Advances in Intelligent and
Soft Computing

160

Editor-in-Chief

Prof. Janusz Kacprzyk
Systems Research Institute
Polish Academy of Sciences
ul. Newelska 6
01-447 Warsaw
Poland
E-mail: kacprzyk@ibspan.waw.pl

T0137824

For further volumes:
http://www.springer.com/series/4240

David Jin and Sally Lin (Eds.)

Advances in Future Computer and Control Systems

Volume 2

 Springer

Editors
Prof. David Jin
International Science & Education
Researcher Association
Wuhan
China

Prof. Sally Lin
International Science & Education
Researcher Association
Guang Zhou
China

ISSN 1867-5662   e-ISSN 1867-5670
ISBN 978-3-642-29389-4   e-ISBN 978-3-642-29390-0
DOI 10.1007/978-3-642-29390-0
Springer Heidelberg New York Dordrecht London

Library of Congress Control Number: 2012934955

Printed on acid-free paper

Springer is part of Springer Science+Business Media (www.springer.com)

Preface

In the proceeding of FCCS2012, you can learn much more knowledge about Future Computer and Control Systems all around the world. The main role of the proceeding is to be used as an exchange pillar for researchers who are working in the mentioned field. In order to meet high standard of Springer, the organization committee has made their efforts to do the following things. Firstly, poor quality paper has been refused after reviewing course by anonymous referee experts. Secondly, periodically review meetings have been held around the reviewers about five times for exchanging reviewing suggestions. Finally, the conference organization had several preliminary sessions before the conference. Through efforts of different people and departments, the conference will be successful and fruitful.

During the organization course, we have got help from different people, different departments, different institutions. Here, we would like to show our first sincere thanks to publishers of Springer, AISC series for their kind and enthusiastic help and best support for our conference.

In a word, it is the different team efforts that they make our conference be successful on April 21–22, 2012, Changsha, China. We hope that all of participants can give us good suggestions to improve our working efficiency and service in the future. And we also hope to get your supporting all the way. Next year, In 2013, we look forward to seeing all of you at FCCS2013.

February 2012   FCCS2012 Committee

Committee

Honor Chairs

Prof. Chen Bin	Beijing Normal University, China
Prof. Hu Chen	Peking University, China
Chunhua Tan	Beijing Normal University, China
Helen Zhang	University of Munich, China

Program Committee Chairs

Xiong Huang	International Science & Education Researcher Association, China
LiDing	International Science & Education Researcher Association, China
Zhihua Xu	International Science & Education Researcher Association, China

Organizing Chair

ZongMing Tu	Beijing Gireida Education Co.Ltd, China
Jijun Wang	Beijing Spon Technology Research Institution, China
Quanxiang	Beijing Prophet Science and Education Research Center, China

Publication Chair

Song Lin	International Science & Education Researcher Association, China
Xionghuang	International Science & Education Researcher Association, China

International Committees

Sally Wang	Beijing normal university, China
LiLi	Dongguan University of Technology, China
BingXiao	Anhui university, China
Z.L. Wang	Wuhan university, China
Moon Seho	Hoseo University, Korea
Kongel Arearak	Suranaree University of Technology, Thailand
Zhihua Xu	International Science & Education Researcher Association, China

Co-sponsored by

International Science & Education Researcher Association, China
VIP Information Conference Center, China
Beijing Gireda Research Center, China

Reviewers of FCCS2012

Z.P. Lv	Huazhong University of Science and Technology
Q. Huang	Huazhong University of Science and Technology
Helen Li	Yangtze University
Sara He	Wuhan Textile University
Jack Ma	Wuhan Textile University
George Liu	Huaxia College Wuhan Polytechnic University
Hanley Wang	Wuchang University of Technology
Diana Yu	Huazhong University of Science and Technology
Anna Tian	Wuchang University of Technology
Fitch Chen	Zhongshan University
David Bai	Nanjing University of Technology
Y. Li	South China Normal University
Harry Song	Guangzhou Univeristy
Lida Cai	Jinan University
Kelly Huang	Jinan University
Zelle Guo	Guangzhou Medical College
Gelen Huang	Guangzhou University
David Miao	Tongji University
Charles Wei	Nanjing University of Technology
Carl Wu	Jiangsu University of Science and Technology
Senon Gao	Jiangsu University of Science and Technology
X.H Zhan	Nanjing University of Aeronautics
Tab Li	Dalian University of Technology (City College)
J.G Cao	Beijing University of Science and Technology
Gabriel Liu	Southwest University
Garry Li	Zhengzhou University
Aaron Ma	North China Electric Power University
Torry Yu	Shenyang Polytechnic University
Navy Hu	Qingdao University of Science and Technology
Jacob Shen	Hebei University of Engineering

Contents

Design of Monitoring System for Aeolian Vibration on Transmission Lines Based on ZigBee

Meng Zhang[1], Beige Yang[2], Hui Xue[2], Yuxiang Lv[1], and Yanling Zhang[1]

[1] College of Physics and Optoelectronics, Taiyuan University of Technology,
Taiyuan 030024, Shanxi, China
[2] Datong Power Supply Company, Datong 037008, Shanxi, China

Abstract. As to accurately grasp the vibration state of the transmission lines, a monitoring system for Aeolian vibration of transmission lines based on ZigBee wireless communication technology and GPRS communication technology is developed, the system is mainly comprised of vibration monitoring module, base station and monitoring center, the design of hardware and software of this monitoring unit are analyzed. This system can monitor the Aeolian vibration of every monitored site on conductor with high efficiency and low power cost, it has a high application value.

Keywords: Transmission line, Aeolian vibration, Monitoring system, ZigBee.

1 Introduction

Aeolian vibration is high frequency motion that can occur when a smooth, steady crosswind about 0.5~10m/s blows on aerial cables, this laminar wind creates vortices, which are detached at regular intervals on the leeward side, alternating from top and bottom of the cable, the detachments creates alternating forces on the cable, when the vortex shedding frequency equals the natural frequency of the cable, a resonance condition occurs[1]. Aeolian vibration is characterized by small amplitude and high frequency, it may cause fatigue failures on the cable where bending strain is no more negligible and hardware components damage, even line breakage or power outage, which seriously impacts the operational security of transmission lines. In this country, transmission line fatigue failures are general exist phenomenon [2], but because of lack of real-time vibration data, we can't accurately grasp the vibration status and forecast fatigue accident. Thus, electricity department is badly needed in a device that can monitor the vibration of transmission lines to master the vibration status and fatigue degree, and to ensure the operational security of transmission lines.

In this paper, in order to achieve the real-time monitoring of Aeolian vibration, a low-cost, low-rate, high efficiency wireless vibration sensors network is build through the current advanced ZigBee wireless communication technology[3], ZigBee network is responsible for data collection, which consists of the coordinator, routers and terminal nodes, the coordinator is responsible for starting and maintaining a wireless network, identifying the device join, a ZigBee network only allows a coordinator, routers as the network repeaters support the network link, the terminal nodes are the executives of the network, which are responsible for collecting and sending data.

2   System Structure

The system is mainly composed of vibration monitoring module, base station and monitoring center. The system structure is shown in Fig.1.

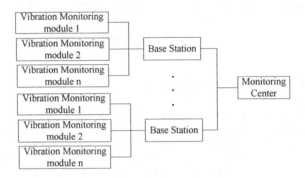

Fig. 1. System structural diagram

The base station is installed on the transmission tower and the vibration monitoring module on the outlet of suspension clamp or anti-vibration devices. Vibration monitoring module(wireless sensor network node) is mostly used to collect lines vibration information, and then transmits the information to ZigBee coordinator of the base station by ZigBee wireless network. Base station sends the packaged message that contains vibration data and meteorological parameters(temperature, humidity, wind speed, wind direction, etc) to the monitoring center by GPRS communication technology. The monitoring center which consists of monitoring center service, database service, printer and other equipments, is mainly responsible for storing and dealing with the data from the base station, it can display the vibration state real-time, and make warning information timely. Furthermore, the monitoring center can also communicate with base station for remote parameter setting.

3   Vibration Monitoring Module Design

3.1   Hadware Design of Vibration Monitoring Module

Vibration monitoring module is installed on the overhead line within the ZigBee wireless network, it is responsible for collecting vibration data of the conductor and the conductor temperature. The data collected is sent to the coordinator of base station through the wireless network.

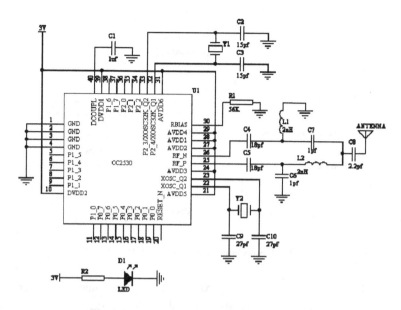

Fig. 2. Hardware schematic of ZigBee module

The system selects the CC2530 of TI as the microprocessor for ZigBee module, CC2530 is a true System-on-Chip solution for 2.4GHz IEEE802.15.4, ZigBee and ZigBee RF4CE application, it integrates a high-performance RF transceiver and a strengthen industry standard 8051 microcontroller core, with 8-KB RAM, 32/64/128/256-KB In-System-Programmable Flash and other features[4-5]. CC2530 has several operating modes, making it highly suited for systems where ultralow power consumption is required. The hardware schematic of ZigBee module is shown in Fig.2.

3.2 Software Design of Vibration Monitoring Module

After the system power supply, vibration monitoring module starts to initialize, then ZigBee nodes join in the wireless network that established by coordinator and enter sleeping mode, when receive the data collection order from the base station coordinator, these nodes begin to work, after sending data to coordinator successfully, the nodes enter sleeping mode again, waiting for the next data collection command. The program flowchart is shown in Fig.3.

4   Base Station Design

4.1   Hardware Design of Base Station

Base station is mainly composed of microprocessor, various sensors, external memory, ZigBee coordinator, GPRS communication module, clock circuit and power management unit. The structural diagram of base station is shown in Fig.4.

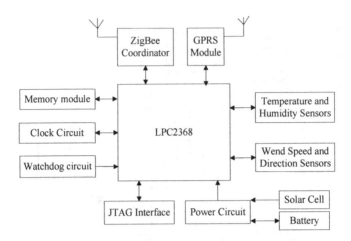

Fig. 4. Structural diagram of base station

Base station uses the LPC2368 with ARM7 kernel as the microcontroller[6], the NXP LPC2368 is an ARM7TDMI-S based high performance 32-bit RISC microcontroller with Thumb extensions, 512KB on-chip Flash ROM with In-System Programming(ISP) and In-Application Programming(IAP); GPRS communication module uses MC52i which has embedded TCP/IP protocol stack; while the base station always works in the wild, the power supply is designed as battery charging from the solar panel under the management of controller.

The base station microcontroller sends command of collecting vibration data to the routes and terminal nodes through ZigBee coordinator and receives vibration data from them, then packages the acquired information and environmental parameters, sends to the monitoring center by GPRS network, meanwhile the base station also accepts the control command that from the monitoring center and executes as the command.

4.2   Software Design of Base Station

In order to improve the efficiency of the system and reduce the length of the main program, modular methods are used in the design of base station software which contains the system initialization program, ZigBee coordinator program, other meteorological data collection program, data storage program, and GPRS communication program, etc.

ZigBee coordinator is responsible for setting up a wireless network, it makes a broadcast to the network when receives the collecting data order from the base station microcontroller, receives the data from the routers or the near nodes through the wireless module, then sends the information to the base station through the serial port. The flow chart of the coordinator is shown in Fig.4.

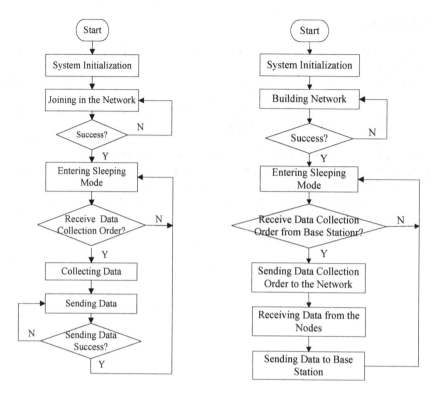

Fig.3. The flowchart of the monitoring module   **Fig.5.** The flow chart of the coordinator

5   Conclusion

The monitoring system for Aeolian vibration on transmission lines based on ZigBee technology has advantages as low power consumptions, auto networking and simple to application, it not only solves the monitoring of Aeolian vibration but also the difficulties of wireless transmission. The system realizes the real-time monitoring of multiple points vibration state on the transmission line and achieves remote transmission of monitoring data through GPRS communication technology, it can make evaluation of the safe operation of the line and give warming information before the fatigue damage, so the system can largely improve the safety and reliability running of transmission lines. Furthermore, it can gather and accumulate atmosphere information and provide base data for the design and maintain of the transmission lines.

References

[1] Shao, T.: Mechanical computation of power transmission conductor. Chain Power Press, Beijing (2003)

[2] Yang, Y.: An on-line measurment system of aeolian vibration of 500KV transmission line crossing. Anhui Electric Power, 29-32 (2006)

[3] Sun, Q., Kong, L., Zhang, Z., Mei, T.: Construction of wireless sensor network monitoring system based on ZigBee. Instrumentation Technology, 7-9,13 (2011)

[4] Fan, J., Wang, J.: Design of wireless monitoring system for enviroment status of museum based on CC2530. Electronic measurement technology, 105-109 (2011)

[5] Texas Instruments, A True System-on-Chip Solution for 2.4GHz IEEE 802.15.4 and ZigBee Application, http://www.ti.com

[6] Zhou, L., Zhang, H.: Head first ARM7. Beijing University of Aeronautics and Astronautics Press, Beijing (2005)

Research on Collaborative Annotation in Semantic Web Environment Based on Ontology Reasoning

XiYong Zhu, BaoChuan Lin, and YiHui Zhou

Institute of Modern Information Technology, Zhejiang Sci-Tech University, 5 Second Avenue,
Xiasha Higher Education Zone, HangZhou, China
zxywolf@163.com, linbaochuan@126.com, zhouyihuityf@yahoo.com.cn

Abstract. Based on current Web collaborative annotation systems, the paper first analyzes the common features of tags in documents as well as context-awareness to extend the concept of tags and make mapping between ontology, then uses ontology reasoning to enrich the semantics of tags, to capture implied semantic information, and to find out the relationships between documents, it finally detects pseudo-relevance among different documents in order to improve the accuracy of knowledge retrieval and knowledge recommendation as well as the flow of knowledge among subjects.

Keywords: Semantic annotation, ontology mapping, ontology reasoning, SWRL.

1 Introduction

In Web2.0, users make collaborative annotation for sharing network knowledge by means of tags so that different knowledge resources may have different features or identifiers and can be easily retrieved. Compared with traditional technologies, annotation reflects not only the contents but also user's understanding of the document and cognitive context awareness. Its semantics is even more divergent. Meanwhile, collaborative annotations among subjects have solved the problems of possible different understandings of the same subject, and have avoided abstract annotations, thus have greatly improved the accuracy of document description.

However, annotation has following limitations: 1) annotation system ignores the correlations that may exist among different tags, because it does not recognize the relationship between related knowledge resources, especially in the case of synonyms [1]. For examples, tags of "student" and "pupil" have the same meaning, but they may be used as two different annotations for the same subject, the system may not recognize their similarity in meanings; 2) annotation system may provide inexplicable correlations between knowledge resources, especially in the case of polysemy. For example, the tag of "mouse" has both the meanings of animal of mice and an object connected to a computer. One may describe it as an animal, while another may refer it to a computer device. Actually there is no relevance to each other, but computer system mistakes that they are correlated with the same tag of "mouse". Therefore, traditional collaborative annotation system lacks necessary semantics.

D. Jin and S. Lin (Eds.): Advances in FCCS, Vol. 2, AISC 160, pp. 7–12.
springerlink.com   © Springer-Verlag Berlin Heidelberg 2012

With ontology and semantics, ontology-based semantic collaborative annotation is gradually replacing the existing tagging technologies. There are two approaches to semantic Web annotation. One of them is to use natural language processing (NLP).

The NLP approach is based upon tags and natural language for knowledge resource description. It first uses syntax (such as subject, predicate, object of a sentence) and the property of a word to find out the corresponding statements for the resource description framework (RFD), then tries to find out the corresponding domain ontology by means of the subject and object in RFD and makes a mapping between predicate and ontology attributes, finally extracts the tags and semantics of annotations [2]. Alani proposed a method of using ontology-related general-purpose language to map relationship between verb and ontology, but found it difficult to implement a mapping from verb to ontology attributes [3]. At the same time, his method applied only to sentences with only one subject or one predicate.

Another approach is using ontology-based semantics annotation prototype. This approach first analyzes the common features of tags, then aggregates tags with similar semantics, and finally maps directly with ontology elements and extracts semantics relations among tags with ontology information. SemTag searched for all possible instances that match the words to be annotated in the TAP project, then constructed respective text vectors according to the context of words and instances [4]. He measured the similarity between words and instances, and finally found out the most matched instances for the words to be annotated.

In this paper, we propose an ontology reasoning-based approach. We first standardize the format of the document tags, analyze the context, and extend the concept groups, then map them with related ontology elements. we use ontology knowledge structure to enrich semantics by ontology reasoning techniques, which make processed tags to be semantic. This approach can distinguish the correlations and detect pseudo-relevance among different documents.

2   Conceptualization of Tags

2.1   Normalization of Tag Format

Users often make subjective errors (such as wrong delimiter, spelling errors, non-compliant symbols, etc.) in describing documents tags, which may cause systems to fail to identify the tags. Therefore it is first necessary to standardize the format of tags and transform them into a word format or a simple statement format which can be interpreted or processed by systems. The processes of standardization can be divided into the following three steps: 1) the first step is to check delimiters errors in tags. Generally, we use semicolon (";") as a separator, because other symbols (such as colon (":") and minus ("-"), etc.) are part of tags, they may result in ambiguity. 2) the second step is to check spelling errors in tags or to detect any misspelled words or wrong orders in tags; 3) the last step is to detect compound words or acronyms in tags. By dividing compound words into separate words or restoring the acronyms into whole phrases, it is possible to determine standardized representation for each tag.

2.2   Context-Awareness Based Tag Processing

Context-awareness is also referred as contextual processing. As different people may have different understandings of the same knowledge, they may use different tags for that knowledge. With the concept of tags only, it seems impossible to provide a comprehensive description of the knowledge features. Therefore, we need to analyze the common features of a certain tag cluster by first defining all tags in knowledge as an initial concept group, and then counting all occurrences in initial concept groups in the system. We can find out the most frequent sets of tags by means of co-occurrences and thresholds among initial groups. The next work to be done is to count up automatically or manually the frequency of tags in resource documents and to determine the accuracy of all tags. By now, we can extend the concept group of tags and use appropriate ontology elements to enrich ontology semantics.

3   Ontology Mapping and Ontology Reasoning Technologies

3.1   Ontology Mapping

By means of concept groups, we search for corresponding domain ontology for each concept through tag-to-ontology mapping and ontology-to-ontology mapping. In this paper, we propose two kinds of mappings: Tag-to-ontology Mapping (TO-Map) and Ontology-to-ontology Mapping (OO-Map).

TO-Map is used to establish a mapping between tags and semantic Web ontology to enrich tag semantics. Ontology-to-ontology Mapping is to present the correlation between ontology in a semantics level. We can use the correlation to achieve interoperability between ontology and make a mapping between entities. Here, we define one as the source ontology, and search for mapping-related ontology within the ontology network and find out more related classes, attributes, and class instances. Ehrig [7] gave an ontology mapping function as follows:

$$\text{map} : O1 \rightarrow O2$$

If sim(p1, p2)> s, then there is map: O1 → O2, in which p1 and p2 are two ontology entities or concepts. Function sim() represents the similarity between the two entities; s is the similarity threshold. It is possibly to measure the similarity of attributes or concepts in one entity respectively.

We use this function to search for the optimal related ontology from source ontology. This process can be divided into following 6 steps.

Step 1: to extract the features (such as concept, attribute names, etc.) to measure similarity according to the features of the source ontology.
Step 2: to select pairs of concepts or attributes from the source ontology and other related ontology entities.
Step 3: to measure similarity.
Step 4: overall similarity calculation. We may start with many indicators to measure the similarity and result in various similarity values, so it is necessary to calculate the overall similarity.

Step 5: optimization: With overall similarity, it usually requires manual operation to determine the degree of correlation between the source ontology and the mapping ontology, and sometimes need adjustments.

Step 6: repeat step 1 to 5 until you find the most related mapping ontology.

By TO-Map and OO-Map, we can find out more ontology attributes, concepts, and entities that are related to the concept groups of tags. These are all very good jobs for the next ontology reasoning technologies.

3.2   Ontology Reasoning

One of the fundamental tasks for ontology reasoning is to acquire hidden or implicit knowledge from specified knowledge. The key to reasoning from ontology is to design a processing mechanism to capture implicit knowledge from the definitions and statements of explicit knowledge.

Ontology reasoning has many applications. One of them is for ontology providers to optimize the expressions and to detect conflict in ontology integration. But its main applications are oriented to ontology users to acquire knowledge from ontology and then to apply knowledge to problem-solving.

Web Ontology Language (OWL) can well describe abstract domain ontology, instances and their relationships, but it is yet unable to describe most logical relationships among ontology entities. In this paper, we introduce semantic Web rule language (SWRL) as well as OWL language. SWRL integrates ontology with rules, as it uses rules in ontology to extend its capability for information description and more powerful logical expressions [8]. One main objective of SWRL is to provide expression capabilities that are not supplied by the OWL while maintaining its compatibility in syntax, semantics and theoretical models with OWL. Therefore, SWRL can be considered as an added way of expression based on user-defined rules.

SWRL has its own set of grammars and symbols. For example, symbol "→" is used to imply logical relationships between premises and conclusions. "?" is a suffix for variables, "^" is to concatenate cited sub-formulas. SWRL also provides many built-in functions in a way similar to method call that returns variable values.

The "uncle" problem is widely discussed in many OWL. In order to determine whether individual A has an uncle of U, A must first find out his parents as set of P, then determine whether there is a brother in the set P. If a brother is found, we can determine that U is an uncle of A.

To address this problem, we define two functions in a ontology: isParent() and isBrother(). Function isParent(x1,x2) indicates that x2 is a parent of x1. Function isBrother(x2,x3) indicates that x3 is a brother of x2. We use SWRL to define the following rule to specify an "Uncle" relationship between x1 and x3:

$$\text{isParent}(?x1,?x2)^\wedge \text{isBrother}(?x2,?x3) \rightarrow \text{hasUncle}(?x1,?x3)$$

We supply the rule with entities data of A, U, and P. If isParent(A, P) and isBrother(P,U) both return "true" value, then isUncle (A, P) will also return the "true" value, which indicates that P is A's uncle.

The above description and correlation discovery cannot be finished by common annotations and must be achieved by means of ontology reasoning. In this approach, we first need to extend the concept of tags to create concept groups and to find out related ontology or mapped ontology, then to make a mapping from SWRL-based descriptions to reasoning rules and put these rules into a reasoning engine, finally to capture hidden semantics in knowledge resources through reasoning.

4   Experimentation

The following experimentation is to demonstrate our theory and approach. We set two types of resource documents shown in Table 1. We standardize tag formats and implemented context-awareness processing.

Table 1. Defintions of experimental resources and documents

Document	Title	Source	User-defined Tags
A	Apple:Magic Mouse-The world's first Multi-Touch mouse	http://www.apple.com/magicmouse/	mouse; electricity; multi-touch
B	mouse	http://en.wikipedia.org/wiki/Mouse	mouse; eliminate; food

In document A, the tag "mouse" is referred as computer-mouse when both tags of "mouse" and "multi-touch" are present. In document B, we delete the tag of "food" and retain two other tags of "mouse" and "eliminate", defining that when both tags of "mouse" and "eliminate" are present, we refer "mouse" to an animal of mice. With ontology system, we have acquired two ontology entities related to "mouse". One is an animal ontology, another is a computer device ontology.

As protégé (an ontology design tool) has no capability to reasoning, we combine it with Jess reasoning engine. We import two ontology entities into Jess library together with their attributes and instances, and set the following reasoning rule.

Rule-1:

hasElectronicProduct(?x1,?y1)^hasKey(?x1,?y2)^isPartOfComputer(?x1,?y3)^has Signal(?x1,?y4)^ swrlb:stringEqualIgnoreCase(?x2,?y5)
→isComputerMouse (?x1)

Through the above reasoning, the semantics of tag groups is obvious. We find that document A is related to computer-mouse in the description of an electronic device, while document B is related to an animal of mice. These two documents are not related to each other even though they both have a tag of "mouse".

We have also conducted similar experiments, the results are shown in Table 2. We have found that semantic collaborative annotation system have enriched the semantics of the tags, it can find out hidden correlations between different documents, and can also detect the pseudo-correlations among documents caused by the diversity of tags in meanings.

Table 2. A comparison of trational annotation and semantics annotation for documents correlation determination

Document Group	Document Title or wiki encyclopedia links	User-defined tags	Correlated between documents(Yes/No)	
			Traditional annotation	Semantics annotation
1	<Protection of Pupil Rights Amendment (PPRA)>	pupil; PPRA	No	Yes
	<International Student Identity Card (ISIC)>	student; ISIC		
2	http://en.wikipedia.org/wiki/Fly	fly	Yes	No
	<Ready to fly>	fly; song		
3	http://en.wikipedia.org/wiki/Fast_repast	repast; fast	No	Yes
	http://en.wikipedia.org/wiki/Fast_food	food; fast		

5   Conclusion

In this paper, we proposed an ontology reasoning-based approach to capturing semantics for documents hidden in tags and enriched the semantics of tags. It helped system find out related tag implicit and explicit knowledge documents, and solved synonyms and polysemy problems caused by traditional annotation technologies. It improved the accuracy for document searching and matching capabilities and could detect peudo-relevance among different documents. Our next work is to use other styles of documents to verify the proposed theory and method.

References

1. Wu, F.: Semantics Enrichment in Collaborative Annotation Systems. Intelligence Magzine 29, 186–188 (2010)
2. Jing, T., Zuo, W.L., Gui, S.J.: Semantic Annotation in Chinese Web Pages: From Sub-clauses to RDF Representations. Computer Research and Development 45, 1221–1231 (2008)
3. Alani, H., Kim, S., Millard, D.: Automatic Ontology-based Knowledge Extraction From Web Documents. Intelligent Systems 18, 14–21 (2003)
4. Dill, S., Tomlin, J.: SemTag and Seeke: Bootstrapping the Semantic Web via Automated Semantic Annotation. In: 12th International Conference on World Wide Web. ACM, New York (2003)
5. Gao, S.L., Shia, H.J.: Research on Ontology Reasoning in Semantic Web. Journal of Huahai Institute of Technology 19, 28–32 (2010)
6. Wwoogle[DB] (September 09, 2009), http://swoogle.umbc.edu
7. Ehrig, M., Staab, S.: QOM – Quick Ontology Mapping. In: McIlraith, S.A., Plexousakis, D., van Harmelen, F. (eds.) ISWC 2004. LNCS, vol. 3298, pp. 683–697. Springer, Heidelberg (2004)
8. Zhang, Y.T., Chen, J.J., Xiang, J.: Research on SWRL-based Ontology Reasoning. Micro Computers Information 26, 182–184 (2010)

Research on Control Strategy of VAV Air Conditioning System

Jiejia Li[1], Rui Qu[1], and Ying Li[2]

[1] School of Information and Control Engineering, Shenyang Jianzhu University, 110168, Shenyang, Liaoning, China
[2] Shenyang Academy of Instrumentation Science, 110043, Shenyang, Liaoning, China

Abstract. Aiming at the characteristics which variable air volume air conditioning system is multi-variable, nonlinear and uncertain, we put forward a recursive wavelet neural network predictive control strategy. Through recursive wavelet neural network predictor on line established controlled object's mathematical model, and using RBF neural network controller on line corrected information we get, thus to improve control effect. The simulation results show that recursive wavelet neural network predictive control has stronger robustness and adaptive ability, high control precision, better and reliable control effect and other advantages.

Keywords: Variable Air Volume Air Conditioning, Recursive Wavelet Neural Network, RBF Neural Network, Predictive Control.

1 Introduction

Variable air volume (VAV) air conditioning system is a full air system of multi-variable and nonlinear. In VAV air conditioning system, when the air conditioning load changes or indoor air parameters' set value changes, it will adjust air flow into the room automatically, and adjust the air conditioning parameters to set value, to meet comfort requirements and production requirements of the personnel indoor. Therefore, this paper puts forward a kind of recursive wavelet neural network predictive control method, to establish reference model online, to complete the identification of system's dynamic characteristics, and using RBF neural network control method to improve control effect.

2 The Working Principle of VAV Air Conditioning System

VAV air conditioning system is mainly composed of VAV air conditioning units and VAV Box. The typical VAV system is shown in Figure 1. Its working principle is: According to control requirements, it regulate new air volume, put air volume into VAV box, and then adjust VAV box to complete control function. So, in order to achieve temperature control requirements, to improve the comfort and energy saving effect, we need good control strategy to meet the needs of people.

D. Jin and S. Lin (Eds.): Advances in FCCS, Vol. 2, AISC 160, pp. 13–18.

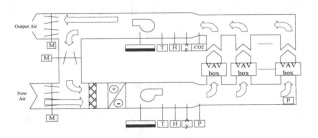

Fig. 1. Typical VAV System Working Principle Diagram

3   Wavelet Neural Network Predictive Control System Design

The wavelet neural network predictive control system structure is shown in Figure 2. The controllers are composed of recursive wavelet neural network predictor and neural network controller. Wavelet neural network predictor uses the characteristics which wavelet neural network can effective control non-linear, time-varying and uncertainty system, and predict object's future output, and then according to the future output to adjust control strategy online, to achieve optimal control of air conditioning.

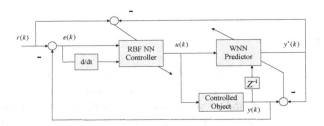

Fig. 2. Structure Diagram of Control System

3.1   Recursive Wavelet Neural Network Predictor Design

In this paper the design of predictor adopts recursive wavelet neural network. The structure diagram is shown in Figure 3. In this network structure, the input of hidden layer is composed of output of input layer and output of associated layer, so recursive network structure has good dynamic characteristics, for VAV air conditioning system which has dynamic, time-varying, non-linear, uncertain system, it has good control effect.

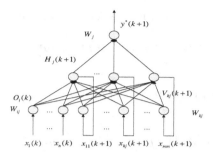

Fig. 3. Structure Diagram of Multi-input, Single Output Recursive Wavelet Neural Network

The input layer of network has n neurons, this paper takes n=3, neurons of input layer are $x_1(k), x_2(k), x_3(k)$, on behalf of output $u(k)$ of neural network controller, actual output delay $y(k-1)$ of controller object and the error $y(k)-y^*(k)$ of object's actual output and predicted output. $O_i(k)$ is input layer's output, $i = 1,2...n$. Hidden layer and associated layer both have m neurons. At k moment, output of hidden layer is $H_j(k+1)$, so input of associated layer is expressed as $x_{kj}(k+1) = \alpha H_j(k), \alpha$ is feedback gain. Associated layer's output is $V_{kj}(k+1)$. So, recursive wavelet neural network's output is:

$$y^*(k+1) = \sum_{j=1}^{m} \omega_j \varphi(\sum_{i=1}^{n} \omega_{ij} O_i(k) + \sum_{k=1}^{m} \omega_{kj} V_{kj}(k+1) - \theta_j) \qquad (1)$$

θ_j is wavelet function's translation coefficient, $\varphi(\bullet)$ is wavelet function. ω_j is the weight between hidden layer and output layer, ω_{ij} is the weight between input layer and hidden layer, ω_{kj} is the weight between associated layer and hidden layer.

Take wavelet function:

$$\varphi(x) = (1 - x^2)e^{-\frac{x^2}{2}} \qquad (2)$$

Define the error performance index function of this network:

$$E = \frac{1}{2}[y(k+1) - y^*(k+1)]^2 \qquad (3)$$

Among them, $y(k+1)$ and $y^*(k+1)$ represent actual output and reference model output of predictor, at $k+1$ moment.

The learning of parameters takes adapt gradient descent method, we can get:

$$\omega_j(k+1)=\omega_j(k)+\eta\Delta\omega_j(k)+\alpha[\omega_j(k)-\omega_j(k-1)] \tag{4}$$

$$\omega_{ij}(k+1)=\omega_{ij}(k)+\eta\Delta\omega_{ij}(k)+\alpha[\omega_{ij}(k)-\omega_{ij}(k-1)] \tag{5}$$

$$\omega_{kj}(k+1)=\omega_{kj}(k)+\eta\Delta\omega_{kj}(k)+\alpha[\omega_{kj}(k)-\omega_{kj}(k-1)] \tag{6}$$

$$\theta_j(k+1)=\theta_j(k)+\eta\Delta\theta_j(k)+\alpha[\theta_j(k)-\theta_j(k-1)] \tag{7}$$

Among them:

$$\Delta\omega_j=-\frac{\partial E}{\partial\omega_j}=-\frac{\partial E}{\partial y^*(k+1)}\bullet\frac{\partial y^*(k+1)}{\partial\omega_j}=[y(k+1)-y^*(k+1)]\sum_{j=1}^{m}\phi(\sum_{i=1}^{6}\omega_{ij}O_i(k)+\sum_{k=1}^{m}\omega_{kj}V_{kj}(k+1)-\theta_j) \tag{8}$$

$$\Delta\omega_{ij}=-\frac{\partial E}{\partial\omega_{ij}}=-\frac{\partial E}{\partial y^*(k+1)}\bullet\frac{\partial y^*(k+1)}{\partial\omega_{ij}}=[y(k+1)-y^*(k+1)]\sum_{j=1}^{m}\omega_j\phi'(\sum_{i=1}^{n}\omega_{ij}O_i(k)+\sum_{k=1}^{m}\omega_{kj}V_{kj}(k+1)-\theta_j)O_i(k) \tag{9}$$

$$\Delta\omega_{kj}=-\frac{\partial E}{\partial\omega_{kj}}=-\frac{\partial E}{\partial y^*(k+1)}\bullet\frac{\partial y^*(k+1)}{\partial\omega_{kj}}=[y(k+1)-y^*(k+1)]\sum_{j=1}^{m}\omega_j\phi'(\sum_{i=1}^{n}\omega_{ij}O_i(k)+\sum_{k=1}^{m}\omega_{kj}V_{kj}(k+1)-\theta_j)V_{kj}(k+1) \tag{10}$$

$$\Delta\theta_j=-\frac{\partial E}{\partial\theta_j}=-\frac{\partial E}{\partial y^*(k+1)}\bullet\frac{\partial y^*(k+1)}{\partial\theta_j}=-[y(k+1)-y^*(k+1)]\sum_{j=1}^{m}\omega_j\phi'(\sum_{i=1}^{n}\omega_{ij}O_i(k)+\sum_{k=1}^{m}\omega_{kj}V_{kj}(k+1)-\theta_j) \tag{11}$$

η is leaning rate, α is inertial parameter, and η,α both in the range of $(0,1)$.

3.2 Neural Network Controller Design

The design of controller adopts RBF neural network structure, the output y_i meets:

$$y_i=f_i(x)=\sum_{k=1}^{N}w_{ik}\phi_k(x,c_k)=\sum_{k=1}^{N}w_{ik}\phi_k(\|x-c_k\|_2)\quad i=1,2,...m \tag{12}$$

$x\in R^n$ is input vector, $\phi_k(\bullet)$ is a function from R^+ to R, $\|\bullet\|_2$ represents Euclidean norm, w_{ik} is the weight between the k th hidden unit and the i th output unit, N is the neurons number of hidden layer, $c_k\in R^n$ is function's center.

Take gauss function of hidden layer's nodes function in RBF neural network, so,

$$y_i = \sum_{k=1}^{N} \omega_{ik} \bullet \exp[-\frac{\|x - c_k\|^2}{\sigma_k^2}] \tag{13}$$

Parameter σ_k controls the width of radial basis function.

For this paper, there are two input values, x_1 is error value e, and x_2 is error variation Δe. The output value is control value $u(k)$.

Define the performance index function:

$$J = \frac{1}{2}[r(k+1) - y(k+1)]^2 \tag{14}$$

In formula, $r(k+1)$ is desired output, $y(k+1)$ is system's actual output.

According to gradient descent method, we get iterative algorithm of output's weight, radial basis function's center and radial basis function's width. The algorithm will skip.

4   Simulation and Conclusion

In order to verity the reasonableness of the control scheme. We made a simulation experiment. In this paper, control object is second-order lag model, its transfer function is: $G(s) = \dfrac{ke^{-\tau s}}{(T_1 s + 1)(T_2 s + 1)}$ .Take $T_1 = 12, T_2 = 5, k = 12, \tau = 10$, set

desired temperature is $25°C$, and add a disturbance, when temperature control at

$1000s$, observe the changes. The simulation comparison diagram of recursive wavelet neural network predictive control and common neural network control is shown in Figure 4.

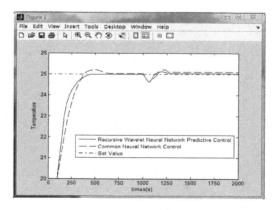

Fig. 4. Comparison Diagram of Recursive Wavelet Neural Network Predictive Control and Common Neural Network Control

Figure 4 shows that: When the input temperature changes from $20°C$ to $25°C$, the adjusting time of common neural network control is at about $600s$, and its overshoot is large. But recursive wavelet neural network predictive control has quick response speed and almost zero overshoot, and its adjusting time is about $450s$. Therefore, recursive wavelet neural network predictive control has better dynamic performance. After disturbance, compared the new method and common neural network, the new method has quicker response speed, smaller error of steady state and it achieves steady state earlier. So, recursive wavelet neural network predictive control has small overshoot, short adjusting time, high control precision, strong robustness and good control performance.

References

1. Yuan, S., Prerz, R.A.: Model Predictive Control of Supply Air Temperature and Outside Air Intake Rate of a VAV Air-Handling Unit. ASHRAE Transaction (S0001-2505) 112(1), 145–161 (2006)
2. Jang, J.S.R., Sun, C.T.: Functional equivalence between radial basis functions and fuzzy interence systems. IEEE Trans. on Neural Networks 4(1), 156–158 (1993)
3. Li, Z., Wang, J., Kang, A.: An adaptive learning algorithm for T-S fuzzy model based RBF neural network. Journal of Lan zhou University of Technology 30(2), 82–85 (2004)
4. Somakumar, R., Chandrasekhar, J.: Intelligent anti-skid brake controller using a neural network. Control Ebgineering Practice (7), 611–621 (1999)

The Type Detection of Mineral Oil Fluorescence Spectroscopy in Water Based on the KPCA and CCA-SVM

JiangTao Lv and QiongChan Gu

Department of Automation Engineering, Northeastern
University at Qinhuangdao,
066004 Qin Huangdao, China
siwenren-yd@163.com

Abstract. The composition of the mineral oil is complex. Especially mineral oil of 3D fluorescence spectra data dimension higher, more complex processing. In this paper, The method of kernel principal component analysis plus canonical correlation analysis (CCA) is used to do classify the spectroscopy data processed by the KPCA. Kernel method is used within principal component analysis, since CCA uses all the kernel principal component of the converted samples from KPCA, no discrimination is lost in the analysis. The experiment results show that it is effective to extract the main feature of the spectroscopy. The identification of the oils can be realized with high discrimination which is 94.11%.

Keywords: KPCA, CCA, SVM, three-dimensional fluorescence spectroscopy recognition, mineral oil.

1 Introduction

The tradition based on the apparent characteristics of statistical features can only reflect the general characteristics of three-dimensional fluorescence spectra of simple components, or a single fluorescence peak sample differential case is practical, to complex water environment pollution in mineral oil identification has great limitations. Especially mineral oil of 3D fluorescence spectra data dimension higher, more complex processing [1]. Kernel principal component analysis has been paid much attention by many researchers. The traditional principal component analysis and kernel principal component analysis is as much as possible to retain the original information, but for the classification problem, without fully considering the maximum classification information, so it cannot be determined after dimension reduction, we can obtain a better composition[2]. On the basis of forefathers of mineral oil type to do a more in-depth study, using KPCA+CCA method to extract mineral oil fluorescence spectral characteristics, based on kernel nearest feature classifier for gender classification. The experimental results show that, this method is very effective.

D. Jin and S. Lin (Eds.): Advances in FCCS, Vol. 2, AISC 160, pp. 19–22.

2 Principle of KPCA and Canonical Correlation Analysis (CCA)

Principal component analysis can only linear mapping, no data of some nonlinear relation, so common that the kernel principal component analysis performance is better than the principal component analysis. Comparison of kernel independent component analysis, elastic component analysis, and kernel discriminant showed better performance [3].

Canonical correlation analysis is the analysis of the correlation degree of two random variables a statistics analysis method, so that after the combination of two groups of correlativity of random variables and reached the maximum. A two set of random variables X and y, CCA purpose is looking for vector to a and b, make the correlation coefficient r(u, v) is maximum of the linear combination $u = x^T a$ and $v = y^T b$ [2]. The CCA is used to analyze two groups of random data inherent correlation analysis, on the view ofNot only sampling, sample, not readily available in the two set of random data, thereforeThe direct use of CCA to handle the analysis of differential problems cannot be achieved. First using sample matrix (1).

$$X = [x_{11}, x_{12}, \cdots x_{1n_1}, x_{21}, x_{22}, \cdots x_{2n_2}, x_{C1}, x_{C2}, \cdots x_{Cn_C},] \tag{1}$$

The upcoming n c dimensional sample X_{ij} arranged by category n x C matrix X, eachLine represents a sample, for simple, will in turn rewritten subscript X = [x_1, x_2, ...X_n]^T. Then according to each sample classification matrix is constructed as (2).

$$Y \begin{bmatrix} 1_{n1} & 0_{n1} & \cdots & 0_{n1} \\ 0_{n2} & 1_{n2} & \cdots & 0_{n2} \\ \vdots & \cdots & \ddots & \vdots \\ \vdots & \cdots & \cdots & 1_{n_{c-1}} \\ 0_{nc} & 0_{nc} & \cdots & 0_{nc} \end{bmatrix} \tag{2}$$

0_{ni} is the elements of all $0_{ni \times 1}$ column vectors. 1_{ni} is the element for all $1_{ni \times 1}$ column vector. This structure of the matrix Y for each row and each matrix X. The sample should be relative, and indicates that the X for each row in a sample in which categories.

3 The Experimental Process and the Interpretation of Result

3.1 Experimental Installation

The spectroscopic data are got by the FS900 high performance steady state fluorescence spectrometer and the complement working software of it produced by the Edinburgh Instruments company of the British. The spectral response dimension of it is 200-900nm. The quartz is selected to be the example sump because the emission fluorescence of the quartz glass is weak.

The experimental installation can be seen from the Fig.1. The experiment containers are as follows: quartz cell pair (2), conical flask (1lx5), wild-mouth bottle (200mlx1), measuring cylinder (50mlx1, 10mlx1), and drip tube dropper (5mlx1).

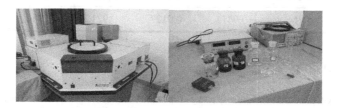

Fig. 1. F920 fluorescence spectrum measurement system

3.2 Experimental Samples

In this paper, the 0# diesel oil, coarse oil, coal oil and 97# gasoline oil are selected to be the experimental example. The SILT of the experimental installation is set to be 0.25mm. The DWELL TIME is 0.1ms. The COOLING TEMPERATURE is -19.7°C. The scanning area of the excitation wavelength (EX) is 250 to 380nm. The step is 10nm. The scanning area of the emission wavelength (EM) is 260 to 530nm. The step is 2nm. The spectroscopic data is converted to the two-dimension form from the EEM matrix form. A line is linked to another line to format a big excitation spectrum. A row is linked to another row to form a big emission spectrum. So, the two-dimension EEM matrix is converted to a one-dimension vector. The one-dimension vector of every mineral oil sample has 1449 digital point. 60 samples are selected. They format a 60x1449 sample matrix. It includes the density information and sort pattern.

3.3 Experimental Consequence Formulas

The Matlab program from the documentation [4] is used to process the samples. The dimension of every characteristic vector is reduced to 50 by KPCA and CCA. The main feature is to be obtained by the CCA. The dimension of it is 20. The method of the parallel coordinates figure is used to make the consequence got by the CCA to be more visual.

Parallel Coordinates [5] is called the Profile plot in the multivariate statistical analysis. The method in the documentation [5] is used to express the conclusion got by the KPCA and CCA. The SVM is used to assort the characteristic got by the KPCA and CCA.

The RBF kernel function of the SVM is selected. The parameter of the RBF kernel function is set to be 0.6[6]. The function in the Matlab is used of the SVM [7].

The cross certification method is used to process the training sample and testing sample. The assorting result is shown in the Tab.1.

We can see from the tab.1 that the discrimination is 94.11%. The better assorting result can be got by this method.

Table 1. The classification result comparison of four classifiers to feature abstracted by the KPCA and CCA

Method	CA	Sens	Spec	AUC
SVM	0.9411	0.9127	0.9212	0.9345

4   Conclusions

The process of the spectral data is very important on the identification mineral oil. In this paper, the merits of two methods are kept and combined to format a new method of spectrum detection and identification. The dimension of the spectrum can be reduced by processed by KPCA. The main information is kept by the KPCA and CCA. The SVM is used to process the data after dimensionality reduction to realize the fast detection and identification. This method is used to realize the effective recognition of the mineral oil.

Acknowledgment. The authors would like to acknowledge financial supports of the Hebei province science and technology plan projects(10276722).

References

1. Tian, G., Shi, J.: Oil identification based on parameterization of three-dimensional fluorescence spectrum. Yi Qi Yi Biao Xue Bao, Chinese Journal of Scientific Instrument 26, 727–728, 735 (2005)
2. Zhang, T., Zhang, J.: Gender Classification of Face Image Based on Kernel Principal Component Analysis plus Canonical Correlation Analysis. Software 32(7), 54–57 (2011)
3. Wan, J., Wang, Y., Liu, Y.: Improvement of KPCA on feature extraction of classification data. Computer Engineering and Design 31(18), 4085–4087 (2010)
4. Zhang, M., Li, T., Zhong, S.: Implementation of principal components analysis method based on Matlab. Journal of Guangxi University 30(s), 74–77 (2005)
5. Xu, Y., Gao, Z., Jin, H.: Principle and overview of parallel coordinates. Journal of Yanshan University 32(5), 389–392 (2008)
6. Ismail, H., Bilal, H., Eyas, E.-Q.: Performance of KNN and SVM classifiers on full word Arabic articles. Advanced Engineering Informatics 22(1), 106–111 (2008)
7. Duin, R.P.W., Juszczak, P., de Ridder, D.: PR-Tools, a Matlab toolbox for pattern recognition (2004), http://www.prtools.org

Effects of NaCl Stress on the Germination Characteristics of Amaranth Seeds

WeiRong Luo and YongDong Sun

School of Horticulture and Landscape Architecture,
Henan Institute of Science and Technology, Xinxiang 453003, China
rowe0803@163.com

Abstract. Effects of NaCl stress on the germination characteristics of amaranth seeds were investigated. The results were as follows: the germination rate, germination potential, taproot length, length of hypocotyledonary axis and proline content of amaranth were increased at lower NaCl concentrations, but then decreased with the increasing of NaCl concentrations. Meanwhile, superoxide dismutase (SOD) activity was significantly increased under NaCl stress, but malondialdehyde (MDA) content was increased under higher NaCl concentration and had no significant change under lower NaCl concentration, compared to the control. These findings indicated that amaranth tolerates NaCl stress through increasing SOD activity and accumulating proline content.

Keywords: NaCl stress, germination characteristics, amaranth, seeds.

1   Introduction

The salinization of soil is one of the most important and widespread environmental problem. Approximately 20% of the world's cultivated lands and nearly half of all irrigated lands are affected by salinity. The salinization of soil has limited the plant growth and crop productivity [1]. In recent years, many major agricultural regions have suffered serious salinity stress, because of climate changes and population growth. Seed germination is a crucial stage in the life history of plants, and is the most sensitive stage to abiotic stress. Salinity stress may result in delayed and reduced germination or may prevent germination completely [2]. Many investigators have reported retardation of germination and growth of seedlings at high salinity [3]. However plant species differ in their sensitivity or tolerance to salt [4]. Amaranth (*Amaranthus mangostanus* L.) is used as animal feed and human food, because of its nutritional value. The objective of the present study was to determine germination characteristics of amaranth seeds under different NaCl concentration treatments.

2   Materials and Methods

2.1   Plant Materials and Treatments

Amaranth was used to study the germination at different NaCl concentrations. Filter paper germination assay was used in the test. The amaranth seeds were disinfected by

D. Jin and S. Lin (Eds.): Advances in FCCS, Vol. 2, AISC 160, pp. 23–27.

soaking with 0.1% $HgCl_2$ for 10 minutes and then washed with distilled water for 3 times. NaCl solutions of 0 (CK), 30, 60, 90 and 120 $mmol·L^{-1}$ were prepared. Glass dishes with the diameter of 9 mm were taken as culture vessels. The prepared different concentration solution of NaCl with the same volume was added to the dishes with two layers of filter papers, respectively. Then 50 seeds were placed uniformly in each dish and cultured in the incubator at $(25±1)°C$, with 3 repetitions in each treatment.

2.2 Growth Measurement

Germination potential of amaranth seeds were calculated after culturing for 2 days. Germination rate, taproot length and length of hypocotyledonary axis were calculated after culturing for 5 days.

2.3 Determination of Physiological Parameters

MDA content was determined using the thiobarbituric acid reaction as described by Sudhakar et al. [5]. Proline content was determined following the method of Sairam et al. [6]. SOD activity was determined according to the method of Meloni et al. [7].

2.4 Statistical Analysis

All data were subjected to analysis of variance (ANOVA test) using the SPSS version 10. 0 statistical package for Windows. Duncan's multiple range tests were applied at the 0.05 level of probability to separate means.

3   Results and Discussion

3.1 Effects of NaCl Stress on the Germination Characteristics of Amaranth Seeds

Growth inhibition is the most common and profound response for most plant species exposed to abiotic stress. In this investigation, germination characteristics of amaranth seeds under different NaCl concentrations were determined and the data were shown in table 1. The four germination parameters of amaranth were increased at lower NaCl concentrations, but then decreased with the increasing of NaCl concentrations. When NaCl concentration was 30 $mmol·L^{-1}$, germination rate, germination potential and length of hypocotyledonary axis reached the top, which were 103.19%, 103.26 and 119.25% of the control, respectively. But at the 60 $mmol·L^{-1}$ NaCl concentration, taproot length reached the top, which was 112.61% of the control. The four parameters reached the bottom under 120 $mmol·L^{-1}$ NaCl concentration, which were 89.36%, 89.13, 79.41% and 89.44% of the control, respectively. These results indicated that lower NaCl concentration treatments promoted seeds germination of amaranth and higher NaCl concentration treatments inhibited that.

Table 1. Effects of NaCl stress on the germination characteristics of amaranth seeds

Treatment (mmol·L^{-1})	Germination rate (%)	Germination potential (%)	Taproot length (cm)	Length of hypocotyledonary axis (cm)
CK	0.94±3.00a	0.92±3.21a	2.38±0.08a	1.61±0.10ab
30	0.97±1.00a	0.95±1.15a	2.64±0.24a	1.92±0.26a
60	0.93±2.65a	0.93±2.89a	2.68±0.06a	1.77±0.18ab
90	0.87±1.53b	0.86±2.52b	2.34±0.29a	1.67±0.18ab
120	0.84±1.53b	0.82±1.73b	1.89±0.11b	1.44±0.03b

3.2 Effects of NaCl Stress on MDA Content of Amaranth

Membrane permeability is a sensitive test to determine abiotic stress and tolerance [8]. MDA is the direct production of lipid peroxidation and its content is often used as an indicator of the extent of lipid peroxidation. It has been used as an indicator of lipid peroxidation and tends to show greater accumulation under abiotic stress [5, 7]. In the present study, MDA content of amaranth increased significantly under higher NaCl concentrations (90-120 mmol·L^{-1}) and had no significant change under lower NaCl concentration (30-60 mmol·L^{-1}), which indicated that a higher degree of lipid peroxidation due to higher NaCl concentration stress (Fig. 1).

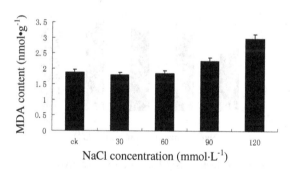

Fig. 1. Effects of NaCl stress on MDA content

3.3 Effects of NaCl Stress on the Proline Content of Amaranth

Effects of NaCl stress on the proline content of amaranth were shown in Fig. 2. NaCl stress progressively increased the proline content. The proline contents of the different NaCl concentration treatments all higher then the control. The proline content at the 60 mmol·L^{-1} NaCl concentration reached the top and was 1.36 fold of the control. Proline can play an important role in the osmotic adjustment of the plant under abiotic stress. Injury caused by NaCl stress was most likely mitigated by accumulating proline to make osmotic adjustments.

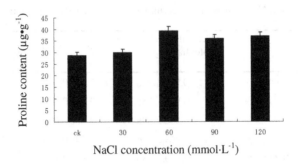

Fig. 2. Effects of NaCl stress on proline content

3.4   Effects of NaCl Stress on SOD Activity of Amaranth

SOD is the key enzymes eliminating reactive oxygen species (ROS). Fig. 3 represented the changes of SOD activities. The results showed that SOD activities under NaCl stress were all higher than the control. In comparison to the control, the SOD activities under the different NaCl concentrations were increased by 41.72%, 30.17%, 43.42% and 52.60%, respectively, which suggested that he antioxidant defense system was enhanced rapidly to scavenge all ROS generated by NaCl stress.

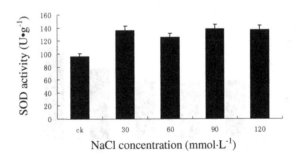

Fig. 3. Effects of NaCl stress on SOD activity

References

1. Bernstein, L.: Osmotic adjustment of plants to saline media. I. Steady state. Am. J. Bot. 48, 909–918 (1961)
2. Yang, Y., Liu, Q., Wang, G.X., Wang, X.D., Guo, J.Y.: Germination, osmotic adjustment, and antioxidant enzyme activities of gibberellin-pretreated Picea asperata seeds under water stress. New. Forests. 39, 231–243 (2010)
3. Sudhakar, C., Lakshmi, A., Giridarakumar, S.: Changes in the antioxidant enzyme efficacy in two high yielding genotypes of mulberry (Morus alba L.) under NaCl salinity. Plant. Sci. 161, 613–619 (2001)

4. Sairam, R.K., Rao, K.V., Srivastava, G.C.: Differential response of wheat genotypes to long term salinity stress in relation to oxidative stress, antioxidant activity and osmolyte concentration. Plant. Sci. 163, 1037–1046 (2002)
5. Meloni, D.A., Oliva, M.A., Martinez, C.A.: The effects of salt stress on growth, nitrate reduction and proline and glycinebetaine accumulation in Prosopis alba. Environ. Ex. Bot. 49, 69–76 (2003)
6. Torech, F.R., Thompson, L.M.: Soils and soil fertility. Oxford University Press, New York (1993)
7. Mansour, M.M.F., Salama, K.H.A.: Cellular basis of salinity tolerance in plants. Environ. Exp. Bot. 52, 113–122 (2004)
8. Shannon, M.C.: Adaptation of plant to salinity. Adv. Agron. 60, 75–119 (1998)

Influence Factors and Solutions of the Campus Electronic Commerce

Xiaomei Shang, Hailan Gu, and Zengtao Xue

School of Electrical Engineering and Information Science, Hebei University of Science and
Technology, Shijiazhuang 050018, China
shangxm@hebust.edu.cn

Abstract. With the Internet in the global business activities of the widely used
and the continuous improvement of the campus network coverage, electronic
commerce is gradually came into the university campus. College students, as a
special consumer groups, has their own features in online shopping. But there
exists some problems in campus e-commerce, such as electronic business
enterprise not paying enough attentions to the market on the campus, college
students' spending power being low, the network security not being guaranteed.
This article puts forward some strategies and advices to speeding up the
construction of campus e-commerce and development through studies.

Keywords: influence factors, solutions, campus, electronic commerce.

1 Introduction

Electronic commerce is a new business model in the rise of the network economy,
representing the business operation the mainstream of development direction. Along
with the network popularization and development, electronic business booms in the
world in a breathtaking pace, and the use of Internet for commercial trade is more and
more common. Campus e-commerce has obtained fast development, and more and
more college students begin to choose online shopping this novel way to shop. The
so-called campus e-commerce, refers to a specific campus e-commerce in which
students exchange goods, service and information online.It differents from the whole
society of e-commerce. Campus e-commerce as a new e-commerce, its development
has gain more and more attention in China.

2 Analysis on the Campus e-Commerce

2.1 The Characteristics of the Ampus e-Commerce

Gender characteristics. Because the popularity of the Internet in boys is higher than
girls, more boys than girls dare to adventure and pursue new things, and boys master
more basic knowledge of the online shopping than girls, the proportion of boys is
higher than girls in online shopping. In recent years, with the maturity of online
shopping environment and the increasing popularity of network knowledge, girls not
only in the traditional way of shopping on the dominant position, while in the
gradually emerging shopping -- online shopping, also occupy a major position.

D. Jin and S. Lin (Eds.): Advances in FCCS, Vol. 2, AISC 160, pp. 29–32.
springerlink.com

Grade characteristics. Through the survey we find that as grade increasing, the proportion of students rapidly increases in online shopping. In the first grade students in online shopping online shopping proportion is 5.7%, the second grade proportion is 9.9%, the third grade net purchase ratio is 12.7%, and the fourth grade net purchase ratio reaches 71.7%. Thus, the increase in network knowledge has a greater impact on online shopping. At the same time as the grade increasing, the proportion of college students who own computers is also higher, which is a powerful online consumer condition.

Psychological characteristics. College students are Ideological and active, sensitive to new things, and have strong ability to accept. In the beginning, because of curiosity they began to shop online. A good impression and a good experience make them soon fell in love with this consumption patterns, and became online store's loyal customers, which gives the development of the campus e-commerce plenty of market space and opportunities .

2.2   The Problems in the Development of the Campus e-Commerce

Electroniccommerce enterprises do not pay enough attention to campus market. Very few companies currently have a special e-commerce market for college students, many companies engaged in e-commerce underestimates the demand and purchasing power of the college students this consumer groups. Some firms think that because of college students' consumption group limited economic conditions, the school's supermarket, dining room can meet their basic necessities of life. Other firms think that undergraduates' consumption is monotonic, namely, they only have strong demand for computers and some digital products in addition to the necessities of life and don't have enough purchasing power for other goods. In fact, the market potential of college students is greater than the enterprises' imagination, their needs coming together, will be a very big potential market. While because of most enterprises' one-sided understanding and judgment, campus of this vast market is not well developed.

*College students' consumption level is low.*With no income sources, college students can only rely on the given living expenses from their parents for life and the economic situation is low. Although many students bring their own computers, the school also provides students with the campus network and other network environment, in general, the students' computer popularity rate is not high, and the campus net hardware conditions are not mature enough. Plus students' academic task which makes them can spend little time in online consumption, these factors will directly restricts the development of the campus e-commerce.

Network security is not guaranteed. Campus e-commerce is different from the traditional consumption mode, and how to guarantee the purchase of goods quality and on time arrival becomes one of the important factors for the campus e-commerce popularity. Besides, in the open network business, how to guarantee the security of data transmission is of concern to students. Therefore, the network security vacuum is still a major factor that influences the development of the campus e-commerce.

2.3   The Strategies and Advice to Accelerate the Development of the Campus e-Commerce

Construct campus e-commerce platform. In view of the needs of the college students and economic condition, the modern college students' online consumption mainly focus on some necessities, school supplies and digital products. While the current online sales of goods are facing the crowd, even for high-income people, the types of goods and the demand of the university students do not match, and the price is on the high side. Enterprises should make accurate orientation to campus market and construct a campus e-commerce platform for school's special service, to let students have more convenient accesses to online shopping.

Improve the credibility and satisfaction degree of online shopping. For students can't ensure enough time online, businesses can only rely on credit degree and customer satisfaction to catch the vast number of consumers. As long as they agree on online shopping, recognize the quality of products, they will be very easy to fall in love with this pattern of consumption and take what he needs conveniently online. Business reputation is built on product quality and service quality, so businesses should always pay attention to product quality and service quality, and enhance consumers' satisfaction with online shopping.

Strengthen the security mechanism and legal system construction. China should further enhance network security and build a complete e-commerce security system. At the same time, we should strengthen the legal construction of electronic business affairs, develop specialized e-commerce laws and regulations and crack down on online fraud, counterfeit, etc. Thus, the consumers' rights and interests have a reliable guarantee. We can form a set of mature and perfect legal system and make our campus e-commerce grow sturdily in a fair and safe environment.

Strengthen the design and construction of shopping site. In the web page design, businesses should consider to meet the needs of different users, allowing users on the site to get a good browsing experience and purchasing experience. Such as the provision of product information and technical information, is convenient for users to obtain the desired product, technical information; to provide product knowledge and the link, is convenient for users to deep understanding of product and get help from other website; through the user mailing list, users can register freely and understand the latest developments of the site and the businesses can release news timely.

Expand the campus market by means of franchising. Setting up sub-stations in various colleges and universities by means of franchising way, network group purchase is apt to achieve economies of scale in a short time. Along with the continuous perfect city planning, University Town is ceaseless emerge in large numbers. The campus group purchase can make full use of the geographical advantage to realize regionalization and localization of the goods and reduce the logistics cost of the network group purchase greatly. The group purchase market locked in the university campus can avoid numerous and miscellaneous business prevalent in the current group purchase website, creating suitable for undergraduate consumption group purchase items, such as: computer group purchase group purchase training, various educational institutions, the holiday travel, hairdressing and fitness and so on.

Provide diversified,and personalized service for users Businesses can make full use of the electronic commerce technology, to meet the needs of students of this special consumer group. Businesses should provide one to one service in the form of 'tailor-made', give full consideration to every potential consumer's personality, and give sufficient attention and satisfaction, so that every college student consumer gets the most satisfactory services.

3 Conclusions

The campus e-commerce development has great potential, but its specialized consumer groups and innovative marketing behavior can also be a test for the businesses whether they can be based on the campus e-commerce development. The businesses only catch the opportunity of global information, recognize campus e-commerce characteristics and existing problems, adopt targeted marketing strategy, make accurate market positioning. Only in this way, college students can get more benefits from campus e-commerce and enterprises can increasingly develop and grow in the expansion of the campus e-commerce market.

Acknowledgments. This work is supported by the Planned Project of Department of Science and Technology of Hebei Province and Planned Project of Department of Education of Hebei Province and the Planned Project of Department of Science and Technology of Shijiazhuang City.

References

[1] Deng, J.: Analysis of factors affecting college students online consumption. Journal of China Market, 101–103 (2010)
[2] Ding, Z.: Discussion on the campus e-commerce development in China. Journal of E-commerce, 40 (2009)
[3] Wu, Q., Wang, Y.: Campus electronic commerce developing problem and countermeasure. Journal of Jiangsu Commercial Forum, 53–55 (2005)
[4] Chen, T.: Undergraduate e-commerce consumer research. Journal of Talents, 72–73 (2010)
[5] Zhang, Y., Cai, R.: The digital campus URP solutions. Journal of Huzhou Normal College, 81–84 (2007)

Application and Design of Artificial Neural Network for Multi-cavity Injection Molding Process Conditions

Wen-Jong Chen and Jia-Ru Lin

Department of Industrial Education and Technology, Changhua University of Education, No.2, Shi-Da Road, Changhua City, Taiwan
vinyan53@gmail.com, apple790606@hotmail.com

Abstract. In this study, an artificial neural network (ANN) with a predictive model for the warpage of multi-cavity plastic injection molding parts. The developed method in this paper indicate that the minimum and the maximum warpage were lower than that of CAE simulation. These simulation results reveal that the optimal process conditions are significantly better than those using the genetic algorithm method or CAE simulation.

Keywords: Artificial Neural Network, Multi-cavity, Genetic Algorithm, Warpage.

1 Introduction

In recent years, the plastic injection molding technology has played an important role in both high-tech and traditional industries. Its diversity, complexity, and production requirements achieved in a short period of time are competitive features essential to the mold industry. To reach the goals mentioned above, the best way is to prevent warpage phenomena. There are many factors affecting warpage problem need to be considered. These are target pressure, mold temperature, melting temperature, packing pressure, packing time, cooling temperature, and cooling time [1]. In this study, an efficient optimization method by coupling artificial neural network (ANN), genetic algorithm (GA) and finite element method (FEM) are applied to minimize warpage of multi-cavity injection molding parts. B.H.M. Sadeghi [2] utilized a back-propagation neural network system to reduce the time required for planning and optimizing of process conditions or operating parameters. In addition, the combination of ANN and GA have also been used in single-cavity mold injection[3][4].

Most of the past studies concentrate on the warpage process parameters proposed for single cavity part. This paper combines the ANN and GA methods to derive the optimal design of the process parameters and perform the runner balance for multi-cavity molds.

2 Methodology

This study first employed SolidWorks 2009 to produce an H-shaped finished part 65 mm long, 45 mm wide, and 10 mm thick. Afterward, the MoldFlow software imported

D. Jin and S. Lin (Eds.): Advances in FCCS, Vol. 2, AISC 160, pp. 33–38.
springerlink.com   © Springer-Verlag Berlin Heidelberg 2012

the file and meshed into 10,348 tetrahedron elements in FE model. In this paper, the properies of the material for parts are as follows (1) Specific heat is 2740 J/kg-C, (2) Thermal conductivity is 0.164 W/m-C, (3) Young's modulus is 1340 MPa, (4) Poisson's ratio is 0.392.

2.1   Contruct BPN Model

The predictive model was completed using BPN as a modeling structure. BPN is currently the most representative and the most popular supervised neural network in ANN learning model [5][6]. It includes the input layer, hidden layers, and the output layer (as seen in Fig.1). In the network infrastructure, there exist transfer functions between the input layer and the hidden layers and between the hidden layers and the output layer [7]. In this paper, the transfer function of the input and hidden layers is the tansig function, while the transfer function of the hidden and output layers is purelin function. The configure diagram in of this model is as seen in Fig.2, during the learning process, the transfer function of the input and hidden layers is the tansig function , as shown in Eq. (1), while the transfer function of the hidden and output layers is purelin, as shown in Eq. (2).

$$f(x) = \frac{e^x - e^{-x}}{e^x + e^{-x}}, \qquad g(x) = ax + c \qquad (1)$$

Where x is net internal activity level, $-\infty < x < \infty$; a and c are constants ;   and are the output of each neuron in different layers.

Equations (2) are then used to calculate the value of the output layer.

$$net_k = \sum_i w_{kh} \cdot h_h - b_k, \qquad y_k = f(net_k) = a \cdot net_k + c \qquad (2)$$

In Eqs. (2),indicates the initial weight values of the hidden layers and the output layer; b_k indicate the bias between the hidden layers and the output layer respectively; h_h indicates the value of the hidden layers. Depending on the error between the expected outputs (target value) and the corresponding outputs of the neurons, indicates the local gradient of the neurons of the output layers, and   indicates the local gradient of the neurons of an arbitrary hidden layer.

The new weight values and can be obtained by adding the original weights to the weighting corrections. They are repeatedly calculated until the results converge. Therefore, the weights revised within the margin of error can be derived through the repeated calculation of the above equations.

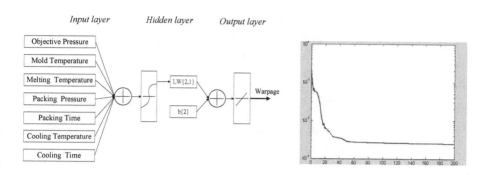

Fig. 1. BPN Structure   **Fig. 2.** Convergence diagram in BPN

3   ANN/GA Step Description

The research flowchart diagram is shown in Fig. 4 and the steps are described as follows,

Step1. Build a multi-cavity model on the SolidWorks 2009 software and export to the Moldflow 2012 software.
Step2. Analyze the multi-cavity model.
Step3. Identify the 36 sets of parameter conditions from the CAE simulation.
Step4. Designate and establish the 30 sets of CAE simulation parameters as training samples in BPN model (as seen in Fig. 3(a)-(b). If this is the case, proceed to Step 4. If not, identify the reasons and make corrections and return to Step 2.
Step5. Obtain the optimal process parameters and runner diameters.
Step6. Finish the contruct of prediction model.
Step7. Stop.

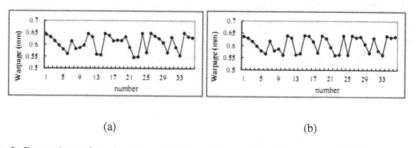

(a)   (b)

Fig. 3. Comparison of results. (a) prediction models and (b) thirty-six sets of CAE simulation.

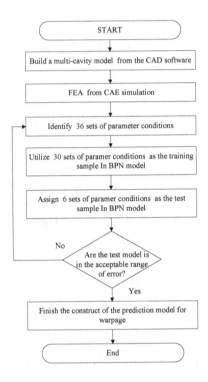

Fig. 4. Research steps of ANN/GA

4 Simulation Results

(1) CAE Simulation. As shown in Fig. 5 and Fig. 6, in CAE Simulation, the objective pressure was 13.6 Mpa, and the runner diameters α and β were 6.89 mm and 4.01 mm, respectively; T_{mold} was 54 °C; T_{melt} was 210 °C; P_{hold} was 68 MPa; t_{hold} was 26 sec; the T_{cool} was 60 °C; and t_{cool} was 12 sec.

(2) ANN predictive model. After performing the same number of calculations and achieving convergence and runner balance through the ANN method, the runner diameters α and β reached 7.04 mm and 4.1 mm, respectively. The objective pressure was 10 MPa; T_{mold} was 47.14 °C; the plastic melting-point temperature was 210 °C; P_{hold} was 80 Mpa; the t_{hold} was 15.71 sec; T_{cool} was 60 °C; and t_{cool} was 15 sec.

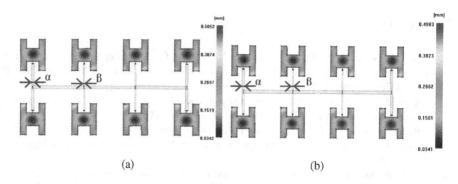

Fig. 5. The warpage distribution of FE simulation. (a) before optimization and (b) after optimization.

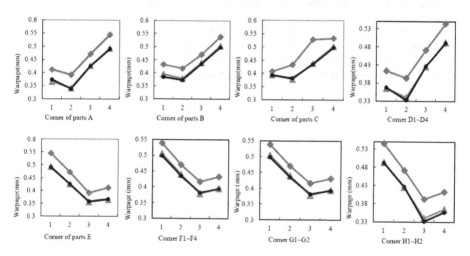

Fig. 6. Warpage of the eight parts in multi-cavity injection mold. (◆ : CAE , ▲ : GA , ● : ANN/GA.)

5   Conclusion

In this study, obviously these results were superior in ANN when compared to the figures obtained through the CAE simulation. They maximum, minimum, and average warpage values were still better than those derived through the GA method. In application of the multi-cavity injection mold with multiple different parts, these results can provide exceptional design methods and increase the credibility of the process parameters.

References

1. Hassan, H., Regnier, N., Pujos, C., Arquis, E., Defaye, G.: Modeling the effect of cooling system on the shrinkage and temperature of the polymer by injection molding. Applied Thermal Engineering 30(13), 1547–1557
2. Sadeghi, B.H.M.: A BP-neural network predictor model for plastic injection molding process. Journal of Materials Process Technology 103, 411–416 (2000)
3. Kurtaran, H., Ozcelik, B.: Warpage optimization of a bus ceiling lamp base using neural network model and genetic algorithm. Journal of Materials Process Technology 169, 314–319 (2005)
4. Ozcelik, B., Erzurumlu, T.: Comparison of the warpage optimization in the plastic injection molding using ANOVA, neural network model and genetic algorithm. Journal of Materials Process Technology 171, 437–445 (2006)
5. Wang, T.-Y., Huang, C.-Y.: Optimizing back-propagation networks via a calibrated heuristic algorithm with an orthogonal array. Expert Systems with Applications 34(3), 1630–1641 (2008)
6. Huang, C.-Y., Chen, L.-H., Chen, Y.-L., Chang, F.M.: Evaluating the process of a genetic algorithm to improve the back-propagation network: A Monte Carlo study. Expert Systems with Applications, Pt. 1 36(2), 1459–1465 (2009)
7. Hagan, M.T., Demuth, H.B., Beale, M.: Neural Network Design, Inc., p. 978

Query Assurance Verification for Outsourced Database

Pengtao Liu

Institute of Information Science and Technology, Shandong University of Political Science and
Law, Jinan, P.R. China
ptwave@163.com

Abstract. In outsourced database model, database faces potential threats from
malicious database service providers. To provide correctness, completeness and
freshness of query result sets, security mechanisms are needed to assure that
queries are executed against the most up-to-date data, and the query results have
not been tempered with and are authentic with respect to the actual data owner.
As an improvement of the existing authenticated-data-structure-based method
called chain embedded signature, a new integrity checking scheme is proposed.
The scheme redefines the predecessor and successor tuples so that it can handle
the situation with duplicate attribute values. Cryptographic dynamic
accumulator is used to ensure freshness of query results.

Keywords: database security, outsourced database integrity, query authentication,
freshness, dynamic accumulator.

1 Introduction

Outsourced database mode(ODB)[1] is a new type of database application form. In
ODB, there are three major participants: the database owner, a third-party database
service provider and database users. Database service provider offers adequate
software, hardware and network resources to host its clients' databases as well as
mechanisms to efficiently create, update and access outsourced data. A database
owner stores its data at an external, and potentially untrusted, database service
provider. It is thus important to secure outsourced data from potential attacks not only
by malicious outsiders but also from the service provider itself. There are two types of
outsourced database security issues: data confidentiality and integrity protection.
Encryption can be used to protect sensitive data. Integrity protection means that the
database user can verify the query results and can be used to prevent database service
providers' tampering of data and query results. Integrity protection solutions must
include correctness and completeness. Correctness enables secure and efficient
authentication of tuples contained in all possible query replies. Completeness implies
that the querier can verify that the server returned all tuples satisfying query
predicates. Current known schemes that provide query integrity over outsourced
databases cannot provide freshness guarantee which means that all database update
operations have been correctly carried out by the service provider and the query
results come from the up-to-date outsourced data.

In this paper, an improved query result integrity verification method of outsourced
database model is proposed, which embeds the verification object(VO) inside the

D. Jin and S. Lin (Eds.): Advances in FCCS, Vol. 2, AISC 160, pp. 39–44.

ODB tables to reduce the verification workload. Furthermore, it uses the dynamic accumulator to support the freshness of the query results.

2 Preliminaries

Basic Query. Suppose to have a relational table T with attributes $(A_1,...,A_m)$. The primary key is A_1. We always assume that attribute types have orders and comparison operators which is the case in the vast majority of practical situations. We call basic query a selection query in the following form: q_s: SELECT * FROM T WHERE A_i BETWEEN c_l AND c_h, where c_l,c_h are range of A_i, satisfying:$-\infty<c_l \leq c_h<+\infty$.

Query condition "$A_i=z$" is a special case of q_s when $c_l = c_h$. Other types of queries can be converted to the corresponding basic queries.

Accumulator. A secure dynamic accumulator consists of the five algorithms[2]. AccGen(1^k, n) creates an accumulator key pair (sk_A, pk_A), an empty accumulator acc_ϕ and an initial state $state_\phi$; AccAdd(sk_A, i, acc_V, $state_U$) allows the authority to add i to the accumulator. It outputs a new accumulator $acc_{V \cup \{i\}}$ and state $state_{U \cup \{i\}}$, together with a witness wit_i for i; AccUpdate(pk_A,V, $state_U$) outputs an accumulator acc_V for values $V \subset U$; AccWitUpdate(pk_A, wit_i,V_w, acc_V ,V, $state_U$) outputs a witness wit'_i for acc_V if wit_i was a witness for acc_{V_w} and $i \in V$; AccVerify(pk_A, i, wit_i, acc_V) verifies that $v \in V$ using an up-to-date witness wit_i and the accumulator acc_V. In that case the algorithm accepts, otherwise it rejects.

A secure dynamic accumulator under strong RSA assumption in [3] can be applied in our method.

3 Prior Works

In this section, we discuss several approaches to implement query integrity including authenticated data structure based approaches [1] and probabilistic based approaches [4].

In authenticated data structure(ADS) based approaches, completeness guarantees are provided by signing the sorted list of date attribute values. The most common ADSes are Merkle hash tree(MHT)[5] and signature aggregation[6]. MHT[5] is applicable only for static scenario. In signature aggregation [6] method, one signature is based on three consecutive tuples instead of one tuple. However, it has high authentication cost and limited authentication ability. Zhang et al. proposed a server transparent query authentication method called chain embedded signature(CES)[7], which embeds the VO inside the outsourced database. It frees the server from heavy verification tasks, and its time and space overhead are reasonable. However, there are flaws of the precursor tuple and successor tuple definition which makes it unable to deal with duplicate attribute values(we will explain it in the following example) and it could not provide freshness guarantees.

Probability based approach can only provide authentication in part. There have been two probability authentication approaches: fake tuple[4] and challenge-response[8]. The basic idea of fake tuple is as follows: certain percentage of random fake tuples are inserted into outsourced real tuples. Through authentication on fake tuples, authentication on real tuples is inferred with certain probability. However, fake tuple approach brings additional storage cost at database service provider and clients. In challenge-response based approach, for each batch of client queries, the server is challenged to provide a proof of query execution which is then checked at the client site as a prerequisite to accepting the actual query results as accurate. However, it could not provide completeness authentication and freshness guarantees.

To provide freshness guarantees, some methods are proposed. In [9], Xie et al. add timestamps to data signatures in authenticated data structure based scheme. Pang et al use bitmap summary of the records and timestamp in tuples to provide freshness guarantees [10]. Every ρ seconds, the data aggregator issues a certified bitmap summary of the records that have been updated. The periods ρ is short, so the bitmap is likely to be sparse and amenable to compression. However, the number of bitmaps is very large and the compression and decompression workload would be heavy. Furthermore, the short bitmap update period means that data owner is almost on-line all the time which is suitable for the practical application.

4   Our Improved Method

4.1   Redefinition of Precursor Tuple and Successor Tuple

Sort the tuples of table $R(A_1,\ldots, A_m)$ in increasing order according to the attribute value for each searchable attribute A_i ($1 \leq i \leq m$). If two tuples has same A_i values, their order rely on the primary key(A_1) of R. Let *seq* be the Monotone non-decreasing sequence.

Definition 1: Immediate Precursor tuple and Successor tuple[7]. t_b is the ith precursor tuple of t_a ($t_b=P_a^i$) iff it satisfies the following conditions:

1) $t_b. A_i \leq t_a. A_i$; and
2)   there is no tuple t_c ($t_c \neq t_a$, $t_c \neq t_b$), satisfying $t_b.A_i < t_c.A_i \leq t_a.A_i$, or $t_b. A_i = t_c. A_i$ and $t_b.A_1 < t_c.A_1$. t_a is called the ith successor tuple of t_b ($t_a=S_b^i$).

For the first tuple in *seq*, its precursor tuple is itself. For the last tuple in *seq*, its successor tuple is itself. If R has m attributes, then each tuple in R has m precursor tuple and m successor tuple.

Fig 1 gives an example table R which has three attributes (A_1, A_2, A_3) and A_1 is primary key. For tuple t_3, its first precursor tuple and second precursor tuple are t_2, and its third precursor tuple is itself. ($P_3^1= t_2$, $P_3^2= t_2$, $P_3^3= t_3$).

	A1	A2	A3
t1	3	7	10
t2	5	7	9
t3	6	9	5
t4	8	1	8

Fig. 1. An example table R

The ith precursor tuple and successor tuple must be unique. However, for t_2. A_2, as the definition 1, we can get two immediate precursor tuple--t_2 and t_1. So, we should modify the definition of precursor tuple and successor tuple.

Definition 2: Immediate Precursor tuple and Successor tuple. t_b is the ith precursor tuple of t_a ($t_b=P_a^i$) iff it satisfies the following conditions:

1) $t_b.A_i \leq t_a.A_i$; and
2) if $t_b \neq t_a$, there is no tuple t_c ($t_c \neq t_a$, $t_c \neq t_b$), satisfying $t_b.A_i < t_c.A_i \leq t_a.A_i$, or $t_b.A_i = t_c.A_i$ and $t_b.A_1 < t_c.A_1$.
3) if $t_b = t_a$, there is no tuple t_c ($t_c \neq t_a$, $t_c \neq t_b$), satisfying $t_c.A_i = t_b.A_i$, and $t_c.A_1 < t_b.A_1$.
t_a is called the ith successor tuple of t_b ($t_a=S_b^i$).

By this definition, for t_2. A_2, its immediate precursor tuple is t_1.

4.2   Providing Freshness Guarantees by Dynamic Accumulator

Taking each tuple as an element of the dataset, data owner accumulates all the tuples and publishes the signature of the accumulated value. Data service provider calculates the accumulated value once and updates the value with data updating. Data service provider must return both the query results and witness that the query results have been accumulated in the up-to-date accumulated value. The user verifies the witness and ensures that the query results come from the up-to-date data. We describe the method as follows:

(1) Database setup. Similar to [7], VO is embedded in the table as follows:

1) Insert two additional boundary tuples $t_{\text{low}}=(-\infty,\ldots,-\infty)$ and $t_{\text{high}}=(+\infty,\ldots,+\infty)$. t_{low} is the first tuple and t_{high} is the last tuple in R;
2) Find immediate precursor tuples of each tuple t_k as $(P_k^1 \ldots P_k^m)$;
3) Add a new attribute C as VO. $t_k.C=\text{zip}(\ldots, t_k.C_i,\ldots)$, where $t_k.C_i = P_k^i.A_1$;
4) Add a new attribute A_{sign}, $t_k.A_{\text{sign}} = \text{SIG}[h(t_k.A_1\|\ldots\| t_k.A_m\| t_k.C)]_{PR}$;
5) Generate accumulator key by AccGen. The key is shared by data owner and database service provider. Let $T=(t_1,\ldots, t_n)$ be the set of all tuples and get the accumulated value acc_T of T by AccAdd. Data owner publishes the signature of acc_T, $S= \text{SIG}[h(acc_T)]_{PR}$.

(2) Authenticating Query Results. The server returns query results which VO is embedded in. In addition, it also returns the accumulated value acc_T of current dataset and the witnesses wit_is that the result tuples are accumulated in the value (computed by AccWitUpdate).

i) Correctness and Completeness. Tuple signature is used to ensure the accuracy of result. In order to provide the completeness, users need to get the boundary tuples which are just beyond the query range and verify the tuple chain formed upon immediate precursor tuple (in $t_k.C$) from low boundary tuple to high boundary tuple:

Step 1. Boundary verification. Let t_x be the low boundary tuple and t_y be the high boundary tuple. If $t_x.A_i < c_l$, $t_y.A_i > c_h$, then go to Step 2, else it fails.

Step 2. Tuple signature verification. Verify signature of each tuple. If there are corrupted tuples, then their signatures will not match. Otherwise, go to the next step.

Step 3. Form a chain upon immediate precursor tuple (stored in $t_k.C$) from t_y to t_x. The tuple chain proves to the querier that the server has indeed returned all tuples in the query range. Specifically, the querier verifies that the values in the boundary tuples are just beyond the range posed in the query. At the same time, the querier verifies that there are no other tuples in between the boundary tuples and the immediately adjacent tuples which fall within the query range. This is because the boundary values are linked to the first and the last tuple.

ii) Freshness. Check the signature of acc_T to ensure that it is the accumulated value of up-to-date data. Then, for each tuple t_i, check the witnesses wit_i that the tuple has been accumulated in acc_T. If all checks are succeeded, users are guaranteed about freshness of the results.

(3) Database Updates. The most important issue of an update operation is to recalculate embedded VO. As in [7], if the sequence of tuples does not change, just update the corresponding tuple. Otherwise, it needs to update not only current tuple but also at most three relevant tuples.

Meanwhile, data owner needs to efficiently update acc_T by AccUpdate and publishes its new signature. Database server also needs to update acc_T by AccUpdate.

5 Analysis

We now illustrate the performance of our approach.

Query results verification cost. It includes: 1) Communication cost C_{comm}, which is linearly with result tuples number r; 2) Tuple signatures verification cost. There are two boundary tuples and r result tuples, so the cost is $(C_h+C_{verify}+C_{IO})(r+2)$; 3) Accumulated value signature verification cost, $C_h+C_{verify}+C_{IO}$; 4) Witness verification cost, rC_{acc_ver}.

So, query results verification cost is $C_{ver}=C_{comm}+(C_h+C_{verify}+C_{IO})(r+3)+rC_{acc\_ver}$

Updating cost. 1) Insert. If the sequence of tuples does not change, it just needs to modify the C attribute and signature of the inserted tuple's immediate successor tuple, $C_{insert}=2(C_h+C_{sign}+C_{IO})$; otherwise, it needs to modify at most m tuples, so $C_{insert}=(C_h+C_{sign}+C_{IO})(m+1)$ at worst. 2) Update. If the sequence of tuples does not change, it just needs to modify the C attribute and signature of the updated tuple's immediate successor tuple, so $C_{update}=2(C_h+C_{sign}+C_{IO})$; otherwise, it needs to modify at most three tuples, so $C_{update}=3(C_h+C_{sign}+C_{IO})$ at worst.

6 Summary

In this paper, an improved query result integrity verification method of outsourced database model is proposed, which embeds the verification object(VO) inside the ODB tables to reduce the verification workload. Furthermore, it uses the dynamic accumulator to support the freshness of the query results. We only analyze the performance theoretically and we will analyze the actual performance of our approach.

Acknowledgements. This work is supported by the Nature Science Foundation of Shandong Province under Grant No.ZR2010FM042,ZR2011FQ019, and the Research Planning Foundation of Shandong University of Political Science and Law under Grant No. 2011Z02B.

References

1. Li, F., Hadjieleftheriou, M., Kollios, G.: Dynamic authenticated index structures for outsourced database. In: ACM SIGMOD 2006, pp. 121–132. ACM, New York (2006)
2. Benaloh, J., de Mare, M.: One-way Accumulators: A Decentralized Alternative to Digital Signatures. In: Helleseth, T. (ed.) EUROCRYPT 1993. LNCS, vol. 765, pp. 274–285. Springer, Heidelberg (1994)
3. Camenisch, J., Lysyanskaya, A.: Dynamic Accumulators and Application to Efficient Revocation of Anonymous Credentials. In: Yung, M. (ed.) CRYPTO 2002. LNCS, vol. 2442, pp. 61–76. Springer, Heidelberg (2002)
4. Xie, M., Wang, H., Yin, J., et al.: Integrity Auditing of Outsourced Data. In: VLDB 2007, pp. 782–793. ACM, New York (2007)
5. Merkle, R.C.: A Certified Digital Signature. In: Brassard, G. (ed.) CRYPTO 1989. LNCS, vol. 435, pp. 218–238. Springer, Heidelberg (1990)
6. Narasimha, M., Tsudik, G.: DSAC: An Approach to Ensure Integrity of Outsourced Databases using Signature Aggregation and Chaining. In: ACM CIKM 2005, pp. 235–236. ACM, New York (2005)
7. Zhang, M., Hong, C., Chen, C.: Server Transparent Query Authentication of Outsourced Database. Journal of Computer Research and Development 47(1), 182–190 (2010)
8. Sion, R.: Query execution assurance for outsourced databases. In: VLDB 2005, pp. 601–612. ACM, New York (2005)
9. Xie, M., Wang, H., Yin, J., et al.: Providing freshness guarantees for outsourced database. In: EDBT 2008, pp. 323–332. ACM, New York (2008)
10. Pang, H., Zhang, J., Mouratidis, K.: Scalable Verification for Outsourced Dynamic Databases. In: VLDB 2009, pp. 802–813. ACM, New York (2009)

Model and Algorithms on Utilization Plan of Shunting Locomotive for Railway Passenger Station[*]

Changfeng Zhu[**]

School of Traffic and Transportation Engineering, Lanzhou Jiaotong University,
Lanzhou,730070, China
cfzhu003@126.com

Abstract. Utilization plan of shunting locomotive is a complicated system engineering that is influenced by multitudinous factors, such as the time of arriving and departing trains, number of trains and so on. The optimization model was established using stratified sequence theory, and algorithms were proposed to this problem. And optimization goal of shunting locomotive operation was reached by adjusting the passenger carriage of the earliest possible placed-in and the latest possible taken-out time for garage. Meanwhile, tight connecting time and the shortest idle running distances of shunting locomotive were also taken into consideration to meet the requirements of the practical operation where delayed trains and extra trains can depart on time. Finally, a case study is carried out to verify the validity and applicability of this model and its algorithm, from the test results, it is shown that the operation scheme of shunting locomotives is not only balance, but also efficient.

Keywords: Shunting locomotives, Optimization algorithm, System engineering.

1 Introduction

The shunting locomotive scheduling of railway large-scale station is the core of the railway passenger transportation scheduling. Recently, a number of research efforts have sought to address the above problem [1-8]. However, most of the modeling and the algorithm in print aimed at shunting locomotive and parking lines for rolling stock as the research objects and applying the scheduling theory[4], [5], [6]. Meanwhile there are some problems should be carried on further study as

1) Operation balance of shunting locomotive was only considered at the train daily and shift operation plans.
2) The constraint was not considered that the arrival and departure of two neighboring passenger carriages should satisfy the minimum time interval.
3) Harmonization between operation in throat area was ignored.
4) Idling running distance of shunting locomotive was no considered.

When there are some parallel operation existing in throat area and restrictions on the amount of shunting locomotives, capacity of arrival and departure lines becomes

---

[*] Project supported by the Spring sunshine plan of Ministry of education(Z2005-1-62008).
[**] Corresponding author.

especially important. When there are lots of arrival and departure lines, shunting locomotives are very busy. Otherwise, shunting locomotives are relative idle.

2 Establishment of Optimization Model

Based on the existing research, this paper had a combination of equipment and train operation organization, and also the below constraints were considered.

1) Idle running distance of shunting locomotive is placing-in account when passenger carriages and locomotives are not in the same place. Idle running distance during placing-in and taking-out process was considered comprehensively.

2) The time interval between two neighboring passenger trains using the same arrival or departure line should satisfy the relevant requirements for safe. Due to the restricts on arrival and departure lines capacity of railway passenger station and a certain time needed for trains to occupy lines.

3) Considered operation capacity of arrival and departure lines of railway passenger station. When there are some parallel operation existing in throat area, and restrictions on the amount of shunting locomotives, capacity of arrival and departure lines becomes especially important.

4) Optimization adjustment of shunting locomotive operation plan are considered under the abnormal conditions of passenger-flow rush hours such as spring festival, farmers' frenzied hunt for work in cities and facilities, and under the abnormal conditions of facilities faults.

2.1 Problem Statement

Suppose that K_0, K_1, K_2 represent respectively the set of shunting locomotives in railway passenger station, the set of shunting locomotives which are for placing-in and taking-out for garage, f is the amount of arrival and departure lines in passenger station, S_{kj} is the total running distance of k-th shunting locomotives placing-in and taking-out j-th passenger carriage. ω_j is the train level, r_j is the arrival time of passenger carriages, the earliest possible time of placing-in for garage is r_{ij}^{e}, Zdt_j is train arrival time of passenger carriage i. T_{zd} is standard operation time of arrival trains on the arrival and departure lines, standard parking time of passenger carriage on parking lines is T_{zb}, T_{zb}^{j} is servicing work time of carriage J_i on parking lines. d_j is deadline of corresponding job. ct_j^{l} is the latest possible time of taking-out from garage, sft_j is the departure time of passenger carriage i, T is standard operation time of taking-out and placing-in carriages by shunting locomotives, standard operation time of departure train on the arrival and departure lines is t_{cf}.

2.2 Establishment of Optimization Model

Suppose $A_t = (t = 1,2,\cdots,k_1)$ is adjustment time orderly that shunting locomotives placing-in and taking-out carriages which belong to the first processing center. And $B_t = (t = 1,2,\cdots,k_2)$ is the relevant values of the second processing center. Besides, the difference between placing-in and taking-out operation time is small, so variance S_χ^2 of which is the frequency $\chi_k \in \{0 \le \chi_k \le n\}$ of the shunting locomotive E_q occupied, D_j is employed to describe the situation that the trains on scheduling and behind scheduling, where $D_j=1$ represent the train belonging to passenger carriage J_i cannot run on time, $D_j=0$ represent the train can run on time. Passenger carriages taking-out and placing-in operation can be regarded as flow processes of process operation where the minimum total tardy process operation $\sum D_j$ are the first optimization objective. The second objective is to reach the minimum variance S_χ^2 for the balance of shunting locomotives with two processing centers and a limited buffer. The third objective is to reach the minimum total running distance of taking-out and placing-in operation for all carriages. Hence, the improved optimization model of shunting locomotive operation scheme in railway passenger station is denoted as

$$FF_2 / K, f, S_{kj}, \omega_j, r_j, T_{zb}^j, d_j, A_t, B_t, T - W / \sum_{k=1}^{K} D_j, \sum_{k=1}^{K}(\chi_k - \sum_{l=1}^{K} \chi_l / K)^2, \sum_{j=1}^{n}\sum_{k=1}^{K} S_{kj} \quad (4)$$

Where

$$S_\chi^2 = \sum_{k=1}^{K}(\chi_k - (\sum_{l=1}^{K}\chi_l)/K)^2$$

The value of train state ϖ_j is equal to the value of train characteristic φ_j multiplying by train degree $\varpi_j = \varphi_j \times \lambda_j$. The relevant values of departure train(DT), arrival train (AT) and pass by train (PT) are shown in Table1.

Table 1. The relevant values for every category train

	Electric multiple unit			Express train			Fast train			General passenger train		
φ_j	0.6			0.7			0.8			0.9		
λ_j	DT	AT	PT	DT	AT	PT	DT	AT	PT	DT	AT	PT
	0.4	0.3	0	0.4	0.3	0	0.4	0.3	0	0.4	0.3	0
ϖ_j	0.24	0.18	0	0.28	0.21	0	0.32	0.24	0	0.36	0.27	0

3 Solution Algorithm

Step1. For taking-out, let $x'_j = r_j$, If the latest taking-out time $x'_j < y'_j$, then $y'_j = d_j$, otherwise, $y'_j = d_j = t_{day}$, where $t_{day} \in \{t_{day} \ge 1440, t_{day} \in Z\}$.

Step2. If the earliest placing-in time is the same, then carriages are sequenced by time order of the latest taking-out time, $x'_j > x'_{j+1}$ is set up when $y'_j < y'_{j+1}$ and $x'_j = x'_{j+1}$. Otherwise, $x'_j < x'_{j+1}$.

Step3. $2n$ items belonging to sequence I and sequence II are consisted of sequence III in sequence, where the next two items should satisfy any of the below inequalities $y'_j < y'_{j+1}$ or $x'_j > x'_{j+1}$, $y'_j < y'_{j+1}$ or $x'_j < x'_{j+1}$.

Step4. For passenger carriage J_i belonging to sequence III. If J_i belongs to sequence I, then shunting locomotive $E_j (E_j \in Z_1)$ is arranged. If E_j is free and has less operating frequency, with long waiting time, then it can be occupied. Then θ_k and A_t or B_t will be modified, $\chi_i = \chi_{i+1}$. If there are no suitable shunting locomotives for this passenger carriage, then time of placing-in and taking-out will be adjusted, then turns to *step2*. If J_i belongs to sequence II, then $E_j (E_j \in Z_2)$ is arranged, if E_j is free and has less operating frequency, then it can be occupied. Next, the current state and corresponding waiting time θ_k and adjustment time A_t or B_t will be modified, $D_j=0$, $j=j+1$. If there is no suitable shunting locomotive for this carriage ($D_j=1$), then placing-in and taking-out time of passenger carriage will be adjusted. And it turns to *step2*.

Step5. Output placing-in operation scheduling, taking-out operation scheduling and usage of shunting locomotive. Summarize $\sum D_j$ and S_χ^2.

4   Empirical Analysis

From18 o'clock to the same time on the next day, there are 21 departure and arrival passenger trains needed to have placing-in and taking-out operation. The arrival time of arrival trains and the departure time of departure trains is shown in Figure 1.

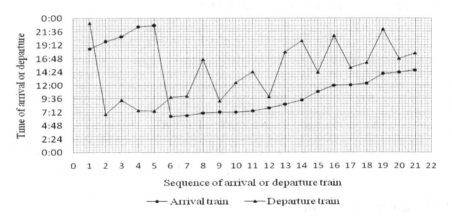

Fig. 1. The arrival time of arrival trains and the departure time of departure trains

The departure operation time of corresponding trains is 40 min, the arrival operation time of corresponding trains is 30 min, the minimum operation time the arrival operation is more than 25 min, servicing work time for carriages is more than 30min, there are 3 shunting locomotives in the stations. At 18 o'clock. D1 was in the garage, D2 was in the arrival yard, and D3 was in the departure yard.

Based on above original data, the optimization results were achieved using simulation calculation platform matlab6.0. And the optimization results of shunting locomotive scheduling of this passenger station during the period which is from 18 o'clock to the same time on the next day was Shawn in Table 2 and Table3.

Table 2. Placing-in operation of arrival passenger trains

					Utilization plan of shunting locomotives					
Train number	K428	7502	7516	T926	T213	K108	K223	K961	118	T21	263
Arrival time at station	18:11	18:46	19:41	20:27	21:38	22:27	23:35	4:06	5:08	5:16	5:56
Arrival time at garage	18:31	19:18	19:11	20:57	22:08	22:57	23:05	4:31	5:38	5:40	6:26
Shunting locomotive	D2	D3	D1	D2	D3	D1	D3	D1	D2	D3	D1

Table 3. Taking-out operation of departure passenger trains

					Utilization plan of shunting locomotives					
Train number	750	756	T920	T207	K110	K223	K967	182	T209	Z56	K859
Departure time from station	18:10	18:37	19:29	20:22	21:49	22:04	22:39	4:10	4:29	5:12	5:30
Departure time from garage	18:40	19:07	19:59	20:42	21:09	22:34	23:09	4:40	5:6	5:45	5:52
Shunting locomotives	D2	D1	D2	D3	D1	D3	D1	D3	D2	D1	D2

The optimization result from table1 and table2 show that the operation scheme of shunting locomotives in this station is balance, and the usage efficiency of shunting locomotives is higher, all trains could run on time which can meet the requirements of practical operation. Based on above research result, we also found that arrival operation time of K428 is 20min, correlative operation time in station of K428 is relative tight. To make sure the full usage of shunting locomotives, operation organization should be paid attention to in practical operation.

5 Conclusions

The shunting locomotive scheduling of railway station is the core of the railway passenger transportation scheduling, and is complicated system engineering. Based on the existing research, influencing factors such as the minimum time interval of two neighboring passenger carriages, harmonization between operation in throat area, capacity of arrival and departure lines, were considered completely. For the larger railway passenger station, this optimization model and algorithms can obtain result more effectively. This model mentioned above has the value to be referenced for the operation scheme of shunting locomotives.

Acknowledgments. This research project is supported by the Spring sunshine plan of Ministry of education Z2005-1-62008. The authors are grateful for the anonymous reviewers who made constructive comments.

References

1. He, Y., Mao, B.: Research on Simulative Calculation of Maximum Assembling of Railway Passenger Station. Journal of System Simulation 18, 213–224 (2008)
2. Zhang, T.: Research on the Assembling Rule of Passengers at Railway Passenger Stations. Journal of the China Railway Society 31, 31–34 (2009)
3. Shi, F., Chen, Y.: Comprehensive Optimization of Arrival Departure Track Utilization and Inbound Outbound Route Assignment in Railway Passenger Station. China Railway Science 30, 108–113 (2009)
4. Zhang, Y., Lei, D.-Y., Tang, B.: Due Windows Scheduling Model and Algorithm of Track Utilization in Railway Passenger Stations. Journal of the China Railway Society 33, 1–7 (2011)
5. Xu, W., Wu, Z.: Mathematical model for the passenger flow in subway station waiting rooms. Journal of Railway Science and Engineering 2, 70–75 (2011)
6. Jia, S., Wen, X.: Study on Safety Assessment Techniques for Subway Operation System Based on Analytic Hierarchy Process. China Safety Science Journal 5, 34–38 (2008)
7. Li, Q.: A System Assurance Study on Bi-directional Evacuation of A Metro Train on Fire. Urban Mass Transit 3, 34–39 (2009)
8. Zhang, Y., Lei, D.: Scheduling Model and Algorithm for Track Application in Railway Station. Journal of the China Railway Society 31, 96–100 (2010)
9. Zhang, Y., Lei, D.: Decision support system of track utilization with CTC at railway passenger station. Journal of Traffic and Transportation Engineering 1, 89–96 (2011)

The Application Research of CRM in e-Commerce Based on Association Rule Mining

Lan Wang[*] and HongSheng Xu

College of Information Technology, Luoyang Normal University, Luoyang, 471022, China
xhs_ls@sina.com

Abstract. Association rule mining in large amounts of data between sets of items found interesting or relevant associated contact is KDD (Knowledge Discovery in Database) is an important research topic. CRM is to achieve the needs of customers and to enhance the strength with customers for enterprise. This article describes the basic theory of association rule mining knowledge. This paper proposes the application research of CRM in e-commerce based on association rule mining. Algorithm analysis and experiments show that the methods proposed consumes less space and time than Apriori algorithm for the dense data.

Keywords: association rule, CRM, data mining.

1 Introduction

Data Mining (Data Mining, DM) is a database of knowledge discovery (Knowledge Discovery in Database, KDD) an integral part. It involves multiple disciplines, including: database technology, artificial intelligence, machine learning, neural networks, statistics, pattern recognition, knowledge-based systems, knowledge acquisition, information retrieval, high-performance computing and data visualization. Through data mining resources from the Internet-related behavior in the extraction, the paper uses of interest or useful patterns and hidden information [1]. Understanding of user behavior is the basis for Internet applications, but also the core. In order to tap more of the contents of the user a sense of Xing; tracking, analyzing the user's usage patterns; to improve the efficiency of network users, to detect the user's buying preferences, provide users with personalized services, to extend the user site dwell time, so it is necessary to use active web mining application mining research.

Association rule mining in large amounts of data between sets of items found interesting or relevant associated contact is KDD (Knowledge Discovery in Database) is an important research topic. With the large amounts of data continuously collected and stored many in the industry for the database from their growing interest in mining association rules. Business transaction records from a large number of interesting associations found that can help many business decisions making, such as the classification design, analysis of cross-shopping and cheap.

---

[*] Author Introduce: Wang lan (1967-), Female, Professor, Master, College of Information Technology, Luoyang Normal University, Research area: Data mining, CRM.

D. Jin and S. Lin (Eds.): Advances in FCCS, Vol. 2, AISC 160, pp. 51–56.
springerlink.com   © Springer-Verlag Berlin Heidelberg 2012

The users, the majority of the e-commerce sites simply try to provide information to the user, without taking into account the needs of users vary widely between, resulting in e-commerce site provides users the information loss. When users download a lot of time wasted on useless information itself, and the need to find a preference for information, its usage of e-commerce sites will be reduced. Users do not like to force users to e-commerce sites or vendors to accept the information they provide such as spam mail, access to certain web pages to force pop-up window, etc.

In this paper, it is about association rule mining. The first leads to association rules, association rules and introduce the basic concepts, basic theory. Association rule mining to find a given data set interesting links between projects. This article describes the basic theory of association rule mining knowledge. This paper proposes the application research of CRM in e-commerce based on association rule mining. The association rules algorithm, the rules limit the conclusions by the properties that appear to reduce the useless association rules, support or confidence through the threshold limit; it can reduce the number of rules.

2   The Research of Association Rule Mining

The so-called association rules, the expression of the project (Item) in the transaction between (Transaction) the relevance of the process. Usually expressed by the following example: A→B.

Association rules is a data mining research is an important research topic. Agrawal, Imieliski and Swami in 1993 for market basket analysis problem first proposed the concept of association rule; its purpose is to find the transaction database, the link between the rules of different commodities. For example, association rule mining can be used to find the transaction database in different commodities (items) the link between, and to find customers to buy goods behavior patterns, and then apply it to goods shelf design, stock arrangements, and buying patterns of the user according to the classification and so on. We usually ask the question, "Buy milk, bread, and customers purchase likelihood?", and "Most customers to buy a computer in which age"? Etc., which are part of data mining association rules need to be resolved. Thus, reflecting the large number of association rules between sets of data items interesting links.

Association rules defined as follows: Let $I = \{i1, i2, ..., im\}$ is the collection of data items (itemset), which is called the elements of the project (item)[2]. Transaction database D is a transaction (transaction) T set, where transaction T is a collection of items, namely: $T \subseteq I$. corresponding to each transaction has a unique identifier, denoted by TID. Let A be $a \subseteq I$ in the collection of items, if $A \subseteq T$, then said transaction T contains A, is shown by equation 1.

$$\left(\sum_{i_j \in X \cup Y} w_j \right) \times Support\,(X \cup Y) \tag{1}$$

An association rule is the implication of the form A⇒B style, here$A \subseteq I$, $B \subseteq I$. Rules $A \cap B = \varnothing$ in the transaction database D, the degree of support (support) that these two items A and B set in the transaction database D, the probability of simultaneously. Denoted as support (A⇒B), that support(A⇒B)=P(A∪B), is shown by equation 2.

$$SC(X) \geq \left[\frac{wminsup \times n}{W(Y,k)} \right]$$ (2)

Rules $A{\Rightarrow}B$ in the transaction database D, the confidence (confidence) is there in the affairs of the itemset A centralized, key set B also probability, denoted by confidence $(A{\Rightarrow}B)$, the confidence$(A{\Rightarrow}B)$=P(B | A), as is shown by equation 3.

$$P(x_{k,m} \mid SUR, SIR, I_1 = 1) = P(x_{k,m} \mid SUR)$$ (3)

Association rule mining problem is to find the transaction database D with a user-given minimum support and minimum confidence min_conf min_sup association rules. Agrawal et al in 1993 designed a basic algorithm (Apriori algorithm) is proposed for mining association rules is an important way - this is a set of ideas based on the frequency of two-stage method, the design of association rule mining algorithm can be decomposed for the two sub-problems.

(1) frequent itemset generation: The goal is to find to meet the minimum support threshold for all itemsets, these items are set called frequent itemsets (frequent itemset), or frequent patterns (frequent pattern). (2) Rule generation: its goal is the frequent itemsets have been found to extract all high confidence of rules is called a strong rule (strong rule). The similarity is calculated as follows equation 4.

$$\hat{x}_{k,m} = \sum_{r=1}^{|r|} rP(x_{k,m} = r \mid SUR, SIR, SUIR)$$ (4)

Application of association rules include analysis of customer shopping, catalog design, commercial advertising mail analysis, additional sales, product design shelves, storage planning, network fault analysis. For example, in market basket analysis found that the customer's buys patterns, and thus for market planning and advertising planning and buying patterns of customers according to the classification.

Frequent pattern mining is an important field of data mining basic research, since it has been proposed, they were academic attention, is still a hot topic at home and abroad. Most previous studies using Apriori and its improved algorithm to mining frequent itemsets, but these algorithms in the mining process will need to repeatedly scan the database and produce a large number of candidate items, which seriously affect the efficiency of the algorithm [3]. To this end, Jiawei Han, who proposed a set of candidates does not produce the FP-growth algorithm and only need to scan the database twice, experimental analysis shows that, FP-growth algorithm is faster than the Apriori algorithm such as an order of magnitude, as is shown by equation 5.

$$P(x_{k,\ m}|SUR, SIR) \equiv P(x_{k,m}|\{p_{k,m}(x_{a,b})|x_{a,b} \in SURUSIR\})$$ (5)

Association rule mining problem is to find the transaction database D with a user-given minimum support and minimum confidence min_conf min_sup association rules. Agrawal et al in 1993 designed a basic algorithm (Apriori algorithm) is proposed for mining association rules is an important way - this is a set of ideas based on the frequency of two-stage method, the design of association rule mining algorithm can be decomposed for the two sub-problems.

Algorithm 1: Based on the association rule subtree mining frequent itemsets
Input: According to the candidate generated using iterative layer by layer to find frequent itemsets support count min_count.
  Output: All weighted collection of frequent itemsets L in the process output
  (1) L_1=find_frequent_1-itemsets(D);
  (2) **for** $i = max\ order$ **down to** 1 **do**
  (3) for(i=1;i≤Size;i++)
  (4) output { $item\ (i)$ } and its support $count\ [\ i\ ]/\ n$;
  (5) C_i=subset(C_k,t);//get the subsets of t that are candidates
  (6) if $l_1[1]=l_2[1]\wedge l_1[2]=l_2[2]\wedge \ldots \wedge l_1[k-2]=l_2[k-2]\wedge l_1[k-1]<l_2[k-1]$ then {
  (7) if (there is an non-root node in $S\ T\ (\ i\ )$)
  (8) $FP[\ ++\ length\ ] = item\ (\ i)$;
  (9) Select p.item$_1$,p.item$_2$,…,p.item$_{k-1}$,q.item$_{k-1}$
  (10) return $L=\cup_k L_k$;

The calculus of a single Item start the database started, find out the frequency higher than the threshold value of the Large 1 Itemset, then join generated by Candidate 2 Itemset, and find the frequency higher than the threshold value of the Large 2 Itemset. And so on, and gradually expanded to more than one Itemset search. Although the Apriori algorithm can reduce the production of non-relevant data items, but the drawback for the search process must be repeated constantly back to the original database search, the efficiency is too low. Defined in a user's maximum forward reference column, all the elements are not repeated. Therefore, when access to web pages have been viewed, the time of browsing behavior to stop adding forward reference list, and output maximum forward reference column, and the frequency higher than the lowest level of support that the maximum forward reference to dig out of the association rules. The following diagram, as is shown by figure1.

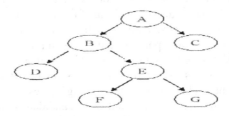

Fig. 1. The result of data mining based on association rule

Therefore, the use of Apriori and maximum forward reference can identify the user's browsing path, and dig out the website of the most preferred group, and further providing the website as an important reference for recommended system.

3   Application of CRM in e-Commerce Based on Association Rule Mining

E-commerce sites, due to inability to identify a specific customer base station, thus providing personalized service and product sales; we can only spend millions, braving the risk of the user objectionable, volunteer information and product sales. The force provided by the so-called e-commerce site for users of the information is simply for the visual and auditory pollution, waste. FP-growth algorithm for mining frequent patterns effectively opened up a new way. However, the FP-tree model and the conditions must be two-way tree traversal, the node will need more pointers; and in the need for mining frequent itemsets to generate a large number of conditions for recursive tree model, dynamic production and release of these conditional pattern tree will cost a lot of time and space, so the algorithm's time efficiency is still not high enough.

Association rule mining task is to dig out of D, according to all of the strong association rules [4]. Strong association rules $Y \Rightarrow X$ corresponding item sets $(X \cup Y)$ is a frequent itemset must be frequent itemset $(X \cup Y)$ derived association rules $Y \Rightarrow X$ of confidence by the frequent item sets X and $(X \cup Y)$ calculate the degree of support.

Candidate can be compressed using hashing k-itemsets $(k > 1)$. For example, the candidate 1 - itemset C1 produces frequent 1 - itemsets L1, to scan the database to calculate the degree of support for the itemset.

Let I = {i1, i2, ..., im} is a collection of items, D = {t1, t2, ..., tn} is the transaction database, where each transaction tj (j = 1,2, ..., n) contains itemset is a subset of I, that I, tj⊆tj has a unique identifier TID.

Algorithm 2: Application of CRM in e-commerce based on association rule mining

Input: Construct a novel, compact data structure FP-tree. It is an extended prefix tree structure that stores all the frequently needed information mining.

Output: Function Generate_C1 scan the transaction database D to generate a one-dimensional candidate set C1, and according to the weight in descending order to the project.

(1) length = 0;
(2) $L_k=\{c \in C_k | WSup(c) \geq wminsup\}$
(3) if (there is an non-root node in $S\,T\,(\,i\,)$)
(4) output { $item\,(i)$ } and its support $count\,[\,i\,]/\,n$;
(5) for each transaction $t \in D\{//$scan D for counts
(6) for each $(k-1)$-subset s of c
(7) if (S T (k1 ,···,km).count [i] > =min_count)
(8) for all (k-1)-subsets s of c do;
(9) (SC, C_1)=Counting(D,W);
(10) output FP and its support $S\,T\,(k_1, ···, k_m).count\,[\,i\,]/\,n$;

Rules support and confidence are the two rules interestingness measures, which are found in the rules to reflect the usefulness and certainty. Support association rule analysis of 20% means that 20% of all purchase transactions and financial management computer software. 50% confidence level means that 50% of the purchase of computer customers also purchases financial management software. If the association rules are in order to meet the minimum support threshold and minimum confidence threshold, then the association rule is interesting, as is shown by figure 2.

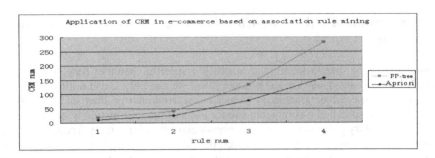

Fig. 2. The result of Application of CRM in e-commerce based on association rule mining

4   Summary

Association rule mining in large amounts of data between sets of items found interesting or relevant associated contact is KDD (Knowledge Discovery in Database) is an important research topic. This paper proposes the application research of CRM in e-commerce based on association rule mining. The association rules algorithm, the rules limit the conclusions by the properties that appear to reduce the useless association rules, support or confidence through the threshold limit; it can reduce the number of rules.

Acknowledgement. This paper is supported by not only Henan Science and Technology Agency science and technology research in 2010 (Key Project) under Grant no. 102102210472, but also Education Department of Henan Province Natural Science Research Program (2010A520030).

References

1. Liu, B., Hsu, W., Ma, Y.: Mining association rules with multiple minimum supports. In: Proc. KDD 1999, San Diego, CA, USA, pp. 337–341 (1999)
2. Feldman, R., Hirsh, H.: Mining associations in text in the presence of background knowledge. In: Proceedings of the Second International Conference on Knowledge Discovery and Data Mining (KDD 1996), pp. 343–346 (1996)
3. Lu, S.F., Lu, Z.: Fast mining maximum frequent itemsets. Journal of Software 12(2), 293–297 (2001)
4. Song, Y.Q., Zhu, Y.Q., Sun, Z.H., Chen, G.: An algorithm and its updating algorithm based on FP-Tree for mining maximum frequent itemsets. Journal of Software 14(9), 1586–1592 (2003)

Access Control Unit with Embedded Operating System µC/OS-II Based on LPC2214

Zhigang Lv

School of Electronics and Information Engineering Xi'an Technological University
Xi'an City, China
gangji780807@sina.com

Abstract. Access Control Unit (ACU) system is one of advanced security facilities accepted by modern buildings, which is regarded as an important symbol of modern building's intelligence. Embedded operating system of µC/OS-II equipped with TCPIP protocol is transplanted into LPC2214 controller, from which two kinds of communication types (TCPIP & RS485) are achieved in ACU. The ACU has not only normal access control functions, but also a series of complicated functions such as anti-trail, anti-hide, anti-intimidate, remote open door and so on, which can satisfy different customer's requirement.

Keywords: ACU, LPC2214, µC/OS-II, TCPIP.

1   Introduction

Access control unit (ACU) is a kind of digit system for managing staff's entrance, which is widely used in factory, school, company, and so on. Access control unit can be divided into several types, password control, IC or ID card control, fingerprint or iris control, RS485 or TCPIP communication and so on.

A kind of access control unit, with embedded operating system µC/OS-II, based on LPC2214 processor was designed in this paper. It provided brushing IC card, inputting password, remote control and other methods of opening door. From memory on the board, it could store 5000 card numbers and 40000 record of entrance, which were not lost when power was turned off. All the information of between PC and ACU was transmitted from TCPIP. Sketch map of ACU is shown as Fig.1.

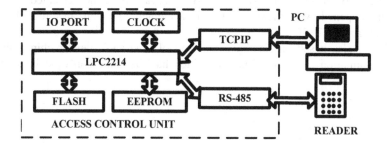

Fig. 1. Sketch map of ACU

D. Jin and S. Lin (Eds.): Advances in FCCS, Vol. 2, AISC 160, pp. 57–61.
springerlink.com   © Springer-Verlag Berlin Heidelberg 2012

2 Hardware Design

Hardware of the system is composed of CPU unit, data storage unit, clock unit, IO management unit, SRAM unit, communication unit and so on, which will be introduced in detail.

2.1 CPU Unit

LPC2214 is designed as the CPU unit of ACU.LPC2214 is based on a 32 bit ARM7TDMI-STM CPU with real-time emulation and embedded trace support, together with 256kB of embedded high speed flash memory.

With their compact 144 pin packages, low power consumption, and up to 9 external interrupt pins, LPC2214 is particularly suitable for industrial control, medical systems, Access Control Unit and point-of-sale and so on. 256 Kb on-chip Flash Program Memory is used to store application program and embedded operating system μC/OS-II. With In-System Programming (ISP) and In-Application Programming (IAP) via on-chip boot-loader software, it is easy to finish software design.

2.2 Data Storage Unit

There are two kinds of memory chip: EEPROM and FLASH, which are designed to save record, parameter, rules and so on.

The former is SST39VF160, a kind of Flash with 1M*16 bits, which can be written, read, erased easily. Data stored in SST39VF160 can be saved for over than 100 years when it loses power. So SST39VF160 is regarded as the main storage media, which is used to store valid card number, entrance rules, records, and so on.

The latter is FM24CL64, a kind of EEPROM with 64K-bits based on I2C bus, which occupies only 2 pins of LPC2214. As the secondary media, it is used to store reader's status, intimidating password, alarm time, network set value and other parameters. Data storage unit is given as Fig.2.

2.3 Clock Unit

In access control unit, date and time is a very kind of information for entrance management. When the current time of brushing card is in the permitted time, the door will be opened. Also, there must be a time element in the entrance record, which will be got from PCF8563.

PCF8563, also based on I2C bus, is a kind of time and calendar chip with low power consumption and multi-function, which has the same hardware interface as FM24CL64.

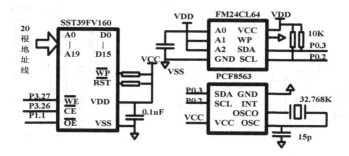

Fig. 2. Data storage unit of block diagram

2.4   IO management Unit

In order to control 8 doors, TLP521 and ULN2803 are designed in the system to finish all the IO management. From TLP521, 24 input signals of on-off are connected to LPC2214 to finish detecting buttons, gate sensors, infrared-sensors other alarm sensors. From ULN2803, 10 relays can be managed to control doors and alarms.

In the design, alarm sensors are connected to interrupt pins of LPC2214 finally. That is to say, interrupt technology is used to detect alarm incidence to ensure respond quickly. From detecting gate sensors, not only door status can be got, but also anti-trail can be achieved. IO management unit is given as Fig.3.

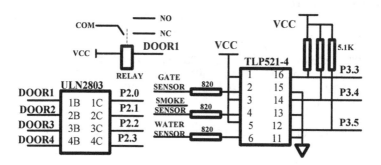

Fig. 3. IO management unit of block diagram

2.5   Communication Unit

RS485 and TCPIP communication technology is applied in the design to communicate with reader and PC.MAX3485 is used to form RS485 interface to get card information from reader. RTL8019 is designed to form TCPIP interface to exchange information between PC and ACU. Communication unit is given in Fig.4.

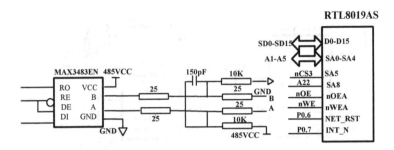

Fig. 4. Communication unit of block diagram

3 Software Design

Embedded operating system of μC/OS-II equipped with TCPIP protocol is applied in software design, whose code is open and programmed by standard ANSIC language. In order to make every unit of ACU run harmoniously, system software is divided into several tasks, every one of which accomplishes specific function.

3.1 Start_task

Start_task is designed to initialize all the chips of ACU and establish other tasks, which is made by main function.

```
int main (void)
{
    ......
    OSInit();      // initialize
    OSTaskCreateExt(Start_Task, (void *)0,
    &task0_stack[2999], TART_PRIO,
     START_ID, &task0_stack[0], 3000, (void *)0, 0);//
establish task
    ......
    OSStart();     //start multi-task operation
} ;
```

3.2 Other Tasks

PC_Comm_Task is designed to communicate with PC through TCPIP interface, which makes data exchange safely and quickly.

Reader_Comm_Task is set to communicate with reader to get card number, worker number, passwords, and so on through RS485 interface.

IO_Manage_Task is applied to manage all kinds of input and output signal of on-off.

ACU_Main_Task is the role task of system, from which all the entrance management is accomplished.

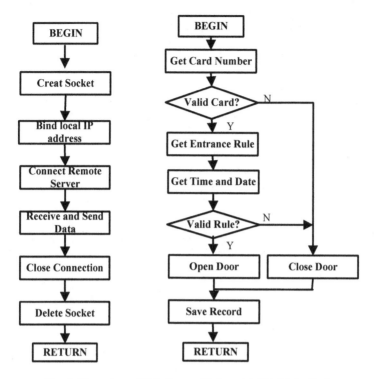

Fig. 5. Flow chart of PC_Comm_Task and ACU_Main_Task

4   Conclusions

A kind of multi-functional ACU with embedded operating systemµC/OS-II based on LPC2214 is introduced in the paper, which has been manufactured in batch. The ACU has many advantages, such as fast working, large storage, high efficiency of communication, and so on, which will be widely applied in the society.

References

1. Shao, B.: Embedded Real-time Operating System *µC/OS-II*, pp. 25–50. Beihang University Press, Beijing (2003)
2. Zhou, L.: Basis and Practice for ARM Controller, pp. 101–135. Beihang University Press, Beijing (2003)
3. He, Y.: Design of Entrance-guarding Controller Based on *µC/OS-ll*. Electronic Engineering, 75–77 (2007)
4. Zhang, C.: System of Defending to Rob by Ttailing Behind of Double Gate by Chip. Computer Engineering and Applications, 79–81 (2007)

The Performance Improvements of SPRINT Algorithm Based on the Hadoop Platform

TianMing Pan

Department of Information Science and Engineering,
East China Normal University, China
pantimmy110@gmail.com

Abstract. The emergence of cloud computing provide the medium enterprises many low-cost mass data analysis solutions. Decision tree algorithm in which one of the biggest problems is its computational complexity is proportional to the size and training data, resulting in a large number of computing time in constructing Data Set. The article aim at the SPRINT algorithm based on the Hadoop platform, presenting a parallel method of constructing a decision tree and then solving the parallel problem in Hadoop platform.

Keywords: Data Mining, SPRINT, Hadoop, Parallel.

1 Introduction

Many companies including the IBM, Google has paid attention to the value of cloud computing applications. The future of the cloud computing will affect the development of computer applications like the Industrial Revolution. Currently, cloud computing in the early stages of research and application[1],and cloud computing is just around the corner out of the lab into the commercial. In this paper, for the SPRINT on the Hadoop platform, we present a parallel algorithm for constructing a decision tree method.

2 Hadoop and Mapreduce

Hadoop is an open source software under Apache, it is the first project as an open source search engine based on Nutch platform. As the project progressed, Hadoop is developed as a separate project.

2.1 Hadoop

The core of the Hadoop is the Mapreduce and HDFS. The idea of the Mapreduce is mentioned by the a paper of Google and then widely circulated. A simple word to explain the MapReduce is task decomposition and a summary of the results. HDFS is the abbreviations of Hadoop Distributed File System and it provides the underlying support for distributed computing[2].

D. Jin and S. Lin (Eds.): Advances in FCCS, Vol. 2, AISC 160, pp. 63–68.
springerlink.com   © Springer-Verlag Berlin Heidelberg 2012

2.2   Mapreduce

We can see Literally that the Mapreduce has two parts: map and reduce. Map aim to break down one task into multiple tasks. Reduce aim to gather the results of multiple tasks and then get final results. This is not a new idea, in fact we can find the shadow of the idea multi-threaded, multi-task design can be found in the shadow of this idea. Whether the real world, or in the program design, one task for can often be split into multiple tasks. The relationship between tasks can be divided into two class: One is not related tasks and can execute in parallel, the other is a mutual dependency between the tasks and the order cannot be reversed, such a task is not parallel processed. In distributed systems, machine cluster can be seen as the hardware resource pool, and it split the task in parallel and then handed over to a spare machine resources for processing. This can greatly improve the computational efficiency and provide the best guarantee for the expansion of computing cluster design. After task decomposition process, it needs to be handled later and then pooled the results and it is the work of reduce. Fig.1 shows the operating mode of MapReduce, map responsible for decomposition of tasks, reduce responsible for merging the task.

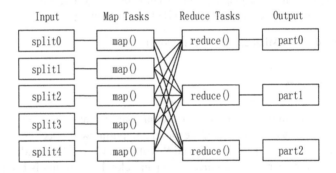

Fig. 1. Operating mode of MapReduce

3   The Improvement of SPRINT[4]

SPRINT algorithm was used for data mining classification very early, it has high value in the data mining[3]. SPRINT algorithm has a distribution features under the cloud computing, in contrast to other algorithms, we borrow the SPRINT's features to improve it and then use the improved algorithm in cloud computing for massive data mining.

3.1   The Concept of Decision Tree

Decision tree is a tree structure, it start from the root for testing the data sample. Depending on the results, it divided the data sample into different subsets, each data sample constitutes a subset of child nodes. Through a series of rules to classify the data, it provides rules that under what conditions we would get what value. Most decision tree algorithms include two phases: phase of tree construction and tree pruning stage. In the stage of tree structure, we construct a tree by recursively calling the classification algorithm. Tree pruning phase aims to cut branches which was too adaptable to the data set. Here we mainly study the phase of the tree structure.

3.2   The Basic Idea of Improved SPRINT

Improved SPRINT algorithm defines two data structures, namely the attribute table and histogram. Property sheet has three parts: attribute value, class identification, the index of data records. All recorded data cannot reside in memory, the list of attributes can be stored on the hard disk. Attribute table divided with the node expansion and it affiliated with the corresponding child nodes. Histogram attached to nodes, it is used to describe the type of an attribute node distribution. When describing the numerical properties of the class distribution, the nodes associated with two histograms Cbelow and Cabove. The former describes the types of samples have been dealing with the distribution, which describes the type of distribution of untreated samples, the value of the two operations are carried out with the update. When describing the attributes of the class of discrete distribution, there is only one node on the histogram. SPRINT pruning uses minimum description length principle.

3.3   Choose the Best Split Attribute

Mitotic index is a measure the pros and cons of the splitting rules. Gini index method can effectively search for the best split point.The division provides the smallest Gini index with the largest information gain was selected as the best segmentation. SPRINT algorithm used the Gini index method, which is essential to generate a good decision tree. Gini index method can be summarized as follows:

(1) If the set T contains m records in n class, then the Gini index is:

$$G\,in\,i(T\,) = 1 - \sum_{j=1}^{n} P_j^2 \qquad (1)$$

Which Pj is the frequency of class j.

(2) If the set T is divided into two parts T1 and T2, respectively, corresponding to the m1 and m2 records, then the Gini index for this division is :

$$Gini_{split}(T) = \frac{m_1}{m} Gini(T_1) + \frac{m_2}{m} Gini(T_2) \qquad (2)$$

Find the best split attribute and split point: For a numeric attribute A, pre-ordering the attribute first, if the result is sorted as v1,v2,...,vi,...,vn. Because the split occurs only between two adjacent points, so there are n-1 possibilities. It's split form has two parts

A≤vi and A>vi, so we usually choose the midpoint $\dfrac{v_i + v_{i+1}}{2}$ as the candidate split point

and compute the Gini values of all the candidate points from small to large. After getting the results we choose the minimum point as the best candidate split points. For a discrete attribute B, if it has m different values, the way is to divide m into two

sets.There are 2^n possible classificatiomethn od, so we calculate the Gini value of

each division to find the best candidate split points. Based on the above methods, we

can get the best division point of all properties, and the best candidate split point with

the lowest Gini value are the best split points. The corresponding property of the best

split point is the best current split property.

4   Parallel Process of SPRINT

The huge amounts of data under the cloud usually parallel processing the data.we
should find the way to handle the problem and reduce fault tolerance of the data.

4.1   Data Structures

In addition to the attribute table and histogram, we need a hash table data structure to
store the data records on both sides of the split point and make the basis for parallel
nodes of SPRINT algorithm.

   The value of the i-th records in the hash table means the tree node number that the
i-th records of the original data divided into. Hash table is divided into two parts:
(NodeID, SubNodeID). NodeID represent the tree node number, SubNodeID represent
the son node of the current tree node. The default SubNodeID is 0 and it means records
is in the tree node's left child node. When SubNodeID is 1,it means is in the tree node's
right child node.

4.2   Parallel SPRINT Algorithm

```
makeTree(Ti, A, M)
(Ti: Training sample set of the i-th sub-site, A: set of
classification attributes, M: number of sub-site)
Initialize the sample set Ti and  generate  the list of
attributes and histograms, then create the node queue and
put the queue into N.
while(The queue is not null)
{
Get the first node in the queue
If N is 0 then mark up as the leaf node, continue
for i=1 to M compute Gini index and the corresponding
optimal splitting parameters of properties Ti, and sent to
the central site; central site to select the best split
point * F, N grow a branch section N1, N2 and put them into
the queue. Splitting the list of attributes of split point
```

```
and use the index of the list to generate a  hash table of
list of records .Sub-site split the other attribute list
according to the hash table, while generating properties
histogram.
}
```

5   Achieve SPRINT in Hadoop [6] [7]

After the above improvements on the SPRINT algorithm, the algorithm can be ported
to MapReduce framework of cloud computing platform for distributed synthesis
process.

5.1   Vertical Division Combined with the MapReduce and SPRINT

In the input stage of the vertical division of SPRINT algorithm, the input of each map is
different, and ultimately the input with the same key is assigned to only one reduce.

A, B, C and D represent the key pairs which key is the attribute category. After the
distribution of tasks through the map, assigning attribute with the same key to the one
reduce. So that we can calculate the minimum Gini points each can reduce the column
for each attribute value and the best solution to split the minimum Gini points, and
generate the hash table.

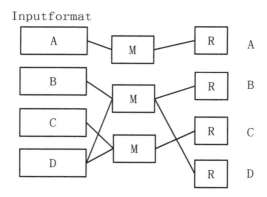

Fig. 2. Map assigned the same key to the same reduce

6   Conclusion and Outlook

This paper introduces the Hadoop and MapReduce programming model, and for the
MapReduce programming model and its parallel program execution, we propose a
parallel SPRINT decision tree-based data mining algorithms. Finally, we give the
algorithm description for transplanting the SPRINT algorithms to the MapReduce
programming model. The next we can level with the SPRINT and MapReduce
algorithm, and compared to vertical integration and horizontal integration.

References

1. Miller, M., Lei, J., Ruizhi, S., Yong, X., et al.: Cloud computing. Mechanical Press, Beijing (2009)
2. Dean, J., Ghemawat, S.: MapReduce: Symplified Date Processing on Large Clusters, vol. 51(1), pp. 107–113. ACM, New York (2009)
3. Campbell: Data mining concepts and techniques. Mechanical Industry Press, Beijing (2010)
4. Shafer, J., Agrawal, R., Mehta, M.: SPRINT: A Scalable Parallel Classifier for Data Mining. IBM Almaden Research Center, U.S
5. Zhu, Z.: Massive data processing and application based on Hadoop model. Beijing University of Posts and Telecommunications, Beijing (2010)
6. Liu, Y., Wang, L.: Improvement of SPRINT Algorithm. Computer Engineering (2008)
7. Pen, C., Lou, K.: Improving the Method Used by SPRINT Algorithm to Find the Best Split Point of Continuous Attribute. Computer Engineering and Application (2009)

Design of Vehicle Electrical System Based on CAN Bus with Analysis of Scientific Materials

Xiao-Guang Li and Zhen-Feng Qu

Department of Electrical Engineering and Automation, Luoyang Institute of Science and Technology, Luoyang, 471023, Peoples R. China
lxg@lit.edu.cn, lylonger@163.com

Abstract. with the growing demand for the safety, reliability and efficient transmission of the vehicle electrical system of the cars. The traditional bale of wire has unable to meet the needs of the tradition. An auto electrical system based on CAN bus is introduced, which used a core controller S3C2440 based on KC-6 DFSTN2000 experiment platform. And the hardware and software design of CAN-bus control system is designed in detail. The experimental results show that the system has high accuracy and stability relatively and the system design is reasonable and feasible.

Keywords: AT90CAN128, Electrical system, CAN-bus, Can Network.

1 Introduction

With the rapid development of information technology, and the people all aspects of the requirements of the car is growing, car digital degree more and more is also high. Now automotive electronic control is "the multi-objective synthesis control and intelligent control" the direction of development. To achieve this goal, the car using network technology to solve distributed control is an inevitable trend; such not only can reduce the harness, and be able to improve each control the operation of the system reliability. Controller LAN (CAN) car interior network of interconnected is an important standard, CAN bus structure was put forward a revolutionary improvement.

2 CAN Bus Profile and the Car CAN Network System Design

CAN bus is more than a host bus, any node on the network are active to other nodes send information [1]. Network node can press system real time requirement into different priority, once produce bus conflict, it will reduce the arbitration bus of time. The CAN bus dual serial communication mode, inspection wrong ability is strong, CAN be easily agreement, realize in the electromagnetic interference environment, long-distance real-time data transmission of reliable, and hardware cost is low. Due to the use of many new technology and unique design, compared with general communications bus [2], CAN bus data communication with prominent reliability, real-time and flexibility.

D. Jin and S. Lin (Eds.): Advances in FCCS, Vol. 2, AISC 160, pp. 69–73.
springerlink.com   © Springer-Verlag Berlin Heidelberg 2012

Generally, constitute the hardware CAN node semiconductor devices mainly have: independent type CAN controller, CAN single-chip microcomputer and CAN transceiver [3]. CAN bus application module in build on the commonly there are two scheme, the first kind is independent CAN periphery: MCU (SCM) + CAN controller + CAN transceiver + DC voltage stabilizing the power modules + peripheral circuit; the second is the single chip microcomputer integrated CAN controller CAN periphery: MCU CAN bring their own controller + CAN transceiver module + + DC voltage stabilizer outer circuit [4]. For the above two schemes for, each has its advantages and disadvantages, which CAN be used to realize independent CAN of different CPU interface adapt to, this design adaptability is stronger. Integration CAN controller in data transmission speed faster, in the compact structure and reliability than independent CAN controller is superior [5]. Comprehensive all aspects of consideration, the second option integrated peripheral structure.

Based on the KC-6 DFSTN2000 experimental environment and realizing body on the experiment the basic information of the monitoring, lamps, failure monitoring and control function of the lights. System from the function from the control and display, data acquisition and processing, the front of the data acquisition and processing parts.

This system from the structure, were set up front and rear, central Taiwan place three CAN node formed a CAN bus network, distance is relatively far node connected by a CAN bus, where each node of the relative device connected to the side with short connected. Central Taiwan setting a master node, CAN realize the function of state control and display, while the front and back of the car and in two according to the needs of the electrical control system were set up two CAN node, such, within the system the functions of the parts of the node is responsible for implementation, through the bus transfers data and interactive information.

The2000model experiments Taiwan Santana car real electrical components [6], the second development and become, by power supply system, starting system, light signal system, control system, ignition system, electric window lift system, the collection control locks, audio systems and instrument experiment plate etc. The utilization of the configuration of experiments on board, testing and setting fault components.

In the front of the car body CAN node is responsible for collecting and sensor signal and the anterior headlight failure monitoring information, and through the CAN bus realization and central node information interaction. In place at the back of the car body CAN node, for monitoring tail lights and the fault information through the relay to realize the before and after two parts headlight switch control. At the same time through the CAN bus back the node fault information to a central control lights node, and receive from the central node control command. Central node in place of the vehicle in central Taiwan is responsible for handling from before and after the two nodes of data. But at the same time, through this node keyboard gives the headlight switch control and command. In considering vehicle control system actual environment is bad, electromagnetic interference is serious. In the signals into the before and after the node and CAN access all through the light coupling between chip segregated, prevent electromagnetic signal and noise to micro controller, improve the stability and reliability of the system.

3   Hardware Design of the CAN Network System

The system CAN node in the MCU adopted ATMEL company AT90CAN128 solutions. In AT90CAN-128 in the integration CAN controller, this simplified the development links and further ensure that CAN interface in the stability of the communication. Based on AT90CAN128 designed on the basis of the power circuit module, node system RESET module, communication module, CAN serial communication module, LCD display module, A/D module, ISP download module, relay control module, board level resource module.

In the system of the node PCB design process, the layout is an important link. In PCB design layout tools in the way point's two kinds, one kind is interactive layout, and the other is the automatic layout. In this system on the layout of the main use interactive layout method, in order to AT90CAN128 chips as the core, peripheral circuit function division unit, put the same as possible to realize the function of the IC put together, such not only beneficial to check, still can reduce interference between components. Each device in the circuit board layout should conform to the corresponding design principle: crystals should try to close to the chip, chip filter capacitor should try to close to the power source of the chip input of the nearly as far as possible; Try to widen power supply, ground width, it is better than the power cord wire and the wide (landlines > the power >cord signals). The final PCB printed circuit board figure as shown in Fig.1.

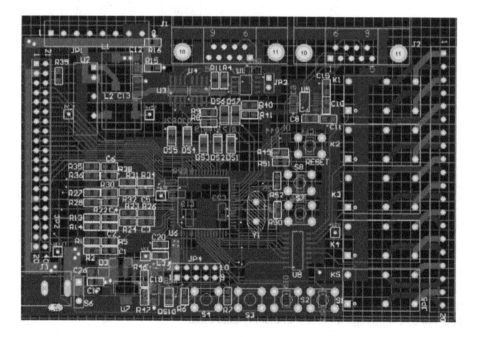

Fig. 1. PCB Figure

4   Software Design of the CAN Network System

The writing of the system software is based on the function of each node and somewhat different distinguish: located in the body of the central control node to realize human-machine interaction design (including the command issued, LCD display, control of lights); The front of the car body node to realize the monitoring of light and through the CAN bus realization and central node information interaction; The node mainly realizes the lights on the control. The core of the system is to the three nodes CAN interface effective control. Fig.2 control is receiving data CAN node flow chart, the other part of the design idea and is consistent.

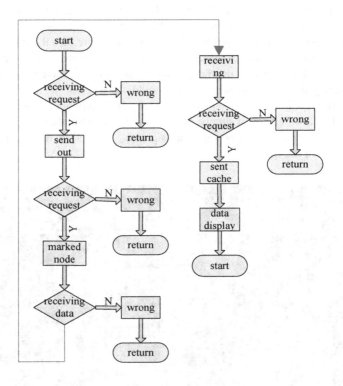

Fig. 2. CAN node receiving data flow chart

In order to validate the whole system can realize the function of the reservation; it needs to write test software system debugging. This system mainly for the test and LCD module CAN module testing. Test procedures including bottom initialization, accept data module, send data module and display data module.

5 The CAN Network System Anti-interference Design

Interference exists in every link of automobile electronic products, in addition to the influence of the physical environment changes often outside; still have other all sorts of electromagnetic interference, such as ignition system electromagnetic field interference, horn, and air conditioning system produced by the perceptual load of induction pulse current interference. The interference produced must have three basic factors, namely the interference sources, coupling channels, jamming signal to sensitive circuits. This article mainly from the two aspects, the hardware of anti-interference design, including the power cord wire and the design, design, emc design, inhibit reflection interference, reasonable decorate element position and its decoupling capacitor configuration and other measures

6 Conclusion

This paper based on the KC-6 DFSTN2000 type car test bench, design the CAN bus solutions, system respectively by arrangement in the front of the car, central Taiwan department at the back of the car and three CAN node component, realize the communication between nodes, formed a complete body CAN network. And to the electrical system design anti-interference and good results have been achieved.

References

1. Piller, S., Jossen, M.P.A.: Methods for State-of-change determination and their applications. J. Power Sources (96), 113–120 (2001)
2. Moore, T.C.: HEV Control Strategy: Implication of Performance Creteria, System Configuration and Design, and Component Selection. In: Proceddings of the American Control Confenence, Albuquerque, New Mexico (1997)
3. PHILIPS Semiconductors. CAN Bus Specification Parts A and B, Eindhoven (1991)
4. Fu, J.-F.: Design of Vehicle Electrical System. J. Auto Electric Parts (01), 1–5 (2010)
5. Van Mierlo, J.J.: Views on Hybrid Driver train Power Management Strategies. EVS-17CD. Beijing (1999)
6. Baldwin, T.: Directional Ground-Fault Indicator for High-Resistance Grounded Systems. IEEE Transaction on Industry Application (39), 325–332 (2003)

A Framework for Mobile Business Intelligence Based on 3G Communication Environment

XiYong Zhu and Yuan Huang

Institute of Modern Information Technology, Zhejiang Sci-Tech University,
5 Second Avenue, Xiasha Higher Education Zone, HangZhou, China
{zxywolf,huangyuan2009}@163.com

Abstract. Enterprise business intelligence has solved many problems of massive data analysis for business decision-making. With the growing popularity of 3G mobile and its communication devices, it is impossible for businesses or other organizations to implement mobile business intelligence. As current hardware and software facilities for most traditional business intelligence systems cannot be simply applied to mobile appliances, this paper first proposes a lightweight framework for mobile business intelligence under 3G communication standards, then describe in detail its physical and logic architectures both in the server side and mobile terminals, finally provides its implementation solutions by using Pentaho and Android platform tools.

Keywords: 3G, business intelligence, mobile business intelligence system, Android, Pentaho.

1 Introduction

With the accumulation of information resources, more and more enterprises plan to use multi-dimensional data analysis and data mining (DM) to support business activities. Business Intelligence (BI) is an application technology which includes the data warehouse (or data mart), reporting, data analysis and data mining components. It supports real-time enterprise decision-making and other business activities that can quickly response to market changes [1].

However, traditional BI systems are faced with many challenges, as they are often deployed in local servers, and cannot provide real-time, accurate and comprehensive information and knowledge. In recent years, much has been achieved in the research on new generation of BI systems. Richard Hackathon proposed a model to understand the value of real-time BI systems [2]. He pointed out that data latency would result in the loss of potential business opportunities. Jin Zhou established a relationship between knowledge management (KM) and business intelligence [3]. She proposed a knowledge management business intelligence model which integrated BI and KM to support management. Wei Shao and Liang Cao proposed a framework for real-time BI to integrate businesses activities [4]. They believed that real-time BI systems and integrations with other enterprise applications (such as ERP, SCM, CRM) are the trends for BI development.

D. Jin and S. Lin (Eds.): Advances in FCCS, Vol. 2, AISC 160, pp. 75–81.
springerlink.com   © Springer-Verlag Berlin Heidelberg 2012

Mobile Business Intelligence (MBI) is a new research field for real-time and integrated BI systems. It is a combination of BI with wireless communication technologies and Internet standards. With the third Generation (3G) wireless communication, it is possible to overcome the limitations of traditional mobile communication network, such as low data transmission speed, quality and security. 3G can provide extremely high speed and a variety of mobile multimedia services, which can greatly improve the MBI systems.

This paper proposes a lightweight framework for mobile business intelligence under 3G communication standards. The framework is based on collaboration model. It uses Pentaho to build a BI system in the server side, and uses Android to present reports and charts in the mobile terminals.

2   3G Mobile Business Intelligence Framework

The lightweight architecture of MBI systems is based on collaboration among various departments in an enterprise. In the collaboration model, enterprises integrate data into centralized databases, which help different departments share the same data, information, or analysis results. The lightweight BI architecture allows various departments in same Web servers to have their own BI modules, which can directly separate data analysis and mining results from Web pages by means of the broadcast and data sharing mechanism from mobile terminals. It can greatly reduce the data processing and transmission burdens on servers.

2.1   Physical Communication Architecture of 3G MBI Framework

Collaboration model, 3G wireless communication technologies, together with smart phone operating system (e.g., Android, Ios) make the 3G mobile business intelligence systems possible. Its physical communication architecture is shown in Figure 1.

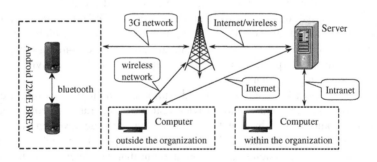

Fig. 1. Physical communication architecture of 3G MBI system framework

In the architecture, mobile devices use 3G wireless network to access the Internet, and communicate with enterprise application servers. All users can not only use wired communication devices to access enterprise BI systems but also use their mobile devices anytime, anywhere to access and operate BI systems. With mobile device's operating system mechanism (e.g., the broadcasting and service mechanisms in

Android system), users can use ETL (extract, transform, load) tools to extract, transform, or load structured and semi-structured data (e.g., pictures, SMS messages, etc.) into data warehouses for real-time analysis of important information and data [5].

2.2   Server-side Logic Architecture of 3G MBI Framework

The lightweight BI framework uses Web Services to integrate and deploy enterprise BI systems in the server side. BI applications can be packaged as Web Services by WDSL and can be published to the UDDI registration center. Users can use UDDI API to access to the Web Services and retrieve WDSL files. End users or system administrators can use Web Services, Web browser, system monitors, and sometimes desktop applications to access, operate and maintain BI systems [6]. The logic architecture for BI systems integration based on Web Services is shown in Figure 2.

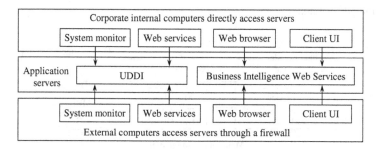

Fig. 2. A server-side integration architecture for BI systems based on Internet/Intranet

The BI systems in the server side may have many complex business processes. Figure 3 is an illustration of a BI system integrated with ERP systems.

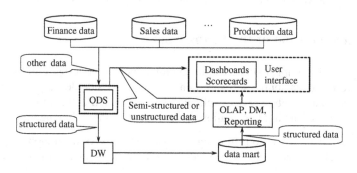

Fig. 3. Server-side function modules and business processes in BI systems

In the BI system, data is stored in data warehouses (DW) in the form of metadata. Daily operation and management data is stored in Operational data store (ODS), and can be appended to the DW on an on-demand basis to meet the requirements of real-time BI systems [7]. Data within or outside the enterprises will be extracted or loaded from enterprise applications, and can be transformed into useful information or knowledge by ETL tools. Knowledge can be displayed in many forms (such as dash boards or scorecards). Some data in simple reporting forms may be directly sent to 3G mobile terminals. The BI systems use Web Services and XML to deliver data, which helps to solve the problems of heterogeneity [8].

2.3   Mobile Device's Logical Architecture of 3G Mobile BI Framework

The 3G-based BI architecture for mobile terminals is shown in Figure 4. With 3G wireless network, terminal devices can retrieve data from servers or other users' mobile devices by using their own imbedded components. These devices can use imbedded operating systems to define the screen layouts for displaying data and knowledge [9]. Sometimes, mobiles can customize BI output by re-mining or re-processing data. In order to improve the processing efficiency and data transmission speed, important data and information can be stored in the user's mobile terminals through metadata management.

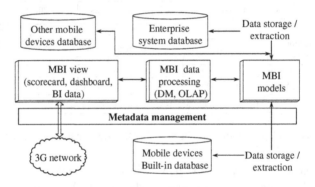

Fig. 4. Business intelligence logic processing in mobile terminals

4   Implementation of the Framework

Here, we give an example to demonstrate the implementation of the framework. In the example, we use open source software Pentaho as server-side development platform, use Android and Web Services client library KSOAP2 in mobile terminals development tools. The implementation process includes the following three steps:

1) Web Services definition. In order to access the servers from mobile terminals, we publish Web Services in the server side. Part of the services.xml file is as follows:

```
<service name="analysisServices">
  <description>
    Annual Revenue Analysis
  </description>
```

```
<parameter name="ServiceClass">
   ServiceRevernueAnalysisClass
</parameter>
<messageReceiver mep=http://www.w3.org/2004/08/wsdl/in-only
     Class="org.apache.axis2.rpc.receivers.RPCInlyMessageReceiver"/>
</service>
```

2) Communications between mobile devices and Web servers. The following is part of program for Android system to access the Web Services by using KSOA2.

```
String serviceUrl = "http://127.0.0.1:8080/axis2/services/analysisServices?wsdl";
String methodName = "overallDashbord" ;
SoapObject request = new
SoapObject("http://ws.apache.org/axis2",methodName);
SoapSerializationEnvelope envelope = new
SoapSerializationEnvelope(SoapEnvelope.VER11);
envelope.bodyOut = request;
HttpTransportSE ht = new HttpTransportSE(serviceUrl);
try{
     ht.call(null,envelope);  }
......
```

3) Information displays in mobile devices. After receiving data from BI systems in the server side, mobile terminals use an Android plug-in chart component to draw chart, part of this program is as follows:

```
String[] titles = new String[] { "Sales for "+a1.getString("time"), "Sales for "+ a2.
getString("time"), "Sales in "+a1.getString("time")+" compared with "+
a2.getString("time")};
List<double[]> values = new ArrayList<double[]>();
values.add(getdoubles(a1.getJSONArray("salesInMonth")));
values.add((getdoubles(a2.getJSONArray("salesInMonth")));
int length = values.get(0).length;
double[] diff = new double[length];
for (int i = 0; i < length; i++) {    diff[i] = values.get(0)[i] - values.get(1)[i];   }
values.add(diff);
...
return
ChartFactory.getCubicLineChartIntent(context,buildBarDataset(titles,values),renderer,
0.5f);
```

Figure 5 is an illustration of the sales in 2003 compared with 2004 in the mobile terminal (mobile phone) after calling Web Services and data warehouses in the server side. It uses bar charts and line charts to present multi-dimensional data analysis.

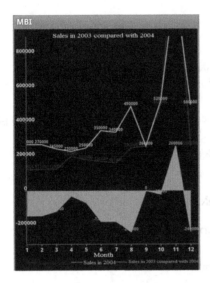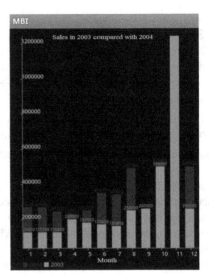

Fig. 5. An example of a BI system for sales analysis in Android simulators

5 Conclusion

Business intelligence systems are widely used in enterprises, traditional BI software packages (such as SAP Business Intelligence, IBM Cognos Business Intelligence) have been frustrated by many challenges. Mobile BI system based on 3G wireless communication devices and mobile intelligent terminals can be used to overcome current limitations. With advances in knowledge management, data mining and mobile communication technologies, mobile BI systems will be an important part of enterprise applications. In this paper, we first proposed a lightweight architecture for 3G-based mobile business intelligence systems, then described in detail it's physical communications infrastructure, server-side logic architecture, and mobile terminal architecture. We finally illustrated its implementation by an example with platforms of J2EE, Pentaho, Android and Web Services. Our next work is to make researches on the metadata management and BI interface layouts in mobile terminals.

Acknowledgements. This research is supported by the National Nature Science Foundation of China (No. 71071144).

References

1. Hackthorn, R.: The BI Watch: Real-Time to Real Value. DM Review 14, 24–43 (2009)
2. Zhou, J., Huang, L.-P.: Research on Event-driven Business Intelligence Based on Enterprise Real-time Decision-making. In: Proceedings of 3rd Chinese Management Annual Meeting (2008)
3. Zhou, J., Huang, L.-P.: Research on Knowledge Management and Business Intelligence. Science of Science and Management of S. & T. 3, 100–104 (2009)

4. Shao, L., Cao, W.: Research on Real-time Business Intelligence Based on Business Activity Management. Logistics Technology 6, 130–133 (2008)
5. Lu, X.-S., Tian, G.-J.: The Summary of the Third Generation of Mobile Communication. Fire Control and Command Control 35, 1–3 (2010)
6. Wang, W.P., Xu, H.-F.: Web Services-based Business Intelligence Network. Computer Systems & Applications 7, 16–19 (2007)
7. Ma, M., Zhao, Y.-C.: A Framework Of Real-Time Business Intelligent And Its Technical Analysis. Computer Applications and Software 26(10), 130–132 (2009)
8. Injun, C., Minseok, S.: An Integration Architecture for Knowledge Management Systems and Business Process Management Systems. Computers in Industry 58, 21–34 (2007)
9. Pentaho Corporation. Pentaho Open Source Business Intelligence Platform Technical White Paper, http://www.pentaho.com

A Hybrid Internal Model Control Method for Switched Reluctance Motor Drives

Wanfeng Shang

School of Mechanical Engineering, Xi'an University of Science & Technology,
Xi'an, 710054, Shaanxi, China
shangwanfeng@gmail.com

Abstract. This paper presents a hybrid internal model control (HIMC) for SRM drive system. The HIMC is integrated with an internal model, feedback linearization technique, and filter technique, such that the control system of SRM is convergent and robust against parameter disturbances and plant-model mismatches. The internal model of SRM is established based on the estimated derivative of speed and torque, so the speed ripple can be reduced. Simulation results demonstrate that the HIMC in SRM drive system can effectively reduce speed ripple, and it is highly robust to parameter variations.

Keywords: Switched reluctance motor, internal model control, speed ripple.

1   Introduction

SRM has emerged and gained increasing popularity for industrial drives with the development of power electronics and high-speed processors [1-3]. However, it has disadvantages of torque ripple, which further leads to speed ripple, because of magnetic saturation and doubly salient structure of SRM,.

People have proposed many advanced methods for SRM control to provide satisfactory performances as required by variable-speed drive system. Fuzzy controllers were developed for speed control of SRM and verified better performance for disturbance torque [4]. Sliding mode control (SMC) was used for SRM speed regulation due to its tough robustness and then was proved effectiveness [5]. Various artificial neural networks (ANN) were employed for online modeling to recreate parameters for controlling SRM and achieved better stable performance, yet for the high speed application, the real-time training-time of ANN need to be further reduced [6]. In fact, for these researches mentioned above, some used a closed loop only for speed control without considering the torque information. While the others utilized a closed loop only for torque without concerning speed control in outer loop, although they tried many advanced control algorithms to reduce speed and torque ripple. Husain [7] used a cascade double-loop structure involving an inner-loop torque control and outer-loop speed control except inner current control, called full closed-loop control. Also, the cascade double-loop control was widely adopted by several scholars [8,9] to solve the simultaneous control of speed and torque in a work process.

D. Jin and S. Lin (Eds.): Advances in FCCS, Vol. 2, AISC 160, pp. 83–89.
springerlink.com   © Springer-Verlag Berlin Heidelberg 2012

However, its control parameters are coupled each other in such full closed-loop control, and then, in practice, are very difficultly adjusted.

IMC is a process control strategy widely used in industry due to its many desirable properties, in particular good robustness against disturbances [10]. Although researchers have employed the IMC principle in the drive system of AC and DC motor for speed control [11], few groups have reported its application to control SRM before Ge [12] proposed a nonlinear IMC based on a suitable commutation strategy for torque control of SRM. However, for Ge's SRM internal model, the torque derivative component with respect to phase current is not easily calculated in actual discrete digital control system because there is no direct mathematical relationship between phase torque and current. Therefore, in the paper, the derivative information of torque with respect to a sample time is acquired for establishing the plant-model of SRM drives and compensating load torque. Furthermore, the HIMC law integrating with the plant-model, feedback linearization and filter technique is deduced.

2   Dynamics Model of SRM

Assuming that mutual inductances between stator phases are negligibly small, the mathematical model of generalized m-phase SRM model is obtained [1], as follows,

$$\frac{d\psi_j}{dt} = u_j - i_j r_j \quad j = 1,...,m \tag{1}$$

$$J\frac{d\omega}{dt} = \sum_{j=1}^{m} T_j - T_L - D\omega \tag{2}$$

where u_j is the voltage applied to the stator terminals of the jth phase and i_j is the current flowing in the jth phase, r_j and T_j are the resistance and electromagnetic torque of the jth phase, respectively. ω is the rotational speed of rotor, T_L is the load torque, and D and J are the viscous friction coefficient and the inertia moment of the load, respectively. The flux linkage ψ_j is periodic in θ with period 2π and shifted by the angle $(j-1)2\pi/m$. Thus the flux-linkage can be expressed as

$$\psi_j(\theta, i_j) = \psi_{sat}(1 - e^{-i_j f_j(\theta)}) \tag{3}$$

The function $f_j(\theta)$ in general can take the form

$$f_j(\theta) = a + b\sin[N_r\theta - (j-1)2\pi/m] \tag{4}$$

Here, ψ_{sat} is the saturated flux-linkage, N_r is the number of rotor poles, and a and b are constants. The torque produced by a single phase can be derived from the D'Alembert principle [9] as

$$T_j(i_j, \theta) = \frac{\partial \int_0^{i_j} \psi_j(\theta, i_j)di_j}{\partial \theta} = -\frac{1}{2}\psi_j^2 \frac{dR_j(\theta)}{d\theta} \tag{5}$$

where $R_j(\theta)$ is the air-gap reluctance of the machine for the jth phase and its derivative with respect to θ is

$$\frac{dR_j(\theta)}{d\theta} = -\frac{f_j'(\theta)}{\psi_{sat} f_j^2(\theta)} \tag{6}$$

3   HIMC Scheme for SRM Drive System

The Eqs. (1) and (2) are rewritten as

$$\frac{d\omega}{dt} = \frac{(T_e - T_L - D\omega)}{J} = \alpha \tag{7}$$

$$\frac{d\alpha}{dt} = \frac{1}{J}[-\sum_{j=1}^{m} \psi_j \frac{dR_j(\theta)}{d\theta}(-ri_j + u_j) - \frac{1}{2}\omega \sum_{j=1}^{m} \psi_j^2 \frac{d^2 R_j(\theta)}{d\theta^2} - D\alpha] \tag{8}$$

In this case, assume the load torque T_L to be constant. The $u_j(t)$ can be written as $u_j(t)=k_j \times u(t)$ where k_j is the electronic commutation signal. If the phase is to be turned on, $k_j =1$ for $i_j>0$, otherwise $k_j=0$ for $i_j=0$ and $k_j=-1$ for $i_j <0$.

Let us consider a SRM drive system as nonlinear multivariable control problem. Suppose that a model of the SRM drive system is available,

$$\left.\begin{array}{l} \dot{x}_1^M = x_2^M \\ \dot{x}_2^M = f^M(x^M) + g^M(x^M)u \\ y_\omega^M = h(x^M) = x_1^M \end{array}\right\} \tag{9}$$

where $x^M=[x_1{}^M, x_2{}^M]^T=[\omega^M, \alpha^M]^T$, and $f^M(\bullet)$ and $g^M(\bullet)$are respectively shown by

$$g^M(x^M) = -\frac{1}{J^M}\sum_{j=1}^{m} \psi_j \frac{dR_j(\theta)}{d\theta}k_j \tag{10}$$

$$f^M(x^M) = \frac{1}{J^M}(r^M \sum_{j=1}^{m} \psi_j \frac{dR_j(\theta)}{d\theta}i_j - \frac{1}{2}x_1^M \sum_{j=1}^{m} \psi_j^2 \frac{d^2 R_j(\theta)}{d\theta^2} - D^M x_2^M) \tag{11}$$

As Eq.(9) shows, it is an affine nonlinear system. The origin of the system is an equilibrium point of the unforced system, i.e. $f(0)=0$ and $g(0)=0$. When $L_g L_f{}^{n-1}h(x^M)\neq0$, which is the Lie derivative of $h(x^M)$ in the directions of $f^M(x^M)$ and $g^M(x^M)$, the relative order of the SRM drive system is n. In this case, $n=2$.

The plant-model output error is further obtained,

$$e^{PM} = y_\omega^P - y_\omega^M \tag{12}$$

Since discrepancies between the model and plant behavior usually occur at the high-frequency end of the system response, a low-pass filter is usually added to attenuate the mismatch. Hence, an asymptotically stable filter is designed by

$$\left.\begin{array}{l} \dot{x}_1^F = x_2^F \\ \dot{x}_2^F = -\boldsymbol{a}^F \cdot \boldsymbol{x}^F + a_1^F e^{PM} \\ y_\omega^F = x_1^F \end{array}\right\}$$
(13)

where $\boldsymbol{x}^F = [x_1^F, x_2^F]$ and $\boldsymbol{a}^F = [a_1^F, a_2^F]$ is such that $\lambda^2 + a_2{}^F \lambda + a_1{}^F$ is strictly Hurwitz. The plant output is to track a smooth reference speed y_ω and then converges exponentially to a constant. Motivated by the objective, we use a control law that linearizes a global input-out relationship. Les us firstly define auxiliary variables

$$\left.\begin{array}{l} \dot{y}_1 = y_2 \\ y_2 = \dfrac{dy_\omega^M}{dt} + \dfrac{dy_\omega^F}{dt} - \dfrac{dy_\omega}{dt} \end{array}\right\}$$
(14)

Here, for $\boldsymbol{y} = [y_1, y_2]^T$, there exists a real vector $\boldsymbol{b} = [b_1, b_2]$ such that $\lambda^2 + b_2 \lambda + b_1{}^F$ is strictly Hurwitz, leading to a linearization relationship

$$\dot{\boldsymbol{y}} = \begin{bmatrix} 0 & 1 \\ -b_1 & -b_2 \end{bmatrix} \boldsymbol{y}$$
(15)

Using Eqs.(9)-(15), we can achieve the HIMC law.

$$u = -\frac{f^M(\boldsymbol{x}^M) - \boldsymbol{a}^F \boldsymbol{x}^F + a_1^F e^{PM} - d^2 y_\omega / dt^2 + \boldsymbol{b}\boldsymbol{y}}{g^M(\boldsymbol{x}^M)}$$
(16)

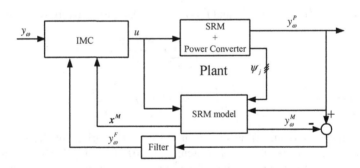

Fig. 1. Control scheme for SRM drive system

Therefore, the control scheme for SRM drive system as shown in Fig.1 is composed of an IMC block, an internal model and a linear filter shown by Eq.(13). This strategy uses an internal model in parallel to the plant of SRM drive system, feeding back the difference between the outputs of the plant and the model, which leads the plant output to follow a constant reference with zero steady state error even if the model is different from the plant.

4   Simulation Results

In this section, the HIMC scheme for SRM drive system is simulated using SIMULINK 7 under MATLAB 7.6. The specification of tested motor is a 12/8 SRM with parameters shown in Table 1.

Table 1. Parameters of SRM

Parameters	Value
Power	15 kW
Stator poles	12
Rotor poles	8
DC voltage	514 V
Rated velocity	1500 r/min
Stator resistance	0.95 Ω
Moment of inertia	0.0708676 kg*m^2
Coefficient of viscous friction	0.01
Aligned phase inductance	175.24 mH
Unaligned phase inductance	16.94 mH

The model of the tested SRM prototype can be given by (9)-(11). In (9), define

$$g^M(\bullet) = -14.1108\sum_{j=1}^{m}k_j c_j(i,\theta) \tag{17}$$

$$f^M(\bullet) = 0.1411\omega - 7.0554\omega\sum_{j=1}^{m}d_j(i,\theta) + 13.4053i\sum_{j=1}^{m}c_j(i,\theta) \tag{18}$$

where $c_j(i,\theta)$ and $d_j(i,\theta)$ displayed in Fig.2 are deduced by

$$c_j(i,\theta) = \psi_j(i,\theta)\frac{dR_j(\theta)}{d\theta} \tag{19}$$

$$d_j(i,\theta) = \psi_j^2(i,\theta)\frac{d^2R_j(\theta)}{d\theta^2} \tag{20}$$

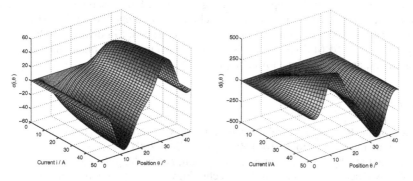

Fig. 2. Parameter surfaces of SRM model

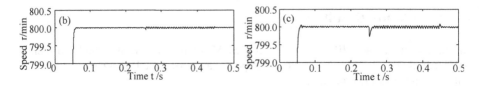

Fig. 3. Speed response to varying model parameter D^M: (a) output of HIMC, and (b) output of PID controller with $D^M=D^P$ for 0-0.25s, $D^M=10D^P$ for 0.25-0.45s, and $D^M=0.1D^P$ after 0.45s

Fig.3 gives speed response to varying model parameter D^M with SRM running at 800 rpm under 2.6Nm. Although some mismatches occur between plant and model of SRM drive system, the output speed of plant can stable at 800rpm using HIMC and PID controller, and the amplitude of speed ripple gets lower with D^M decreasing. However, comparing Fig.3(a) and Fig.3(b), we obtain the HIMC can achieve less speed ripple than PID controller Therefore, the HIMC has better robustness against model mismatches and lower speed ripple.

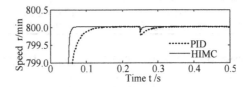

Fig. 4. Speed response to load torque disturbance

Furthermore, an instant disturbance of load torque is applied to the SRM with unloaded torque at around 0.25s. For a great disturbance, the torque observer work to compensate the drive system and then the stability of HIMC can be assured. Fig.4 has obviously two spikes in speed curve where 2Nm load torque occurs, and the spike from HIMC is smaller than one from PID controller. Hence, the HIMC has better robustness against disturbances.

5   Conclusions

The work described in the paper is to design a robust HIMC scheme for SRM drive system. Feedback linearization and filter technique are integrated into the HIMC such that the drive system can converge to the reference speed. Simulation results confirm that the HIMC of SRM drive system is robust against the model-plant mismatches and disturbances, and can achieve lower ripple for speed.

References

1. Krishnan, R.: Switched reluctance motor drives: modeling, simulation, analysis, design, and applications, Boca Raton, FL, USA (2001)
2. Krishnamurthy, M., Edrington, C.S., Emadi, A.: Making the case for applications of switched reluctance motor technology in automotive products. IEEE Transactions on Power Electronics 21(3), 659–675 (2006)

3. Shang, W., Zhao, S., Shen, Y.: Application of LSSVM with AGA Optimizing Parameters to Nonlinear Modeling of SRM. In: 2008 ICIEA, pp. 775–780 (2008)
4. Paramasivam, S., Arumugam, R.: Hybrid fuzzy controller for speed control of switched reluctance motor drives. Energy Conversion and Management 46(9-10), 1365–1378 (2005)
5. Inanc, N., Ozbulur, V.: Torque ripple minimization of a switched reluctance motor by using continuous sliding mode control technique. Electric Power Systems Research 66(3), 241–251 (2003)
6. Rahman, K.M., Gopalakrishnan, S., Fahimi, B.: Optimized torque control of switched reluctance motor at all operational regimes using neural network. IEEE Transactions on Industry Applications 37(3), 904–913 (2001)
7. Husain, I., Ehsani, M.: Torque ripple minimization in switched reluctance motor drives by PWM current control. IEEE Transactions on Power Electronics 11(1), 83–88 (1996)
8. Sahoo, S.K., Panda, S.K., Xu, J.: Indirect torque control of switched reluctance motors using iterative learning control. IEEE Transactions on Power Electronics 20(1), 200–208 (2005)
9. Shang, W., Zhao, S.: A Sliding Mode Flux-Linkage Controller with Integral Compensation for Switched Reluctance Motor. IEEE Transactions on Magnetics 45(9), 3322–3328 (2009)
10. Alvarez, J., Zazueta, S.: An Internal-Model Controller for a Class of Single-Input Single-Output Nonlinear Systems: Stability and Robustness. Dynamics and Control 8(2), 123–144 (1998)
11. Mohamed, H.A.F., Yang, S.S., Moghavvemi, M.: Design and Real-Time Implementation of an Internal Model Speed Control for an Induction Motor. In: 2008 ICRAM, vol. 1(2), pp. 735–740 (2008)
12. Ge, B., Wang, X., Su, P., Jiang, J.: Nonlinear internal-model control for switched reluctance motor. IEEE Transactions Power Electronics 17(3), 379–388 (2002)

Research for Control System of Soccer Robot Based on DSP

Chuan Wang[1,2] and Sheng-Jie Zhao[2]

[1] School of Information Engineering, Wuhan University of Technology,
Wuhan, 430063, Peoples R. China
[2] Henan Normal University, Xinxiang, 453007, Peoples R. China
{wangchuan20112012,shengjiezhidao}@163.com

Abstract. Aiming at the problem that the intelligence is not high of the soccer robot control system, this paper not only simplifies the design of the soccer robot control system but also enhances considerably its reliability and performance, which included a core controller based on TMS320F2812, introduced BLDCM control circuit and driver circuit with energy materials, designed a fuzzy PID controller. Experiments show that the new soccer robot controller features a quick response, high control precision and high servo rigidity.

Keywords: Fuzzy PID Controller, DSP, BLDC, Soccer Robot.

1 Introduction

Robot soccer is a typical intelligent robot system, as a mobile robot soccer robot, a family special members[1][2], integration of the control theory, the neural network, communication, computer, digital signal processing, mechanical institutions to learn and artificial life, and other high-tech areas, the system can be divided into visual, decision-making, communication and control these four subsystems[3]. The control subsystem is each subsystem to normal work foundation platform which decided the basic properties of the whole system[4].The performance of the robot control system is mainly by low-level decision, research and development, integration and high control accuracy reliability of the control system which has important significance.

Based on TMS320F2812 model controller DSP as the core[5], it has the introduction of brushless DC motor control circuit and drive circuit, and it completed the double closed loop control of BLDCM basic modules, it takes the digital incremental PID control and fuzzy control method, and based on DSP F2812 of soccer robot control system software design, the use of MATLAB/Simu link software FIS editor realized the incremental fuzzy PID controller design, and through the simulation experiment platform to validate the control system it can actually improve the performance of soccer robot control subsystem.

D. Jin and S. Lin (Eds.): Advances in FCCS, Vol. 2, AISC 160, pp. 91–95.
springerlink.com   © Springer-Verlag Berlin Heidelberg 2012

2   The Hardware Design of the Control System

The hardware design of the control system includes detection subsystem, control subsystem and other subsystems. Which detection subsystem includes ultrasonic ranging and temperature compensation unit module; Control subsystem include motor control module, motor driver module, brushless dc motor phase detection module, electric current detection module and speed detection module; Other subsystems including power supply circuit, crystals circuit and serial interface communication module, etc.

Main processor. TI company [6] TMS320F2812 which belongs to C2000 series DSP products is designed for control of the design of a 32-bit digital signal processor, processing speed of 150 MHz. In the current process control in the field, it is TMS320 series of second generation product that compared with the traditional single chip microcomputer, it has the function of strong, rich in resources, low power consumption of outstanding performance. It integrates 128 kB flash memory, 16 road and 12 of the high speed A/D converter, high performance SCI module, optimized event manager including 16 PWM output interface, and general way timer and capture decoder interface.   DSP control circuit of electric motor. The using of BLDC realize a DSP of BLDCM control at least need if the signals achieve full bridge type driving all of the pulse signal PWM1, 2,3,4,5,6, BLDC rotor position which is used to determine the phase, and then it realizes the control of the three phase shifting road hall sensor detection signal H1, 2, 3, in view of the above mentioned all kinds of signals, the circuit diagram with energy materials is shown in Fig.1.

Among them, PWM1 ~ 6 in again 100 Europe respectively after resistance by photoelectric isolated, driving circuit of BLDCM drive to corresponding interface, the driver circuit control. And cap1 ~ 3 is receiving BLDCM hall sensors to listen to the phase signals.

Phase detection circuit. After the hall sensors to collect to BLDCM motor rotor in the phase information with energy materials, it can go through the H1, H2, H3 three out of line. As the chart shows, when Hx for high, diode separate it and behind the connection circuit, + 3 v high voltage through the leds to make triode after conduction photoelectric isolation of the output into low voltage, finally the schmidt inverter to let the capx with high voltage; Instead, when Hx for low, + 3 v voltage through the diode conduction, it leads for low voltage not emit light, transistor quarantined, schmidt inverter through the resistance from + 3 v power supply for high voltage, low voltage output eventually.the phases detected circuit is shown in Fig.2.

Driving circuit. Drive circuit using IR International Rectifier company launched IR2133 series, IR2133 was designed for high voltage, high speed power MOSFET and IGBT drive and the grid of the design of integrated circuit chip. The internal integration independent of each other three groups of high side (HIN1, 2, 3) and low side (LIN1, 2, 3) bridge gate drive circuit arm, so it is especially suitable for three-phase motor driver with energy materials.

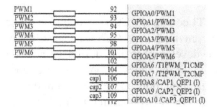

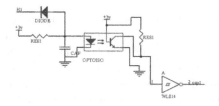

Fig. 1. DSP control circuit to the BLDC motor   **Fig. 2.** Phases detected circuit

Inverter circuits and electric current detection circuit. IR2133 from out of control signals HO1, 2, 3 and LO1, 2, 3 respectively is as shown in Fig.3 such as indicated from six MOS tube, the discretion of the IGBT in one side take on motor driving voltage, the six control signal as the MOS tube can be the switch signals the operation of the drive motor. Both sides meet in MOS tube one-way diode to prevent the reverse flow through the MOS tube was burnt out.

On the drive motor IGBT in process, the system needs to current loop feedback control,the system in order to prevent the flow of current burn out devices monitoring, therefore, can be in driving circuit for flow resistance in to meet with the current in the circuit testing. As shown in figure, all the way through the current detection to roll into the LM358 current amplifier resistance, to adjust to the current 0~3.3V, then the corresponding value to the ADC circuit transmission DSP. DSP detect the current value after processing, storage in ADC corresponding conversion results register. When the system is to use these data, just from the transformation results in a register will be read out current value. The current detected circuit is shown in Fig.4.

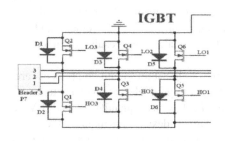

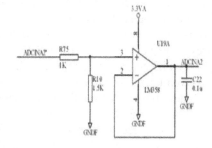

Fig. 3. Inverter circuit of the Three-phase BLDCM Control   **Fig. 4.** Current detected circuit

3   The Software Design of the Control System

The software of this system design is based on TI company's CCS3 (Code Composer Studio Version 3 abbreviation) programming environment, CCS3 is specifically targeted at DSP high level language program development and design of a Code adjusted emulators, and Code design package, it provides strong, mature core functions, easy, configuration and graphic visualization tools, make the system design faster. Using C language and assembly language can mix programming realize. The software is designed in the integration of the CCS3.3 DSP software, and through the hardware of simulation, test program. The software of the system design is divided into four big step: ultrasonic detection, the wheel conversion and motor control.

The running of BLDC motor. BLDC motor control with energy materials mainly including PWM produce, commutation and speed adjustment. It is the use of PWM waves with T1 timer which is used to include the Settings of the dead zone time produce, the speed of the motor by photoelectric encoder of a pulse through the pulse decoding QEP unit receive, calculated, commutation work is mainly refer to capture the phase signals back for the corresponding PWM mouth output Settings. The program of main work including T1 timer, general/special function interface setting, the cycle of PWM, accounting for space ratio and trigger an interruption of the initialization; On the basis of testing phase get phase number, by setting up PWM1 ~ 6 of high and low level of the adjustment of the need to get phase; Adjust the comparator value of change than get speed adjustment of empty.

Phase detection. BLDC motor by the analysis of the three phase and DSP event manager application, it is known that the phase detection main work including, of general motors/special function interface initialization set; Read interface level information and decoding; It will get the phase of the information in the corresponding phase number.

Electric current detection. In order to realize closed loop control of motor, it need to motor inverter circuits of the access of the current collection. Current detection of main job is defined, and the clock information; Initialize ADC clock, and the conversion channel and channel number, start signal; ADC results obtained the corresponding conversion current value.

Fuzzy PID control algorithm. The fuzzy control algorithm's main components are: by a given target detection of environment, from the photoelectric encoder to motor speed detection as input, through calculated E, EV and quantify E and EV to MoHuLun field as fuzzy controller two input signal, design and inquires the fuzzy control table to get the blur KP, KI, KD precision control quantity, the motor control signal output and converting it into a PWM control register of quantitative five sections.

Through the simple language program which gets the PID fuzzy quantity, three parameters through gravity method of the blur can be got clear PID parameters values. Fuzzy controller output set the PID parameters, procedures and into the PID control algorithm procedures. With the traditional PID control algorithm robot right wheel engine, it can put the traditional PID control by feedback and set quantitative amount in the difference in value Set point Error as input of reference, a Proportion by control parameters Proportion, Integral parameter Integral, differential parameters of Derivative three decided to rOut as the output value of control.

4   The System Test and Results Analysis

By using of MATLAB tool of the module soccer robot control system design and simulation. In adopted which based on FIS and DSP football robot control system, it watches the performance of the robot.In tests, the first robot football above insert two sign, a rigid steel wire with a rubber, a small red flag is a rigid steel wire,which use the rigid steel wire rubber test robot speed regulation and steadiness, it use the rigid steel wire with a flag to test its operation and steadiness. From the operation of the robot, it is known that the flag small jitter, smooth operation, when soccer robot close to football, rubber mark had a slightly swings, but overall stable operation, thus it can be seen, the new system of the robot localization accurate and stable operation.

5   Conclusion

In this paper, the incremental chip TMS320F2812 digital PID fuzzy control method, double closed loop control module BLDCM soccer robot control system to the bottom out the design, the design scheme simplified the hardware and software design, It's a good way to solve the complicated calculation and the bottleneck problem of the algorithm; It has the high speed control accuracy, good dynamic quality; The intelligent realize PID adaptive control; Which can also be convenient for the expansion of the function. Experiments show that the system has high accuracy and stability, and system design is reasonable and feasible.

Acknowledgement. In this paper, the research was sponsored by the Soft Science Project of Henan Province(Project No.112400440062,112400450305)and the Nature Science Foundation of China (No.61173071).

References

1. Qu, X.-L., Cao, Y.-F.: Design of Flight Control System for Aerial Robot Based on ARM. Industry and Automation 36(1), 41–46 (2010)
2. Wu, C.-Y., He, L.-Y.: Design and Realization of Instructional RPPR-Robot. Research And Exploration In Laboratory 26(10) (2007)
3. Deng, J.-H., Du, Y.-X.: Linux Transplantation Based on Processor of S3C2410. Microcomputer Applications 25(8), 53–55 (2009)
4. Mayingqing, ChengFu, ZhaoChen: Hardware Design of Soccer Robot Control System. Moderm Ectronics Technique 30(13), 118–120 (2007)
5. Yang, F., Xiao, H., Wang, X., et al.: A robot controller for teaching based on ARM and $\mu c/o$ II . Control & Automation 22(17), 33--35 (2006)

A Method for Web Service Discovery and Dynamic Composition Based on Service Chain

Dong Yang and Lei Liu

College of Computer Science and Technology
QianJin Street 2699, ChangChun, China
{yangdong,liulei}@jlu.edu.cn

Abstract. This paper proposes a dynamic web combined method supporting service quality restraint based on service chain, which selects all involved services as alternative service collection through data stream analysis according to the real time requirements provided by the users. After that, this method can construct reversal structure service chain based on above to meet user's requirement. In the process of combination, this method considered the supports to service quality, recorded the invocation dependence of service in usage, and outputted all the services used and topological structure after running. So the dynamic combination of web service with service quality restraint is realized. This method can make sure the feasible solution and optimal solution be found, and also can handle with the combination structure of branch, polymerization and recycle well. Simulation and model test verified the method mentioned in this paper.

Keywords: web service, service composition, service chain, call chain.

1 Introduction

In the current booming Internet, Web services as a standard component, with its easy development, deployment, interoperability and good properties, have been widely used. With the increasing complexity of the actual demand and the large size of the demand for services, a single Web services are increasingly difficult to meet the requirements of users, therefore, it is necessary to combine Web services to form a larger particle size, more powerful user demand component.

The current combination of Web services technology is mainly divided into two categories: combination methods based on template of fixed business processes and combination methods based on techniques to solve real-time tasks.

The former is represented by BPEL4WS, in which, business process in reality are used to generate the template, just as in the combination process every step of the template select the appropriate set of available services and can help. It is easier to implement such methods, but over-reliance on real-world business processes to generate the template, although other mechanisms can make use of community service to achieve a certain degree of dynamic service selection and runtime binding of user-submitted tasks, and flexible real-time workflow is still very difficult to handle.

The latter is according to the user's real needs, select a few atomic services from all the appropriate ones, and automatically combine them into a larger particle size of the service implementation as a whole. Because of the large number of alternative services, the great flexibility of combination , and involving the mapping between the parameters of atomic services , the processing is more difficult.

In recent years, not only require the users functional requirements are met, and more concerned about the quality of Web services. Web Quality of Service (Quality of Service, QoS) usually refers to its non-functional attributes (eg reliability, response time, etc.). Combination methods with quality of service constraints or quality-driven service composition technologies has been a research focus.

First Zeng et al [1] proposed a combination of services with Qos constraints. Its main concern is how to achieve the best overall combination performance according to the individual atomic services Qos parameters.

The Web service composition problem With Qos constraints is NP-complete problem. Researchers usually convert it into other forms of the problem to solve. Some researchers convert service composition problem with Qos constrains into multi-dimensional 0-1 knapsack problem or the shortest path problem with the multiple constraints [2]. Zeng et al pointed out that a combination of services can be transformed into one or more directed acyclic graph [3].

Also some researchers proposed a the reverse service combination algorithm based on semantic matching and service chain. The method views the service combination as a chain of service call sequence, In the sequence, the head node and tail node, respectively, match the user's input, the user's expect output, and the other nodes can match their adjacent ones[4]. To a certain extent, the method can flexibly handle the combination of services, but it can only handle a serial sequence. The algorithm complexity is too high, and it does not support Qos constraints.

Other researchers proposed a dynamic forward-chain synthesis and optimization algorithms [5].This method can form the combination needed, and all the atomic services used in the combination, but except how to form the combination. In particular, its result includes a lot of useless information.

Some researchers have proposed a further non-back dynamic synthesis method based on the reverse chain [6]. The method by calculating the input closure to reduce the scale of the problem, can find the set of services used to form the combination needed at little cost, but it did not give how to combine. And this method also does not take into account the constraints of the quality of service.

At present, the various methods of dynamic combination can usually handle the serial structure well, but they usually don't support the branch, aggregation and loops structure or the support is very poor, In addition, these combination methods generally do not support quality of service constraints.

Based on other researchers' work , this paper proposes a dynamic web services combination method supporting service quality restraint based on service chain, In the first step of this method, filter out all the relevant services according to users' requirement through data flow analysis to reduce scale of the problem.

Then the method view the user input and user expectation outputs of as a starting point, generate the forward parse tree and the combination reversely.

In the process of combination, supports of service quality have been considered, all the services used in combination and their dependencies are recorded. All the used services and topological structure of combination will be output after running, so the dynamic combination of web service with service quality restraint is realized.

2 Definitions

In this paper we use the following definitions:

Definition 1: WS

In this paper, we formally define web service as a triplet:
    WS = (I, O, C) Here, I is input of web service, O is output of web service and C is constraints

Definition 2: interface parameters

Parameter = < I, O > here I is the input parameters collection of the web service, WS.I. WS.I = $\{i_1, i_2, \ldots i_m\}$, m is the number of input parameters of the WS.O is the output parameters collection of the web service.

WS.O = $\{o_1, o_2, \ldots, o_n\}$, n is the number of output parameters of the WS.

Definition 3:

For any two concepts c1 and c2, if in the domain ontology, c1 and c2 are "sameAsOf" relationship, called c1 and c2 completely equivalent, denoted by c1 = c2,; if c1 and c2 are "subClassOf" relationship, called c2 included in c1, denoted by c2 < c1.

Definition 4:

For any two concepts c1 and c2, if in the domain ontology, there are c1 = c2 or c1 <c2, c1 and c2 semantic concept called compatible, denoted by c1 $\leq$ c2.

Definition 5:

For any two sets of concepts C1 and C2, if $\forall ci \in C1$, $\exists cj \in C2$ and $ci \leq cj$, called C1 and C2 are semantic compatible. denoted by C1 $\subseteq$ C2.

Definition 6:

For any two Web services WS1 and WS2, if WS2.I $\subseteq$ WS1.O, called WS2 input depends on the WS1, denoted by WS1 $\Rightarrow$ WS2.

Definition 7:

If the common output of atomic Web services WSi1, WSi2, ... WSik can meet the requirements of the input parameters of atomic Web services WSj called WSj co-dependent on WSi1, WSi2, ... WSik.
    denoted by WSi1 $\wedge$ WSi2 $\wedge$ ...... $\wedge$ WSik $\Rightarrow$ WSj.

Definition 8: *chain of services*
If a sequence of services WS$_1$,...,WS$_i$, WS$_{i+1}$,...,WS$_n$ exists and complies the following conditions:

1. The user's input match WS$_1$.Input ;
2. WS$_i$.Output satisfies WS$_{i+1}$.Input (1≤i<n);
3. WS$_n$.Output satisfies the output of the client needs.
Then this sequence WS$_1$,...,WS$_i$, WS$_{i+1}$,...,WS$_n$ is a *chain of services*.

3 Calculus

3.1 Single Atomic Service Qos Attribute Value Calculus

Qos attributes usually refers to non-functional properties of Web services, such as cost, execution time, the credibility and reliability, etc. According to their role in evaluation, researchers often make them dimensionless and map them into [0,1].

3.2 Web Services Combination Qos Attribute Value Calculus

In this paper, we view web service combination as a chain of services . The Qos attributes value of all the node of the service chain constitute the overall Qos attribute value of the whole chain and The Qos attributes value of all atomic services in the node constitute.

3.2.1 Adjacent Nodes Qos Attribute Value Calculus

For any two of any two adjacent nodes N and N +1 in the services chain, assumed that their own Qos vector value were QosVal (N) and QosVal (N +1), the overall Qos value from the beginning node to node N is QosTotal (N), then in the call sequence from the beginning node to node N +1 , the overall Qos value QosTotal(N+1) can be calculated as follow:

$$QosTotal(N+1) = QosTotal(N) \oplus QosVal(N+1)$$

Here $\oplus$ is compound operation. For response time and cost , it means addition, while for reliability , it represents multiplication.

3.2.2 Single Node Qos Attributes Value Calculus

For any node N in the services chain, assumed that its own Qos vector value is QosVal (N). Node N contains k single atomic web services: WS_{N1}, WS_{N2},.....WS_{Nk} , their Qos vector value are noted by QosVal(WS_{N1}), QosVal (WS_{N2}), ... , QosVal(WS_{Nk})

When k is one , node N contains only one atomic web services and now the overall Qos vector value of the node is its own Qos vector value.

When k is bigger than one, node N contains more than one atomic web services and now the overall Qos vector value of the node can be calculate as follow:

$$QosVal(N) = QosVal (WS_{N1}) \oplus QosVal (WS_{N2} \oplus ...... \oplus QosVal (WS_{Nk})$$

Here $\oplus$ is compound operation. According to the type of Qos attributes value, $\oplus$ represents different operation.

a. minimax type

Because the relationship of atomic web services in the same node is parallel, many Qos attributes value, for example, execution time, should be calculated according to the maximum one.

b. cumulative type

Because the relationship of atomic web services in the same node is parallel, the execution of the node depends on the execution of all the member atomic web service in the node. So many Qos attributes value, for example, execution cost, should be calculated according to the sum . Here $\oplus$ represents addition.

c. multiplication type

Because the relationship of atomic web services in the same node is parallel, the successful execution of the node depends on whether every member atomic web service in the node can be executed successfully. So many Qos attributes value, for example, reliability, should be calculated according to the product of all the member atomic web services' reliability.

3.3 Formalized Representation of the Web Service

For any web service with m input, n output WSj (I,O,C), It can be converted into the form of Generative grammar. Assumed that all its m input are denoted by $I_1, I_2, \ldots\ldots, Im$, and all its n output are denoted by $O_1, O_2, \ldots\ldots, On$, thus this web service can be transformed as follow:

$$P_j.1 \qquad I_1 \wedge I_2 \wedge \ldots \wedge Im \xrightarrow{\ C\ } O_1$$

$$P_j.2 \qquad I_1 \wedge I_2 \wedge \ldots \wedge Im \xrightarrow{\ C\ } O_2$$

...

$$P_j.n \qquad I_1 \wedge I_2 \wedge \ldots \wedge Im \xrightarrow{\ C\ } On$$

Here C represents the constrains of web services , and $P_j.n$ is the serial number of the web services.

4   Method and Algorithm

Because of the large number of candidate web services, To reduce the scale of the problem, we filter out all the relevant web services according to the user input and requirement through data stream analysis. After that, we get the candidate web service set which contains all the potential candidate web services to construct the needed combination. If the services number in the set is 0, it means that the requirement of user can not be met. Then we can construct the combination needed.

```
Construction algorithm
Variable Description :
            SetUserInput              set of user input grammar
            SetAvailableInput         set of currently available grammar
            SetAvailabelService       set of currently available web services
            SetService                set of candidate web services
{ SetAvailabelService = ∅  ;
  SetAvailableInput = SetUserInput   ;

Repeat
    {   1. select currently executable web services from set   SetService.
        2. put it into set   SetAvailabelService
        3. put the Corresponding Generative grammars into set SetAvailableInput
    } Until   no currently executable web services in set SetService
        4. Compare set   SetAvailableInput with set SetUserInput
        5. select the best web services combination
        6. output the combination and its topological structure
}
```

5   Simulation Experiment and Analysis

Due to the lack of a unified base pool of standard tests for web services, we generated a total of 1000 random services, including various types of composite structure. The tests show:

a. This method can make sure the feasible solution and optimal solution be found, and also can handle with the combination structure of branch ,polymerization and recycle well.

b. To reduce the time complexity, we can use TOP(K) method , but in this case, there is no guarantee that the optimal solution can be found.

6   Conclusions and Future Work

This paper, based on previous research works, proposes a method which can make sure the feasible solution and optimal solution be found, and also can handle with the combination structure of branch ,polymerization and recycle well.

Future work will mainly focus on further reduce time complexity.

References

1. Liu, Y., Ngu, A.H., Zeng, L.Z.: QoS computation and policing in dynamic web service selection. In: International World Wide Web Conference, pp. 66–73 (2004)
2. Yu, T., Lin, K.-J.: Service Selection Algorithms for Composing Complex Services with Multiple qoS Constraints. In: Benatallah, B., Casati, F., Traverso, P. (eds.) ICSOC 2005. LNCS, vol. 3826, pp. 130–143. Springer, Heidelberg (2005)
3. Zeng, L.Z., Benatallah, B., Ngu, A.H.H., et al.: QoS- Aware Middleware for Web Services Composition. IEEE Transactions on Software Engineering 30(5), 311–327 (2004)
4. Fu, Y.-N., Liu, L., Jin, C.-Z.: Service chain-based approach for Web service composition. Journal on Communications 28(7) (2007)
5. Thakkar, S., et al.: Dynamically composing Web services from on-line sources. In: 2002 AAAI Workshop on Intelligent Service Integration, Edmonton, Alberta, Canada (2002)
6. Non-Backtrace Backward Chaining Dynamic Composition of Web Services Based on Mediator. Journal of Computer Research and Development 42(7), 1153–1158 (2005)
7. Ma, J., Fan, Z.-P., Huang, L.-H.: A Subjective and Objective Integrated Approach to Determine Attribute Weights. European Journal of Operational Research 112(2), 397–404

Passive Optical Network Based on Ethernet Access Technology

Zhang ZhanXin[1], Liu Tao[1], and Zhu Kaiyu[2]

[1] College of Science
[2] College of information Technology, Hebei United University, Tangshan 063009, China

Abstract. This paper focuses on the different user community EPON(Ethernet Passive Optical Network) access scheme design, first introduced the PON access, PON access technology according to the different data link layer technology mainly divided into APON(ATM+PON), EPON (Ethernet+PON) and GPON (ATM/GEM+PON), EPON can carry data services, voice services, IPTV, CATV as a professional, so we adopt EPON technology as the area of the access technology. According to the area of different user's demand for bandwidth and different access methods, selection of optical network terminal, beam splitter and optical network unit equipment, the final design of the area network access scheme, and in the area of optical power loss making a budget. Finally the scheme used by the equipment and its configuration command is introduced in detail in this paper.

Keywords: EPON, APON, ONU, OLT.

The application of EPON technology on one hand can make the user community for voice, data and high quality access to services, and can use its passive, resembling a multiplex, with business the characteristic of diversity, for operators to save a lot of cost, but also occupy the market, so EPON technology has become the telecom, Netcom fixed telecom operators to promote access network technology, and to be in the high-end area of large-scale promotion and application of.

1 EPON Working Principle

EPON system using WDM technology to realize the bidirectional transmission (compulsory)

In order to separate the same fiber on the number of users and the direction of the signal, uses the following twomultiplexing technology

Downlink data stream using time-division multiplexing (TDM) technology

From the local terminal interface unit transmits the signal time division multiplex is sent to each subscriber unit.

In Figure 2 downlink data by radio from a OLT to a ONU, according to IEEE802.3 protocol, the data packet is not a fixed length, the length can be up to 1518 bytes. Each packet carrying the letter head that uniquely identifies the data to arrive at ONU1, ONU2 or ONU3, in addition there are a number of packets is sent to all ONU, called a broadcast packet, and packet is sent to a group ONU, called the multicast packet. The

D. Jin and S. Lin (Eds.): Advances in FCCS, Vol. 2, AISC 160, pp. 103–107.
springerlink.com   © Springer-Verlag Berlin Heidelberg 2012

data stream is passed through the beam splitter is divided into 3 independent signals, each signal containing specific ONU packets sent to all. When the ONU receives a data stream, extracting only to its own data packet, would be given to the other ONU data packet discard. For example, in Figure 2 ONU1 received packets 1, 2, 3, but only 1 data packets sent to the terminal user.

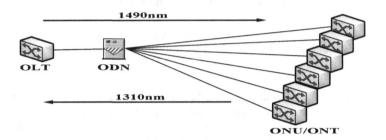

Fig. 1. EPON Working principle

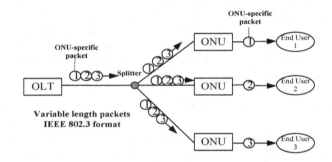

Fig. 2. EPON downstream data flow

1.1   Uplink Data Stream by Using TDMA Technology

Figure 3 shows the uplink data streams using time-division multiplexing, each ONU is assigned a time slot of transmission. These slots are synchronized, so when the data packet is coupled to a fiber, different ONU data packet does not generate the interference. For example, ONU1 in a first time slot data packet for transmission in 1, ONU2 in second without occupying the time slot data packet for transmission in 2, ONU3 in third without occupying the time slot transmission packet3.

ONU to OLT on business flow using time-division multiplexing, each ONU in a particular time slot transmission transmission, to avoid conflict.

The Hebei province Tangshan city a lot of high-grade residential district, Tangshan telecommunication company plan in the area takes the lead in undertaking FTTx broadband pilot project, the project can satisfy the household at least more than 10M broadband accesr.

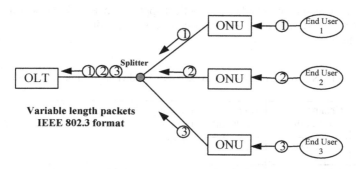

Fig. 3. The EPON of uplink data stream

This area can be divided into three buildings in a residential area:

A: luxury independent villa area, area of a total of 20 buildings, a total of 20 households:

B: multilayer District, the district consists of 15 buildings, each building two unit 10 layers, each layer of a household, a total of 300 households.

C: high-level area, the district has 6 buildings, each building three unit 20 layers, each layer of two, a total of 720 households.

A total of 1040 households village. Residential property within the relatively centralized area, has completed the building of sound and data cables integrated wiring access, and the area within a telecommunications access computer room, community access room from the city core room 4.5KM.

2   Technical Scheme Design

Each floor to the optical fiber distribution box cloth placed 8 core user cable, according to different levels of customers within the community broadband demand and price, district 20 independent villa high-end customers using FTTH access mode provided broadband services, the other general customers using relatively inexpensive FTTB provide business. Because the domestic ZTE, Huawei and other major equipment manufacturers can now provide a ONU device user port maximum is 24, according to each area of high-rise has 120 access points to calculate, then every high-rise inside need to install 5 ONU equipment. But because the ONU cabinet large volume, inconvenient installation in the corridor, and taken into account in a period of time the new business opening rate will remain at low levels, the uplink bandwidth is relatively abundant, so, in the high-level and multi-storey uses FTTB+LAN means to provide broadband business is the highest price of an FTTx access scheme. If in the future user bandwidth is not enough, can be used to increase and decrease the ONU splitter optical divide ratio method to speed.

3   Bandwidth Calculation

With GPON technology as an example to introduce the calculated bandwidth. According to the PON uplink / downlink according to 1.25G / 2.5G computing, if online ratio 1: l, each PON port can be assigned to each user maximum theoretical bandwidth value as shown in table 3-2, practical application, users consider online than business characteristics, business model, user access bandwidth value should be greater than the table 1.

Table 1. Each PON port can be assigned to each user bandwidth maximum theoretical value

Technol ogy	Channel bandwidth	Effective bandwidth	Light splitting ratio	Downlink bandwidth/ door	Uplink bandwidth/ door
	(up / down)	(up / down)			
GPON	1.25G/2.5G	1.1G/2.3G	1:64	36M	17M
	1.25G/2.5G	1.1G/2.3G	1:128	18M	8.6M

The overall scheme design

In real life, all users of the community at the same time can not be online, is to user bandwidth demand is 10M and 60% of online users as an example to illustrate the cell EPON access scheme.

A area villa area has a total of 20 villas, using FTTH access mode provided broadband service. Because of the OLT device in each EPON interface can provide 1G bandwidth, the villas with 1:32 optical splitter are respectively connected to 20 villas.

B multilayer district has 15 buildings, a total of 300 users. Because the B total 2G bandwidth, i.e. two EPON interface access to the area. Each user requires 6M of bandwidth, Each EPON interface can be assigned to a 160 user using, if the GE downward by 1:8 optical splitter to access the 8 corridor exchanger (B zone, because each building has two units, so the B zone adopts distributed light (1x4/1x2) is more appropriate), each corridor switch average access 160/8=20users. Considering the user community is relatively centralized state, it recommended a downlink 24 x FE corridor switch more appropriate, to sum up, B total of splitter 2, ONU equipment 15.

C District high-rise district has 6 buildings, a total of 720 users. Because the C total 5G bandwidth, i.e. 5 EPON interface access to the area. Each user requires 6M of bandwidth, each EPON interface can be assigned to a 160 user, C area using the same 1:8 optical splitter to access the 8 corridor exchanger (using the convergent light is more appropriate), each corridor switch average access 160/8=20 users, so still using descending 24 x FE port corridor exchanger. C zone of each unit has 40 households per cell, 2 ONU equipment, a total of 18 units, to sum up, C total of splitter 5, ONU equipment 36.

In the planning area, is divided into three areas: the villa, multilayer area, high-rise District, villas users to access network bandwidth demand is bigger, so the use of the relatively high cost of the FTTH (fiber to the home) access mode, the multi-storey and high-rise area user to access network bandwidth demand for smaller, the relatively low cost of the FTTB (fiber to the building) access mode. According to the user and the required bandwidth, developed the community access program, OLT equipment selection of ZTE's ZXA10 C220; villa zone splitter using 1:32 optical splitter, multilayer and high-rise district with 1:8 optical splitter; villa area of the ONU ZXA10 D400 equipment selection, multilayer zone and the upper area using 24 ZXA10 F822 interface. The following are also calculated by the area EPON network access scheme of optical power loss, meet the technical requirements of the standard ODN.

References

1. Liu, S.: EPON network planning and implementation. Telecommunications Network Technology 3(2) (2009)
2. Angelopoulos, J.D., Boukis, G.C.: EPON-a next generation access network. Today Science of Telegraphy 12(6) (2004)
3. Lu, J., Xu, G.: EPON technology in FTTx access network application. 12(12) (2009),
   `http://10.16.114.12/KNS50/Navi/Bridge.aspx?LinkType=Bas` `eLink&DBCode=cjfd&TableName=cjfdbaseinfo&Field=BaseID&V` `alue=DXWJ&NaviLink=%e7%94%b5%e4%bf%a1%e7%bd%91%e6%8a%80` `%e6%9c%afit _blank`Telecommunications NetworkTechnology
4. Marelli, P., Consonni, E.: Indoor Cables for FTTH Applications Allowing Quick and Simple Subscriber Connection. Photonie Network Communications 1(2) (2010)

Research on Agricultural Information Retrieve Based on Ontology

Xuning Liu[1,*], Genshan Zhang[1], Weihua Chen[2], Hong Tian[3], and Liying Duan[1]

[1] Department of Computer, Shijiazhuang University, Shijiazhuang, Hebei, China
[2] Department of Information Engineering, Hebei Software Institute, Baoding, Hebei, China
[3] Department of Information Engineering, Beijing University of Technology Tongzhou College, Beijing, China
sjzhei@163.com

Abstract. In order to solve the difficult problems of agricultural information retrieve and improve the efficiency of intelligent retrieve, the paper studied the application of ontology, introduced principles of ontology construction and ontology process, then developed the retrieve system with the help of intelligent searching technology, which can be integrated with the traditional retrieve systems, greatly improving the semantic support and performance of retrieve system, developing the implicit knowledge in the mind of the users, standardizing and improving the information demand of the user. The practice has proved that the ontology technology in the agricultural information greatly promotes the accuracy and reliability of the agricultural information retrieve.

Keywords: ontology, agricultural information, retrieve.

1 Introduction

With the rapid development of information technology, the requirement of agricultural information resources that the users acquire are growing, but traditional information retrieve tools only help the user to retrieve by keyword searching interface, in many cases the true intentions of users are difficult to express clearly with a few keywords, which leads to the low accuracy of existing retrieve systems. To improve the accuracy of existing retrieve systems, information retrieve must be based on keywords from the current level to the knowledge level, and rational design of intelligent agricultural information retrieve system is made based on the semantic level of information organization and representation [1].

China is influent in agricultural information resources, and daily life of farmers has become increasingly close relationship with circulation of agricultural products information. Therefore the accurate and complete retrieve of agricultural information resources has great significance and social value for sustainable agriculture and rural development. The knowledge retrieve based on ontology will greatly improve the efficiency and accuracy of agricultural information retrieve [2].

---

* Corresponding author.

D. Jin and S. Lin (Eds.): Advances in FCCS, Vol. 2, AISC 160, pp. 109–113.
springerlink.com   © Springer-Verlag Berlin Heidelberg 2012

2   Ontology Concept

Ontology is defined as the system description of objective world, and artificial intelligent regards ontology as basic terms and relation which made vocabulary of related fields, and the definition of extension of the rules that use relationships and these terms is made. As ontology is regarded as conceptual model tools which descript knowledge of information system in the semantic and knowledge level, since the introduction of domestic is made, then foreign researchers pay great concern on it, and has been widely used in many areas of computer, which has good concept hierarchy structure and support of logical reasoning, and thus the ontology is used in semantic retrieve of information retrieve, it is intelligent search which is based on the conceptual understanding level and thinking search [3]. The goal of ontology is to capture the relevant knowledge in the field, providing common understanding of knowledge of this area, determining the agreed-upon terms of areas and gives these words and the clear definition of terms of the interrelationship from different levels of formal model, through the relationship of concept the semantics of concept is described.

3   Ontology Technology

3.1   Construction Theory of Ontology

For different ontology are applied in different areas, so now a standard method and rule does not fit for ontology construction, generally more influent factors are the following proposed principles: the clarity and objectivity, integrity, consistency, the largest single scalability and the principle of least constraint. These five principles are universal, for the construction of agricultural ontology, four basic principles should be followed: the principle of standardization, the principle of maximum reusability, the principle of participation, collaboration of domain experts and combination principle of construction and evaluation[4].

The clarity and objectivity is that the ontology should use natural language terms to give a clear, objective definition of the semantics, and should explain the meaning of the terms which is defined, and which is independent of all backgrounds should be defined by natural language and be explained; the principle of the definition of integrity is intact, and can express the meaning of specific terms; the principle of consistency is that the knowledge reasoning of the conclusions does not contradict with the meaning of the term itself; The maximum principle to a one-way scalability is that existing content knowledge is not modified while general or specific terms are add to ontology; the principle of least constraint is that model can treat the object as little as possible to limit the constraints list; The principle of standardization is that standardized terminology and standardized equipment is made full use; Standardized construction processes will contribute to improve the standardization of ontology sharing, and can effectively prevent and avoid the generation of information silos; The principle of maximum reuse is that agriculture ontology can be classified based on the concept, and multi-level inheritance mechanisms is used to minimize and avoid duplication of definitions, redundant construction; The principles of participation and collaboration of experts is that agricultural knowledge is used to make ontology of the agricultural sector, some problems also involve areas such as agro-processing, rural economy and management and other related areas such as agro-processing, active participation and collaboration is essential; The principle of construction and evaluation is that agricultural ontology construction is huge one, is the formal process of knowledge of agriculture, experts in the field need active participation and collaboration, also are inseparable from staff for their hard work, the combination of domain knowledge and knowledge is not an easy thing, the combination of assessment helps to promote the cooperation of related personnel and will contribute to speed up the ontology construction progress [4].

3.2   Construction Method of Ontology

The construction method of ontology is showing as following:

(1)Firstly we need make the purpose and domain of the ontology;
(2)Ontology analysis: the meaning of all terms and relationships of ontology are defined, the step requires the participation of experts in the field, understanding the more about this area, the more perfect ontology is built;
(3)Ontology expression: the generally semantic models are used to express ontology;
(4)Ontology Inspection: the basic standard of ontology construction is clarity, consistency, integrity and scalability. Clarity is that the ontology terms should be unambiguously defined; Consistency is that the relationship between the terms should be logically consistent; Integrity is that ontology concepts and their relationships should be complete, and should contain all the field concepts, but often are difficult to achieve, and require continuous improvement; and the scalability is that ontology should be able to expand, and growing in the field when adding a new concept;
(5)The ontology is constructed by capturing concepts and relations in the domain, encoding them and integrating all concepts and relations to the whole, evaluating ontology and documentation [5].

3.3   Application of Ontology in Information Retrieve

Because the ontology has a good concept hierarchy and the support of logical reasoning, and has been widely used in the search. In general, the idea of information retrieve is based on ontology:

(1) The help is from experts in the field, the related areas of the ontology are made;
(2) The data are collected from information sources, and the collected data are stored in the base by the required format according to the established ontology data;
(3) The search query is collected from user interface, then the query converter transforms the query into the format ontology according to the provisions, with the help of the ontology the base matching data sets are used to meet the conditions;
(4) The results are returned to the user through the customization process[5].

4   Intelligent Retrieve

While collecting and indexing web page, traditional searching engines apply different technologies, and there are disadvantages of different data formats for different data sources, for the same query, different searching engines will produce different results, which makes the search results are not comprehensive and accurate. Intelligent searching is that these searching engines are combined and strengthened, so making distributed and multi-level searching systems. This searching system takes something that user need find to distribute the next level of one or more retrieve agent system according to some algorithms or strategy, next retrieve agent system distributes the next level, until the leaf level is end, leaf level of searching engine directly find information in their corresponding database.

This searching system uses distributed load-balanced structure, each retrieve agent system is responsible for various types of metadata in different regions of the database, which not only improve coverage of the retrieve system, but also improve the performance of retrieve system, at the same time the distribution of resource library is free, and the creation, classification and management are more simple and easy, and the robustness, maintainability, scalability of system has also been enhanced. The results can be obtained from different searching engines, and irrelevant, redundant webs are threw away, a unified user interface is made, therefore only a user can retrieve the entrance to a different area of the repository. As a single retrieve system the return of data accuracy is not high, a classification based on neural network technology is used in the searching methods, which uses neural network based classification algorithm to determine the relevance of these results, and searching results are sorted by the proper order, in general, the search results page which are classified as associated will be placed on the list above, unrelated website will be put the list below, or be removed, so that we can make the user acquire what he really wants faster and easily.

5   Implementation of Agricultural Information Retrieve

In order to overcome the limitation that query is too small while the key words are used to search, intelligent search algorithms based on ontology is proposed. The intelligent query allows the user to input natural language searching style approach to information retrieve, so the user provides query environmental information and increases the number of query keywords, which helps system understand the need of user query. The advantages of intelligent query based on ontology mainly are two points: First the system uses link of multiple keywords in the user query to the field of ontology concepts which exists on the semantic ambiguity of the concept of keyword ambiguity elimination, which can ensure the correctness of returned document; Second the system uses task of ontology and domain ontology to understand the user search needs, and reason according to relationship that exists keyword search query of user type, then answer user questions and acquire the real needs of users.

When users make search requests, the main interface will submit the query to management query, the expression parse uses ontology to transform a query problem that the user submits the smallest semantics canonical form that the computer can understand, and the ontology is used to regulate the retrieve of information, after the specification retrieve of information is submitted to inference engine, then inference engine receives the related terminology, according to the variety of agricultural ontology model which are stored, the application of formal ontology content is used to reason, the content-type user information needs and submit the results and find terminology which is related to the concept, then returns to the remote resource management and information retrieve system, resource management and information retrieve system is responsible for library management and user access to information queries, and index in accordance with the terms ontology model, user query request is saved and accessed relevant information, semantic annotation and metadata extraction system is responsible for the content of the document analysis characteristics, and make the framework design semantics of the domain ontology retrieve system explicit.

6   Conclusion

This paper analyzed the deficiencies of traditional information retrieve methods, and proposes a framework for intelligent information searching system based on ontology, solve low rate of recall and precision of traditional information retrieve, then carry out implementation technology of intelligent information searching system, and specific areas of ontology are made. The framework of design achieves different areas of intelligent information retrieve on the network, and improves high precision retrieve of agricultural information resources.

Acknowledgments. This work is supported by China Agricultural University, additional facilities and materials are kindly provided by the Information and Electrical Engineering at China Agricultural University.

References

1. Chen, K., Wu, G.: Research on information search technique based on ontology. Chinese Information Journal, 19, 51–57 (2004)
2. Xu, J.: Ontology and information search. China Information Review 3, 57–58 (2001)
3. Chang, C.: The construction and transformation of ontology in the agricultural information management, pp. 45–78. China Agricultural Academic, Beijing (2005)
4. Chen, X., Chang, Z.: The theory and intelligent search engine. Computer Applications 23, 479–480 (2003)
5. Deng, Z., Tang, S., Zhang, M., et al.: Research concept on ontology. Perking University Journal (Science Edition) 38, 73–78 (2002)

Improvement on Parameter Estimation Method in ARMA Based on HOS and Radon Transform

Zhen-Feng Qu and Xiao-Guang Li

Department of Electrical Engineering and Automation, Luoyang Institute of Science and Technology, Luoyang, 471023, Peoples R. China
lylonger@163.com, lxg@lit.edu.cn

Abstract. For the traditional ARMA parameter estimation method in practical applications, there will be too computationally intensive solution non-unique, and the estimated algorithm unstable, fuzzy system identification using higher order statistics on the estimated MA model parameters, then disconsolation algorithm to estimate the original image improvement algorithms; the same time, the introduction of the Radon transform two-dimensional image projection of the one-dimensional projection of the image under a certain angle, effectively reducing the algorithm computational complexity, simulation experiments show that the algorithm achieved good results.

Keywords: Blind image restoration, Radon transform, ARMA PSF, Scientific Materials.

1 Introduction

The existing image blind recovery method is fuzzy identification method(The iterative blind disconsolation [1] ,simulated annealing algorithm[2], and NAS-RIF algorithm [3] belong to this category.), the prior parameters limited support recovery technology, zero level domain separation technology, ARMA model parameters estimation method.etc.This paper discusses the key parameter estimation method is ARMA. The traditional ARMA parameters estimation method in practical application, will appear calculation is too big, not the only solution and estimation algorithm unstable problem. In this paper based on the existing technology, the introduction of higher order statistics, Radon transformation and sliced theory method, the algorithm greatly reduce the computing complexity, healing effect also are improved obviously.

2 The Traditional ARMA Model

ARMA model identification is mostly to complete respectively AR model and MA model identification method based, is real image will with a 2D regression (AR) model, fuzzy function also is the point spread. Function (PSF) with a two-dimensional moving average (MA) model to describe [4]. The recovery process is basically identification model parameters, and then can using classical method for image restoration of the estimated image.

D. Jin and S. Lin (Eds.): Advances in FCCS, Vol. 2, AISC 160, pp. 115–119.
springerlink.com © Springer-Verlag Berlin Heidelberg 2012

Identification model parameters have many methods, but the methods of main limitation: when the parameters are too much number of parameters of the variation of parameters on a single not sensitive; In addition, the convergence process may arise in the local minimum problem.

3   Improvement on Parameter Estimation Method in ARMA

In this paper the characteristics of Radon transformation dimension reduction, make the image in every Angle projection is 1D signal, and we can adopt appropriate estimate 1D signal of the methods to estimate the image projected sequence, and finally to 1D projection image re-projection sequence to domain can obtain estimate of the image. The basic flow chart of this algorithm is shown in Fig.1.

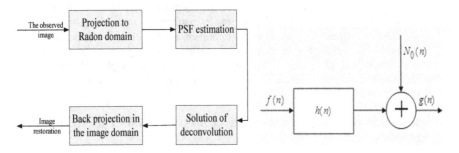

Fig. 1. The flow chart of this algorithm   **Fig. 2.** Image degradation model

Radon transformation tθ can be understood as the image in the space projection, tθ space every bit of corresponding to a straight line, and the Radon transformation is image pixels in each a line integral. Therefore, the image of high grey value line will be in a tθ space form light spots, and low grey value of straight in tθ space will form the dark spots. According to the definition of Radon transformation type, Radon transformation test line belongs to a kind of global operator, and to template as the basic algorithm model of convolution than local operators, Radon transformation method of linear theory more robust.

Multidimensional signal from the projection will recover out of the various methods of basic theory is on center slice theory; it points out $N-1$ the dimensions of the projection transformation N is ant derivative spectrum. It provides leaves the dimensions of the leaf transform itself spectrum. It provides with certain Angle of a section of [5]. So in 2D, $R_\theta(t)$ set to $f(x,y)$ the Radon transformation, then according to the center slice theory, one dimension, the transformation $R_\theta(t)$ is spectrum $f(x,y)$.it provides 2D spectrum. It provides transform in the Angle of the leaf θ in a slice. Therefore, the projection of one dimension is equal to transform spectrum. It provides 2D ant derivative spectrum. It provides transform in the Angle of leaves in a slice. This theorem can be easily extended to N dimensional. Then through the Filter back-projection theory [6], can will Radon domain to image re-projection Projection domain to get images.

According to the characteristics of Radon transformation convolution, Radon transformation can not only will the image to 1D projection Radon domain, and still retain the basic characteristics of the convolution. Figure 2 shows the Radon domain in the image degradation model so, if degradation in image is composed of two signals (the f (x, y) and point spread function (h (x, y)) of the convolution, then the degradation of the projection is still image is the equivalent of two signals of the projection of convolution respectively. In addition, through the Radon transformation from 2D images to 1D reduces the calculate necessary data, saving calculation time, making the complex two-dimensional calculation into a simple one dimensional calculation. Fig. 2 shows the Radon domain in the image degradation model.

Written expression form as follows:

$$g(n) = f(n) * h(n) + N_0(n) \tag{1}$$

$h(n)$ projection of point spread function is the $h(x,y)$ projection, $g(n)$ is observed $g(x,y)$ image $f(n)$, $N_0(n)$ projection of additive noise is the $N_0(x,y)$ projection.

Through the above Radon transformation, get R_i. R_i is about a group of t_j discrete data. PSF will use a non-Gaussian and is independent of the distribution of the MA with (q) described, projection sequence remember as (n) R_i. By converting output R_i sequence (n) of the PSF estimated 1D projection. The process as follows:

Tectonic output projection sequence R_i (n) the accumulation of the matrix C_1 and C_2:

$$C_1 = \begin{bmatrix} C_{kR_i}^{k-1}(q,q) & & & 0 \\ C_{kR_i}^{k-1}(q,q-1) & C_{kR_i}^{k-1}(q,q) & & \\ \vdots & \vdots & \ddots & \\ C_{kR_i}^{k-1}(q,1) & C_{kR_i}^{k-1}(q,2) & \cdots & C_{kR_i}^{k-1}(q,q) \end{bmatrix} \quad C_2 = \begin{bmatrix} C_{kR_i}^{k-1}(q,0) & C_{kR_i}^{k-1}(q,1) & \cdots & C_{kR_i}^{k-1}(q,q) \\ & C_{kR_i}^{k-1}(q,0) & \cdots & C_{kR_i}^{k-1}(q,q-1) \\ & & \ddots & \vdots \\ 0 & & & C_{kR_i}^{k-1}(q,0) \end{bmatrix}$$

Will point spread function (namely degradation system h (n) as FIR system, so make:

$$h = \begin{bmatrix} h(0), h(1), \cdots, h(q) \end{bmatrix}^T \tag{2}$$

And continue to tectonic cumulate matrices c_1 and c_2

$$c_1 = \begin{bmatrix} C_{kR_i}(-q,0)C_{kR_i}^{k-3}(q,0)C_{kR_i}(q,q) \\ C_{kR_i}(-q+1,0)C_{kR_i}^{k-3}(q,0)C_{kR_i}(q,q) \\ \vdots \\ C_{kR_i}(-1,0)C_{kR_i}^{k-3}(q,0)C_{kR_i}(q,q) \end{bmatrix} \quad c_2 = \begin{bmatrix} C_{kR_i}(0,0)C_{kR_i}^{k-3}(q,0)C_{kR_i}(q,q) \\ C_{kR_i}(1,0)C_{kR_i}^{k-3}(q,0)C_{kR_i}(q,q) \\ \vdots \\ C_{kR_i}^{k-3}(q,0)C_{kR_i}(q,q) \end{bmatrix}$$

Above all, q is the order number of parameters for MA. K is the order number of representatives cumulate, general is equal to or greater than 3, in the type: $C_{ckRi}(m, n)$

=$C_{kRi}(m, n\ddot{}...,n)$ representative cumulate a special section, and can prove equation (3) is the only solution. According to the literature [7] simultaneous equations

$$
\begin{bmatrix} C_1 \\ C_2 \end{bmatrix} h = \begin{bmatrix} c_1 \\ c_2 \end{bmatrix}
\tag{3}
$$

In the Radon projection domain, its basic iterative formula can be expressed as

$$
f_{k+1}(n) = f_k(n) \sum_i \left[\frac{h(i-n)}{\sum_j h(i-j) f_k(j)} \right] g(i)
\tag{4}
$$

In the type iteration times, the iterative k for initial value $f_0(n)$ greater than zero, initial value $f_0(n)$ to R-L iteration algorithm has a little influence on the results, after several iterations, it convergence to the original image in the projection of Radon domain $f(n)$. Estimate the projection f (n), through the Radon inverse transform, $f(n)$ will the projection to image domain, get the estimate $\hat{f}(x, y)$.

Solution matrix equation (3), can estimate the point spread function h (n) at θi in place of Radon domain projection. We construct of the equation is just contain cumulate matrices, that is to say this is a kind of only using higher-order cumulate of MA parameters estimation method [8], it can apply to any additive Gaussian colored noise.

After giving MA parameter or point spread function Radon domain projection $h(n)$, want to recover the original image $f(x, y)$ of Radon domain projection $f(n)$, it must be solution convolution, This paper uses the R-L algorithm.

4   The Experimental Results and Analysis

The aim of the experiment is to verify that the proposed based on higher order statistics and the Radon transformation blind image restoration of the efficiency of the algorithm, but at the same time, recover the obtained result and ML maximum likelihood method to do further comparison analysis.

To compare the performance of the algorithm, the image is used test degradation by original image and a given point spread function do convolution, at the same time, the Gaussian noise superposition must have. Point spread function for a 7x7 of the Gaussian filter, the size of the images for 256 x256.

From the Fig.3 (a) to Fig.3 (f) can see whether maximum likelihood method or the method in this paper can be on certain level to improve degradation image visual effect, but the proposed algorithm in the rebound effect and reserve the detail of the image is better than maximum likelihood method.

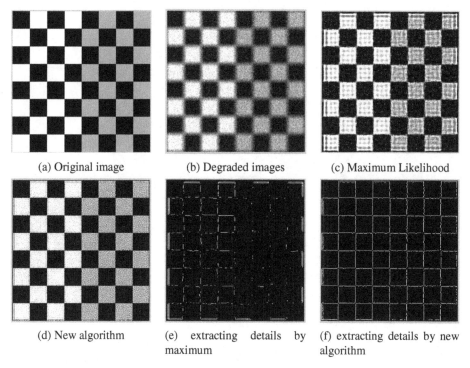

(a) Original image (b) Degraded images (c) Maximum Likelihood

(d) New algorithm (e) extracting details by maximum (f) extracting details by new algorithm

Fig. 3. Effects of various algorithms to compare

References

1. Ayers, G.R., Dainty, J.C.: Iterative blind disconsolation method and its application. Optics Letters. 13(7), 547–549 (1988)
2. McCallum, B.C.: Blind disconsolation by simulated annealing. Optics Communications 75(2), 101–105 (1990)
3. Kurdur, D., Hatzinakos, D.: Blind image disconsolation revisited. IEEE Signal Processing Magazine 13(6), 61–63 (1996)
4. Telalp, A.M., Kaufman, H.: On statistical identification of a class of liner space-invariant blurs using non-minimum-phase ARMA models. IEEE Transactions on Account Speech and Signal Processing 36(8), 1360–1363 (1988)
5. Zhao, H., Liu, Y., Li, J.: Effects on spatial resolution of CT image caused by nonstationary axis. Journal of Tsinghua University 42(8), 997–1000 (2002)
6. Zhang, P., Zhang, Z.: Research and Comparison on Several CT Reconstruction Algorithms. Computerized Tomography Theory and Applications 10(4), 4–10 (2001)
7. Zhang, X.D., Zhang, Y.S.: FIR system identification using higher order cumulates alone. IEEE Trans. Signal Processing 42(10), 2854–2858 (1994)
8. Feng, H.-J., Tao, X.-P., Zhao, J.-F., Li, Q., Xu, Z.-H.: Review and Prospect of Image Restoration with Space-variant Point Spread Function. Opto-Electronic Engineering 36(1), 1–7 (2009)

CTMC-Based Availability Analysis of Cluster System with Multiple Nodes

Zhiguo Hong[1], Yongbin Wang[1], and Minyong Shi[2]

[1] School of Computer, Communication University of China, Beijing 100024, China
hongzhiguo1977@yahoo.com.cn, ybwang@cuc.edu.cn
[2] School of Animation and Digital Arts, Communication University of China, Beijing 100024,
China
myshi@cuc.edu.cn

Abstract. The availability of cluster system with multiple nodes was investigated by taking common mode failure (CMF) into account. Firstly, the availability of system with one cluster node was analyzed. Then on the basis of analysis, contraposing to the cluster system with two nodes and three nodes, four continuous time markov chain (CTMC) models in the case of No CMF and in the case of CMF were constructed respectively. Moreover, numerical comparison of steady availability for these two cases was conducted by solving related linear equations and the conclusion on the priority and inferiority comparison of different cases is derived thereby, which offers theoretical foundation for establishing cluster systems.

Keywords: availability, markov process, common mode failure, cluster.

1 Introduction

With the ever-increasing Internet-based applications, cluster system has been widely used to accelerate and shorten the computation time. Take computation-intensive rendering for instance, due to the stronger ability of computation the rendering time of a high-performance cluster node for a complex still frame would be much shorter than that of an average PC. Consequently, the requirement of having high availability and reliability for cluster systems is also critical in offering efficient rendering services for 3D scene producers. Nowadays, methodologies of performance analysis mainly take three ways, which are direct measurement [1], actual experiments and tests for real platform[2],mathematic modeling [3], and simulation. However, if performance bottlenecks are detected, adjustment and redeployment should be conducted during running phase of multimedia systems, it would be a waste of manpower and material resources. So, how to construct a considerably precise model for cluster system and analyze its high availability is an urgent and important issue, which presents instructive qualitative foundations for establishing cluster system.

Recently, there are some popular modeling methodologies for analyzing availability of cluster systems, which are Reliability Block Diagrams (RBD), markov process [4] and stochastic Petri nets [5]. An important guiding principle for designing cluster system is that the probability of catastrophic failure should be avoided to the fullest

D. Jin and S. Lin (Eds.): Advances in FCCS, Vol. 2, AISC 160, pp. 121–125.
springerlink.com
© Springer-Verlag Berlin Heidelberg 2012

extent. In the case of just one cluster node in the cluster system, the fault or error in hardware or software would lead to crash. As a result, such cluster system would be unavailable to offer expected services, which could affect users' experiences and suffer commercial loss. Usually, the availability of cluster systems can greatly upgraded with parallel connections among multiple nodes. Nevertheless, although system with parallel connections has higher availability than system with a single node, the occurrence of common mode failure (CMF) does exist which means the simultaneous failure of multiple components follows the same mode.

For this reason, to simulate systems with several nodes and model such systems more precisely, this paper studied the availability of system comprising multiple cluster nodes in the case of CMF. Firstly, the performance of system with a single cluster node was cited. Then, contraposing to the systems with two cluster nodes and three cluster nodes, four continuous time markov chain(CTMC) models were proposed respectively. By solving related equations, numerical results of system's availability were derived. Finally, analysis by comparison of these cases is discussed.

2 Availability Analysis of System with One Cluster Node

In the case of system with one cluster node, the availability of system depends totally on that of the cluster node. If such cluster node fails or crashes, system won't work until the cluster recover and resume. We assume the time to failure of cluster node to be a random variable with the corresponding distribution being exponential with rate λ, also the time to repair to be exponentially distributed with rate μ. Fig. 1 shows the CTMC model of system with one cluster node.

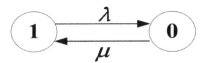

Fig. 1. ON-OFF Model

As Fig.1 demonstrates, state "1" represents availability of the system, and state "0" symbolizes un availability of the system. Numerically, the chance of system's availability equals the probability when the CTMC model is in the state of "1". So we can get the equations of describing the long-term, steady probability for cluster system as follows [6]: $\lambda(1-\mu(0=0$ and $(1+(0=1$. By solving this equation, we can have the availability, i.e. the steady probability of CTMC model being state "1" is: $(1=\mu/(\lambda+\mu)$. For the parameters of $\lambda=1\times10$-3hr-1 and $\mu=1$hr-1, the availability can be further calculated:$A1=(1=1/1.001\approx0.999001$.

3 Availability Analysis of System with Two Cluster Nodes

Similarly, we take the system with two cluster nodes as study object and model it using markov process. Two CMTC models for cluster system in the case of CMF and in the case of No CMF are constructed as Fig. 2 and Fig. 3. Here, state "2" represents both cluster nodes are in available state, state "1" represents only one cluster node is available and state "0" represents two cluster nodes are unavailable for the time being. We further consider the system to be functioning as long as one of two cluster nodes is working [7].

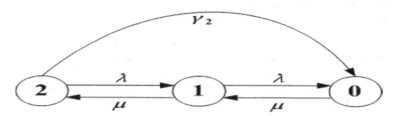

Fig. 2. CTMC Model of System with Two Cluster Nodes in the case of CMF

As Fig.2 demonstrates, in the case of CMF, we can express the CTMC model by using the following equations:

$dP2(t)/dt=-\lambda P2(t) -\gamma2\,P2(t)+ \mu P1(t)$
$dP1(t)/dt=-\mu P1(t) -\lambda P1(t)+ \mu P0(t)+ \lambda P2(t)$
$dP0(t)/dt=-\mu P0(t) + \lambda P1(t)+ \gamma2\,P2(t)$

We concern the steady state and get: $dPk(t)/dt=0$, k=0, 1, 2. The steady probability can be got by evaluating the limits of these equations. Here we symbolize A2 for the availability in this case. Then for the similar parameters of $\lambda=1\times10$-3hr-1, $\mu=1$hr-1, $\gamma2=5\times10$-4hr-1 and combining the constraint of "(0+(1+(2=1", we can calculate: A2=(1+(2=10005015/10020015≈0.9985029963.

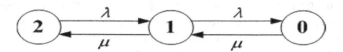

Fig. 3. CTMC Model of System with Two Cluster Nodes in the case of No CMF

As Fig.3 shows, in the case of No CMF, we can obtain the steady probability from solving similar equations. In the same way, we denote B2 for the availability in this case. We can calculate: B2=(1+(2=1001000/1001001≈0.9999990010.

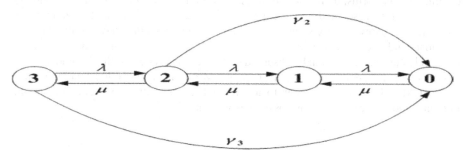

Fig. 4. CTMC Model of System with Three Cluster Nodes in the case of CMF

As Fig.4 demonstrates, in the case of CMF, we can express the CTMC model by using the following equations:

$dP3(t)/dt = -\lambda P3(t) - \gamma 3 \, P3(t) + \mu P2(t)$
$dP2(t)/dt = -\lambda P2(t) - \mu P2(t) - \gamma 2 \, P2(t) + \lambda P3(t) + \mu P1(t)$
$dP1(t)/dt = -\mu P1(t) - \lambda P1(t) + \mu P0(t) + \lambda P2(t)$
$dP0(t)/dt = -\mu P0(t) + \lambda P1(t) + \gamma 2 \, P2(t) + \gamma 3 \, P3(t)$

We concern the steady state and get: $dPl(t)/dt=0$, $l=0, 1, 2, 3$. The steady probability can be got by evaluating the limits of these equations. Here we symbolize A3 for the availability in this case. Then for the similar parameters of $\lambda=1\times10\text{-}3\text{hr-}1$, $\mu=1\text{hr-}1$, $\gamma2=5\times10\text{-}4\text{hr-}1$, $\gamma3=2\times10\text{-}4\text{hr-}1$ and combining the constraint of "(0+(1+(2+(3=1", we can calculate: A3=(1+(2+(3=≈0.9997992207.

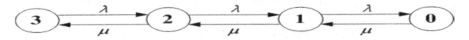

Fig. 5. CTMC Model of System with Three Cluster Nodes in the case of CMF

As Fig.5 shows, in the case of No CMF, we can get the steady probability from solving similar equations. In the same way, we denote B3 for the availability in this case. We can calculate: B3=(1+(2+(3=1001001000/1001001001≈0.9999999990.

4   Analysis By Comparison

In sum, we collect the numerical approximations for different cases, i.e. A2≈0.998502 9963, B2≈0.9999990010, A3≈0.9997992207, B3≈0.9999999990. It can be seen the differences between each other are hard to distinguish, but for one year the total differences of time could not be neglected. Assume 365 days for one year, then we can deduce the average time of unavailability for each cases. For the state of being unavailability, cluster systems are in the process of crashing or repairing and unable to offer services. To exhibit the enhancement of availability with the addition of nodes' number and to demonstrate the situation in the case of CMF more obviously, the unavailable time for different cases is 786.8251mins, 0.5251min, 105.5296mins and 0.0005 min respectively. It can be seen that for time-critical services, such as commercial platforms, unavailability for several minutes would cause great economic losses. For this case, additional expense of increasing cluster nodes' number is cost-effective. However, for small-scaled cluster system with the unavailability being a secondary index compared with its output, merely adding new cluster nodes is not a preferable option. Also, it can be observed that for cluster system with the same number of nodes, the unavailability time in the case of CMF is much longer than that in the case of No CMF. Therefore, in order to assure high availability of system, we should minimize the chance of CMF as possible as we can.

5 Summary

In order to model and simulate cluster system more precisely, we investigated the availability of cluster systems in the case of CMF. By citing CTMC model for one cluster node, we constructed four (CTMC) models for cluster system with two nodes and three nodes in the case of No CMF and in the case of CMF respectively. Furthermore, numerical comparison of steady availability for these two cases was conducted by solving related linear equations and the conclusion on the priority and inferiority comparison of different cases was derived thereby. On the basis of current work, further research on performance of cluster systems can be carried on.

Acknowledgement. The paper is supported by the High Technology Research and Development Program of China (2011AA01A107); Engineering Planning Project for Communication University of China (XNG1126).

References

1. Kambo, N.S., Deniz, D.Z., Iqbal, T.: Measurement-Based MMPP Modeling of Voice Traffic in Computer Networks Using Moments of Packet Interarrival Times. In: Lorenz, P. (ed.) ICN 2001. LNCS, vol. 2094, pp. 570–578. Springer, Heidelberg (2001)
2. Gooding, S.L., Arns, L., Smith, P., et al.: Implementation of a distributed rendering environment for the teraGrid. In: Challenges of Large Applications in Distributed Environments (CLADE), pp. 13–22. IEEE Computer Society, Washington (2006)
3. Jin, Y., Zhou, G., Jiang, D.C., et al.: Theoretical mean-variance relationship of IP network traffic based on ON/OFF model. Science in China Series F: Information Sciences 52(4), 645–655 (2009) (in Chinese)
4. Liu, X.: Research of Dependability Evaluation of Cluster System. Master thesis, Graduate School of Chinese Academy of Sciences, Beijing (December 2003) (in Chinese)
5. Leangsuksun, C., Shen, L., Liu, T., et al.: Dependability Prediction of High Availability OSCAR Cluster Server. In: Proc. of 2003 International Conference on Parallel and Distributed Processing Techniques and Applications (PDPTA 2003), Las Vegas, Nevada, USA, June 23-26 (2003)
6. David, M.D.: QoS and Traffic Management in IP and ATM Networks. McGraw-Hill Professional Publishing, New York (2000)
7. Sahner, R.A., Trivedi, K.S., Puliafito, A.: Performance and Reliability Analysis of Computer Systems: An Example-Based Approach Using the SHARPE Software Package. Kluwer Academic Publishers (1996)

Research on Control Technology of Point Type Feeding in Aluminum Electrolytic Process

Jiejia Li[1], Rui Qu[1], Hao Wu[1], and Ying Li[2]

[1] School of Information and Control Engineering, Shenyang Jianzhu University,
110168, Shenyang, Liaoning, China
[2] Shenyang Academy of Instrumentation Science, 110043, Shenyang, Liaoning, China

Abstract. Aluminum electrolysis is a process of nonlinear, time varying and large time delay, and it is difficult to control, high energy consumption. Therefore, aluminum electrolysis control system's hot issue is to save electric energy, to improve current efficiency. We proposed composite fuzzy neural network control method which combined neural network control and PID control, through tracking parameters of cell resistance which reflected alumina concentration, to adjust control strategy of controller, to control feeding quantity of alumina feeding device, so that we can control alumina concentration in ideal range. Experimental results show that: this method has good control performance and energy-saving effect.

Keywords: Aluminum electrolysis, Fuzzy neural network, Composite control.

1 Introduction

Aluminum electrolysis is a complex industrial process, and it has strong interference and large energy consumption. So, the research on energy saving control technology of aluminum electrolysis has become a hot issue of academic research. Nowadays, the control methods for aluminum electrolysis mostly adopt adding alumina to electrolytic cell, because of the differences of each aluminum electrolytic cell, control effect is not ideal. So, this paper uses fuzzy neural network composite control strategy, add feeding on the basis of needs, in order to make each electrolytic cell to best working state, to improve control effect and reduce energy consumption.

2 Hardware System Design

System is composed of management level, monitoring level and control level.

Control level's main function is to complete data acquisition of analog and switch value. According to the different working states of electrolytic cell, we use different control strategies to complete the control of alumina feeding, and to make electrolytic cell in the best working state, to meet the requirements of technical indexes. Monitoring level is mainly to complete system's supervisory control function, to monitor electrolytic cell's state. Upper managing computer's main function is to receive data information.

D. Jin and S. Lin (Eds.): Advances in FCCS, Vol. 2, AISC 160, pp. 127–132.
springerlink.com   © Springer-Verlag Berlin Heidelberg 2012

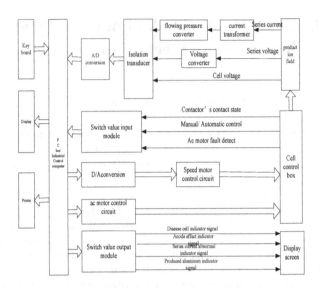

Fig. 1. Structure diagram of hardware system

Process control level is mainly composed by control host, analog input module, switch value input module, analog output module, switch value output module and so on. Its hardware system structure is shown in figure 1.

Main controller adopts IPC-90 industrial control computer. It has real-time processing ability, intelligent modular, multiprocessor system structure and other characteristics.

In analog input module, analog signal isolates transmitter and A/D conversion template through voltage, collects real-time data. And according to electrolytic cell's different working states, we will use different control strategies to let it in best working state.

Switch input channel detects system's state, monitor the reliability of switch device.

The output control signal of analog output module will control alumina feeding device which is the key technology of aluminum electrolysis. Control strategy for feeding is completed by software control algorithm.

Output control signal of switch output channel alarms for all kinds of fault, displays electrolytic cell's working state.

3   Composite Fuzzy Neural Network Controller Design

The schematic diagram of control system is shown in figure 2.The controller includes fuzzy neural network controller and PID controller. Outputs of controller are feeding rate and time of feeding device. The controller adopts composite time sharing control strategy, set error is $\Delta R(k)$, when $\left|\Delta R(k)\right|$ > set error's range, it will use PID control to complete quick adjustment, when $\left|\Delta R(k)\right|$ < set error's range, it will use fuzzy neural network control to ensure high control precision and good real-time performance.

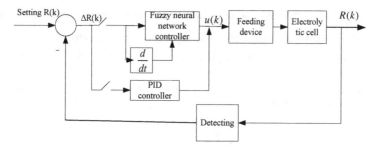

Fig. 2. Schematic diagram of fuzzy neural network control system

The fuzzy neural network controller is shown in figure 3.

The controller includes: input layer, fuzzy layer, rule layer and output layer. They achieve that perform allocation, fuzzification, fuzzy computing and defuzzification. The model parameters of network model are fuzzy center m and width σ of fuzzy layer's each neuron, and weight W is rule layer to output layer, weights of other parts all set them 1.

Y is the output of neurons, the right upper corner is on basis of the layer neurons in, the right lower corner is on basis of order number.

The first layer is input layer, it can pass external inputs to next layer directly.

$$y_m^{(1)} = u_m^{(1)} = x_m \tag{1}$$

x_m is the m th component of external input, is the change rate of cell resistance.

The second layer is fuzzy layer, neurons fuzzy up input data by fuzzy function.

$$y_{mi}^{(2)} = \exp\left(-\frac{\left(u_{mi}^{(2)} - m_{mi}^{(2)}\right)^2}{2\left(\sigma_{mi}^2\right)^2}\right) = \exp\left(-\frac{\left(x_m - m_{mi}^{(2)}\right)^2}{2\left(\sigma_{mi}^2\right)^2}\right) \tag{2}$$

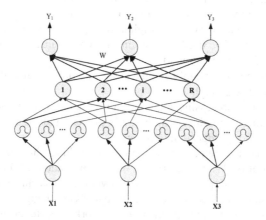

Fig. 3. Structure diagram of fuzzy neural network controller

$u_{mi}^{(2)}$, $y_{mi}^{(2)}$ are the input and output from the m th neuron of the first layer to the i th neuron of the second layer. $m_{mi}^{(2)}$, $\sigma_{mi}^{(2)}$ are fuzzy center and width of those neurons.

The third layer is rule layer, it completes logic and operation to fuzzy results of the second layer.

In rule layer, we use product operation to achieve "and" operation in inference rules. The expression is shown below:

$$y_r^3 = \prod_{k=1}^{K_r} u_{r_k}^{(3)} \tag{3}$$

Among them, Kr is the number of all input variables which are owned by the third layer's the r th neuron.

The fourth layer is output layer, we use weighted square method to achieve defuzzification.

$$y_i = \frac{\sum_{r=1}^{R} w_{nr} y_r^3}{\sum_{r=1}^{R} y_r^3} \tag{4}$$

Among them, y_i is output of the ith fuzzy neural sub-network, this controller's output is used as control variable of feeding device.

4 System's Software Design

The software main program diagram of control system is shown in figure 4. It mainly includes: data acquisition module, fault processing module, produced aluminum processing module and predictive module and so on.

Data acquisition module mainly includes: analog data acquisition and switch value data acquisition these two signals. Through collecting the information of analog series current, series voltage and cell voltage, digital filtering and computing, and then it will send them to computer. The data acquisition of switch value includes acquisition of contactor's state and manual/automatic switch.

Fault processing module mainly includes: the monitoring and alarming of anode effect, heat cell and disease cell.

Produced aluminum processing module mainly includes: monitoring of produced aluminum, controlling of produced aluminum's process and the new control strategy which will be used after produced aluminum.

Control module is the core part of main program, includes that fuzzy neural network control and PID control. When the error is big, we will adopt PID control. When the error is small, we will adopt fuzzy neural network control, and use variable speed device to control alumina feeding device, so that it can change the rate of feeding, adjust the concentration of alumina, let it in best working state, ensure the output and quality of produced aluminum and achieve energy saving and efficient target.

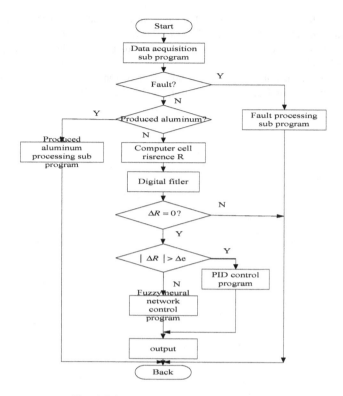

Fig. 4. Diagram of software main program

5   Conclusion

In this paper, based on the characteristics which aluminum electrolytic process is nonlinear, strong interference, time-varying and large time delay, we proposed that composite fuzzy neural network control method, this method combined neural network control technology and PID control technology, according to the changes of electrolytic cell's parameters, adjusting alumina feeding device to control concentration of alumina and to ensure it in setting range and electrolytic cell in best working state. The experimental results show that this method has high control precision, good dynamic performance and good application value.

References

1. Zhang, Y.F., Ma, B., Zhang, J.L., Chen, M., Fan, Y.H., Li, W.C.: Study on fault diagnosis of power-shift steering transmission based on spectrometric analysis and SVM. Guang Pu Xue Yu Guang Pu Fen Xi 30(6), 1586–1590 (2010)
2. Caccavale, F., Digiulio, P., Iamarino, M., Masi, S., Pierri, F.: A neural network approach for on-line fault detection of nitrogen sensors in alternated active sludge treatment plants. Water Science and Technology 62(12), 2760–2768 (2010)

3. Cheng, Y.L., Huang, J.C., Yang, W.C.: Modeling word perception using the Elman network. Neurocomputing 71(16/18), 3150–3157 (2008); 11(5), 25–28 (2006)
4. Haykin, S.: Neural networks-A Comorehensive Foundation, 2nd edn. Tsinghua University Press, Beijing (2001)
5. Ren, X.: Recurrent neural networks for identification of nonlinearsystems. In: Proceeding of the 39th Conference on Decision and Control, pp. 2861–2866 (2000)

The Method of Normalization Maximal Energy Spectrum for Sequence Stratigraphy Division

Yunliang Yu, Ye Bai, Xin Wang, and Wenbo Li

College of Earth Sciences, JiLin university, ChangChun, China
{yuyunliang,baiye,wangxin,liwenbo}@jlu.edu.cn

Abstract. Sequence stratigraphy division is important steps in petroleum exploration and development, logging data is the reflection of lithology and physical properties, which contains a large amount information of periodic changes in the sequence stratigraphic. In this paper, we analyze the logging signal by wavelet transform and spectral analysis, based on the law of optimal scale and optimal scaling factor corresponding to the energy spectrum on the frequency band that occupies the maximum proportion of the total energy, the method of normalization maximal energy spectrum is put forward. Use this method to select the optimal scale factor and wavelet coefficients corresponding to it after transforming the logging data by wavelet, and then divide the sequence stratigraphy according to the cyclicity of this wavelet coefficients. The method has further improved the objectivity and accuracy of sequence stratigraphy division.

Keywords: wavelet transform, normalization maximal energy spectrum, logging signal, sequence stratigraphy, wavelet coefficient.

1 Introduction

The logging signal is consists of a series signals with wide frequency range, which is the reflection of lithology and physical properties[1], it contains many informations about sequence stratigraphy boundaries[2]. From the viewpoint of Fourier series expansion and the energy spectrum, the optimal scale and optimal scaling factor corresponding to the energy spectrum on the frequency band should occupy the maximum proportion of the total energy, then the method of normalization maximal energy spectrum is put forward. The method can search the maximum energy within decomposition scales of wavelet transform, and the scale corresponding to maximum energy is optimal scale factor. We get the wavelet coefficients corresponding to the optimal scale factor, and divide the sequence stratigraphy according to the cyclicity of this wavelet coefficients.

2 The Principle of Energy Spectrum Calculation with Wavelet Transform

The energy of signal can be obtained through wavelet coefficients of different decomposition scales after discrete wavelet transforming[3], and according to the way

D. Jin and S. Lin (Eds.): Advances in FCCS, Vol. 2, AISC 160, pp. 133–137.
springerlink.com

of spectrum to show the signal energy got by wavelet decomposition is called wavelet energy spectrum, then the energy distribution of signal can be displayed clearly.

One-dimensional discrete wavelet transform algorithm is obtained by mallat algorithm[4]. The signal is decomposed by high-pass filter and low-pass filter at increment levels. We get high frequency detail information by high-pass filter, and low frequency approximation information by low-pass filter, decrease the sampling rate of low-frequency approximate information to half, and repeat the process above until reaching the decomposition level predefined[5]. The decomposition process is shown in figure1.

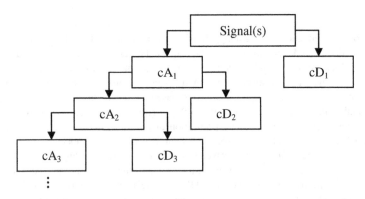

Fig. 1. The decomposition process of 1-D discrete wavelet transform

From the figure1 we can see, the original signal $s(n)$ can be calculated by formula as follows:

$$s(n) = cD_1(n^{2^{-1}}) + cD_2(n^{2^{-2}}) + cD_3(n^{2^{-3}}) + \ldots\ldots + cD_j(n^{2^{-j}}) + cA_j(n^{2^{-j}})$$

$$= \sum_{i=1}^{j} cD_i(n^{2^{-i}}) + cA_i(n^{2^{-i}}) \approx \sum_{i=1}^{j} cD_i(n^{2^{-i}}) \tag{1}$$

$i = 1,2,3....j$ is scales of wavelet decomposition, $cD_i(n^{2^{-i}})$ is wavelet coefficients at different scales.

Signal energy can be obtained by square of the wavelet coefficients at different scales, the formula as follows:

$$E_j = \left| cDj(n^{2^{-i}}) \right|^2 \tag{2}$$

The energy spectrum obtained through ranging the signal energy at each level can reveal the inherent characteristics of the signal more intuitive, more sophisticated, and more accurately.

3 Use the Method of Normalization Maximal Energy Spectrum to Extract Optimal Scales

From the viewpoint of Fourier series expansion and the spectral energy, the optimal scale and energy spectrum corresponding to it on the frequency band should occupy the maximum proportion of the total energy, the method of normalization maximal energy spectrum is put forward. Firstly, select a wavelet, and transform the logging signal by this wavelet, then calculate the squared modulus of wavelet coefficients. Secondly, eliminate the DC component by differential, normalize the wavelet coefficient spectrum, and select the optimal scale that has maximum energy within the scope of wavelet transform scales, the most important, we can get wavelet coefficients corresponding to the optimal scale. At last, the sequence stratigraphy is divided according to the cyclicity of the wavelet coefficients.

The formula of normalization maximal energy spectrum as follows:

$$\arg \max_{a \in (1,S)} [\sum_{i=1}^{N} |W_f(a,i)|^2 - \max_{1 \le i \le N} |W_f(a,i)^2|] / \sum_{i=1}^{N} |W_f(a,i)|^2 \tag{3}$$

N is sampling points of logging signal, S is upper bound of the wavelet decomposition scales, and the wavelet coefficients can be calculated as formula (4):

$$W_f(a,i) = FFT(|CWT(a,t)|) / |FFT(|CWT(a,t)|)|^2 \tag{4}$$

Through comprehensive analysis and description above, the optimal scale in wavelet transform can be summarized as follow steps:

(1)select a wavelet, and transform the logging signal by the wavelet;
(2)calculate the squared modulus of wavelet coefficients, and eliminate the DC component by differential;
(3)calculate the normalization maximal energy spectrum by formula(3);
(4)search the scale of maximum energy in the spectrum as optimal scale, and get wavelet coefficiens corresponding to optimal scale using formula (4);
(5) divide the sequence stratigraphy according to the cyclicity of the optimal wavelet coefficients.

4 Case Study

Log Wu5 located in the Shahe jie Formation of Kongnan area of Huanghua depression of China, we divide the sequence stratigraphy with the logging data(shown in figure 4) by Morlet wavelet. The time-frequency chromatogram of wavelet coefficients is shown in figure 2, and the energy spectrum is shown in figure 3. The centre of energy group can imply the boundary of sequence stratigraphy, the distribution of sampling points of boundary ranges among $1-500$, $500-1400$, $1400-1800$ and $1800-2100$.

In figure 2, the scales between 420-500 present large-scale sedimentary cycles, which can be used to identify subsequence sets, the scales between 200-400 present small-scale sedimentary cycles, which can be used to identify parasequences, and the scales below 100 correspond to high-frequency sedimentary cycles. We finally select the scale of 320 to identify the boundary of parasequences, and the scale of 480 to identify the boundary of subsequence sets by the method of normalization maximal energy spectrum.

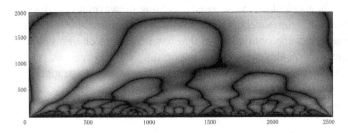

Fig. 2. The time-frequency chromatogram of wavelet coefficients of log Wu5

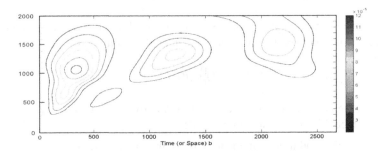

Fig. 3. The scalogram percentage of energy for each wavelet coefficient of log Wu5

The result of automatic division is shown in figure 4:

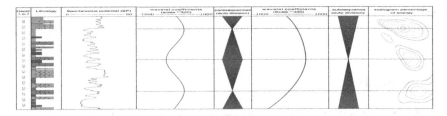

Fig. 4. Division result of log Wu5 by the normalized maximum spectral energy method

The energy group centers of optimal scale correspond to upper boundary of falling half cycle, the boundary is the period of strongest energy. According to the cyclicitys of wavelet coefficients of optimal scales, we can divide two subsequence sets at optimal scale 480, and four parasequences at optimal scale 320. In figure 4, we can see that the optimal scale factor and wavelet coefficients corresponding to it obtained by the method of normalization maximal energy spectrum is reasonable. The cyclicity of wavelet coefficients is fully consistent with boundary of sequence stratigraphy, which proves the correctness and accuracy of this method.

5 Conclusion

Use the method of normalized maximum spectral energy to select the optimal scale factor and wavelet coefficients corresponding to it, and then divide the sequence stratigraphy according to the cyclicity of this wavelet coefficients.

The wavelet coefficients of optimal scale extracted by the method has simple curve shape, and obvious cyclicity, which can be the final basis for sequence stratigraphy division.

References

1. Lee, S.H.: Multiscale data integration using Markov Random Fields. SPE 76905, pp. 68–78 (2002)
2. Mirowski, P., Seleznev, N., McCormick, D.S., et al.: New Software For Well-To-Well Correlation Of Spectroscopy Logs. AAPG Computer Applications in Geology (2005)
3. Zhu, H., Kwok, J.T.: Improving de-noising by coefficient de-noising and dyadic wavelet transform pattern recognition. In: Proceedings of International Conference on Pattern Recognition, New York, vol. 2, pp. 273–276 (2002)
4. Duan, T.Z., Cedric, M.G., Sverre, O.J.: High frequency sequence stratigraphy using syntactic methods and cluster applied to the upper limestone coal group of Kincardine Basin, United Kingdom. Mathematical Geology 33(7), 825–843 (2001)
5. Foster, I., Kesselman, C., Nick, J., Tuecke, S.: The Physiology of the Grid: an Open Grid Services Architecture for Distributed Systems Integration. Technical report, Global Grid Forum (2002)

Access Control Model in Migrating Workflow System with Site Service Alliance

Wei Su[1,2] and Guangzhou Zeng[1]

[1] School of Computer Science and Technology, Shandong University
[2] School of Management Science and Engineering, Shandong University of Finance and Economics, Jinan Shandong 250061, China
echorainbow@msn.com, gzzeng@sdu.edu.cn

Abstract. The migrating workflow is a new direction within the workflow management area. The access control model in a migrating workflow system is one of the most important areas in the research. In the traditional access control model the instance needs to apply permission in each site. In this article, we present the conception of sites' service alliances (SA) into the migrating workflow system to reduce the cost of access control. In the system, there are several s. The sites which belong to the same provide same service and resource to the instance. With conception of sites' service alliances we adapt the migrating workflow system model and present a new access control model on it. In the new access control model, the instance applies permission for a sites' service alliance. So the instance does not need to apply permission in each site. The cost of access control is reduced largely.

Keywords: Migrating Workflow System; Site Service Alliance; Access Control.

1 Introduction

The migrating workflow (defined in reference [5]) is a new direction within the workflow management area. It was proposed by Andrzej Cichocki etc to handle the distributed and dynamic character of business processes. A migrating workflow transfers its code (specification) and its execution state to a site, negotiates a service to be executed on its behalf, receives the results to finish a task, and moves on. Thus, the actual workflow instance is defined during the run-time. In a migrating workflow system there are four entities. They are migrating instance, site, event and log. The site is the workplace of the instance. The instance can get service or resource from the site.

In a migrating workflow system, when an instance arrives at a site and applies resource or service from the site, the system needs to decide whether allow the application or not. So the access control model in a workflow system is one of the most important areas in the research. The researchers have done much work about it. In the reference [4], WANG present a weighted role and periodic time access control model of workflow system. In another paper of us (reference [7]), we present a RBAC model considering the user reliability in workflow system. In the traditional access control model, the instance needs to apply permission in each site. The system cost in access control is large.

D. Jin and S. Lin (Eds.): Advances in FCCS, Vol. 2, AISC 160, pp. 139–145.
springerlink.com

In this paper we present the conception of sites' service alliances (SA) into the migrating workflow system. The sites in the same SA provide same or similar service and resources to the instance in the workflow system. With the conception of SA we propose a migrating workflow system model named MWSA. In this model, all the sites in the workflow system belong to one of the SA. In MWSA system, the access control model can be adapted to reduce the system cost. So in the latter part of this article we present an access control model in MWSA system named AC_MWSA. In this model, the instance does not apply permission for a single site. The instance applies permission for sites' service alliance. When a migrating instance applies service or resources for a site, the system can just find the SA which the site belongs to, and make the decision according to whether the instance has the permission in the SA. In this way the cost of the system on access control can be reduce largely.

2   MWSA Model

The migrating workflow model has been defined on reference [5] and reference [10]. There are four entities in a migrating workflow system. They are migrating instance, site, event and log. We introduce the conception of sites' service alliances (SA) to the migrating workflow system to get a new model MWSA (migrating workflow system with sits' service alliance). Before define the model, we give a definition of the sites' service alliances (SA) at first. Then we give the definition of MWAS.

2.1   The Site Service Alliance

There are many sites in a migrating workflow system. The site is the workplace of the instance. The instance can get service or resource from the site. Some sites in the system provide same or similar service or resource to the instance. We propose a new conception the sites' service alliance (SA) to make the sites which provide same or similar service or resource belong to the same SA. Then we give the definition of SA.

Definition 1. SA (SAid, SAS, SAR) is an alliance of sites in the system. SAid is the identification of the alliance. SAS is the set of service that the sites in the alliance can provide. SAR is the set of resources that the sites in the alliance can provide.

2.2   The MWSA Model

With the new conception of SA, we adapt the traditional migrating workflow model to propose a new migrating workflow model named MWSA. In the MWSA model, the definition of the four entities is must changed. Then we give the definition of the MWSA model.

Definition 2. MWAS (MI, S, E, L, SA) is a migrating workflow system with sits' alliance, there are five entities in the system. MI is a migrating instance in the system. S is a site that can provide service and resource to the instance. E is an asynchronous notification that the instance send to the workflow, or to the sites. L is the log that the instance carries. SA is an alliance of sites in the system.

Definition 3. MI (MWid, MIid, L) is a migrating instance in a migrating workflow. MWid is the identification of the migrating workflow that the instance belongs to. MIid is the identification of the instance. L is the log that the instance carries.

Definition 4. S (Sid, SS, SR, SAid) is a site that can provide service and resource to the instance. Sid is the identification of the site in the migrating workflow system. SS is the set of service that the site can provide. SR is the set of resources that the site can provide. SAid is the alliance that the site belongs to.

Definition 5. E (Eid, EK, MIid, Sid, Tid, Bt, Ft) is an asynchronous notification that the instance send to the workflow, or to the sites. E is the set of events in the system. Eid is the identification of the event. EK is the kind of the event. MIid is the identification of the instance who is in this event, Sid is the identification of the site which the event happen on, Tid is the identification of the task that the instance want to fulfill in this event, Bt is the beginning time of this event, Ft is the finishing time of this event.

Definition 6. L (Lid, TS, IP) is the log that the instance carries. Lid is the identification of the log. TS is the specification of the instance, which describes the service path and task of this instance. IP is the instance's permission set in the system.

3   AC_MWSA Model

In the MWSA system, there are alliances of sites, so the access control model on the system can think more about the alliances to reduce the system cost. In the next of this article we present an access control model on this system named AC_MWAS. In the AC_MWAS model, when an instance get to a site, the system will check the instance's TS in its log to find the task that is to be finished in this site. Then the system will check the T to find the service and resources which are needed to finish the task. At the last, the system will check the instance's IP in the log to find the permission that the instance have get. If the permission has been already assigned to the instance for the SA which the site belongs to, the instance can get the service or resource from the site directly. Otherwise the instance will apply the permission to finish the task in the site. The system will decide whether allow the application according to the local rule just as in the traditional workflow system. We get the AC_MWSA model as the fig.1 show.

In the AC_MWSA model, there are five entities: instance, role, permission, event and task. Then we will give the definition of the model. In this model the definition MI, S, SA and E have been given in the definition 2 to definition 5.

Definition 7. P(Pid, Rs, Op, Sv) is a permission on a instance to a site alliance in the system, Pid is the identification of the permission, Rs is the resource that the instance access in the site alliance, and Op is the operation that can do on those resource, Sv is the service that the instance can get from the site alliance.

Definition 8. T(Tid, Rs, Op, Sv, Bt, Ft) is a task in the system, the task is minimal which can not be divided any more. Tid is the indemnification of the task, Rs is the resource that is needed to fulfill this task, Op is the operation that is needed to do on the resource, Sv is the service that is needed to fulfill the task, Bt is the earliest begin time when the task can begin, Ft is the latest finish time when the task must be finished.

Definition 9. MIPSAA is a relationship. $MIPSAA \subset MI \times P \times SA$. If $(MI_i, P_j, SA_k) \in MIPSAA$ then assign the permission P_j to the instance MI_j for the sites' alliance SA_k.

Definition 10. MITA is a relationship $MITA \subset MI \times T$ If $(MI_i, T_j) \in MITA$ then the task T_j is assigned to the instance MI_i.

Definition 11. SAA is a relationship. $SAA \subset S \times SA$. If $(S_i, SA_j) \in SAA$ then the site S_i is belong to the site alliance SA_j.

Definition 12. TO is a partial order relation. $TO \subset T \times T$. If $(T_i, T_j) \in TO$ then the task T_j must be finished before the task T_i is began to do.

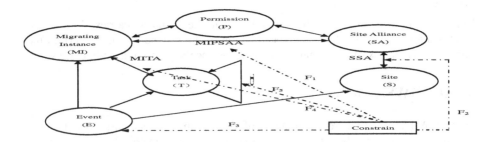

Fig. 1. AC_MWSA model

In the MW_RBAC model the relationships are constrained by several rules.

Rule 1: If a site belongs to an alliance, then the service and recourse which the site can provide belong to the set of the service and recourse of the alliance.

$$F_1 = \left\{ \; \forall (S_i, SA_j) \in SSA \left| \begin{array}{c} S_i.SS \subseteq SA_j.SAS \\ S_i.SR \subseteq SA_j.SAR \end{array} \right. \right\} \tag{1}$$

Rule 2: If permission is assigned to a instance for a SA, there must be a site belonging to the SA. The permission's service and resource belong to the site's service and resources.

$$F_2 = \left\{ \forall (MI_i, P_j, SA_k) \in MIPSAA \left| \begin{array}{c} \exists (S_l, SA_k) \in SSA \\ P_j.Rs \subseteq S_l.RS \\ P_j.Sv \subseteq S_l.SS \end{array} \right. \right\} \tag{2}$$

Rule 3: In an event, there are instance, site and task. The system will ensure that the site belongs to the site alliance that the instance has the permission. The event is begun after the earliest begin time of the task, and is finished before the task's latest finish time.

$$F_3 = \left\{ \forall E_i \left| \begin{array}{l} \exists P_l, T_m, MI_n, S_j \in SA_k \\ E_i.Sid = S_j.Sid, E_i.Tid = T_m.Tid \\ E_i.MIid = MI_n.MIid, \\ (MI_n, P_l, SA_k) \in MIPSAA \\ T_m.Bt \le E_i.Bt, T_m.Ft \ge E_i.Ft \end{array} \right. \right\} \qquad (3)$$

Rule 4: When a task is assigned to an instance, the system will ensure that the instance has the permissions that allow the instance to get service and access the resource and do some operations. They must fulfill the formula 3.

$$F_4 = \left\{ \forall (MI_i, T_j) \in MITA \left| \begin{array}{l} \exists (MI_i, P_k, SA_l) \in MIPSAA \\ T_m.Rs \subseteq P_l.Rs, T_m.Op \subseteq P_l.Op \\ T_m.Sv \subseteq P_l.Sv \end{array} \right. \right\} \qquad (4)$$

Rule 5: Constrain on the TO is same to the RBAC model in workflow, as in reference [2].

4   The Algorithm Based on The MW_RBAC Model

In this part, we will give out the algorithm based on the AC_MWSA model.

Algorithm 1:

```
Begin
Initialize the MI and the S, SA, P, and T in every site
according to the workflow system;
while (MI!= )
{
when MIi arrives at the site Sl
If(MIi.L.TS!= )
{
```
for ($T_j \in MI_i.L.TS$)
```
{
```
find $P_k \in P$
where $T_j.Rs \subseteq P_k.Rs, T_j.Op \subseteq P_k.Op, T_j.Sv \subseteq P_k.Sv$
if $(\exists (MI_i, P_k, SA_m) \in MIPSAA \wedge S_l \in SA_m)$
```
{
then the instance MIi can access to the site Sl to get
resource and service to finish the task Tj.
}//if
else
```

```
{
The system decides whether permit the application
according to the local rule just as in a traditional
workflow system.
 if the application is permitted
  if (S∈SAm)
  then add a new MIPSAA(MIi, Pk, SAm) to the system
}//else
set E (Eid, EK, MIid, Sid, Tid, Bt, Ft)
  if T is fulfilled, MI.L.TS= MI.L.TS - T;
}//for
}//if
}// while
End
```

5 Conclusions and Future Work

The migrating workflow is a new direction within the workflow management area. In this article we present the conception SA (site alliance) to the system. The sites in the system are divided into different alliances. The sites in one alliance can provide the same or similar resource and service to the instance. With the conception of SA we present a migrating workflow system named MWSA, and build an access control system named AC_MWSA in MWAS system. In this model, the instance applies permission for SA. According to the traditional access control model, AC_MWSA model does not assign the permission to a sign site. So the system cost for access control is reduced largely. In the future work, several aspects need further study in our scheme. The consistency analysis should be thought more about it. The difference of the sites in one alliance should be considered. All the questions will be discussed in our future research.

Acknowledgement. This paper is supported by the National Natural Science Foundation of China under Grant No.60573169 and the Natural Science Foundation of Shandong Province under Grant No.ZR2009GM021.

References

1. Sandhu, R., Coyne, E., Feinstein, H., Youman, C.: Role-Based Access Control Models. Computer 29, 38–47 (1996)
2. Wang, X.M., Zhao, Z.T., Hao, K.G.: A weighted role and periodic time access control model of workflow system. Journal of Software 14, 1841–1848 (2003)
3. Bertino, E., Ferrari, E., Atlur, V.: A flexible model supporting the specification and enforcement of role-based authorization in workflow management systems. In: Proceeding of the Second ACM Workshop on Role-Based Access Control, pp. 1–12 (1997)
4. Castano, S., Casati, F., Fugini, M.: Managing workflow authorization constraints through active database technology. Information Systems Frontiers 3(3) (2001)
5. Cichocki, A., Rusinkiewicz, M.: Migrating workflows. In: Advances in Workflow Management Systems and Interoperability, pp. 311–326 (1997)

6. Qiu, J., Tan, J., Zhang, S., Ma, C.: Workflow Dynamic Authorization Model with Role Based Access Control. Journal of Computer-Aided Design & Computer Graphics 7(16), 992–998 (2004)
7. Su, W., Zeng, G.-Z.: Dynamic RBAC Model Considering the User Reliability. Computer Engineering 15(31), 84–86 (2008)
8. Shafiq, B., Joshi, J., Bertino, E., Ghafoor, A.: Secure interoperation in a multidomain environment employing RBAC policies. IEEE Transactions on Knowledge and Date Engineering 17(11), 1557–1577 (2005)
9. Su, W., Zeng, G.-Z.: A RBAC Model Considering the User Reliability in Workflow System. In: Proceeding of ICIS 2010, pp. 426–430 (2010)
10. Wang, R., Zeng, G.: An efficient service recommendation using differential evolutionary contract net for migrating workflows. Expert Systems with Applications 37(2), 1152–1157 (2010)
11. Wu, X., Zeng, G.: Goals description and application in migrating workflow system. Expert Systems with Applications 37(12), 8027–8035 (2010)

Galvanic Coupling Type Intra-body Communication Human Body Implantable Sensor Network

Xiao-fang Li and Shuang Zhang

The Engineering & Technical College of Chengdu University of Technology,
Leshan, 614000, China
41441499@qq.com, zhangshuanghua1@126.com

Abstract. With the deep research of galvanic coupling type intra-body communication, the development of sensor technology, using a sensor in the service of mankind is a very important subject. According to the quasi static galvanic coupling characteristics, the sensors distributed in the tested signals site, detected biological signals, converts the signal into a galvanic signal; the sensor as the transmitting electrodes to transmit signal to human. The improved PDA contact with the body in any part of the body, will become the received sensor signal, thereby restoring signal. Sensor, PDA formed body communication network in the connections of human body.

Keywords: galvanic coupling, sensor, sending electrode, receiving electrode, intra-body communication.

1 Introduction

Before twentieth Century, the ordinary human is difficult to access to medicine, medical resources are mainly used for teaching, which is regarded as the doctor's secret. People can only go to the traditional medical system to check their physical condition, medical diagnosis only stay in the physical examination, treatment is only dealing with the structural lesions of organ and combination. In twentieth Century, with the rapid development of electronic technology, have the very big impact to the new medical technology development. In 1903, William Einthoven invented the first string galvanometer, used for measuring the bioelectricity, this is the early electrocardiogram machine. In 1929, Hans Berger invented the first EEG machine, used for measuring and recording the electrical activity of the brain. In 1935, human use electronic amplifier to confirm the electrical activity of the brain cortex has a specific rhythm. In 1960, William chardack and Wilson Greatbatch use electronic amplifier, invented the first implantable cardiac pacemaker. Implantable electronic officially entered into our life.

In 1995 the United States Massachusetts Institute of Technology T.G.Zimmerman [1],[2] first put forward the concept of intra-body communication, scientists of all over the world made a lot of research in intra-body communication, but also obtained a lot of achievements. Intra-body communication is a new short distance" wireless" communication mode, it see the human body as a weak electrical signal transmission medium, so as to realize the data transmission and share in the human body surface,

D. Jin and S. Lin (Eds.): Advances in FCCS, Vol. 2, AISC 160, pp. 147–152.
springerlink.com   © Springer-Verlag Berlin Heidelberg 2012

internal, around etc the all human contact electronic devices. The implantable sensor using this characteristic, the collected organ signal convert into different frequency electrical signal, through human medium, spread over the whole body. In order to achieve the sensor data and PDA transmission, and between the sensor network.

2   Galvanic Coupling in Intra-body Communication

Galvanic coupling intra-body communication works as shown in figure 1. Communication system consists of the transmitter, receiver, body composition. The transmitter and receiver are respectively provided with two electrodes. Its working principle is by the sending electrode attaching on the surface of the body to inject safety alternating galvanic to human body, another section the detection electrode differential receive voltage signals, to achieve signal transmission [3][4]. Does not require additional connection, does not need to the coupling of the ground.

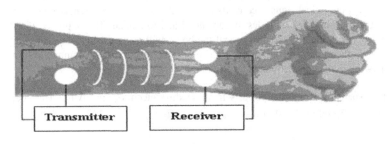

Fig. 1. Galvanic coupling intra-body communication schematic diagram type

The galvanic coupling model is still in its initial stage, most of the achievements are experiment measured data and the prototype data. Quasi static conditions in intra-body communication theory model are relatively lack. The existing achievements, University of Macao's Pun Sio Hang and Gao Yue Ming of the Fuzhou University [6][7] [8][9]ues anatomy theory and the electromagnetism theory, abstract the human arm as a homogeneous medium consisting of a multilayer cylinder volume conductor, from the aspect of space enriches the volume conductor theory. In the theoretical analysis, the article is composed of a cylinder Maxwell equation:

$$\nabla^2 V = \frac{1}{r}\frac{\partial}{\partial r}(r\frac{\partial V}{r}) + \frac{1}{r^2}\frac{\partial^2 V}{\partial \phi^2} + \frac{\partial^2 V}{\partial z^2} = 0 \qquad (1)$$

By using the separation of variables method, obtained the analytic solution of the equation:

$$V(r,\phi,z) = \sum_{n=0}^{\infty} \left[A_n Z_n^{(1)}(kr) + B_n Z_n^{(2)}(kr) \right] \left[C_n \cos(n\phi) + D_n \sin(n\phi) \right] \left[E_k \cos(kz) + F_k \sin(kz) \right]) \qquad (2)$$

among, A_n、B_n、C_n、D_n、E_k、F_k are undetermined constants, while $Z_n^{(1)}(kr)$, $Z_n^{(2)}(kr)$ can take the Bessel function $J_n(kr)$ and $Y_n(kr)$, or modify the Bessel function $I_n(kr)$ and $K_n(kr)$。The solution in intra-body communication, according to the different boundary conditions, simplification, can get each layer model analysis.

3   Implantable Device

An implantable device is an electronic device can be implanted within the body. Implantation of the device is usually divided into two categories: one is used to maintain or help the body for normal work, this is referred to as an implantable therapeutic device; such as: cardiac pacemakers, cochlear. Another is used to monitor and collect the mechanism information device, this is referred to as an implantable sensor; such as: swallowing type radio capsule.

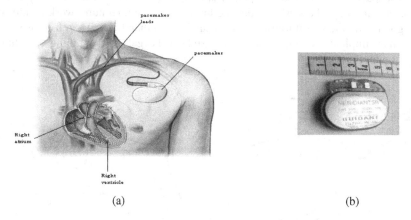

(a)   (b)

Fig. 2. Treatments of Slow Heart Rhythms (a) Implantation of Cardiac Pacemaker.(b) Cardiac Pacemaker

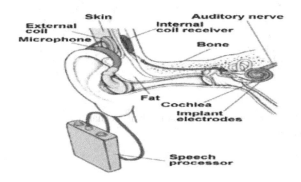

Fig. 3. Implantation of Cochlear Prostheses

With low power consumption, high reliability of the integrated circuit technology is developing rapidly, the diversification electron device in vivo energy supply, no toxicity, the research biocompatibility and excellent performance of biomaterials; various implantable sensor also gradually emerge as the times require, and services for human. But at present all kinds of sensor applications most stay cable, surface test. However, body sensor on the body measurements are subject to many environmental interference, the signal will be extremely inaccurate because of the environmental impact, and complex transmission line network make users feel very inconvenient. Therefore, we put forward human implantable sensor network based on galvanic coupling type human body communication.

4 Galvanic Coupling Type Human Body Implantable Sensor Network Communication

Body sensor due to many environmental factors, the information is often not very stable, noise impact is big. The implantable sensor located environment is human constant space. Regardless of temperature, or appropriate almost without any changes in the human body, the sensor under the human body temperature, work condition is very good. And the signal transmission is stable in human body. Therefore, we design the human implantable sensor network based on the galvanic coupling type human body communication.

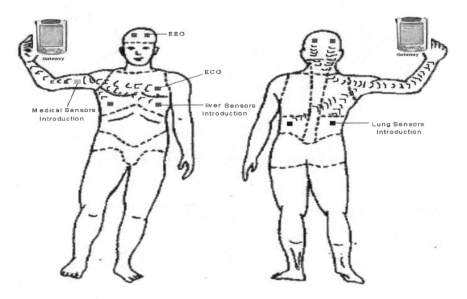

Fig. 4. Designated sensor implanted and human body signal transmission network structure Implantable specify sensor

Different parts of the body, to convey different biological signals. By measuring these biological signals, can observe the different physiological state of human body. Due to various parts form the signal is different, therefore requiring different biosensors to detect. Therefore implant the designated biological sensor to the human body to be detected, to detect signal. The detected signal is converted into a specified frequency electrical signal, and transmission in the human body.

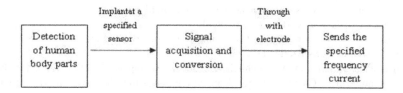

Fig. 5. sensor signal acquisition and transmission Senser network

In the specified sensor, it has two functions: one with specific induction chip, detect the determination of the biological signal, the sensor has the efficacy of signal receiving function. The detected signal convert into a specified frequency galvanic, so that PDA testing, the sensor have the sending the signal function. In the galvanic coupling type intra-body communication, it has the same function of transmitting electrodes.

The human body as a multilayer filled coaxial communication wires, the electric signal emited by sensor will transmit through every part of human body. Through the PDA USB excuse, in PDA surface arranged a multi channel signal detection sensor. This sensor is used to detect human body and receives specified frequency galvanic signal. Transmission the electrical signals to the PDA, to restore the distribution the detect signal of various sensors in vivo. In the galvanic coupling type intra-body communication, it has the same function of receiving electrodes.

The implantable sensor, can real-time, stability monitor human body signal. Because the human body is an approximate constant temperature, constant pressure, constant humidity stable environment. Sensors in the body temperature environment, can work well. With the human body is a quasi static electromagnetic field, electrical signal interference is very weak; the conductive performance is very good, so the use of an implantable sensor for intra-body communication will provide good protection for real-time medical monitoring system.

5 Conclusion

With the research of the quasi static galvanic coupling type intra-body communication developing, the development of sensor technology, the future of biological signal measurement and transmission will be more and more convenient. Many public hand in hand use a player to listen to music, many people share the same PDA detection in human body together; information, resources will be able to realize the sharing. Also in the human body implant specific biological signals, using a specific technique can achieve human encryption, so as to achieve the specified user identification function.

References

1. Zimmerman, T.G.: Personal Area Networks (PAN): Near-Field Intra-Body Communication. PhD thesis. Massachusetts Institute of Technology (1995)
2. Zimmerman, T.G.: Personal area networks: Near-field intrabody communication. IBM Systems Journals 35(3 & 4), 609–617 (1996)
3. Wegmuller, M.S.: Intra-Body Communication for Biomedical Sensor Networks. [PhDThesis]. ETH, Switzerland (2007)
4. Wegmueller, M.S., Kuhn, A., Froehlich, J., et al.: An Attempt to Model the Human Body as a Communication Channel. IEEE Transactions on Biomedical Engineering 54(10), 1851–1857 (2007)
5. Gao, Y.M., Pun, S.H., Du, M., et al.: A preliminary two dimensional model for Intra-body Communication of Body Sensor Networks. In: International Conference on Intelligent Sensors, Sensor Networks and Information Processing (ISSNIP 2008), Sydney, pp. 273–278 (2008)
6. Gao, Y.M., Pun, S.H., Mak, P.U., Du, M., Vai, M.I.: Preliminary modeling for intra-body communication. In: 13th International Conference on Biomedical Engineering (ICBME 2008), Singapore, pp. 1044–1048 (2008)
7. Pun, S.H., Gao, Y.M., Mak, P.U., Vai, M.I., Du, M.: Quasi-Static Field Modeling of Human Limb for Intra-Body Communication with Experiments. IEEE Transactions on Information on Biomedical Engineering 15(6), 870–876 (2011)
8. Lindsey, D.P., Mckee, E.L., Hull, M.L., Howell, S.M.: A new technique for transmission of signals from implantable transducers. IEEE Transactions on Biomedical Engineering 45(5), 614–619 (1998)
9. Konstantas, D., van Halteren, A.T., Bults, R.G.A., Wac, K.E., Jones, V.M., Widya, I.A.: Body Area Networks for Ambulant Patient Monitoring Over Next Generation Public Wireless Networks. In: 14th IST Mobile and Wireless Communications Summit (2004)
10. Sukor, M., Ariffin, S., Fisal, N., Yusof, S.K.S., Abdallah, A.: Performance Study of wireless Body Area Network in a Medical Environment. In: Proc. 2nd Asia International Conference on Modeling & Simulation, AICMS (2008)

Ontology-Based International Education Exchange Web Platform Model

Jing Tian[1], ZongLing Zheng[2], and Yun Zeng[3,*]

[1] College of Continuing Education and Vocational Education, Yunnan Agricultural University,
Kunming 6500201, China
[2] College of Economics and Management, Yunnan Normal University,
Kunming 650500, China
[3] Department of Engineering Mechanics, Kunming University of Science and Technology,
Kunming 650051, China
jingtian2003003@163.com, zling75@yahoo.com.cn,
zengyun001@163.com

Abstract. Educational exchange is an important part of diplomatic cultural affairs, which occupy an important position in international relations, and it is an important part in one country's foreign policy. In the context of economic globalization, Traditional international educational exchange and cooperation can not meet the need; we must seek new areas of development. Ontology-based Educational Exchange Web Platform (OEEWP) is an important tool to obtain, share and use of educational resources. This paper analyzes the origins and meaning of ontology, and application of domain ontology in educational exchange. Based the research of ontology and domain ontology at home and abroad, OEEWP model is been constructed.

Keywords: Ontology, Educational exchange, Web platform model.

1 Introduction

International education exchange has become an important part of China's foreign strategy, which greatly contributed to national security, foreign relations, and improved national international environment. To develop educational exchanges is a significant measure of national prosperity. Ontology is one of the most popular terms of information science in recent years. Ontology is a philosophical concept, used to describe the nature of things. In the last 20 years, Ontology has been used in the computer field for knowledge representation, knowledge sharing and reuse. To construct the model of OEEWP, its purpose is to enable users in different countries via the Internet for educational exchanges, and promote education, science and technology, cultural exchange and development in the world, increase mutual understanding and trust, thereby reduce the dispute caused by the ethnic and cultural differences.

---

* Corresponding author.

D. Jin and S. Lin (Eds.): Advances in FCCS, Vol. 2, AISC 160, pp. 153–157.
springerlink.com   © Springer-Verlag Berlin Heidelberg 2012

2   The Origin Meaning of Ontology

From the view of the history of Western philosophy, ontology as a science originated in the questioning of all things primitive. The term ontology has been applied in many different ways, the word element "onto", "logia" comes from the Greek, and it is a philosophical term. Ontology is the philosophical study of the nature of being, existence or reality as such, as well as the basic categories of being and their relations. So philosophical ontology is real objective description to any area of the world, and this description may not be entirely based on the existing knowledge base, but also includes the process of 'truth'. The core meaning within computer science is a model for describing the world that consists of a set of types, properties, and relationship types. Exactly what is provided around these varies, but they are the essentials of an ontology. There is also generally an expectation that there be a close resemblance between the real world and the features of the model in an ontology.[1] What many ontologies have in common in both computer science and in philosophy is the representation of entities, ideas, and events, along with their properties and relations, according to a system of categories. In both fields, one finds considerable work on problems of ontological relativity. [2] The concept of ontology has been introduced in the field of artificial intelligence, knowledge engineering, and library and information to solve issues about the information extraction, knowledge conceptual representation and knowledge organization systems. According to Gruber (1993), "An ontology is an explicit specification of a conceptualization". [3] Studer, Benjamins and Fensel further refined and explained this definition in 1998. In their work, they defined an ontology as: "a formal, explicit specification of a shared conceptualization. Formal: Refers to the fact that an ontology should be machine-readable. Explicit: Means that the type of concepts used, and the restrictions on their use are explicitly defined. Shared: Reflects the notion that the ontology captures consensual knowledge, that is, it is not the privilege of some individual, but accepted by a group. Conceptualization: Refers to an abstract model of some phenomenon in the world by having identified the relevant concepts of that phenomenon" [4]. Although there are many different ways to define Ontology, but from the content, different researchers have a unified understanding to Ontology, which is been regarded as a field (the field can range from specific applications, can also be a wider range), and there are different subjects (people, machines, software systems, etc.) in this field, those subjects can communicate (talk, interoperability, sharing, etc.) each other, and a consensus has been provided by the Ontology.

3   The Application of Domain Ontology

Domain Ontology Application The emphasis of ontology research is different due to application domain. Ontology related to particular discipline domain is called as domain ontology. The domain ontology aims to catch relative domain knowledge, provide common comprehend on it, determine common admissive concept in it, give explicit definition of these concepts and their correlations from different hierarchic formalization model, and provide main theory and principle for domain actions. In education exchange domain, there are three field applications based-ontology, education, knowledge index, information integration, and so on. In education field, ontology is used to describe remote user live and action condition. In distributed system, the ontology is an intermediation in charge of information exchange between agent and surroundings. The everyday vocabulary and framework designed to teaching system can provide formalization education task in proper and abstract. Learner can find itself issue in the procedure of interviewing teaching system, meanwhile the system can continuously adjusted itself to suit for learner, so the ontology teaching system is a kind of self-adapting teaching-learning system. In knowledge field, the ontology system upgrades knowledge index from traditional index based-keyword to based-semantics. Its basic idea is that relative domain ontology is built at firstly, and information collected by ontology is classified. Index demand of user may be transformed into standard pattern and easy to index eligible information to user. In information integration, the ontology system can partly solve semantics isomerism issue existed in traditional distributed information integration. [5]

4 OEEWP Model

The performance of a search engine can be improved by the use of an ontology. In its conventional form an ontology accounts for the representation of shared concepts in a domain by specifying a hierarchy of terms facilitating communication among people (collaboration) and applications systems (integration of tools).[6] On the basis of the central role of ontology in knowledge sharing and knowledge reuse, most knowledge system has adopted ontology as a skeleton. Usually when the system was established, according to the system application requirements and application environment to construct the appropriate ontology, then, the system would be designed with comprehensive application by ontology as the center. There are Many question in the Educational exchanges, such as the variety of educational resources, low haring level, interoperability poor, keyword search results poor, the lack of intelligent human-machine interface and human support and other issues. Based on the above problems, we need build a model of OEEWP, (see Fig.1).

This model specifically includes the following modules: First, the data source. Educational resources have Variety classes, according to their degree of data structures can be divided into unstructured, semi-structured and structured. There are different ways to access to resources in platform, which may submit by the user to share with others, can get artificial platform, and can automatically obtain and so on. Second, ontology extraction, the case library of ontology can be edited complete to manual, but when the numbers of instances are massive, the artificial writing would be a challenging task. Therefore, we can use machine learning technologies to achieve semi-automatic ontology instances extracted. Different structured degrees using different extraction methods. Third, ontology library, Ontology and ontology instance store. Ontology is the concept of sharing knowledge and educational resources. Ontology instance is a description of educational resources. Ontology is the core of the platform. Fourth, ontology management, Ontology is dynamic and must be managed and maintained. Otherwise, there will be incompatibilities and inconsistencies. This work is accomplished generally by the education change experts, that they like database administrators to manage databases. Fifth, the inference engine, Inference engine can be completed examples checking. Therefore, the inference engine can optimize the retrieval mechanism to support intelligent human-machine interface. Sixth, the use of resources, use of resources means users browse, search, and users with intelligent agents interact with other systems. However, different from the traditional way, with the supported of Ontology, the platform can provide more intelligent user interface. At the same time, the platform can also be easily interact with other systems by intelligent agents, and achieve cross-platform sharing of resources.

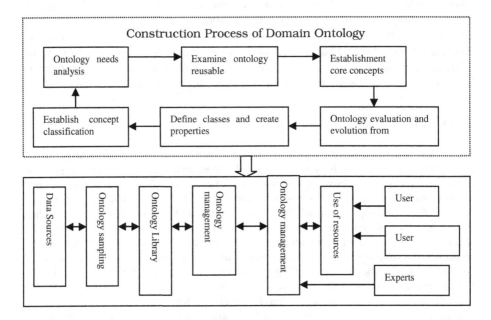

Fig. 1. OEEWP model

5   Summary

To build international educational exchange platform may base on some distributed, heterogeneous data sources, if we see these distributed, heterogeneous data sources as an ontology, then the unified integration platform by access to resources is a question about integration of these ontologies. As the real world is constantly changing, the content, structure, form of knowledge has been changed, the meaning of certain terms also has been changed, and the needs of users is changing, while, and these changes are unpredictable when ontology is established. The source consistency between fixed ontology and changing knowledge may be damaged; the ontology can no longer accurately reflect the new state of knowledge sources. How to make ontology adapt to dynamic changes in the external world, and make timely adjustments based on the change of external knowledge sources, which is a significant content in ontology research. In order to adapt to this change, domain ontology must have evolved features that ontology could accurately reflect the dynamic development of education.

Acknowledgement. The research reported here is financially supported by the Humanity and Social Science Funds of Education Ministry of China Grant NO. 10YJC880112. Part works is supported by the Science Research Foundation of Yunnan Education Department Grant No. 2010Z081. And part works is supported by the National Natural Science Foundation of China under Grant No. 51179079.

References

1. Garshol, L.M.: Metadata? Thesauri? Taxonomies? Topic Maps! Making sense of it all (2004), http://jis.sagepub.com/content/30/4/378.abstract
2. Sowa, J.F.: Top-level ontological categories. International Journal of Human-Computer Studies 43(5-6), 669–685 (1995)
3. Gruber, T.: A translation Approach to portable ontology specifications. Knowledge Acquisition 5, 199–220 (1993)
4. Studer, R., Benjamins, V.R., Fensel, D.: Knowledge engineering: principles and methods. IEEE Transactions on Data and Knowledge Engineering 25(1-2), 161–197 (1998)
5. Zhao, J.G., Xu, D.Z., Luo, Q.Y.: Ontology and its application. Journal of Sichuan University of Science & Engineering: Natural Science Editton 20(6), 105 (2007)
6. Aldea, A., Bañares-Alcántara, R., Bocio, J., Gramajo, J., Isern, D., Kokossis, A., Jiménez, L., Moreno, A., Riaño, D.: An Ontology-Based Knowledge Management Platform, http://arnetminer.org/viewpub.do?pid=387346

Research on Interactive Teaching Platform System in Remote International Chinese Education

Jing Tian[1], ZongLing Zheng[2], and Yun Zeng[3,*]

[1] College of Continuing Education and Vocational Education, Yunnan Agricultural University,
Kunming 650201, China
[2] College of Economics and Management, Yunnan Normal University, Kunming 650500, China
[3] Department of Engineering Mechanics, Kunming University of Science and Technology,
Kunming 650051, China
jingtian2003003@163.com, zling75@yahoo.com.cn, zengyun001@163.com

Abstract. One of the important changes of Chinese international promotion is turn face to face paper-based teaching methods to take advantage of modern information technology, multimedia network teaching. Currently, non-native Chinese learners are increase year by year, and the more and more learners have high-level Chinese ability, which demand a higher level quality of teaching Chinese language. By combine modern network technology and traditional teaching methods, we should explore new teaching platform system to adapt to the new situation requirements of Remote Chinese International Educational (RCIE). This paper analyzes the concepts of "RCIE", "Interactive" and "Interactive Teaching". Based on the characteristics of the non-native learners of Chinese, the design principles of Interactive teaching of RCIE were been analysis. Teaching Platform System of RCIE were been construct in view of the interaction theory.

Keywords: International Chinese Education, Teaching Platform System, Interactive Theory.

1 Introduction

We must pay attention to the trends of use of modern means in second language teaching. With the development of language teaching theory and computer technology updates, computer-assisted second language teaching has gone from the simple to complexity gradual. Garrison and Anderson reputed that the distance education thinking of the first generation Characterized by correspondence based on the behavioral theory. Correspondence with radio and television, audio and video and computer technology as part of the second generation of distance education is development with cognitive learning theory development. Based on electronic communications and computer technology, the third generation of distance education with the characteristics of two-way communication reflects the constructivist learning theory. [1] Two-way interaction in an online environment is an important factor that

---

* Corresponding author.

influences the success or failure of a program. Moore considers interaction as "vitally important" in the design of distance education [2]. Interactive theory provides a broad framework for distance education and teaching, and allows the integration of other theories, and which is one of the core theories of modern distance education. By examining the relationship between dialogue dimensions, structural dimensions and autonomy dimensions, the theory provides such a framework composed the unit of infinite number of learner characteristics, instructional features and design characteristics. [3] Based on the interaction theory, this paper explore teaching platform system of RCIE in network environment, to further enhance the quality of teaching of RCIE and to guide further reform of teaching.

2   Define the Concept

2.1   RCIE

"Chinese International Education (CIE)" is a new concept in recent years. The first time proposed this term can be traced back to March 30, 2007; the State Council Academic Degrees Committee examined and adopted "the Setting Program of Chinese International Education Master Degree" in the twenty-third meeting. The ICE in program is defined in: "ICE is a Chinese teaching for non-native Chinese speakers of overseas". [4] RICE is a teaching methods by develop online teaching-learning resources and link non-native Chinese language learner through the network, which is use of network technology, multimedia technology, computer technology and information technology tools to enable teachers to use online teaching resources and materials, It allows learners according to their needs, through autonomous, cooperative learning and real-time and non real-time interaction, receive Chinese language knowledge by targeted training, get the Chinese communicate ability in a virtual network environment.

2.2   "Interactive" and "Interactive Teaching"

Interactive was a computer term originally that refers to the system receives input from the terminal, then processing, and return the results to the terminal, namely the man-machine dialogue; it is an important characteristic in the computer. In the Macquarie Dictionary, "Interactive" is defined as action on each other, which means an act with the person or the other things affect each other. In this way, "interactive" is used to describe the various interactions events, refers to an interaction situation of environment, individuals and behavior in an environment. Interactive in teaching and learning should be interact ability or features supporting by media. [5]   From the view of communication point, interactive is an information exchange between communicator and recipient. From the view of pedagogical point, interactive is one of the most basic features of teaching activities. Interactive teaching is information's two-way, multi-directional flow between the teaching and learning, which emphasis interaction between teachers and students.

3   Design Principles

To build a good teaching platform, first of all, we have to "people-oriented", the "people" are the non-native speakers of Chinese, and only analyze and understand their learning characteristics can better grasp the design principles of teaching platform of RICE, can put together a variety of resources to effectively achieve the purpose of optimizing selection. The individual of the non-native learners of Chinese are differences in gender, age, hobbies, etc, they also has the following learning characteristics: different purposes; different Chinese levels; backgrounds and so on. Based on these characteristics, to Building the interactive teaching platform system should follow some principles: First, the principle combined with content interaction and interpersonal interaction. Learning content is the core of teaching; content interactive of distance learning is based on the cognitive processes of learning resources, which is a direct way to achieve learning goals. The main task of interpersonal interactions is to ensure effective results and ongoing cognitive activity. Therefore, platform design should base on content interaction, and supplement by the interpersonal interaction. Second, the principle combined with synchronous interaction and asynchronous interaction. The interaction between teachers and students can carry on anytime, anywhere in technical supporting, real-time synchronous interaction increase the realism of interactive scenarios and timeliness interactive information. With the development of broadband technology and the emergence of new tools, real-time interactive applications will be more extensive, but the most important online teaching is non-real-time interaction. Therefore, platform design should base on asynchronous interaction, supplemented by synchronous interaction. Third, the principle combined with teacher control and learner control. In recent years, the concept of "learner-centered" has been widely accepted for everyone. Using this concept in interactive design fields is been promoted by the network's self-organization and the open nature. Necessary teachers control will not become an obstacle to self-study, but can be make self-learning process smoother, so platform design should promote self-control-oriented learner, supplemented by teacher control. Fourth, the principle combined with the network means and the traditional means. Because there is a big dependent to computer and communication lines in the network interaction, if the network fails, the interaction will be immediately interrupted. For this problem, platform design should base on network means, and supplemented by traditional means. [6]

4   Teaching Platform of RCIE Based on Interactive Theory

According to Moore, the development of the transaction is influenced by three basic factors: (i) the dialogue developed between teacher and learner, (ii) the structure that refers to the degree of structural flexibility of the programmed and (iii) the autonomy that alludes to the extent to which the learner exerts control over learning procedures. [7] Transactional distance is one of the results of teaching (action) and structure, autonomy and dialogue are mechanisms of transactional distance. In short, Moore's theory provides a Multi-dimensional framework, which allow coexist with fragmentation and a systematic, comprehensive, holistic view in distance education. (See Fig.1.).

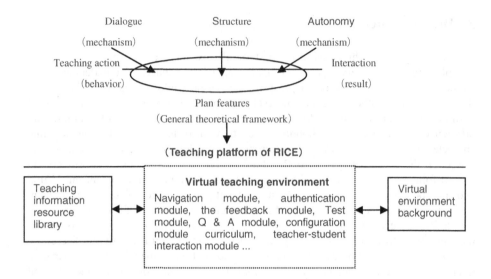

Fig. 1. Teaching platform of RCIE based on interactive theory

To build teaching platform of RCIE need to construct virtual learning environment, teaching information resource library, a virtual environment background. If a learner is the first time to join this platform, he would test His Chinese level by the Chinese level assessment module, According to test results, the system would determine which learning resources would present to the learner. This action is the spontaneous movement in virtual learning environment, the purpose is optimizing selection. With the learner's action, teaching system will be issued the appropriate action in a virtual learning environment by the navigation module, Q & A module, feedback module, test module and other modules. This action is feedback action in virtual learning environment.

5 Summary

With the development and progress of science and technology, the form and content of RCIE are changed with the diversification based on the characteristics and requirements of learners. With the Chinese to the world, the main battlefield of CIE is also from China to the world, the network platform for Chinese language teaching has become a focus of the reform of international teaching methods. Interactive teaching becomes a new content in development of RCIE. But because of the separation temporal and spatial distance, the difficulty of interactive teaching has also greatly increased. Therefore, some teaching facilities and the environment need be provided by teaching management side, in order to effectively improve the teaching quality of RCIE.

Acknowledgement. The research reported here is financially supported by the Humanity and Social Science Funds of Education Ministry of China Grant NO. 10YJC880112. Part works is supported by the Science Research Foundation of Yunnan Education Department Grant No. 2010Z081. And part works is supported by the National Natural Science Foundation of China under Grant No. 51179079.

References

1. Tian, J.: Enlightenment and extending of Transactional Distance Theory in Distance Education. China Educational Technology 9, 47–51 (2010)
2. Moore, M.G.: Three types of interaction. The American Journal of Distance Education 3(2), 1–6 (1989)
3. Xi, J.H., Zhang, X.M.: Research and challenge of the international distance education: Interview with renowned international experts in distance education. Dr. Michael Moore. Open Education Research 51(5), 4–5 (2004)
4. State Council Academic Degrees Committee. A notice about "set up Master Degree of Chinese International Education program",
   http://news.china-b.com/jyxw/20090310/714879_1.html
5. Li, C.P.: The content and features of interactive teaching. Educational Theory and Practice 27(2), 45–47 (2004)
6. Ge, W.J., Ma, L.: Design and development research of Interactive distance learning system. China Adult Education 6, 120–121 (2007)
7. Giossos, Y., et al.: Reconsidering Moore's Transactional Distance Theory [EB/OL],
   http://www.eurodl.org/?article=374

Evaluation of the Energy-Saving Potential of Pilot Projects for Renewable Energy Buildings

XiaoChang Yang and Jian Yao*

Faculty of Architectural, Civil Engineering and Environment, Ningbo University, China
yaojian@nbu.edu.cn

Abstract. The aim of this paper is to evaluate the energy-saving potential of pilot projects for renewable energy buildings integrated with solar water heating systems or ground source heat pump systems. A simple equation based manual calculation method was adopted. Results show that renewable energy application in buildings contributes to a great potential in reducing energy consumption and greenhouse gas emissions.

Keywords: Renewable energy, building energy consumption, solar energy.

1 Introduction

There is a growing concern about building energy consumption in China since the building sector is a large consumer of energy and a significant contributor to greenhouse gas emissions. Lots of studies have been conducted on the improvement of building energy efficiency. Some of them focused on passive strategies in building design and urban planning such as daylighting and ventilation [1-7]. Other measures have also been carried out to reduce building energy consumption and associated environmental pollutions [8-12]. A higher share of renewable energy would be another effective measure in reducing building energy use. Solar energy and geothermal utilization in buildings has a great potential in China and some pilot projects have be conducted in some cities including Ningbo, a southern city in China. This paper provides a simple evaluation of the energy-saving potential of pilot projects for renewable energy buildings integrated with solar water heating systems or ground source heat pump systems.

2 Energy Savings Evaluation

2.1 Solar Water Heating Systems

The area of solar collector of pilot projects for renewable energy buildings is 121474.6 m². According to the building code DB33/1034-2007 "regulation of design, installation and acceptance of solar hot water systems in residential buildings", the following equation can be used to calculate the energy-saving and carbon emission-reduction potential of buildings applying solar hot water systems: $4585 \times 0.50 \times 0.75 = 1719$ MJ/(m²·a). Where 4585 is annual solar radiation, 0.5 is the

---

* Corresponding author.

efficiency of solar collector, 0.25 is heat loss factor. Then the total energy generated from solar hot water systems is: 121474.6×1719=208814837.4MJ. According to the equation provided by "Technical Guidebook for Solar Water Heating System of Civil Buildings", the energy-saving potential can be evaluated. Carbon emissions reduction is 208814837.4/29.3×0.866/0.9/1000×44/12=25144 tce and Energy saving is 25144/2.65=9488 tce.

Fig. 1. A residential building with solar water heating systems

Fig. 2. An office building with ground source heat pump systems

2.2   Ground Source Heat Pump Systems (GSHP Systems)

The total area of GSHP systems is 1032726.77 m^2 (for soil systems) and 579541 m^2 (for water systems). Therefore, the cooling load is 220000 kW and the heating load is 160000 kW (10000 kW for soil systems and 6000 kW for water systems) when the cooling load is 140 kW/m^2 and heating load is 100 kW/m^2.

Assuming the cooling and heating periods are 150 days and 90 days, respectively, and HVAC will be operated 10 h per day. Then the energy-saving and carbon emission-reduction potential of buildings applying GSHP systems can be calculated as shown in the following tables.

Table 1. Cooling energy for GSHP

Load factor	Days	Calculation	Cooling energy (KWh)
100%	2.3%	150d×2.3%×10h/d×100%×220000KW	7590000
75%	41.5%	150d×41.5%×10h/d×75%×220000KW	102712500
50%	46.1%	150d×46.1%×10h/d×50%×220000KW	76065000
25%	10.1%	150d×10.1%×10h/d×25%×220000KW	8332500
total	150		194700000

Electricity consumption : 194700000÷4.0（EER）÷1000=48675 kWh

Table 2. Heating energy for GSHP (Soil systems)

Load factor	Days	Calculation	Heating energy (KWh)
100%	2.3%	90d×2.3%×10h/d×100%×100000KW	2070000
75%	41.5%	90d×41.5%×10h/d×75%×100000KW	28012500
50%	46.1%	90d×46.1%×10h/d×50%×100000KW	20745000
25%	10.1%	90d×10.1%×10h/d×25%×100000KW	2272500
total	90		53100000

Electricity consumption : 53100000÷3.45（EER）÷1000=15391 kWh

Table 3. Heating energy for GSHP (Water systems)

Load factor	Days	Calculation	Heating energy (KWh)
100%	2.3%	90d×2.3%×10h/d×100%×60000KW	1242000
75%	41.5%	90d×41.5%×10h/d×75%×60000KW	16807500
50%	46.1%	90d×46.1%×10h/d×50%×60000KW	12447000
25%	10.1%	90d×10.1%×10h/d×25%×60000KW	1363500
total	90		31860000

Electricity consumption : 31860000÷3.2（EER）÷1000=9956 kWh

Table 4. Cooling energy for regular HVAC systems

Load factor	Days	Calculation	Cooling energy (KWh)
100%	2.3%	150d×2.3%×10h/d×100%×220000KW	7590000
75%	41.5%	150d×41.5%×10h/d×75%×220000KW	102712500
50%	46.1%	150d×46.1%×10h/d×50%×220000KW	76065000
25%	10.1%	150d×10.1%×10h/d×25%×220000KW	8332500
total	150		194700000

Electricity consumption : 194700000÷2.5（EER）÷1000=77880 kWh

Total electricity consumption for regular HVAC systems is 77880+42480=120360 kWh. Therefore, Electricity reduction is 120360-74022=46338 kWh.

Table 5. Heating energy for regular HVAC systems

Load factor	Days	Calculation	Heating energy (KWh)
100%	2.3%	90d×2.3%×10h/d×100%×160000KW	3312000
75%	41.5%	90d×41.5%×10h/d×75%×160000KW	44820000
50%	46.1%	90d×46.1%×10h/d×50%×160000KW	33192000
25%	10.1%	90d×10.1%×10h/d×25%×160000KW	3636000
total	90		84960000
Electricity consumption : 84960000÷2.0（EER）÷1000=42480 kWh			

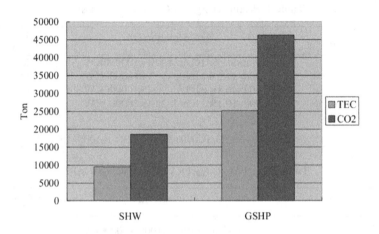

Fig. 3. Energy-saving potential of pilot projects for renewable energy buildings

3   Conclusions

Through simple calculations, this paper gives an evaluation of the energy-saving potential of pilot projects for renewable energy buildings integrated with solar water heating systems or ground source heat pump systems. The results show that renewable energy application in buildings has a great potential in reducing energy consumption and greenhouse gas emissions.

Acknowledgments. This work was supported by projects in Ningbo University (XYL11004 and XKL11D2073).

References

1. Koo, S.Y., Yeo, M.S., Kim, K.W.: Automated blind control to maximize the benefits of daylight in buildings. Building and Environment 45, 1508–1520 (2010)
2. Dai, L., Cai, L.: Analysis on the Characteristics of Modern Urban Residence in Ningbo: A Case Study of the Residence at Daici Alley and Deji Alley in Jiangbei District. Journal of Ningbo University (Natural Science & Engineering Edition) 24, 128–132 (2011)

3. Yao, J., Yuan, Z.: Study on Residential Buildings with Energy Saving by 65% in Ningbo. Journal of Ningbo University (Natural Science & Engineering Edition) 23, 84–87 (2010)
4. Yao, J., Zhu, N.: Evaluation of indoor thermal environmental, energy and daylighting performance of thermotropic windows. Building and Environment 49, 283–290 (2012)
5. Wang, W.L.: Master Plan for Small Towns in Perspective of Urban and Rural Planning Act. Journal of Ningbo University (Natural Science & Engineering Edition) 22, 288–292 (2009)
6. Hu, X.-Y., Xu, J., Zhang, J.-Q.: Study on the Modern Rural Residence Planning with Regional Features of Fanshidu Village. Journal of Ningbo University (Natural Science & Engineering Edition) 24, 110–114 (2011)
7. Cai, L., Dai, L.: Analysis on Characteristics of the Residence of Plain Areas in Ningbo: A Case Study on Zoumatang Village. Journal of Ningbo University (Natural Science & Engineering Edition) 22, 430–434 (2009)
8. Yao, J., Yan, C.: Evaluation of The Energy Performance of Shading Devices based on Incremental Costs. Proceedings of World Academy of Science, Engineering and Technology 77, 450–452 (2011)
9. Yao, J., Yan, C.: Effects of Solar Absorption Coefficient of External Wall on Building Energy Consumption. Proceedings of World Academy of Science, Engineering and Technology 76, 758–760 (2011)
10. Gao, H.-S., Yang, X.-X., Chen, Y.-H., et al.: Water Pollution Control in Rural Villages: Case Study. Journal of Ningbo University (Natural Science & Engineering Edition) 23, 74–78 (2010)
11. Zhou, Y., Ding, Y., Yao, J.: Preferable rebuilding energy efficiency measures of existing residential building in Ningbo. Journal of Ningbo University (Natural Science & Engineering Edition) 22, 285–287 (2009)
12. Yao, J., Zhu, N.: Enhanced Supervision Strategies for Effective Reduction of Building Energy Consumption-A Case Study of Ningbo. Energy and Buildings 43, 2197–2202 (2011)

A Simulation Analysis of the Thermal Performance of West-Facing Rooms

Chang Guo and Jian Yao[*]

Faculty of Architectural, Civil Engineering and Environment, Ningbo University, China
yaojian@nbu.edu.cn

Abstract. The purpose of this paper is to investigate the energy and thermal performance of west-facing rooms using building simulations. Results show that solar shading for west-facing room is extremely important and should be taken prior to other design options. Heating energy can also be decreased by lowering the operating set point of air-conditioners without reducing occupant thermal comfort.

Keywords: Thermal performance, building energy, solar shading.

1 Introduction

To achieve an energy-efficient world, the Chinese government has to transform the building sector, which today accounts for about 40% of China's energy use. A large number of measures are necessary to reduce energy consumption in new and existing buildings. A wide range of studies have been conducted on the improvement of the thermal performance of building envelopes. For example, many researchers suggested that daylighting and ventilation should be included in building design and urban planning [1-7]. Other energy efficiency measures have also been studied to save energy and reduce emissions [8-12]. However, the thermal performance of west-facing rooms has rarely been investigated. This paper tried to give an insight into the energy and thermal performance of west-facing rooms through a series of building simulations.

2 Methodology

A residential building in Ningbo was modeled in this paper. The thermal design for building envelope of this model, such as the U-values of walls (1.45W/m²K), roof (0.97W/m²K) and windows (3.2W/m²K), complies with the design standard for the energy efficiency of residential buildings in hot summer and cold winter zone. Energy and thermal performance simulations were carried out with the program DOE-2. The residential building model was illustrated in Fig.1 and the investigated west-facing room is living room R3.

---

[*] Corresponding author.

D. Jin and S. Lin (Eds.): Advances in FCCS, Vol. 2, AISC 160, pp. 171–175.
springerlink.com   © Springer-Verlag Berlin Heidelberg 2012

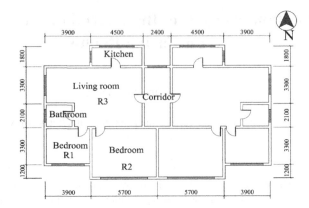

Fig. 1. The building plan

3 Results and Discussion

Fig. 2 gives the monthly cooling and heating energy consumption of the living room R3. It can be seen that the cooling energy consumption mainly occurs in June to September with the maximum one in July, while the heating is needed from December to March. The cooling energy demand in July and August is higher than heating energy demand in other month because of the location of the room and the facing of the window. Therefore, enhanced cooling measures including solar shading should be adopted to reduce the cooling energy requirement.

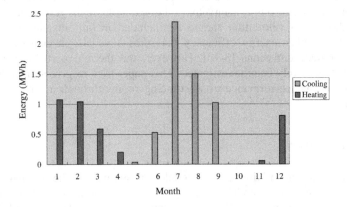

Fig. 2. The cooling and heating energy consumption of the building in each month

Fig. 3 presents the monthly cooling and heating peak loads of R3. The situation in this chart is similar to that in Fig. 2. Cooling load is dominant with the maximum value of more than 8 kW. The maximum heating load occurred in February which is

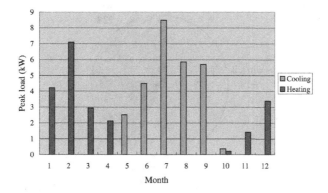

Fig. 3. The cooling and heating loads of the building in each month

more than 7 kW. This means air-conditioners installed should have a size of more than 8 kW for cooling, which is a big investment for occupants.

Hours of cooling and heating in each month are shown in Fig. 4. Although cooling energy demand and cooling load are higher than those of heating, cooling hours are close to heating hours. According to the above two charts we can infer that the base temperature is not far lower than the set point of 18 °C. Therefore, the heating energy and running hours can be lowered if thermal comfort in winter is not as high as cold climate zones. In fact, people in this climate region can tolerate cold in winter and heating energy reductions are achievable.

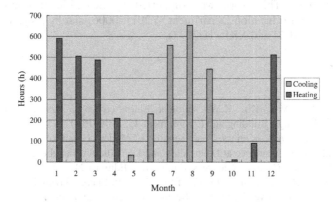

Fig. 4. Hours of cooling and heating in each month

Fig. 5 illustrates the breakdown of heat gains through the building envelope in summer. It can be that solar radiation through windows account for roughly 50% of heat gains, while walls only account for 22%. This means solar shading should be taken prior to other insulation measures.

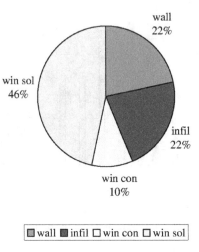

wall
22%

win sol
46%

infil
22%

win con
10%

☒ wall ■ infil ☐ win con ☐ win sol

Fig. 5. Breakdown of heat gains through the building envelope in summer

4   Conclusions

This paper investigated the energy and thermal performance of west-facing rooms using building simulations. Results show that solar shading for this room is extremely important and should be taken prior to other options. Heating energy can be reduced by lowering the set point of air-conditioners when operating.

Acknowledgments. This work was supported by Scientific Research Fund of Zhejiang Provincial Education Department (Y201120006) and by projects in Ningbo University (XYL11004 and XKL11D2073).

References

1. Koo, S.Y., Yeo, M.S., Kim, K.W.: Automated blind control to maximize the benefits of daylight in buildings. Building and Environment 45, 1508–1520 (2010)
2. Dai, L., Cai, L.: Analysis on the Characteristics of Modern Urban Residence in Ningbo: A Case Study of the Residence at Daici Alley and Deji Alley in Jiangbei District. Journal of Ningbo University (Natural Science & Engineering Edition) 24, 128–132 (2011)
3. Yao, J., Yuan, Z.: Study on Residential Buildings with Energy Saving by 65% in Ningbo. Journal of Ningbo University (Natural Science & Engineering Edition) 23, 84–87 (2010)
4. Yao, J., Zhu, N.: Evaluation of indoor thermal environmental, energy and daylighting performance of thermotropic windows. Building and Environment 49, 283–290 (2012)
5. Wang, W.L.: Master Plan for Small Towns in Perspective of Urban and Rural Planning Act. Journal of Ningbo University (Natural Science & Engineering Edition) 22, 288–292 (2009)
6. Hu, X.-Y., Xu, J., Zhang, J.-Q.: Study on the Modern Rural Residence Planning with Regional Features of Fanshidu Village. Journal of Ningbo University (Natural Science & Engineering Edition) 24, 110–114 (2011)

7. Cai, L., Dai, L.: Analysis on Characteristics of the Residence of Plain Areas in Ningbo: A Case Study on Zoumatang Village. Journal of Ningbo University (Natural Science & Engineering Edition) 22, 430–434 (2009)
8. Yao, J., Yan, C.: Evaluation of The Energy Performance of Shading Devices based on Incremental Costs. Proceedings of World Academy of Science, Engineering and Technology 77, 450–452 (2011)
9. Yao, J., Yan, C.: Effects of Solar Absorption Coefficient of External Wall on Building Energy Consumption. Proceedings of World Academy of Science, Engineering and Technology 76, 758–760 (2011)
10. Gao, H.-S., Yang, X.-X., Chen, Y.-H., et al.: Water Pollution Control in Rural Villages: Case Study. Journal of Ningbo University (Natural Science & Engineering Edition) 23, 74–78 (2010)
11. Yan, Z., Yong, D., Jian, Y.: Preferable rebuilding energy efficiency measures of existing residential building in Ningbo. Journal of Ningbo University (Natural Science & Engineering Edition) 22, 285–287 (2009)
12. Yao, J., Zhu, N.: Enhanced Supervision Strategies for Effective Reduction of Building Energy Consumption-A Case Study of Ningbo. Energy and Buildings 43, 2197–2202 (2011)

Research of College Science and Technology Incentive Measures Based on Scientific and Technological Statistics Index

Nan Xie, Xiaohong Shen, Yijia Zhu, and Zhengshen Shi

Zhejiang Water Conservancy and Hydropower College, Hangzhou, P.R. China 310018
xienan@zjwchc.com

Abstract. Scientific and technological (S&T) statistics takes more and more important part in college S&T daily working. According to the statistical requirements of annual report on the college S&T statistics, this paper proposes college three S&T statistical assessment levels in colleges. It also gives some index of statistical data about a college in recent five years and puts forward some charts and analysis results based on the relevant statistical data. From the analysis charts, we can analyze in detail the college S&T development state, level and trend, and provide a scientific reference in college S&T management while drawing up more objective and realistic incentive measures, even promote college S&T management innovation.

Keywords: S&T statistics index, college S&T management, incentive measures.

1   Introduction

Scientific and technological (S&T) statistics as a branch of statistics refers to one statistical field while quantitatively determining size and structure of national and regional scientific and technological activities. It is one basic work when the S&T management departments decide scientific ways or strengthen macro-management. And it brings spread and application in the size, status, structure of a national S&T activities, and quantitatively measures economic and social development impact. Additional, it has an important role in promoting S&T management innovation. After the 1960s, S&T statistics has happened, and has gradually formed a system of S&T statistics. In the late 1980s, we have collected a more systematic statistical information and promoted S&T statistics development in order to establish a more comprehensive S&T survey system and a index system of S&T statistics and make the S&T management be more standardized and scientific [1,3].

College science and technology statistics is very important daily work for college S&T department and it is quantitative measurement for the college S&T aspects, such as S&T size, level, speed, structure and other so on. Also, it reflects the S&T status and development trend. More then, it is an important component for a state S&T statistic work.

D. Jin and S. Lin (Eds.): Advances in FCCS, Vol. 2, AISC 160, pp. 177–182.
springerlink.com   © Springer-Verlag Berlin Heidelberg 2012

2 Index of College Scientific and Technological Statistics

S&T statistics is a systemic and scientific work. The index of S&T statistics is relatively independent and intrinsically linked, and it is quantitative for S&T work while doing S&T activities.

However, the index system of S&T statistics is complex and its operation is more difficult, the authenticity of statistical data is difficult to guarantee, thus S&T statistics affects the important reference on S&T management and decision-making is affected by the S&T statistics [3]. According to the requirements of the college S&T statistics report (shown in Table 1), the index of college S&T statistics could be divided into three levels. The first level index is mainly measured by technological base, S&T input and S&T achievements. The second level index is generally evaluated by the S&T manpower, S&T conditions, funding, published papers, technology transfer and intellectual property rights, and so on. The third level index is a refinement of the second [2,4].

Table 1. Evaluative index of college S&T statistics

First level index	Second level index	Third level index
S&T bases	S&T manpower	Number of college senior title, doctors
	S&T conditions	Research institutions, national or ministerial key laboratories, S&T exchanging activities
S&T input	S&T funding	The total and annual growth about S&T funding
	S&T projects	S&T projects, the national proportion in all projects
S&T achievements	Knowledge innovation	Papers published (including three-index); S&T works
	Technology transfer and intellectual property rights	Patents application and authorization; patents contacts money
	S&T award	Number of national and provincial awards

Table 2 ensamples a part of S&T statistic main index of one college in recent five years (2007-2011). From Table 2, we can reflect some conditions about the college, such as operating condition, the ability of scientific research and the S&T developing level.

Table 2. Part of S&T statistics index about one college from 2007 to 2011

Year	2007	2008	2009	2010	2011
S&T Personnel	326	359	417	432	609
Number of research projects	105	151	110	138	279
Number of provincial projects	4	7	6	9	15

Table 2. (*continued*)

Funds of R&D(ten thousand Yuan)	108.9	497.8	476.3	608.2	889.7
Funds of crosswise projects (ten thousand Yuan)	101.5	293.6	259.1	341.8	457.3
Core papers	56	84	103	126	149
Papers recorded ed by three-index	14	20	28	37	51
Number of S&T award	8	12	10	14	18
Number of provincial awards	3	4	4	6	10
Number of patents applied	4	5	4	3	5
Number of patents authorized	0	1	2	2	5

3 Data Analysis of S&T Statistic Index

According to quantization data in Table 2 and three-level index of college S&T statistics in Table 1, we could analyzes in detail developing level and trend of the college mentioned before. The statistics of Fig. 1~ Fig. 6 displays up developing trend of the college S&T level. Fig. 1 is compare on S&T personnel, and we can see the S&T manpower is more and more strong, including advanced titles and doctors. Also Fig. 1 could preferably reflect college scientific scale and team of one college. Fig. 2 is compare on the number of scientific research projects each year and it rises fine except Year 2009. Moreover the national and provincial project ratio is more and more high. So it could display well scientific research levels. From Fig. 3, the funds of R&D are increasing steeply and the ratio of crosswise projects is decreasing. So it well explains the S&T scale and level of college scientific research. Fig. 4 is column chart about core papers published and recorded by three-index. It also informs the teachers' enthusiasm and wideness of participating in S&T activities, and we can know that the college S&T troops are more and more big. All in all, from Fig. 1 to Fig. 6, we can know a well S&T management and a beneficial policy are very important to college S&T developing.

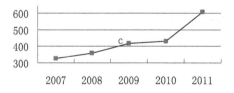

Fig. 1. S&T persons percent year

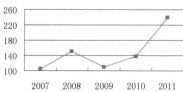

Fig. 2. S&T projects percent year

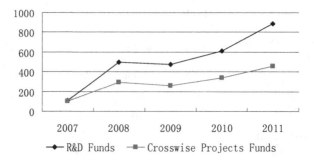

Fig. 3. Compare of all and crosswise R&D Funds

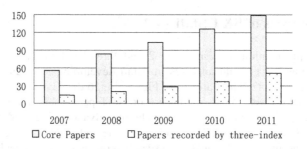

Fig. 4. Compare of Core and recorded papers

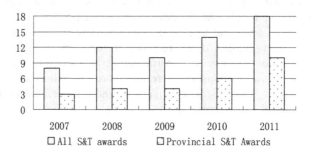

Fig. 5. Annual distribution of S&T awards

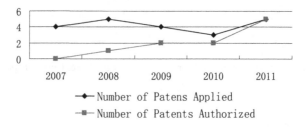

Fig. 6. Annual distribution of Patents applied and authorized

4   S&T Incentive Measures

S&T statistics is a part of S&T management. And its details are from daily S&T management. Now the emphasis of S&T management is diverted to modernization information management mode. S&T manager could track every S&T activities and change all S&T management information to information resource and resource sharing in order to guarantee S&T statistics resource [2]. To build college S&T management and strengthen S&T teams and expand college S&T scale, we put forward some related S&T incentive measures.

① Broadening college S&T channels and increasing college S&T input. We must enhance teachers' enthusiasm, and endeavor to national or local S&T projects, and participate in national or local S&T projects. And college S&T management department should support college S&T personnel and provide some S&T funds and improve scientific research conditions. Additional, the S&T management department should encourage teachers acquiring more and more projects been from enterprise or popular organization or community team.

② Accelerating steep on college S&T talents teams. Firstly, the college S&T manager should make scientific and reasonable talents developing planning and managing means, and perfect S&T talents incentive measures in order to foster excellent talents and create a good requirement about talents flowing. In addition, the college S&T management department should strengthen training and select subjects' leader and establish a well obsolete system on S&T talents and carry out the talents flowing management based on high starting point and strict claims. Especially, the college S&T management department should strengthen young scientific research talents and provide a well stage to put their abilities to good use.

③ Formulating object and reasonable S&T award measures for different scientific research levels and achievements (including projects, papers, prizes, and so on).

④ Increasing scientific research checking in annual college teacher checking system.

⑤ Protecting patents of college teachers S&T achievements. The S&T management department should respect and protect teachers' right who invent patents.

5   Conclusion

College statistics results can reflect a college S&T comprehensive strength such as scientific research projects, S&T development or S&T achievements conversion. And the meaning of S&T statistics proclaims the changing regulars of scientific research activities and provides decision basis on examining or adjusting scientific research policy. S&T statistics index analysis is an important part among scientific research working. It also have decisive effect that how to make scientific research management standardize, scientific, systemic and innovated.

Acknowledgments. The authors wish to thank the project supporting of the foundation of education department of Zhejiang province of China for contract Y201122779. Also, this work was financially supported by 2011' project of Zhejiang Higher Education Association (Project CODE: KT2011208), under which the present work was possible.

References

1. Ma, Y.: Design of Index System for Science and Technology Statistics Based on Management Innovation of Technology Management. Statistics and Decision, 68–69 (2009)
2. Liu, J.: Point, Line, Plane, Volume in Science and Technology Statistics – Studying the Index of Science and Technology Statistics and Improving Science Research Management. Science and Technology Management Research, 215–218 (2011)
3. Wu, M.: Discussion on Design of Index System for Scientific and Technological Statistics. Enterprise Manangement. Guangdong ShipBuilding 28, 62–63 (2010)
4. Altshuller, G.: The Innovation Algorithm. Technical Inmovation Center, Worcester (1999)

A Research on the Relation between China's Household Savings and Social Security Expenditure Based on Dynamic Panel Data Model

Liting Fang

Department of Statistics, School of Economics,
Xiamen University, Xiamen 361005, P.R. of China
flt749@yahoo.com.cn

Abstract. This paper analyzes the relation between China's social security system and household savings using dynamic panel model. Empirical results show that, there is a positive relationship between pension expenditure and household savings, but there exists non-significant negative correlation between unemployment insurance, medical care insurance, maternity insurance and household savings. In addition, empirical results also show that, income elasticity and inertia of household savings in China are large, and the relation between ratio of education and medical expenses and the savings is significant negative, however, there exists only weak positive correlation between economic growth and resident savings.

Keywords: Social Security, Household Savings, Dynamic Panel Model.

1 Introduction

As we know, savings plays an important role in economics growth theory, many foreign theoretical research about economic growth are almost directly related to savings. Therefore, to study the factors affecting the behavior of household savings is of great practical significance. In addition to the explanation of causes of the high savings rate, there are also more people who propose improving the level of social security to reduce savings. As the social security is important for people's livelihood and social long-term development, this article focuses on five areas in social security (pension insurance, unemployment insurance, medical care insurance, work injury insurance, maternity insurance) to make a basic model analysis about how they affect household savings.

All the time, Domestic and foreign scholars have analyzed the factors affecting household savings in different aspects. Implementation of the social security system also has an important impact on household savings. According to traditional life cycle theory, the implementation of social security schemes makes the private savings decrease due to the substitution effect of the wealth. Foreign scholar Feldstein (1974) showes that the implementation of American social security schemes makes the private savings decrease by 30%-50% by estimating the consumption function in which the social security wealth is one of the independent variables. However, There are also researches illustrating that the negative relationship between the social security schemes and private savings is not exist or at least not very obvious. For example, Kruz (1980) finds that there is no effect of social security wealth on private savings, but the effect of the private pension fund on savings is exist. Domestic scholar Shi(2010) shows that PAYG pension insurance system has a significant positive impact on China's household consumption. Yue (1997) analyzes the impact of the pension insurance fund on savings and investment and proposes that different funding model means different transforming problems among the savings, consumption and investment. Wan (2005) also analyzes China's urban household savings from the perspective of pension insurance, and finds that the urban household pension insurance system which is not perfect enough is a major cause of high savings rate. In addition, there are more researches about the savings rate with reference to the factors of social security when analyzing the reasons of household savings.

2   Research Design

2.1   Variable Selection and Data Description

This paper mainly study the impact of the five kinds of insurance expenditure in the structure of social security on urban household per capita savings (S_t). So the dependent variable in the regression analysis is the level of urban household per capita savings (S_t).In the relevant variables affecting S_t , we mainly select four groups of variables. The first group is the lag urban household per capita savings (S_{t-1}) which can represent the inertia of household savings. The second group contains five insurance expenditures: pension expenditures (ylb), medical care insurance expenses (med), work injury insurance expenses (gsb), maternity insurance expenses (syb), and unemployment insurance expenditures (une).The third group is uncertain factors including disposable income (inc), proportion of education and medical expenses (Ratio of income) (edu) and price index (cpi).The fourth group is macro-environmental factors, including GDP growth rate (gro) and fixed asset investment (inv).

We use China's provincial-level panel data from 1999 to 2009. The data including the household savings, income levels, spending levels and the level of investment in fixed assets are all from "China Statistical Yearbook", and the data of the five kinds of insurance expenditure are from "China Statistical Yearbook" partly and also from "Chinese Yearbook of Labor Statistics and Social Security" partly.

2.2   Model Specification and Estimation Methods

Unlike the traditional savings model, We not only consider some uncertain factors including household income, price level and interest rate, macroeconomic conditions and other common independent variables, In particular ,we also introduced five kinds of insurance expenditure into the regression models. Therefore, the dynamic panel data model is set as follows:

$$\ln S_{i,t} = \alpha + \beta_0 \ln S_{i,t-1} + \beta_1 \ln(\text{every insurance expenditure})$$
$$+ \beta^T \cdot (\text{control variable}) + \mu_i + \varepsilon_{i,t} \tag{1}$$

where $S_{i,t-1}$ refers to the lag urban household per capita savings, μ_i refers to the provincial-level effect, $\varepsilon_{i,t}$ is the random disturbance, α is the constant term, $\beta_i (i = 0,1)$ is the variable regression coefficient, β is the coefficient vector of control variables.

As the lag urban household per capita savings was included in the estimated regression model, there would be endogeneity problems when estimating the model. In this case, parameter estimators may be biased and inconsistent if we still use the general panel data regression methods (such as OLS method).We must use the appropriate instrumental variables and adopt difference GMM method to estimate the model in order to get the consistent estimators (Arellano, 1991). When we assume that the residuals are zero mean and current independent, the first-order differential residuals ($\Delta\varepsilon_{i,t}$) are not associated with all the household per capita savings ($S_{i,t}$) and all the other independent variables (at ($t-2$) time and before). This means that we can use all of these values as instrumental variables of $\Delta S_{i,t-1}$ and adopt GMM method to estimate model in order to obtain the more efficient estimators.

3   Empirical Results Analysis

We use five models to estimate the effect of each insurance on household savings separately, model1 to model5 are the estimated functions of pension insurance, unemployment insurance, medical care insurance, work injury insurance and maternity insurance respectively. We take the logarithmic form of the variables except the ratio variables. The results are showed in table1.

We can see from table1 that, in all models P values of Hasen test are large, which illustrates the instrumental variables are all efficient. Since P values of AR(1) test in all models is small and is large in AR(2) test, the residual series are all accord with the basic assumption in GMM estimating of dynamic panel model. This indicates that the result of our estimating is consistent and credible.

The coefficients of lag lns in all models are minus, this shows savings has a strong inertia, but the other 4 insurances have no significant impact on savings except pension insurance. Pension insurance has positive effect on household savings, which demonstrates reduction in precautionary savings due to pension insurance is not enough to reduce total savings. The coefficients of other 4 insurances are all minus though not significant, this result at least reveals that unemployment and health care may be potential factors affecting household savings.

Estimating results of control variables shows that the income elasticity of savings is very strong, and the growth rate of household medical care and education expenditure are much larger than income growth rate, moreover, the effect of GDP growth rate on household savings is very small.

Table 1. Estimating Results

	Model 1	Model 2	Model 3	Model 4	Model 5
1.lns	0.272*** (6.65)	0.271*** (7.85)	0.286*** (6.97)	0.095** (2.06)	0.089** (-2.31)
lnylb	0.083** (2.04)	——	——	——	——
lnune	——	-0.030 (-1.54)	——	——	——
lnmed	——	——	-0.005 (-0.14)	——	——
lngsb	——	——	——	0.022 (1.47)	——
lnsyb	——	——	——	——	-0.023 (-1.50)
lninc	0.707*** (-3.48)	0.555*** (9.50)	0.678*** (-3.62)	1.019*** (8.09)	1.277*** (14.34)
lninv	0.253*** (3.60)	——	0.317*** (3.72)	——	——
cpi	-0.003 (-1.43)	-0.001 (-1.25)	-0.003* (-1.72)	0.004*** (8.90)	0.003*** (3.93)
edu	-0.165*** (-9.58)	-0.096*** (-17.76)	-0.166*** (-10.71)	-0.037*** (-10.73)	-0.065*** (-10.71)
gro	0.028*** (10.75)	0.021*** (11.08)	0.025*** (9.56)	0.004** (2.19)	0.006** (1.96)
Hasen test P value	0.377	0.282	0.493	0.589	0.790
AR(1) test P value	0.002	0.062	0.005	0.007	0.089
AR(2) test P value	0.407	0.273	0.418	0.360	0.667

Notes1: " ***" ," **" ," *"means refusing the null hypothesis under 1%,5% and 10% significant level respectively, t value is in the parentheses.

4   Conclusion

In this paper, we use the dynamic panel model to analyze the relationship between China's social security system and household savings .The empirical results show that, four other insurances do not have a significant impact on household savings except pension insurance. However, there is a positive correlation between pension insurance expenditure and household savings .The possible reason is that China's house price level has continued to soar in recent years. As a result, the uncertain future expenditures of residents is expected to increase. In addition, the estimated coefficient of unemployment insurance, medical care insurance and maternity insurance is negative, this shows that the improvement of the social security system is a potential factor to promote people to reduce precautionary savings. In conclusion, we know that to continue to establish a perfect social security system is an important premise for China's government to solve the problem of inadequate consumption.

References

1. Arellano, B.: Some Tests of Specification for Panel Data: Monte Carlo Evidence and an Application to Employment Equations. The Review of Economic Studies 58(2), 277–297 (1991)
2. Feldstein, M.: Social Security, Induced Retirement and Aggregate Capital Accumulation. Journal of Political Economy 82(5), 905–926 (1974)
3. Kruz, M.: Social Security and Capital Formation. The President's Commission on Pension Policy (1980)
4. Li, H.: Theoretical and empirical analysis on the influence of social security on resident saving. Economist 6, 87–94 (2010) (in Chinese)
5. Li, X., Zhu, C.: Social Security and China's Household Saving Rate: An Empirical Study Based on China's Dynamic Panel Data at the Provincial Level. Journal of Xiamen University (Arts & Social Sciences) 205(3), 24–31 (2011) (in Chinese)
6. Shi, Y., Wang, M.: The Effect of PAYG Pension Insurance on Savings. The Journal of Quantitative & Technical Economics 3, 96–106 (2010) (in Chinese)
7. Wan, C., Qiu, C.: The Welfare Analysis of Optimal Voluntary Saving Ratio—Based on China's Pension System. The Journal of Quantitative & Technical Economics 12, 61–70 (2005) (in Chinese)
8. Yue, Y.: The Impact Analysis of Pension Insurance Fund for Savings and Investment. Journal of Shanxi Finance and Economics College 2, 50–52 (1997) (in Chinese)

The Development of Video Monitoring System Based on JPEG2000

Qiang Guo, Guangmin Sun, Dequn Zhao, Shengyu Liao,
and Fan Zhang

Department of Electronic Engineering, Beijing University of Technology,
Beijing, 100124, China
gmsun@bjut.edu.cn

Abstract. The high stability is required in the process of target detection system. A video monitoring system based on JPEG2000 for target detection is developed in the paper. Firstly, the image-based video compression method and streaming-based video compression method have been compared. And then a comparison of still image compression algorithm based on JPEG2000 standard and traditional JPEG standard has been made. The experimental results show that the video monitoring system based on still image compression has higher stability and low error influence than based on video streaming compression. And JPEG2000-based compression algorithm has better performance than traditional JPEG-based compression algorithm in the condition of low bit rates.

Keywords: Video Monitoring System, Image Compression, Stability.

1 Introduction

With the increasing of the network bandwidth, computer processing speed and storage capacity, a variety of video processing technologies have been appeared. Advantages of the full digital and network video monitoring systems are becoming apparent increasingly. Video monitoring system has lots of information and data. Therefore more bandwidth and storage space is needed. It requires to be compressed before transmission and storage.

There are two ways to realize video monitoring. One is streaming-based video compression method such as H.264. It has a smooth image quality in the case of high compression ratio. H.264 uses multi-frame prediction to improve data compression. It has great advantages when the network transmission bandwidth is limited. But when P-frame disturbed or predicted based on erroneous I-frame, the video decoding error will occur. It's the worst when there is something error in the I-frame for the actual application of the interference. The next P-frame would be greatly affected. Another method for video monitoring is the single frame image compression method such as JPEG-based single image compression algorithm. Compared with the streaming-based video compression, there is only I-frame and no P-frame in image-based video compression. So the compression ratio is small. But it can guarantee the system stable and the number of error frame as less as possible. When one frame of image is

D. Jin and S. Lin (Eds.): Advances in FCCS, Vol. 2, AISC 160, pp. 189–195.
springerlink.com © Springer-Verlag Berlin Heidelberg 2012

disturbed, it is only need one frame to convergence to the correct image. It was far better than the streaming-based video compression method.

According to the high stability of target detection, one video monitoring system based on single-frame image compression is proposed in this paper. The error accumulation will be reduced as far as possible when the error frame happens. Compared with JPEG, JPEG2000 used the discrete wavelet transform and its compression quality could be 10% -- 30% higher than JPEG. And JPEG2000's image is more smoothly. At low bit rates, JPEG2000 image of the "slow" form could be applied to remove the inherent block distortion in JPEG coding method. And the subjective image quality is greatly improved. In this paper, still image compression algorithm is based on JPEG2000 standard instead of traditional JPEG standard. It will not only keep the stability of system, but also meet the requirements of image quality at low bit rates.

The paper is organized as follow. Section 1 describes the background of the video monitoring system and program selection. Section 2 introduces the System Architecture. Section 3 gives the details key technology of the system. The experimental results and data analysis is presented in section 4. Section 5 is conclusions.

2   System Architecture

The system is divided into three parts: video collection, video transmission client and video transmission server.

Fig. 1. System Architecture

The part of video collection uses the digital webcam to get the real-time bitmap. The resolution of bitmap is 320*240 and it is saved in the root directory of the video transmission client software.

First of all, video transmission client compresses the image information with the JPEG2000 algorithm. The local client tries to connect to the remote server. The packaged frame data will be sent once the connection successes.

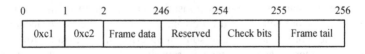

Fig. 2. Frame Format

Video transmission server is the key part of the video monitor system. Its main functions are analyzing the received data stream and data-processing. There are two threads in video transmission server: the receiving thread and decoding thread. Firstly the receiving thread starts TCP/IP transmission server in order to detect whether there is any client connection at any time. Secondly the received frame data is processed. The receiving thread checks and extracts the valid frame data stream and remove the header and other additional frame information. Then determine whether the data obtained is completely or not. If it is confirmed that the whole data of one image is obtained, the image is decoded by JPEG2000. Otherwise it will be considered as the wrong information and not be treated.

In addition, there are other operations and treatments to the received data stream, including preserving original data stream, displaying and saving the data of image.

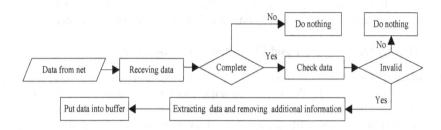

Fig. 3. Receiving Thread

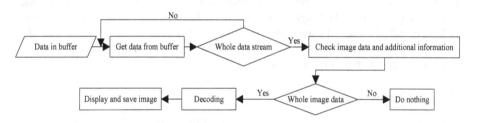

Fig. 4. Decoding Thread

3   The Key Problems and Solutions

Through the introduction of the last section, the basic framework of the system is implemented, the key problems in the system is presented in this section.

3.1   Parameter Acquisition in Video Transmission Server

The image parameters packaged in the frame is sent to the receiver. So the receiver could obtain the image parameters. The additional parameters of the image are added to the image data stream. And then the image data stream packaged into frame structure could be sent out. Each packet contains 256 bytes. Image data is packaged into multiple frames. Before packaging, the image detection flag and data size information are added behind the image data. Each image frame contains the valid data with the certain length and the effective image data.

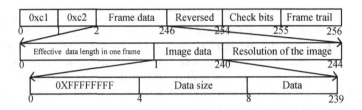

Fig. 5. Data Encapsulation Format

One 256-byte frame includes three parts: the length of valid data, image data and image resolution. Image data contains additional information and image content.

3.2   Getting One Complete and Correct Image at Decoder

The received data saved in circular buffer is detected bit by bit. At the receiver a circular buffer is defined. When received data, check the validity of the data frame. The header, the frame tail and the detection flag of each frame data would be checked. If it is correct and complete frame data, extract the valid data inside the frame and put it into the circular buffer. An effective and complete image data by comparison data in circular buffer is extracted bit by bit. The method in this paper is added the key bits 0xFFFFFFFF and the size of image information before each image data. Firstly find keywords and calculate the length of valid data between two keywords. Then compare the length with the size of image information. If it is equal, it means that obtain the valid image information and the next step is decoding operation. Otherwise, it is the wrong message and not to be deal with.

3.3   Synchronization

There are receiving thread and decoding thread in the receiver. One circular buffer is defined to save the temporary data. Two threads have access to buffer data separately. The operation to the buffer should pay attention to thread synchronization. The solution is to define a thread switch to make two threads use the buffer interchangeably. Lock buffer when the receiving thread is in writing state so that other threads could not use it. When the writing is finished, unlock the buffer. So do the decoding thread.

All of above are the key problems encountered in the process of system. After solving the above problems, the function of video monitor is realized and the stability of the system could be guaranteed.

4   Experimental Results and Data Analysis

JPEG, JPEG2000 and H.264 algorithms are used for comparing in order to verify the performance of the system. Frame loss rate (FLR) is defined as:

$$FLR = \frac{N_{\text{lose frame}}}{N_{\text{transmitted frame}}} \tag{1}$$

Where $N_{\text{lose frame}}$ is the number of degraded frames, $N_{\text{transmitted frame}}$ is the total number of transmitted frames.

The loss information can lead to the degradation of frame. During H.264 decoding, if the I-frame error, then all subsequent frames will be affected until the next I-frame appears. And if an error occurs in the P-frame, then the error will be accumulated even if there are no new errors. In this paper, if the PSNR of frame is below 27dB, it is considered as a wrong frame.

H.264 reference software JM11.0 is used as the experimental platform to encode video sequences using IP ... PIP ... P's coding structure. One I-frame is encoded every 30 frames. There are five reference frames and QP is 28. The first 100 frames of standard image sequence with QCIF format are encoded. By error concealment, the I-frames are not losing and the macro-block losing rate of P-frame is 10%. Then test the PSNR of decoder. Results are shown as follows.

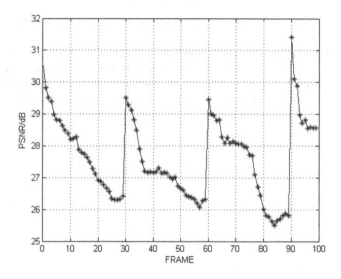

Fig. 6. PSNR in the Case of 10% Macro-block Losing Rate of P-frame

For two kinds of single-frame image compression algorithm based on JPEG and JPEG2000, the error appears in one frame will not affect the next frame, so only one image will be affected. Results are listed as Table 1.

By comparison of data from the table 1, it can be seen that when there are disturbed or dropped frames, video compression algorithms based on multiple reference frames need more frames to convergence to the correct image than image-based video compression method. Therefore, image-based method has a better stability in the system.

Table 1. Total Error Frames With Different Compression Algorithms in the Same Frame Loss

FLR(%)	Different Algorithms	Error Frames
10	H.264	33
10	JPEG	10
10	JPEG2000	10

The main objective evaluation of image quality is peak signal to noise ratio (**PSNR**) defined as:

$$PSNR = 10\lg\left(\frac{255^2}{MSE}\right)$$ (2)

The mean squared error (**MSE**) is computed pixel by pixel in the whole image between original and the reconstructed. **MSE** is defined as:

$$MSE = \frac{\sum_1^{Framesize}(I_i - P_i)^2}{Framesize}$$ (3)

Where I is the original image pixel value and P is the pixel value after the reconstruction.

A comparison of the compression quality between JPEG and JPEG2000 at low bit-rate is made. For the JPEG compression algorithm, Huffman coding is used and with sampling rate of 4:2:0. For the JPEG2000 compression algorithm, Kakadu algorithm is used. In the same compression ratio, the PSNR results for two algorithms are shown below:

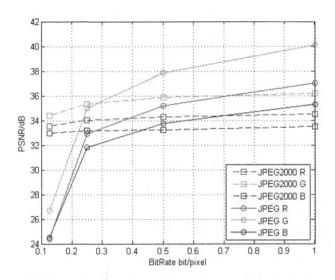

Fig. 6. PSNR of JPEG2000 and JPEG at Low Bit Rates

From the results we can see that in the case of high bit rate, the quality of JPEG is significantly better than JPEG2000. But for the case when bit rate is lower than 0.25 bit, JPEG has begun to obviously appear the visual mosaics while JPEG2000 image is blurred and the basic image information can be recognized.

5   Conclusions

A video monitoring system based on JPEG2000 is proposed in this paper. It is suitable for target detection in the network communication with limited bandwidth. The small amount of data, high stability and higher image quality can be achieved in the system. Experimental results show that the image-based compression algorithm needs fewer frames for recovering to the correct image than video-based algorithm when the error happened. And in the image-based compression algorithms, JPEG2000 compression algorithm has significantly better results than traditional JPEG-based algorithm in low bit rates.

References

1. Adams, M.D.: The JPEG 2000 Still Image Compression Standard. ISO/IEC JTC 1/SC 29/WG 1N 2412 (December 03, 2005)
2. Xiao, R.-H.: Difference between JPEG 2000 and JPEG & Their Practical Application. Computer Development & Applications 22(9), 44–46 (2000)
3. Li, G., Ye, X.-M., Li, J.-Y.: Application of JPEG2000 in digital library. Journal of Zhejiang University of Technology 36(5), 523–526 (2008)
4. Xu, S.-F.: The Analysis of the Core of JPEG and JPEG200. Journal of Huzhou Vocational and Technological College, 91–94 (March 2003)
5. Jiang, A.-D., Hu, S.-H.: Image compression based on JPEG2000. Journal of Zhengzhou College of Animal Husbandry Engineering 29(4), 34–49 (2009)
6. Chen, L.: Scout System Based on JPEG2000 in Internet. Computer Knowledge and Technology 5(9), 2370–2378 (2009)
7. Wang, Z.-S., Ma, Y.-D.: Thinking Based on JPEG and JPEG2000, http://www.paper.edu.cn
8. Zhou, X.-Y., Wang, J.-C.: Multi-Scalability Mode of Still Image Compression Standard JPEG and JPEG200. Computer Technology and Development 17(1), 12–17 (2007)
9. Wang, B., Ge, W.-L.: Design of Intelligent Based on Video Surveillance. Computer Knowledge and Technology 7(22), 5420–5421 (2011)
10. Ji, W.-P., Shen, L.-S.: Error concealment method for H.264 video. Computer Engineering and Applications 43(10), 11–14 (2007)
11. Naman, A.T., Taubman, D.: JPEG2000-Based Scalable Interactive Video (JSIV). IEEE Transactions on Image Processing 20(5) (May 2011)
12. Francesc Aulí-Llinàs, F., Bilgin, A., Marcellin, M.W.: FAST Rate Allocation Through Steepest Descent for JPEG2000 Video Transmission. IEEE Transactions on Image Processing 20(4) (April 2011)

Design of Shortwave Broadband OFDM Communication System Based on Raptor Code

Shengyu Liao, Guangmin Sun, Dequn Zhao, and Qiang Guo

Department of Electronic Engineering, Beijing University of Technology,
Beijing, 100124, China
gmsun@bjut.edu.cn

Abstract. A design scheme for shortwave broadband OFDM system based on Raptor code is proposed. Raptor code is utilized to improve the channel capacity and erasure performance, as well as increasing the bit-rate. In addition, OFDM is applied to increase the spectrum efficiency of the shortwave broadband communication system. The experiment results demonstrate that the proposed system has the better performance on error resilience and the transmitted bit-rate.

Keywords: Raptor Code, OFDM, Shortwave Communication System.

1   Introduction

SW (short wave) is transmitted by electromagnetic wave, whose density and height are easily affected by weather or some other factors. Consequently the shortwave communication has its inevitable drawbacks of noise and instability. In the meantime, compared to other communication modes such as satellite communication, shortwave communication has no advantages in transmission rate. But its superiority lies in the simple equipment, which is featured as flexible. More significantly, the ionosphere, served as the shortwave transmission medium, is out of the reach of human damage. Therefore in the event of an emergency, especially some serious natural disasters or wars, shortwave communication may be the best even the only communication mode.

At present, the highest reliable data transfer rate of most transceivers is no more than 2400bps[1] with a limited anti-interference ability, which badly restricts the application of shortwave transceivers. In order to overcome all the shortcomings, lots of new technologies are applied to modern shortwave communication system, including modern shortwave channel technology[2], modern shortwave communication terminal technology, the digital communication equipment and network technology, aiming to enhance the communication system to its highest level with the limited transmitting power and comparatively hostile environment considered. Raptor code, with the merit of adapting to time-varying channel finely and not relying on feedback and retransmission, is just the right cure of the shortwave communication system. As a result, the Raptor code will greatly improve shortwave communication efficiency on the basis of transmission reliability. This paper, on the foundation of lucubrating short-wave channel characteristics and shortwave

D. Jin and S. Lin (Eds.): Advances in FCCS, Vol. 2, AISC 160, pp. 197–203.
springerlink.com   © Springer-Verlag Berlin Heidelberg 2012

communication key technologies, structured a practical and Raptor code based short wave broadband OFDM communication system, which realized the efficient and trustworthy transmission of files with a high speed. The paper structure is as follows. Chapter 2 focuses on digital fountain code technology, Chapter 3 shows the framework of proposed system, Chapter 4 proposes the design and realizing schemes of system protocols, Chapter 5 tests the system under real condition and gives the results. And finally in Chapter 6, the correspondent conclusion is presented.

2   Raptor Code

Digital fountain code is a new type of erasure code technology. It can produce infinite output symbols, and flexibly control the code rate. Since each encoded symbol is independent and random, the fixed bit rate is not implemented. Fountain code can be called rateless code [4]. The main characteristic of fountain code , compared to the traditional RS code and LDPC code is that there is no definition of the code length in fountain code, that is, the code length of fountain code tends to be infinite[5]. Thus, in the practical application,, a unidirectional asynchronous transmission mechanism with low delay and high efficiency is used. It shows strong adaptability to the time-varying channel.

2.1   Encoding and Decoding Principle of LT Code

LT code is the first practical fountain code. The encoding and decoding method is simple and the decoding cost is low[6]. Encoding process of LT code is described as follows. Firstly, the initial data is equally divided into K packets, an integer d which ranges from one to k is chosen according to the coding degree distribution. Then d packets are randomly chosen from the original data. A coding packet is generated with XOR over the d packets. Repeat the two steps above to generate enough coding packets.

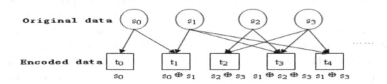

Fig. 1. Encoding Process of LT Code

It is shown in Fig. 1 that the encoding method of LT code is very intuitive. For the given input data, each encoded symbol for output is obtained through independent and random generation.

A kind of iterative decoding algorithm is used for LT Code. In each iteration , the decoder searches for the packets whose degree is 1. Then the packets which connect with the 1-degree packets could be decoded directly. Then, the XOR will be done between the decoded packets and their conected packets. Then, the connection relation between decoded results and the original encodeing packets will be deleted.

This process will be repeated until there's no 1-degree packets. It's a successful decoding progress if all the packets are restored, or the decoding failed. The algorithm is called the BP decoding algorithm of fountain code. The decoding process is shown in Fig. 2.

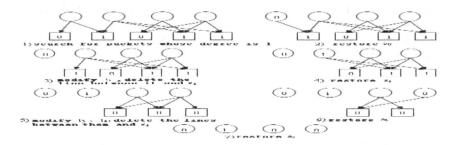

Fig. 2. Frame Format

2.2 Encoding and Decoding Principle of Raptor Code

Although LT code have excellent performance, the complexity of coding and decoding is nonlinear. In the decoding process, when the unit number of receiving coding information is close to the original input unit number, LT code can't decode with the cost of fixed time and space. Furthermore, only all the original input units are restored, it can be decoded successfully. In order to solve these problems, Shokrollahi puts forward the Raptor code technique [7]. By concatenated coding method, Raptor code efficiently keeps balance between the complexity of LT code and transmission efficiency.

In the encoded process for Raptor code, the original data is divided into k packets. Firstly, LDPC. is utilized for first encoding. Then weakened LT code is used for second encoding. Weakened LT code means that its generated code packets do not contain high-connection-degree packet and can't fully restore the original data. In the decoding process for Raptor code, with BP method to decode LT code, we only restore a certain proportion of the intermediate coding calibration unit. Then the rest packets can be decoded by traditional error correcting code.

Raptor code is suitable for time-varying channel and doesn't need feedback retransmission. This advantage overcomes the effects of time varying and fading shortwave channel with strong interference. So it substantially improves the efficiency of shortwave communication system and ensures the reliability of the transmission.

3   Framework of the Proposed System

Short-wave OFDM communication system based on Raptor code is a point to point communication system. PC and the radio are linked with serial communication interface of RS232 standard. Centralized control technology is used to control the system in order to make the transmitting and receiving integrated. The communication and control functions are required. It is a real-time process when voice communication works. The range of communication rate is between 2.0kbps to 20kbps.

3.1   System Framework

The proposed system is composed by PC and radio. UI, Master control module, protocol module, modulation module, demodulation module, radio control module, AD/DA conversion module and voice processing module are run on PC. The block diagram is presented in Fig. 3.

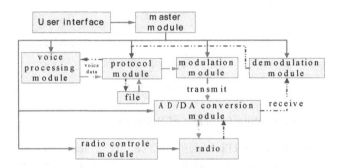

Fig. 3. System Framework

In protocol module, Raptor code is used to encode and decode the data in order to ensure data efficiency and reliability. In modulation module and demodulation module, OFDM techniques are applied to modulate and demodulate data in order to improve spectrum efficiency of the system.

3.2   Design of Module Functions

Master control module is responsible for transferring instructions from UI to each module and status and feedback of each module to the interface. Radio control module is used to control the radio status and send commands to the radio. Modulation and demodulation module is based on OFDM technology. In reference to many foreign literatures about short-wave radio system using OFDM technology, the parameters of the system are set as follows. [8] The number of sub-carriers is 64. Sub-carrier spacing frequency is 62.5Hz. Modulation type is QPSK. Channel coding method is convolution code.   In the frame structure, there are three OFDM symbols each frame and the symbol period is 20 ms. Guard interval is 4ms, and the scattered pilots are spread in both time and frequency directions. There is one pilot out of 12 sub-carriers in the frequency direction and one pilot out of 4 sub-carriers in the time direction [9]. When data transmission instructions are received, protocol module will be responsible for packing, sealing frame and encoding the data to be transmitted with Raptor code. Then protocol module sends the code to the modulation module. When data received, the demodulation module will send the receiving data to protocol module. Protocol module will decode the receive data with Raptor code. If received data is real-time voice data, the protocol module sends the data to the voice processing module. If received data is a file, protocol module saves the file to the current path.

4   Protocol Design of the Proposed System

A uniform scheme of protocol design is employed in the system. It contains packet header, instruction/data and parity bit. Each sub-protocol is used according to different functions. Packet header contains the information on ID of transmitters and receivers, as well as the length of instruction/data. The controlling and state instruction and packaged data which need to be transmitted will be checked by CRC to ensure the accuracy.

4.1   File Transmission Protocol

The transmitted files are coded with Raptor code. Due to the linear relationship between the complexity and the code size of Raptor code, the costing time correlates with the code size linearly. If the number of data is so huge, the transmission efficiency will be affected. Thus, the original files need to be divided into several data blocks and coded for each block. The coded data is packaged and formed as data frame. Each frame is composed of several packets.

File transmission is performed as unidirectional mode and bidirectional mode. For unidirectional transmission, the receivers do not send the feedback information. The Raptor code is utilized in the transmitters in order to keep the data received accurately. For bidirectional mode, a certain number of data is transmitted and the transmitter waits for the feedback. Then, the receiver modifies the frame length, bit-rate and the transmission time. If the feedback is lost, it is proven that the channel is quit noisy and it needs to re-select the transmission frequency.

4.2   Control Protocol

Transmitter or receiver will control the modules of this station in order to perform well in efficiency and stability. Control protocol contains control and state instructions. The control instructions are employed to control all modules in order to make sure all modules run orderly. The master control module control the state of all modules. When transmitting file, the master control module sends the information of file to be transmitted to protocol module. Next, the protocol module controls relevant module according to the type of the file to be transmitted: if it is real-time voice data, the protocol module will send control instructions to processing module to instruct it to process voice data; if it is file data, the protocol module will send control instructions to modulation module, DA module and audio control module to process file data. When receiving file, the protocol sends control instructions to demodulation and AD module. All modules should respectively send state constructions to master control module to tell it their running state.

5 Experiment and Evaluation

In this paper, Bandwidth for the proposed system is 10KHz. System has been setup with a short wave radio whose transmit power is 1kw. The system has run under the real condition. It is used to transmit and receive data. The transmit rate is higher than 2.4kbps during testing. The result is given as Fig. 4.

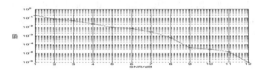

Fig. 4. SNR-BER Curve

The BER of traditional short wave system is required to be lower than 10-4 when the transmit bit-rate of the system is 2.4kbps and the SNR is 9dB. As figure4 indicts, the proposed system has reached the standard when SNR is 8.5dB. In addition, the BER of proposed system has is 10-5 when the transmit bit-rate 9.6kbps. The results demonstrate that the proposed system has the better performance on error resilience and the transmitted bit-rate is also improved. It indict that the proposed scheme of shortwave broadband OFDM system based on Raptor code is feasible.

6 Conclusions

A design scheme for shortwave broadband OFDM system based on Raptor code is proposed in the paper. Raptor code and OFDM are applied to improve the quality of short wave channel which is affected by time-varying, multi-path effect, fading and Doppler shift. It has enhanced the noise immunity and has improved the spectrum efficiency of the system. In addition, a protocol scheme which is suitable for short wave communication has been proposed in the paper. The testing result shows that the proposed system performs better than traditional system under real condition. The traditional channel estimation utilized in proposed system may be replaced by an advanced algorithm in order to improve the system performance.

References

1. Ma, M., Yang, J.: Orthogonal Frequency Division Multiplexing (OFDM) Technology and Its Application in HF Communication, pp. 36–39 (October 2006)
2. van Nee, R., Prasad, R.: OFDM for Wireless Multimedia Communications, pp. 1–5. Artch House, Boston (2002)
3. Watterson, C.C., Juroshek, J.R., Bensema, W.D.: Experimental confirmation of an HF channel model. IEEE Transaction on Communication Technology 18(6), 792–803 (1970)
4. Li, L., Wu, Z., Wang, W.: Codec Research and,Analysis on Fountain Codes. China New Telecommunications, 41–46 (July 2010)

5. Li, H.-L.: Research on Coding and Decoding of Fountain codes, pp. 11–12. Master degree thesis of Harbin Institute of Technology, Harbin (2009)
6. Yao, W.-D., Li, H., Chen, L.-J., Xu, H.-G.: Study on the fountain codes technology in deep space communications. Systems Engineering and Electronics 31(1), 40–44 (2009)
7. Shokrollahi, A.: Raptor codes. IEEE Transactions on Information Theory 52(6), 2551–2567 (2006)
8. Li, Y., Huang, H.-G.: Research on OFDM-based Short-wave Radio System. Communications Technology, 23–25 (September 2009)
9. ETSI, "Digital Video Broadcasting (DVB): Framing structure, channel coding and modulation for digital terrestrial television," ETSI EN300744 V1.5.1 (2004)

A Weighted Fuzzy Clustering Algorithm Based on Density

Cuixia Li and Yingjun Tan

[1] School of Software, Zhengzhou University, Zhengzhou, 450002, China
[2] Department of Information Engineering, Henan PolyTechnic College,
Zhengzhou, 450046, China
qyliying@126.com

Abstract. The selection of the initial points influences greatly on the results of traditional partition clustering algorithms. If the algorithm starts with improper points, it will return a local minimal value. A novel weighted Fuzzy C-means algorithm (shorted by DFCM) was proposed to overcome the shortcoming. Its accuracy and effect are improved through the calculation of the relative density differences attributes, using the results of the center to determine the initial method for clustering. The numerical experiments proved that the DFCM algorithm not only can obtain better results steadily but also can distinguish the importance of each attributes.

Keywords: clustering, Fuzzy C-means clustering, attribute, weight, density, misclassification number.

1 Introduction

Clustering analysis is a kind of semi-supervised learning method. Its purpose is to partition a set of objects into clusters such that the objects in the same cluster are more similar than the others in the different clusters according to some criteria. Clustering has been used more and more widely in many fields, such as machine learning, pattern recognition, information retrieval, image recognition and market management. Fuzzy C-means clustering algorithm is one of the most traditional methods. It can discover a partition minimizing the objective function by starting with random initial centers. So the algorithm is influenced greatly by the initial centers. If choosing an improper initial center, the clustering algorithm can converge to a local minimum point. So it is necessary that the algorithm should began with the better results of initial centers [1]. When the attributes of objects to be clustered is so consistent and contributing equally to the cluster results, the precision of clustering will be influenced. So, how to clustering the objects with consistent attributes is another difficult problem.

Variable selection and weighting have been important research topics in cluster analysis[2,3]. A weighted fuzzy c-means clustering algorithm based on density was proposed to solve the above problem. The algorithm firstly computed the data's correlation density to optimize the initial cluster centers. Then with the aid of attributes

D. Jin and S. Lin (Eds.): Advances in FCCS, Vol. 2, AISC 160 pp. 205–211.
springerlink.com   © Springer-Verlag Berlin Heidelberg 2012

weighted which can distinguish the improtance of ecah attribute, the method improved the precision.

2   The Traditional Clustering Methods

The key step of partitional clustering algorithm is to construct the initial clusters so that the objects in the same cluster are more similar than the objects in different clusters. These algorithm maps a function by related parameters of the objects' initial clustering results. Then the category will be obtained by comparing the function's   value with the given criteria.

Suppose that $X = \{x_1, x_2, ... x_n\}$ is an s-dimensional dataset with n points, and is divided into c clusters $V = \{v_1, v_2, ... v_c\}$, each cluster can be represented by its cluster center v_i. By fuzzy set theory, we use membership to denote the relationship between clusters and objects. Then $u_{ik} \in (0,1)$, subject to $\sum_{i=1}^{c} u_{ik} = 1$. Let U be a partition matrix, $1 < m < +\infty$ and $2 \leq c < n$, then the objective function of the FCM is $J(u,v) = \sum_{k=1}^{n} \sum_{i=1}^{c} u_{ik}^m \|x_k - v_i\|^2$ .The iterative process of FCM is similar with that of C-means. The only difference is to compute of partition matrix and cluster center. The iterative formula   of u_{ik} is $u_{ik} = \left(\|x_k - v_i\|^2 \right)^{1/(1-m)} \Big/ \sum_{j=1}^{c} \left(\|x_k - v_j\|^2 \right)^{1/(1-m)}$ and the one of v_i is $v_i = \sum_{k=1}^{n} u_{ik}^m x_k \Big/ \sum_{k=1}^{n} u_{ik}^m$.

3   Weighted Fuzzy C-Means Clustering Algorithm Based on Density

3.1   Choosing Initial Cluster Centers Based on Density

On account of the influence of the initial cluster centers, the better centers should have better representation, make the density of this cluster higher and be closed to the cluster's center. In order to describe the choosing method of initial center, we computed the center based on the attributes' density. The computing method is as follows:

For the data point x in data set and reference similarity $Rsim$, the neighborhood of x is the field with x as center and $Rsim$ as radius. The number of the data in neighborhood is the density of x based $Rsim$. It can be marked as $Den(x, Rsim)$ [4].

For the data point x , similarity $Rsim$ and Threshold $MinD$, if $Den(x, Rsim) \geq MinD$, x is core data, $MinD$ is density threshold. If x is in the neighborhood of y and y is core data, x can be considered direct density reachable from y. If $x_1 = x$, $x_n = y$, and x_i can direct density reach from x_{i+1}, x can direct density reach from y. If exist r making both x and y can direct density reach from r, x and y are density connection。

Adjacent core data: If core data x satisfy $sim(x, y) \geq 2Rsim$ with given $Rsim$ and $MinD$, x, y are adjacent core data. The density character of x,y can be defined:

$$E_x = Rsim_x - \frac{MinD}{2}, \quad E_y = -(Rsim_y - \frac{MinD}{2}) \tag{1}$$

$$E(x, y) = (\frac{2E_x}{Rsim}, \frac{2E_y}{MinD}) = (\frac{2P_x}{Rsim} - 1, 1 - \frac{2P_y}{MinD})$$

After computing $E(x,y)$, the initial cluster's center can be obtained to overcome the shortcoming of classical clustering algorithm.

If the input data set $X = \{x_1, x_2, \ldots x_n\}$, reference similarity $Rsim$, threshold $MinD$ and the number of clusters $K(K$ is a positive integer and $K \leq n)$ have been given, the process can be described as follows:

For any x in given data set X, we computed the neighborhood $N_{\min D}(x)$ according to parameters; If Z ($Z \in X$), the neighborhood $N_{\min D}(z)$ is 1 which shows that the neighborhood of its only includes itself. It is to say that the similarities between z and other objects are all less than R_{sim}, then put the density value into the reference points set CS; If L ($L \in X$) $N_{\min D}(L) \geq MinD$, then put the mean of the objects' density in the neighborhood into the reference points set CS (S_1, S_2, $\ldots S_j$), where $J < n$. If $j > K$, we put the mean of the maximum s_i and minimum s_j into the reference points set CS, delete s_i and s_j until the number of the reference points set is K. If $J<K$, we input the density again and return to the first step. If the point in the reference set is not the original data in sample set, a point in original set which is the most similar one will substitute it[5].

3.2   Data Clustering Based on the Fuzzy Weighted

After obtaining the initial clusters, we can cluster the data set as follows:

Supposed that $X = \{X_1, X_2, \ldots, X_n\}$ is a set including n objects. The object $X_k = \{x_{k1}, x_{k2}, \ldots, x_{ks}\}$ includes s attributes. We also suppose that $W = \{w_1, w_2, \ldots, w_s\}$ is the weights set of s attributes, β is the parameter of attribute's weight w_j, U is a partition matrix, V is a cluster centers set including k centers. The purpose of the algorithm is to partition X into c clusters; minimize the objective function as formula (3):

$$J(u,v,w) = \sum_{i=1}^{c}\sum_{k=1}^{n}\sum_{j=1}^{s} u_{ik}^{m} w_{j}^{\beta}\left(x_{kj} - v_{ij}\right)^{2} \qquad (2)$$

Where $\sum_{j=1}^{s} w_{j} = 1, \sum_{i=1}^{c} u_{ik} = 1, \quad 1 \le i \le c, 1 \le j \le s$.

By Lagrange multiplier's approach, we can obtain the necessary conditions for the minimum of $J(u,v,w)$ as follows:

$$v_{ij} = \sum_{t=1}^{n} u_{ik}^{m} x_{kj} \Big/ \sum_{t=1}^{n} u_{ik}^{m} \qquad (3)$$

$$u_{ik} = \left(\sum_{j=1}^{s} w_{j}^{\beta} \left\| x_{kj} - v_{ij}\right\|^{2}\right)^{1/(1-m)} \times \left(\sum_{t=1}^{c}\left(\sum_{j=1}^{s} w_{j}^{\beta}\left\|x_{kj} - v_{ij}\right\|^{2}\right)^{1/(1-m)}\right)^{-1} \qquad (4)$$

$$w_{j} = D_{j}^{1/(1-\beta)}\left(\sum_{t=1}^{m} D_{t}^{1/(1-\beta)}\right)^{-1}, \text{ where } D_{j} = \sum_{i=1}^{c}\sum_{k=1}^{n} u_{ik}^{m}\left(x_{kj} - v_{ij}\right)^{2} \qquad (5)$$

Consequently, the procedure of the DFCM can be describes as follows:

Step 1: Fix the number of clusters c, the weighting exponent m, the iteration limit *Tcount*, the attribute's weight β and the tolerance ε, initialize the partition matrix U and attribute weight W;

Step 2: Update the cluster center v_{ij} by (3), if convergence criterion is not met, goes to step 3 else stop;

Step 3: Update the membership function u_{ik} by (4), if convergence criterion is not met, go to step 4 else stop;

Step 4: Update the attribute weight W by (5), if convergence criterion is not met, goes to step 2 else stop;

Typical convergence criteria are: the value of J between two successive iterations is less than ε, or arrive to maximum iterations *Tcount*. It is easy to prove that given m and β, the above algorithm(shorted for DFCM) converges to a local minimal solution or a saddle point in a finite number of iterations[6].

4   Experiment Results

In this section, experiment results are used to check the clustering performance. We analyzed the Hessian matrix of (2) which is similar to the methods in [6,7] in order to obtain the valid parameters(m and β).

Experiment 1: The data set used in this experiment called Data is a synthetic one and includes 300 points. The objects in Data have 4 attributes. The attributes x_1, x_3 follows a normal distribution and can be divided into 3 clusters. x_2 and x_4 are random attributes following uniform distribution. Figure 1 plots the points in different two-dimensional subspaces.

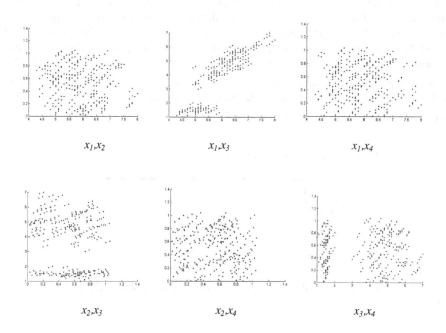

x_1,x_2 x_1,x_3 x_1,x_4

x_2,x_3 x_2,x_4 x_3,x_4

Fig. 1. The Points in Data

In order to finding the better *MinD* and *Rsim*, we can get *MinD*=100 and *Rsim*=0.2 after implementing the algorithms many times. After analyzing the Hessian matrix of objective function, m should be 2.5. So we implement the algorithm many times with m=2.5, the figure 2 shows the weights of different attributes when β ranges from 2 to 10.

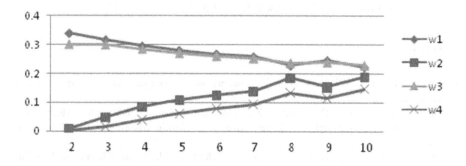

Fig. 2. Attributes' weights

Figure 2 shows that x_1, x_3 contribute more to the cluster results than x_2 and x_4, which is consistent with the real structure.

Experiment 2: A real data set-IRIS is used in this experiment. Maximal mis-classification number, minimal misclassification number and average misclassification number are used to compare the effects of different algorithms. According to some literatures[7,8], FCM algorithm can have a better results when $m=2$. We implemented the FCM algorithm 100 times with $m=2$. WFCM (Attribute-weighted Fuzzy C-means algorithm) is also implemented 100 times with effective parameters $m=2$ and $\beta=2.5$ starting with random initial cluster centers. At the same time, we implement DFCM with $m=2$ and $\beta=2.5$ after choosing the better initial cluster centers. The effects of the above three algorithms are shown in table 1.

Table 1. The average iterative times of FCM, WFCM and DFCM

Algorithms	Data	Iris
FCM	27	21
WFCM	19	14
DFCM	14	10

As shown in table1, the algorithm in this paper is more stable and effective than the others.

5   Conclusions

A new clustering algorithm is proposed based on the density. The initial cluster center choosing methods not only consider the dispersion but also the representation of the centers. After choosing the better initial centers, a weighted fuzzy c-means algorithm which can recognize the different importance of each attribute improve the efficiency and the stability of clustering. As we know, the algorithm are influenced by parameters Rsim and MinD, we only give the choosing methods by experiments. If a theoretical parameters' formula can be given, the effects of this algorithm will be enhanced.

References

1. Han, I., Kamber, M.: Data Mining: Concepts and Techniques, pp. 335–389. Morgan Kaufmann Publishers, Berlin (2000)
2. Huang, J.Z., Ng, M.K., Rong, H., Li, Z.: Automated Variable Weighting in k-Means Type Clustering. IEEE Transactions on Pattern Analysis and Machine Intelligence 27(5), 657–668 (2005)
3. Modha, D.S., Spangler, W.S.: Feature Weighting in k-Means Clustering. Machine Learning 52, 217–237 (2003)

4. Zhang, W., Wu, J., Yuan, X.: K-means text clustering algorithm based on density and nearest neighbor. Journal of Computer Applications 30(7), 1933–1935 (2010)
5. Guan, Y.: Application of Improved K-means Algorithm in Telecom Customer Segmentation. Computer Simulation 28(08), 138–140, 152 (2011)
6. Yu, J.: On the Fuzziness Index of the FCM Algorithms. Chinese Journal of Computers 26(8), 974–981 (2003)
7. Yu, J., Cheng, Q., Huang, H.: Analysis of the weighting exponent in the FCM. IEEE Transactions on Systems, Man and Cybernetics-part B: Cybernetics 34(1), 634–639 (2004)

A Study of the Relation between Financial Development, Urban Facilities and Urban Household Consumption

Kunming Li

Department of Statistics, School of Economics,
Xiamen University, Xiamen 361005, P.R. of China
li968ming@yahoo.com.cn

Abstract. This paper analyzes the relation between financial development, urban facilities and China's urban household consumption using dynamic panel model. The empirical results show that the expansion of the size and efficiency of financial institutions has a delayed impact on household consumption, but the effect of urban facilities is not clear. Except coverage rate of tap water, coverage rate of gas and per capita public green area, the effect of other 3 variables are not significant. Besides, we also find that education and medical expenditure and house price have significant effect on household consumption, but don't find powerful evidence of the effect of macroeconomic and demographic factors.

Keywords: Financial Development, Urban Facilities, Consumption, Dynamic Panel Model.

1   Introduction

In the traditional theory of consumption economics, there are three factors affecting consumer behavior: economic factors, environmental factors and own factors. In fact, especially when the infrastructure is not perfect, consumption convenient condition often becomes an important factor affecting household consumption. Analyzing the relationship between urban residents' consumption and the level of financial development and urban facilities can not only help to improve the consumption theory, but also provide a reference to the development of China's domestic economy. So it has obvious theoretical and practical significance.

The literature on financial development focused on the relationship between financial development and economic growth. There is a representative study from King and Levine (1993a).They constructed a indicator system to measure financial development, and found that financial development is an important prior variable of economic growth. King and Levine (1993b) and Levine (1999) constructed a endogenous growth model to study the relationship between financial relationship and economic growth and found that the development of financial intermediation accelerated economic growth from the perspective of industrial development. The studies from Han(2001), Shi et al (2003), Chen et al (2008), Fan(2011) showed that

D. Jin and S. Lin (Eds.): Advances in FCCS, Vol. 2, AISC 160, pp. 213–218.
springerlink.com   © Springer-Verlag Berlin Heidelberg 2012

the correlation between financial development and economic growth was significant positive. The studies from Liu et al (2010a, 2010b), Zhang et al (2010) and Liu et al (2011) showed that household income was positively correlated with urban facilities and economic growth.

2   Research Design

2.1   Model Specification

General empirical equation of consumption theory is:

$$c_t = \alpha_0 + \alpha_1 c_{t-1} + \alpha_2 y_t + \varepsilon_t \tag{1}$$

Where c_t and y_t are the consumption and income at t period respectively, ε_t is the random disturbance.

To take into account the effect of financial development and urban facilities on consumption, we added some variables referring to financial development and urban facilities and other control variables in equation (1) .Then the final model is:

$$c_t = \alpha_0 + \alpha_1 c_{t-1} + \alpha_2 y_t + \beta X_t + \gamma Z_t + \varepsilon_t \tag{2}$$

Where X_t refers to the core variables representing financial development and urban facilities, Z_t is the control variable. The meanings of each variable can be seen in Table 1.

2.2   Estimation Method

Our target is to estimate model (2), but there is a lagged dependent variable in (2), and the GDP growth rate and medical and education expenditure levels may be the endogenous variable, thus, if we use general panel model to estimate model (2), there will exists endogeneity problem which will result in estimation bias. In order to solve the endogeneity problem we use the instrumental variable method to estimate, in which we take the lagged values of the endogenous independent variables as instrumental variables, In detail, we use GMM method proposed by Arellano and Bond (1991, 1995) to eliminate endogeneity problems.

There may exist the lagged effect of financial development on consumption, thus we will take the first order and second order lagged values of the two indicators of financial development into the model estimating, besides, house prices may be an important indicator for residents to determine the degree of future expenditure uncertainty and residents always judge house prices based on historical data, so we also take the first order and second order lagged values of house prices into the model. In addition, in the selected variables, tap water and natural gas is a major basic livelihood project, and investment of government departments is generally not affected by other factors, thus water penetration and gas penetration can be used as strictly exogenous variables. Taking into account that much more instrumental variable will lead to overfitting of endogenous variables easily, so we only select more than 2 order lagged values of consumption as instrumental variables.

In order to compare and judge the effects of GMM estimation conveniently, we also use the maximum likelihood estimation method to estimate model (2). And we take logarithm of consumption, disposable income and house price to avoid heteroscedasticity.

Table 1. Variables and their symbols

Symbol	Variable	Symbol	Variable	Symbol	Variable
con	household consumption	traf	the number of buses every ten thousand people own	house	real estate sales / sales area
inc	disposable income	road	per capita road area	gro	GDP growth
sca	financial institutions credit / GDP	green	per capita green area	Inf	inflation (CPI growth)
eff	private sector credit / total credit	toil	the number of toilets every ten thousand people own	child	children dependency ratio
wat	water penetration	tran	transfer income / disposable income	old	old-ager dependency ratio
pet	natural gas penetration	edu	medical and education expenditure / disposable income		

3   Empirical Results and Analysis

The maximum likelihood estimation and GMM estimation results of model (2) can be seen in table 2, from which we can obtain the following conclusions:

(1) There is strong consumption inertia. The coefficient of L1.lncon is 0.2462918 and is very significant. This shows that the consumption of urban residents is affected by consumption habits of the past.

(2) The impact of financial development on household consumption will change with time going. The effect of the size of financial institutions on household consumption is negative but not significant. The coefficients of L1.sca and L2.sca are both significant. This shows that the impact of financial development on household consumption is lagging, and China's financial efficiency has no significant current effect on household consumption, but the lagged impact is negative.

(3) The level of urban facilities has a rather vague impact on household consumption. Only the coefficient of water penetration (wat), natural gas penetration (per) and per capita green area (green) is significant.

(4) Medical and education expenditure and house prices as factors affecting consumption growth cannot be ignored. The coefficient of proportion of Medical and education expenditure is significantly 0.015441, and the coefficient of two lagged value of average house price is significantly 0.0694052.

Table 2. Estimated results

Variable	Parameters Estimation	
	MLE	Two-step GMM
L1.lncon	0.2930399***	0.2462918***
lninc	0.7220197***	0.6444396***
sca	0.0000233	-0.0004494
L1.sca	0.0003228	0.0009819*
L1.lncon	0.2930399***	0.2462918***
lninc	0.7220197***	0.6444396***
sca	0.0000233	-0.0004494
L1.sca	0.0003228	0.0009819*
L2.sca	-0.0005357***	-0.0013233**
eff	0.0014185**	0.0001417
L1. eff	-0.0027173***	-0.0053677**
L2. eff	0.0016197**	0.004821
wat	0.0011276***	0.0026899***
per	-0.0010459***	-0.0019895*
traf	0.0031054***	-0.0001801
road	-0.0017943*	0.0049939
green	-0.002126*	-0.0124094*
toil	-0.0005701	-0.0003801
tran	-0.0027738***	0.0016157
edu	0.0181867***	0.015441*
lnhouse	0.0255585	0.0134658
L1. lnhouse	-0.0044987	0.0080947
L2. lnhouse	-0.0247795*	0.0694052**
grow	-0.0003696	0.0005859
inf	-0.0018911**	-0.003173
child	0.0012421**	-0.0027621
old	0.0015765	-0.0011985
cons	-0.4770457***	
P value of Sargan test		0.552
P value of Hansen test		1.000
P value of AR (1) test		0.007
P value of AR (2) test		0.329

Note:(1) ***,** and * denoted that it passed the test at the significance level of 1%, 5% and 10% respectively.(2) The null hypothesis of sargan test and hansen test is equivalent to the effectiveness of instrumental variables, thus the larger P value of the test is, the more effective the instrumental variables are. (3) There were not 1-order and 2-order autocorrelation in the null hypothesis of AR (1) test and AR (2) test respectively, but the consistency of GMM estimates required that there must be 1-order autocorrelation and second order which is not relevant in the residual series.

(5) The impact of macroeconomic factors and demographic factors on household consumption was not significant.

4   Conclusion and Discussion

In this paper, we used the dynamic panel model to analyze the relationship between consumption of urban residents, financial development and urban facilities. Empirical results showed that the impact of the level of financial development on Chinese household consumption will be reversed over time, and the expansion of the size of financial institutions and the improvement of financial efficiency have a lagged effect on household consumption. However, the level of urban facilities has a rather vague impact on household consumption. The coefficients of other three indicators are not significant except water penetration, gas penetration and per capita green area. In addition, we also found that medical and education expenditure and house prices as factors affecting consumption growth cannot be ignored. However, there is no evidence that macro-economic factors and demographic factors affect household consumption.

The empirical results indicate that, there exists no sustainable effect of financial development on the household consumption, and rough-based development model, low level of financial management services, imperfect financial regulatory system and asymmetric market information environment eventually result in this unsustainability. The result above tells us that as to improve China's economic growth and residents' living standards, the administration must face up to the importance of financial development on economic development and people's living and then change the "bigger and stronger" monotonous development model and pay more attention to efficiency, service, and modern management methods to make financial institutions play its special role fully.

The empirical results also show that, the level of urban facilities, especially water and road transport facilities plays an essential role in improving the household consumption. Thus, the relevant government departments need to continue to attach importance to the investment of urban infrastructure, especially in some cities where facilities is relatively backward, since a good urban environment and the complete infrastructure level is necessary to induce the development.

References

1. Chen, W., Zhang, H.: An Empirical Analysis of Financial Development and Economic Growth in China. Modern Economic Science (3), 49–57 (2008) (in Chinese)
2. Liu, S., Hu, A.: Test on the Externality of Infrastructure in China: 1988-2007. Economic Research Journal (3), 4–15 (2010) (in Chinese)
3. Liu, S., Hu, A.: Transport Infrastructure and Economic Growth: Perspective from China's Regional Disparities. China Industrial Economics (4), 14–23 (2010) (in Chinese)
4. Liu, S., Hu, A.: Availability of Infrastructure and Income Growth in Rural China—Based on Static and Dynamic Unbalanced Panel Regression Results. Chinese Rural Economy (1), 27–37

5. Han, T.: Financial Development and Economic Growth: Empirical Models and Policy Analysis. World Economy (6), 3–9 (2001) (in Chinese)
6. Shi, Y., Wu, Z., Zhen, H.: An Empirical Analysis of Financial Sector Development and Economic Growth in China. Forecasting (4), 1–5 (2003) (in Chinese)
7. Zhang, G., Li, X., Chen, G.: The Effects of Public Infrastructure on Employment, Output and Investment. Management World (4), 5–15 (2010) (in Chinese)
8. Arellano, M., Bond, S.: Some tests of specification for panel data: Monte Carlo evidence and an application to employment equations. The Review of Economic Studies 58(2), 277–297 (1991)
9. Arellano, M., Bover, O.: Another look at the instrumental variable estimation of error components models. Journal of Econometrics 68, 29–52 (1995)
10. King, R.G., Levine, R.: Finance, entrepreneurship and growth: theory and evidence. Journal of Monetary Economics 32, 513–542 (1993)
11. King, R.G., Levine, R.: Finance and growth: Schumpeter might be right. The Quarterly Journal of Economics 108, 717–738 (1993)
12. Levine, R.: Law, Finance and economic growth. Journal of Financial Intermediation 8, 8–35 (1999)

The SoC Reconfiguration with Single MPU Core

Liangjun Xiao and Limin Liu

Institute of Embedded Systems, IT School, Huzhou University,
Huzhou, Zhejiang, 313000, China
sjtzzf@hutc.zj.cn, liulimin@ieee.org

Abstract. As the advanced hardware, SoC, System on a Chip, is a research hot point of embedded systems. The reconfigurable SoC, RSoC is a useful solution for some embedded applications with high reliability requirement. In this paper, the organization and design of a RSoC are discussed. The organization is focus of single MCU cores. The design is based on PLD, programmable logic device and hardware description language.

Keywords: SoC, reconfiguration design, embedded systems, PLD.

1   Introduction

As the advanced hardware, SoC, System on a Chip, is an ideal device for embedded system applications [1]. A SoC may integrate some application system on one semiconductor chip. When a system is integrated on a chip, but not one or more PCB boards, the system should be with smaller size, higher reliability, faster operation and lower cost for most applications. Actually, hardware to replace some software is a useful feature for some embedded solutions [2]. For some specific applications, the replacement would obtain better operating efficiency. It is the reason for SoC to become a hot point in embedded designs [3].

Normally a SoC design is concerned to IC design and microelectronics. Its applications are based on hardware and software. For the organization, some functions or parameters of SoC can be modified. The SoC is called reconfigurable SoC, RSoC [4]. With other words, it is the SoC reconfiguration.

In order to guarantee reliable operation of systems, dynamically updating concept for SoC organization or functions can be employed. A new kind of SoC, therefore, Dynamically Reconfigurable System On a Chip, DRSoC, is issued. When the dynamic reconfiguration is to repair some components or functions in SoC, the reconfiguration is a typical self-repairing procedure. There are different ways, such as single MPU or multiple MPU core SoC, to approach DRSoC.

A reconfiguration, especially dynamic reconfiguration, is an important feature for most SoC, System on a Chip, applications. It is useful and helpful for some applications with high reliability demand, such as military, aerospace and others. Therefore, it is a significant direct research of embedded system development, especially for some advanced fields.

D. Jin and S. Lin (Eds.): Advances in FCCS, Vol. 2, AISC 160, pp. 219–222.
springerlink.com   © Springer-Verlag Berlin Heidelberg 2012

2   The Organization of Reconfigurable SoC

With reconfiguration, a SoC can supply some solution for self-repairing systems. And the repairing procedure is not influenced the system operation, that means the system is not turned off [5].

There are various ways to implement SoC dynamic reconfiguration. Here, a method based on SoC with single MPU, microprocessor unit, core is discussed.

An organization of SoC with single MPU core is indicated in Fig. 1. For this SoC, MPU with its instruction set and memory cannot be changed after a design. The reconfiguration may be just concerned to IP, Intellectual Property, modules.

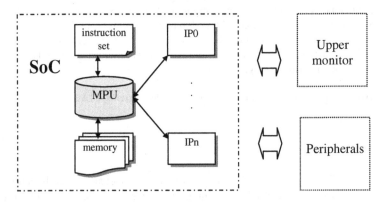

Fig. 1. An organization of SoC with single MPU core

There are some conditions for SoC dynamic reconfiguration. Firstly, the hardware of the SoC is a PLD, programmable logic device, such as CPLD, complex programmable logic device, or FPGA, field programmable gate array. In fact, the SoC is a PSoC, Programmable SoC. For example, users want to add more IPs into the SoC. The relevant hardware resource should be supplied. The PLD is the better choice. Secondly, the SoC has IAP, in-application programming, function. That means the monitor to program some memory of the SoC through serial port in operation. Then there is enough space in the memory.

Actually, the design is a typical cooperative design of hardware and software [6]. PLD is some hardware device. A SoC based on PLD is hardware modeled by software, such as VHDL codes. On the other hand, the operations of the SoC are managed by some application software.

When some peripherals are updated, the control requirement is different for the SoC. The relevant IP module in the SoC may be modified for the updating. The dynamic reconfiguration can supply the better solution. For example, a LCD controller IP is a module programmed with VHDL language, a hardware description language, on a PLD. Normally the LCD is controlled to display information. A new requirement is to display information with backlight in the night. For the dynamic reconfiguration, the monitor programs a new LCD control module on the SoC chip, then switch the new IP into operation. At same time, the original LCD IP is withdrawn. The SoC operates with the new LCD control IP. And the LCD works under new demand.

3   PLD Design of the RSoC

A Reconfigurable SoC, RSoC, is based on PLD devices mainly. A SoC application is a special design for some object. Its requirement about flexibility is not so high. Since it is built on a PLD chip, its hardware cost is the more fixable in some level. Therefore, the disadvantage of hardware is not very obvious for a SoC. For most SoC applications, more hardware functions mean more operation efficiency, and the cost is similar, the flexibility is acceptable. If possible, it is better for SoC hardware to replace software in most functions. However, for a SoC system, the application software is essential yet. Some hardware of PLD can be instead of low level software, but not all. Therefore, cooperative design of hardware and software is a better choice for most SoCs.

Since RSoC is the newer concept, and its design and application are modified frequently. Its design is based on some PLD devices with VHDL language [7-8]. Actually, a RSoC solution is complex. Therefore, the RSoC modeling should be behavior modeling or hybrid modeling, but not structural modeling in VHDL development. It is because that the structural modeling is depended on lower-level components. If without suitable components, the result of structural modeling is not satisfied.

When a SoC is built by FPGA, the updating or reconfigurable procedures are easier. It is because of FPGA is a kind of PLD device to be easily modified. The new FPGA products supply enough resource for RSoC design.

The frame of a RSoC based on VHDL is as follows.

```
Library
Entity RSoC_syst is
Port( );
End SRSoC_syst;
Architecture strt of RSoC_syst is
          Signal
          Constant
Begin
Res: process;
Clk :clk_div;
Handle: process;
  Begin
    Initialization;
    Information Process;
    Efficiency Analysis;
    Reconfiguration;
End process;
End strt;
```

There are some analysis and decision algorithms in memory of a SoC. They are software to be run in MPU core. According to the algorithms, the MPU determines reconfiguration of backup control codes and switches them into operation.

Therefore, a solution of the RSoC with single MPU core must be a cooperation of software and hardware designs. The scheme of dynamic reconfiguration for SoC is easier to be designed. But its updating capability is limited. It is because of that the MPU and instruction cannot be modified. Therefore, this method may not improve more performance for a SoC system. The further work is to increase the range of dynamic reconfiguration of a SoC, for example, to update instruction set of the SoC. If one instruction can replace several instructions, the efficiency of the SoC operation may be improved more.

4   Conclusions

As the advanced hardware, SoC, System on a Chip, is an ideal device for embedded system applications. A RSoC is composed of single MPU core and some IP modules. When the SoC is required updated or repaired, the monitor module control backup model to be modified. The design is based on PLD with VHDL. It is useful and helpful to some embedded applications with high reliability requirement.

Acknowledgments. This research was supported in part by the National Natural Science Foundation of China under grant 60872057, by Zhejiang Provincial Natural Science Foundation of China under grants R1090244, Y1101237, Y1110944 and Y1100095. We are grateful to NSFC, ZJNSF and Huzhou University.

References

1. Claasen, T.M.: An industry perspective on current and future state of the art in system-on-chip (SoC) technology. Proceedings of the IEEE 94, 1121–1137 (2006)
2. Sifakis, J.: Embedded systems design - Scientific challenges and work directions. In: Proceedings of DATE 2009, p. 2. IEEE Press, Nice (2009)
3. Liu, L., Luo, X.: The Reconfigurable IP Modules and Design. In: Proc. of EMEIT 2011, Harbin, China, pp. 1324–1327 (2011)
4. Saleh, R., Wilton, S., et al.: System-on-chip: reuse and integration. Proceedings of the IEEE 94, 1050–1069 (2006)
5. Ostua, E., Viejo, J., et al.: Digital Data Processing Peripheral Design for an Embedded Application based on the Microblaze Soft Core. In: Proc. 4th South. Conf. Programmable Logic, pp. 197–200. San Carlos de Bariloche, Argentina (2008)
6. Liu, L.: A Hardware and Software Cooperative Design of SoC IP. In: Proc. of CCIE 2010, Wuhan, China, pp. 77–80 (2010)
7. Kilts, S.: Advanced FPGA Design: Architecture, Implementation, and Optimization. Wiley, New Jersey (2007)
8. Dimond, R.G., Mencer, O., Luk, W.: Combining Instruction Coding and Scheduling to Optimize Energy in System-on-FPGA. In: Proceedings of 14th Annual IEEE Symposium on Field-Programmable Custom Computing Machines, Napa, CA, USA, pp. 175–184 (2006)

Simulation Model of Spares Demanded and Shared Based on Different Lifetime Distributions

Jing Yang, Fang Li, and Peng Di

Dept. of Management Sci., Naval Univ., of Engineering,
Wuhan 430033, China

Abstract. In term of spares demanded and shared with different lifetime distributions, using the simulation method, simulation model of spares demanded and shared are build. Simulation model of spares demanded solves the problem of demand quantity of spare parts with non-exponent distribution of lifetime distribution, comparing with the tradition mathematics analytical model more simple; simulation model of spares shared solves the problem of demand quantity of spares shared with different lifetime distributions. At the same time the model solves the problem of demand quantity of spares of serial system with different types of parts. The work expands the range of calculating demand quantity of spares. Finally, the results of computer simulation and analytical method validate its effectiveness and correctness.

Keywords: lifetime distribution, spare parts shared, simulation.

1   Introduction

Scientific and proper ration of spares has been a significant problem and it is especially important to properly determine ration of spares and select the correct model to calculate needed amount of spares. Configuring the number of spare parts not only affects equipment maintenance and even affects the operational readiness.Therefore, the scientific and rational solution to the configuration problem of spare parts is an urgent task before us, but a wide range kinds of spare parts and different types of life distribution, how to determine the spare parts configuration needs to be studied.

2   The Simulation Model of Spare Parts Requirements

2.1   The Description and Assumption of Problem

Spare parts requirements refers to determine the support time T and the number of required spare parts supply meeting guarantee degree P of each unit based on the division of tasks and the use Strength of equipment. The definition of guarantee degree P[1] is the probability to provide this kind of spare parts in need of spare parts, its actual meaning is, if P is provided reliability, according to the life distribution type

D. Jin and S. Lin (Eds.): Advances in FCCS, Vol. 2, AISC 160, pp. 223–228.
springerlink.com   © Springer-Verlag Berlin Heidelberg 2012

of spare parts can calculate accordingly the reliable life t_p .Spare parts are unrepairable parts can not be assumed, the replacement cycle is t_p ,within the support time T the number of Spare parts needs to be replaced is the amount of spare parts corresponding equipment should be in the reserves.

In this paper, the assumptions about the simulation model of spare parts requirements are as follows: 1. parts failure does not occur during the reserve period; 2. equipment failure can be excluded by the replacement of spare parts and the time of replacement parts can be ignored.

2.2   The Idea and Process of Simulation

Taking a certain type of lifetime distribution part as an example, consider the number of working part is one and the number of spare parts is n, when the working part is failure, it is replaced by the reserved spare parts until no spare parts is available for the system failure.

The idea of simulation is to first get the sample value of life of the original working parts, with a reserve parts replacement, continue to work; Then get the sample value of the new spare parts replacement and then be replaced; repeated sampling until no spare parts can be available, at that time the sum of (n +1) sample values is sample value of the life of the system, then the value compares with the task time, if the sample value of the life of the system is greater than the task time, we think that the reserved spare parts meet the needs otherwise don't meet.

This process is repeated N times, if the statistics on the number of spare parts to meet the needs is K times, when N is large enough, K / N is the task reliability of systems (i.e. spare parts requirement rate).Simulation process includes:

(1) Input the parameters of the lifetime distribution of the parts, the number of parts is n, the times of simulation is N, the task time is TT.

(2) Variable initialization, the times of the mission success K = 0, the times of simulations S = 0.

(3) Compare the size of S and N, when S $\leq$ N ,go on the following steps, when S> N, go to step(7).

(4) Get the sample value of the lifetime of the working parts t_1 , then turn on the sampling of the spare parts for lifetime, get sample set t_i , calculate the cumulative work time T of the original working parts plus n spare parts, $T = \sum_{i=1}^{n+1} t_i$.

(5) if T $\geq$ TT, the simulation mission is successful, K = K +1, S = S +1, go to step (3) for the next time of simulation.

(6) If T <TT, then the simulation mission is fail, S = S +1, go to step (3) for the next time of simulation.

(7) the end of simulation, get an approximation of the probability of mission success P = K / N, i.e. spare parts requirement rate.

2.3   Examples and Validate Simulation Models

For example the distribution function of parts with two-parameter Weibull lifetime is: $F(t) = 1 - e^{-(\lambda t)^a}$

Corresponding the lifetime sampling formula[2] is: $t_{cy} = [-In(1-\eta)]^{1/a} / \lambda$

Where: η is a random numbers with [0, 1] uniformly distributed.

Assuming parameters of Weibull distribution are $\lambda = 1 / 1000$, $a = 2$, the task time is 5000h, fixed rate μ is 0 (it means that the parts can not repair), find the total number of spare parts to meet P is 0.95.Enter the above conditions into the simulation model and set the times of simulations N = 10000, the result shown in Figure 1, Figure shows we need eight spare parts to meet P more than 0.95.

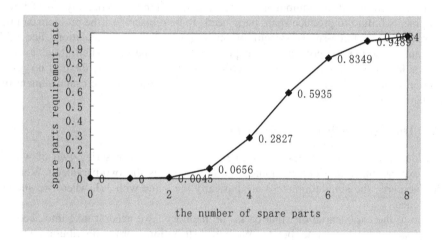

Fig. 1. The result of simulation model

Using the analytical calculation model in paper[3] to verify the simulation results, calculated as follows:

Check the Γ function value table

$\gamma=2$, $\Gamma(1+0.5)=0.8862$, $\Gamma(1+1)=1000$, L=1,

so, m=1000×0.8862=886.2

$\delta^2 = 1000^2 \times [1-0.8862^2] = 1000^2 \times 0.2146$, substituted into the formula

$$S = \left[L \cdot (1-\mu) \cdot \frac{T}{m} + \mu_p \sqrt{L \cdot \frac{T}{m} (1-\mu) \left\{ \mu + (1-\mu) \cdot \frac{\sigma^2}{m^2} \right\}} \right]$$

$$= \left[1 \cdot (1-0) \cdot \frac{5000}{886.2} + 1.65 \sqrt{1 \cdot \frac{5000}{886.2} (1-0) \left\{ 0 + (1-0) \cdot \frac{1000^2 \times 0.2146}{886.2^2} \right\}} \right]$$

$$= 7.64$$

As spare parts can only take integers, so requiring eight parts, the results are the same with this simulation model results.

3   The Simulation Model of Spare Parts Shared

3.1   The Description and Assumption of Problem

The so-called spare parts shared, for example there are many types of equipment with several different types, different work intensity and different work environment, they are equipped with the same type of components, when the component fails and needs replacement, and it requires a certain number of spare parts. Because they use the same type of parts, so parts can be shared between them.

The assumptions of simulation model of parts shared is the same with 1.1, for example, there are three equipments shared the same kind of parts, anyone of three equipments failure to produce spare parts needs to be replaced, so the problem can be abstracted as a three-series system of spare parts demand calculation. For this problem, when the components with exponentially lifetime distributed, it has been a good solution. But for the different lifetime distributions of parts, there is no good solution to this problem. This paper is to address this situation, the establishment of simulation model of spare parts shared.

3.2   The Idea and Process of Simulation

Consider three different types of lifetime distribution parts in series, two of which with exponentially lifetime distributed, the other one is the two-parameter Weibull lifetime parts, the simulation steps are basically the same with 1.2, where we shows difference.

Due to the exponential distribution is "no memory" ,we needn't take into account the impact of working time, but for Weibull distribution, normal distribution and so on, when sample the remaining lifetime, the impact of working time must be taken into account.

The distribution function of parts with exponential lifetime is: $F(t) = 1 - e^{-\lambda t}$

Corresponding the lifetime sampling formula is: $t_{cy} = -In(1-\eta)/\lambda$

The two-parameter Weibull lifetime parts corresponding the lifetime sampling formula[4] is:

$$t_{sy} = (1/\lambda)[(\lambda t_o)^a - In(1-\eta)]^{1/a} - t_o$$

Where: t_o express parts have worked time, η is a random numbers with [0, 1] uniformly distributed.

3.3   Examples and Validate Simulation Models

Assuming parameters of Weibull distribution are $\lambda = 1/1000$, $a = 2$, parameters of exponential distribution are $\lambda_1 = 1/500$、$\lambda_2 = 1/1000$, the task time is 2000h, fixed rate μ is 0 (it means that the parts can not repair), find the total number of spare parts to meet P is 0.95.

Enter the above conditions into the simulation model and set the times of simulations N = 10000, the result shows we need eleven spare parts to meet P more than 0.95 (P=0.9643).

If parts do not share, using the analytical calculation model in paper[3] to calculate, the results are: they need 4,8,5 spare parts, the total is 17, compared with the results of the simulation algorithm more six, forming waste. Currently, the research literature about this problem is almost no such cases; it can not be directly tested on the simulation model.

However, the model can be used to verify a special case, that is, when the components are exponentially distributed, for the k-components in series, n reserve parts, its spare parts requirements can be calculated [5]:

$$P(x \leq n) = \sum_{x=0}^{n} \frac{(k\lambda t)^x \exp(-k\lambda t)}{x!}$$

Where, λ is the exponential distribution of failure rate. Assuming the distribution parameter $\lambda = 1 / 1000$, the task time is 2000h, taking different the number of spare parts, using analytical and simulation to solve the problem, the results obtained are shown in Table 1.

Table 1. Three components in series comparison of the results

the number of spare parts	4	5	6	7	8	9
parts requirement rate analytical method	0.2851	0.4457	0.6063	0.7440	0.8472	0.9161
parts requirement rate simulation method	0.2833	0.4492	0.6079	0.7462	0.8430	0.9127

Comparison of results from the table can be seen that the results of the two methods agree well, indicating that this simulation model of spare parts shared is valid.

4   Summary

In this paper, a simulation model of spare parts shared is established to solve the shared parts with arbitrary lifetime distribution demand for spare parts demand forecasting has strong guidance and practicality.

References

1. Zhang, T., Gao, D.-H., Guo, B., Wu, X.-Y., Tan, Y.-J.: Spare availability model for phased mission systems. Journal of Systems Engineering 21(1), 86–90 (2006)
2. Yang, Y.-M., Sheng, Y.-X.: System reliability digital simulation. Beijing University of Aeronautics and Astronautics Press, Beijing (1990)
3. Chen, Y.-Q., Jin, J.-S., Yu, J.: Calculation Model of Spare Parts Considering the Servicing Time of Components. ShipBuilding of China 46(2), 107–109 (2005)
4. Li, J.-G., Ding, H.-B.: Calculation Models of Spare Part Capacity. Electronic Product Reliability and Environmental Testing (3), 11–14 (2000)
5. Wang, P.-G., Jin, J.-S.: A calculation model and dynamic management for guantity of spare part on warship. Journal of Naval University of Engineering 17(3), 103–106 (2005)

An Adaptive Filtering Approach to Estimate Genetic Regulatory Networks with Time-Varying Delay

Zhenling Wang and Huanqing Wang

Institute of Complexity Science, Qingdao University, Qingdao, 266071, China
Zhenling.eric@gmail.com, Hqwang79@yahoo.com

Abstract. In this paper, a class of nonlinear time delayed genetic regulatory networks (GRNs) is investigated from an adaptive filtering approach. For GRNs with time-varying delays and uncertain noise disturbances, several adaptive laws are derived to ensure the stochastic stability of the error states between the unknown network and the estimated model. Based on Lyapunov stability theory, the proposed adaptive laws are demonstrated to be effective. A simulation example is provided to verify and visualize the theoretical results.

Keywords: Genetic regulatory network (GRN), stochastic stability, adaptive Filtering, system synchronization, time-varying delay.

1   Introduction

During the past decade, significant progress in genome sequencing and gene recognition has been achieved, and a large volume of experimental data is now available. Today, researchers have put forward several theoretical models to describe gene regulatory networks successively, particularly in biological and biomedical sciences. Generally, there are two types of models: the discrete model (Boolean model)[1] and the continuous model[2-4] (differential equation model). In this paper, we will consider the differential equation model.

It is well known that time delay is ubiquitous in biological, chemical, and electrical dynamical systems. The existence of time delay is often a source of instability and poor performance. In [4], an uncertain delayed GRN model was introduced to estimate the unknown system parameters, in which the time delay is constant. However, in reality, time delays in the process of transcription, translation, and translocation processes can not be a constant[3]. Thus, in this paper, a class of nonlinear time-varying delayed genetic regulatory networks is considered under the case the time-varying delay $\tau(t)$ satisfies the condition $\dot{\tau}(t) \leq \beta < 1$.

The outline of this paper is as follows. The next section briefly introduces the GRN model formulation and preliminaries of the model. Then, a GRN model with nonlinear time delay is investigated. In section 3, an adaptive filter is designed to estimate the unknown states and parameters. A numerical example showing the effectiveness of the proposed adaptive filtering scheme is presented in Section 4. Finally, concluding remarks are made in Section 5.

D. Jin and S. Lin (Eds.): Advances in FCCS, Vol. 2, AISC 160, pp. 229–235.
springerlink.com   © Springer-Verlag Berlin Heidelberg 2012

2   Model Formulation and Preliminaries

2.1   Original GRN Model

In [3], the delayed GRN model is described by

$$dm_i(t)/dt = -a_i m_i(t) + G_i(p_1(t-\tau), p_2(t-\tau), \cdots, p_n(t-\tau)) ,$$
$$dp_i(t)/dt = -c_i p_i(t) + d_i m_i(t-\tau), \qquad\qquad (1)$$

where $m_i(t)$, $p_i(t) \in R$ are the concentrations of mRNA and protein of the ith node at time t, whose degradation rates are denoted by $a_i > 0$ and $c_i > 0$, respectively, d_i is the translation rate, and τ is a time delay. The function G_i represents the feedback regulation of the protein on the transcription, which is generally a nonlinear monotonically increasing function, $i = 1, 2, \cdots, n$.

The gene regulation function G_i in (1) is very complex. Usually, SUM logic[5] is applied; that is, each transcription factor acts additively to regulate the ith gene, and the regulatory function is in the form of $G_i = \sum_{j=1}^{n} G_{ij}(p_j(t))$ [3]. If the transcription factor j is an activator of gene i, then

$$G_{ij}(p_j(t)) = \alpha_{ij}(p_j(t)/\xi_j)^{H_j}[1 + (p_j(t)/\xi_j)^{H_j}]^{-1}, \qquad\qquad (2)$$

If transcription factor j is a repressor of gene i, then

$$G_{ij}(p_j(t)) = \alpha_{ij}[1 + (p_j(t)/\xi_j)^{H_j}]^{-1}, \qquad\qquad (3)$$

where H_j are called the Hill coefficients, α_{ij} are scalar transcriptional rates of the transcription factor j to i, and ξ_j are positive constants.

Thus, the delayed GRN model can be written as follows:

$$dm(t)/dt = -Am(t) + Wf(p(t-\tau)) + L,$$
$$dp(t)/dt = -Cp(t) + Dm(t-\tau). \qquad\qquad (4)$$

where $m(t) = [m_1(t), m_2(t), \cdots, m_n(t)]^T$, $p(t) = [p_1(t), p_2(t), \cdots, p_n(t)]^T$, $f(p(t-\tau)) = [f_1(p_1(t-\tau)), f_2(p_2(t-\tau)), \cdots, f_n(p_n(t-\tau))]^T$ with $f_j(x) = (x/\xi_j)^{H_j}/[1 + (x/\xi_j)^{H_j}]$, $m(t-\tau) = [m_1(t-\tau), m_2(t-\tau), \cdots, m_n(t-\tau)]^T$, $A = diag\{a_1, a_2, \cdots, a_n\}$, $C = diag\{c_1, c_2, \cdots, c_n\}$, $D = diag\{d_1, d_2, \cdots, d_n\}$, $L = (l_1, l_2, \cdots, l_n)^T$ with $l_i = \sum_{j \in I_i} \alpha_{ij}$. I_i is the set of all the repressors of gene i, and $W = (w_{ij}) \in R^{n \times n}$ is defined by

$$w_{ij} = \begin{cases} \alpha_{ij} & \text{if gene j is an activator of gene i,} \\ 0 & \text{if there is no link from gene j to gene i,} \\ -\alpha_{ij} & \text{if gene j is an repressor of gene i.} \end{cases}$$

moreover, the function $f_i : R \rightarrow R$ is monotonically increasing and satisfies the Lipschitz condition

$$\left| f_i(x) - f_i(y) \right| \leq h_i \left| x - y \right|, \quad \forall x, y \in R \tag{5}$$

where $h_i > 0$ are the Lipschitz constants for $i = 1, 2, \cdots, n$.

2.2   The Uncertain GRN Model with Time-Varying Delay

In this paper, an uncertain GRN model with time-varying delay is considered

$$\begin{aligned} dm(t) &= [-Am(t) + Wf(p(t - \tau(t))) + L]dt + v(t)d\omega, \\ dp(t) &= [-Cp(t) + Dm(t - \tau(t))]dt + \sigma(t)dv, \\ y &= C_0 x(t) \end{aligned} \tag{6}$$

where y is the output, $x(t) = [m^T(t), p^T(t)]^T$, $C_0 = (I_n 0_n)$ with I_n and 0_n denoting the n-dimension identity and zero matrices, respectively, $v(t)$ and $\sigma(t)$ are external noise intensity functions, $\omega(t)$ and $v(t)$ are two independent one-dimensional Brownian motion satisfying the mathematical expectations $E\{d\omega(t)\} = 0$, $E\{dv(t)\} = 0$ and $E\{d\omega(t)^2\} = 1, E\{dv(t)^2\} = 1$.

Sometimes, system parameters and the concentrations of mRNA and protein can be obtained by direct measurements. However, some system parameters might be unknown. Therefore, assume that in system (6), A, W, L are uncertain matrices and $p(t)$ is unknown but the output $y = m(t)$ can be observed. The objective is to estimate the unknown system parameters and states.

To achieve this goal, an adaptive filtering is designed as follows:

$$\begin{aligned} d\tilde{m}(t) / dt &= -\tilde{A}(t)\tilde{m}(t) + \tilde{W}(t)f(\tilde{p}(t - \tau(t))) + \tilde{L}(t) - K(t)(\tilde{m}(t) - y(t)), \\ d\tilde{p}(t) / dt &= -C\tilde{p}(t) + D\tilde{m}(t - \tau(t)), \end{aligned} \tag{7}$$

where $\tilde{m}(t) = [\tilde{m}_1(t), \tilde{m}_2(t), \cdots, \tilde{m}_n(t)]^T$, $\tilde{p}(t) = [\tilde{p}_1(t), \tilde{p}_2(t), \cdots, \tilde{p}_n(t)]^T$, $K(t) = diag\{k_1(t), k_2(t), \cdots, k_n(t)\}$ is the adaptive feedback matrix, $\tilde{W}(t) = (\tilde{w}_{ij}(t))_{n \times n}$, $\tilde{A}(t) = diag\{\tilde{a}_1(t), \tilde{a}_2(t), \cdots, \tilde{a}_n(t)\}$, and $\tilde{L}(t) = [\tilde{l}_1(t), \tilde{l}_2(t), \cdots, \tilde{l}_n(t)]^T$ are matrix- and vector-valued functions of time t.

Now, subtracting (6) from (7), the following error dynamical system can be obtained.

$$\begin{aligned} de_m(t) &= [-\tilde{A}(t)\tilde{m}(t) + \tilde{W}(t)f(\tilde{p}(t - \tau(t))) + \tilde{L}(t) - K(t)e_m(t) \\ &\quad + Am(t) - Wf(p(t - \tau(t))) - L]dt - v(t)d\omega, \\ de_p(t) &= [-Ce_p(t) + De_m(t - \tau(t))]dt - \sigma(t)dv \end{aligned} \tag{8}$$

where $e_m(t) = \tilde{m}(t) - m(t) = [e_{m1}(t), e_{m2}(t), \cdots e_{mn}(t)]^T$, $e_p(t) = \tilde{p}(t) - p(t) = [e_{p1}(t), \cdots e_{pn}(t)]^T$.

Next, some useful definition and lemmas are introduced.

Definition 1. The two networks (6) and (7) are said to be stochastically synchronous with disturbance attenuation $\gamma > 0$, if

i) network (8) with $v(t)=0$ and $\sigma(t)=0$ is asymptotically stable;

ii) with zero initial conditions, there exists a scalar $\gamma > 0$ such that

$$E \int_0^\infty (\|e_m(s)\|^2 + \|e_p(s)\|^2) ds \leq \gamma^2 \int_0^\infty (\|v(s)\|^2 + \|\sigma(s)\|^2) ds$$

for all nonzero $v, \sigma \in L_2[0,\infty)$.

Lemma 1. For any vectors $x, y \in R^n$ and positive definite matrix $G \in R^{n \times n}$, the following matrix inequality holds:

$$2x^T y \leq x^T G x + y^T G^{-1} y.$$

3   Adaptive Filtering Design

In this section, in order to ensure that the error dynamical system (8) is asymptotical stable, some adaptive laws are given to estimate the unknown system parameters, $\tilde{A}(t), \tilde{W}(t), \tilde{L}(t)$ as well as $K(t)$.

Next, some simple sufficient conditions are given, as follows.

Theorem 1. The two networks (4) and (6) are stochastically synchronous with disturbance attenuation $\gamma > 0$, if

$$\tilde{a}_i = q_i e_{mi} \tilde{m}_i(t), \qquad\qquad i = 1, 2, \cdots, n \qquad (9)$$

$$\tilde{w}_{ij} = -r_{ij} e_{mi} f_j(\tilde{p}(t - \tau(t))), \qquad i, j = 1, 2, \cdots n \qquad (10)$$

$$\tilde{l}_i = -u_i e_{mi}, \qquad\qquad i = 1, 2, \cdots n \qquad (11)$$

$$\tilde{k}_i = \eta_i e_{mi}^2 \qquad\qquad i = 1, 2, \cdots n \qquad (12)$$

and

$$\gamma > \frac{1}{\sqrt{2(c - \mu)}} \qquad\qquad (13)$$

where q_i, r_{ij}, u_i and η_i are positive constants, $c = \min_{1 \leq i \leq n}\{c_i\}$, and μ is a positive constant satisfying $\mu < c$.

Proof: Consider the Lyapunov function candidate:

$$V(t) = \frac{1}{2}[e_m^T(t)e_m(t) + e_p^T(t)e_p(t)] + \sum_{i=1}^n \frac{1}{2q_i}(\tilde{a}_i - a_i)^2 + \sum_{i=1}^n \sum_{j=1}^n \frac{1}{2r_{ij}}(\tilde{w}_{ij} - w_{ij})^2$$

$$+ \sum_{i=1}^n \frac{1}{2u_i}(\tilde{l}_i - l_i)^2 + \sum_{i=1}^n \frac{1}{2\eta_i}(\tilde{k}_i - k)^2 + \frac{1}{1 - \beta} \int_{t - \tau(t)}^t [\delta_1 e_m^T(s)e_m(s) + \delta_2 e_p^T(s)e_p(s)] ds$$

where k, δ_1 and δ_2 are positive constants to be determined.

From the Itô formula, the following equation holds:

$$dV(t) = \mathcal{L}V(t)dt - e_m^T(t)v(t)d\omega - e_p^T(t)\sigma(t)d\nu.$$

The weak infinitesimal operator $\mathcal{L}$ of the stochastic process is given by

$$
\begin{aligned}
\mathcal{L}V(t) \leq\ & e_m^T(t)[-\tilde{A}(t)\tilde{m}(t)+\tilde{W}(t)f(\tilde{p}(t-\tau(t)))+\tilde{L}(t)+Am(t)-Wf(p(t-\tau(t))) \\
& -L-Ke_m(t)]+e_p^T(t)[-Ce_p(t)+De_m(t-\tau(t))]+\sum_{i=1}^n (\tilde{a}_i-a_i)e_{mi}\tilde{m}_i(t) \\
& +\sum_{i=1}^n\sum_{j=1}^n (\tilde{w}_{ij}-w_{ij})e_{mi}f_j(\tilde{p}_j(t-\tau(t)))-\sum_{i=1}^n(\tilde{l}_i-l_i)e_{mi}+\sum_{i=1}^n(\tilde{k}_i-k)e_{mi}^2 \\
& +\frac{\delta_1}{1-\beta}[e_m^T(t)e_m(t)-e_m^T(t-\tau(t))e_m(t-\tau(t))] \\
& +\frac{\delta_2}{1-\beta}[e_p^T(t)e_p(t)-e_p^T(t-\tau(t))e_p(t-\tau(t))]+\frac{1}{2}v^T(t)v(t)+\frac{1}{2}\sigma^T(t)\sigma(t) \\
=\ & -e_m^T(t)[A+(k-\frac{\delta_1}{1-\beta})I_n]e_m(t)+e_m^T(t)W[f(\tilde{p}(t-\tau(t)))-f(p(t-\tau(t)))] \\
& -e_p^T(t)(C-\frac{\delta_2}{1-\beta}I_n)e_p(t)-\frac{\delta_1}{1-\beta}e_m^T(t-\tau(t))e_m(t-\tau(t))-\frac{\delta_2}{1-\beta} \\
& \times e_p^T(t-\tau(t))e_p(t-\tau(t))+\frac{1}{2}v^T(t)v(t)+\frac{1}{2}\sigma^T(t)\sigma(t).
\end{aligned}
\tag{14}
$$

According to Lemma 1 and the Lipschitz condition (5), the following inequations hold:

$$e_m^T(t)W[f(\tilde{p}(t-\tau(t)))-f(p(t-\tau(t)))] \leq \frac{1}{2\theta}e_m^T(t)WW^Te_m(t)+\frac{\theta h^2}{2}e_p^T(t-\tau(t))e_p(t-\tau(t)),$$

$$e_p^T(t)De_m(t-\tau(t)) \leq \frac{\zeta}{2}e_p^T(t)DD^Te_m(t)+\frac{1}{2\zeta}e_m^T(t-\tau(t))e_m(t-\tau(t)). \tag{15}$$

where $h=\max_{1\leq i\leq n}\{h_i\}$, θ and ζ are positive constants.

From (14), (15) and choose $\delta_1=(1-\beta)/(2\zeta)$, $\delta_2=\theta h^2(1-\beta)/2$, then we can obtain that

$$
\begin{aligned}
\mathcal{L}V(t) \leq\ & -e_m^T(t)[A+(k-\frac{1}{2\zeta})I_n-\frac{1}{2\theta}WW^T]e_m(t) \\
& -e_p^T(t)(C-\frac{\theta}{2}h^2I_n-\frac{\zeta}{2}DD^T)e_p(t)+\frac{1}{2}v^T(t)v(t)+\frac{1}{2}\sigma^T(t)\sigma(t) \\
\leq\ & -[a+k-\frac{1}{2\zeta}-\frac{1}{2\theta}\lambda_{\max}(WW^T)]e_m^T(t)e_m(t) \\
& -[c-\frac{\theta}{2}h^2-\frac{\zeta}{2}\lambda_{\max}(DD^T)]e_p^T(t)e_p(t)+\frac{1}{2}v^T(t)v(t)+\frac{1}{2}\sigma^T(t)\sigma(t)
\end{aligned}
\tag{16}
$$

where $a=\min_{1\leq i\leq n}\{a_i\}$, $c=\min_{1\leq i\leq n}\{c_i\}$, and $\lambda_{\max}(M)$ denotes the maximal eigenvalue of the symmetric matrix M.

As we know, values of θ and ζ can be selected to be sufficiently small, so that $c - \theta h^2 / 2 - \zeta \lambda_{max}(DD^T)/2 > 0$, and let $k = -a + 1/(2\zeta) + \lambda_{max}(WW^T)/(2\theta) + 1$, then it is easy to verify that the condition i) in Definition 1 is satisfied.

For $\gamma > 0$, set

$$J(t) = E \int_0^t [e_m^T(s)e_m(s) + e_p(s)^T e_p(s) - \gamma^2(v^T(s)v(s) + \sigma^T(s)\sigma(s))]ds \tag{17}$$

From (14), (17) and under zero initial conditions, for $\rho = 2\gamma^2 > 0$, one has

$$J(t) \le E \int_0^t 2\gamma^2 [-(a+k - \frac{1}{2\zeta} - \frac{1}{2\theta}\lambda_{max}(WW^T))e_m^T(s)e_m(s) \tag{18}$$
$$- (c - \frac{\theta}{2}h^2 - \frac{\zeta}{2}\lambda_{max}(DD^T) - \frac{1}{2\gamma^2})e_p^T(s)e_p(s)]ds$$

let $\mu = \frac{\theta}{2}h^2 + \frac{\zeta}{2}\lambda_{max}(DD^T)$, and choose k as mentioned above, from (18), one has

$$J(t) \le E \int_0^t 2\gamma^2 [-e_m^T(s)e_m(s) - (c - \mu - \frac{1}{2\gamma^2})e_p^T(s)e_p(s)]ds \tag{19}$$

thus, from (14) and (17), it is easy to verify that $J(t) < 0$. Consequently

$$E \int_0^t (e_m^T(s)e_m(s) + e_p(s)^T e_p(s))ds \le \gamma^2 E \int_0^t (v^T(s)v(s) + \sigma^T(s)\sigma(s))ds \tag{20}$$

therefore, the condition ii) in Definition 1 is satisfied. Thus, the proof of the theorem is completed.

4   A Simulation Example

In this section, we will give an example to justify the theoretical analysis in this paper.

Consider a time varying delayed GRN that the kinetics of the system is determined by six coupled first-order differential equations. Taking into account stochastic disturbance, the following unknown GRN model is considered

$$dm(t) = [-Am(t) + Wf(p(t - \tau(t))) + L]dt + v(t)d\omega,$$
$$dp(t) = [-Cp(t) + Dm(t - \tau(t))]dt + \sigma(t)dv, \tag{21}$$
$$y = m(t)$$

where, $A = diag\{0.4, 0.36, 0.48\}$, $D = diag\{2, 2, 2\}$, $C = diag\{0.2, 0.5, 0.6\}$, $w_{13} = w_{21} = w_{32} = -1.4$, other entries of W is zero. $L = (1.5\ \ 1.5\ \ 1.5)^T$, $\sigma(t) = e^{-0.1t}$, $v(t) = 0.1e^{-0.05t}$, $\tau(t) = 0.5(1 + \sin t)$.

Assume $w_{13} = -1.4$ and $l_1 = 1.5$ are unknown, and the adaptive filter (7) and (9)-(12) is designed. The error states e_m and e_p in (8), and the adaptive laws $\tilde{w}_{13}$ and $\tilde{l}_1$ are shown in Fig.1, respectively, where $r_{13} = u_1 = 10$.

From Fig.1, it is clearly seen that parameters $w_{13} = -1.4$, $l_1 = 1.5$ can be effectively estimated by $\tilde{w}_{13}$ and $\tilde{l}_1$.

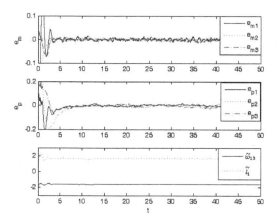

Fig. 1. Error states e_m and e_p and adaptive laws $\tilde{w}_{13}$ and $\tilde{l}_1$.

5   Conclusion

In this paper, an adaptive filtering approach has been developed to estimate the unknown genetic regulatory networks with time varying delays. The adaptive laws are demonstrated to be effective, and it is easy to apply these sufficient conditions to the real networks. Finally, we give a simulation to justify the obtained results.

References

1. Huang, S.: Gene Expression Profiling, Genetic Networks, and Cellular States: An Integrating Concept for Tumorigenesis and Drug Discovery. J. Molecular Med. 77, 469–480 (1999)
2. Wang, Z., Gao, H., Cao, J., Liu, X.: On delayed genetic regulatory networks with polytopic uncertainties: Robust stability analysis. IEEE Trans. NanoBiosci. 7(2), 154–163 (2008)
3. Ren, F., Cao, J.: Asymptotic and robust stability of genetic regulatory networks with time-varying delay. Neurocomputing 71, 834–842 (2008)
4. Yu, W.W., Lü, J., Chen, G.R., Duan, Z.S., Zhou, Q.H.: Estimating Uncertain Delayed Genetic Regulatory Networks: An Adaptive Filtering Approach. IEEE Trans. Automat. Control 54, 892–897 (2009)
5. Li, C., Chen, L., Aihara, K.: Stability of Genetic Networks with Transcriptional Regulators. Nature 403, 335–338 (2000)

Intelligent Interaction System of Distance Education Based on Natural Language Matching

Lian-duo Yan[1], Xiu-lan Ma[2], and Li-ling Wang[1]

[1] Department of Ideological and Political, Hebei Normal University of Science & Technology,
066004 Qinhuangdao, China
[2] College of Continuing Education, Hebei Normal University of Science & Technology,
066004 Qinhuangdao, China
{lianduoyan,xiulanma,lilingwang}@163.com

Abstract. Rapid development of distance education enables intelligent interaction as an important part to utilize resources in knowledge database of distance education network and to automatically answer questions from students. It can also improve teaching efficiency and quality. Therefore, intelligent interaction system also becomes a research focus. The proposed system represents relation among words by establishing natural language thesaurus database. It returns the closest answer by computing similarity among questions. Key techniques as natural language processing, semantic analysis and fuzzy matching to implement intelligent interaction system were also presented.

Keywords: distance education, intelligent interaction, natural language processing.

1   Introduction

With the development of distance education, intelligent interaction system becomes a research focus currently. Researches on intelligent interaction system mainly address on improving existing interaction mode and adequately using resources in distance education network to automatically answer questions from students, so as to achieve teaching part of interaction and improve teaching efficiency. Seen from existing interaction system, as the natural language process based on English, text searching, data mining, knowledge representation and knowledge reasoning technology have been developed for long time, the interaction system has high intelligence, the recall and precision of which are also high [1, 2]. For limitations in complexity and semantic recognition of Chinese word, the researches on intelligent interaction are relatively backward and there is little widely used and acclaimed interaction system.

The needed intelligent interaction system should provide friendlier interface. The students can question with natural language and system performs word processing, similarity computation, semantic understanding and fuzzy matching till arriving at answer. However, the existing interaction systems mainly base on keywords matching [3-5], which is merely match between keywords. In the matching process of problems and materials, the semantics of sentence will not be considered, which significantly affect on comprehensiveness and accuracy of answer. The proposed system not only provides user interface to present problem with natural language, but also improve semantic understanding on problems and materials. Obviously, it will bring out improvement on interaction manner. The paper is organized as follows: section 2 gives flow of intelligent interaction system; section 3 presents overall design on the system; section 4 gives detail analysis and design on related modules and section 5 concludes our work.

2   Flow of Intelligent Interaction System

The flow of intelligent interaction system oriented to distance network education is as following:

Step 1: In system establishment, experienced teacher recorded large amount of common problems and easy points into question interaction system. Meanwhile, new emerged words were divided into nouns, verbs, adjectives, pronouns. The similarity among words is determined to assign same attribute value o a set of synonyms.

Step 2: If a student has question, it submit problem described with natural language to system.

Step 3: The word processing module deal with problems, divide word sentence and mark out the words of speech. Then, it stores word into local database for searching, so as to arrive at attribute value of each word and obtain relative information of word for next step analysis.

Step 4: Carry out Step 3 on problems in the database and match the obtained word information with word in users' problem. Exclude irrelevant questions according to certain rules and type value to find all candidate questions.

Step 5: For each candidate question, compute similarity between answer and target question based on similarity of nouns, verbs and describe in candidate question and user problem.

Step 6: According to computation result of Step 5, obtain the candidate question with maximum similarity.

Step 7: Submit corresponding answer in candidate question obtained from Step 6 to students. If the student think question is consistent, the interaction process ends.

Step 8: If student do not think the problem is inconsistent, we can consider there is no relevant problem in the database. The system uploads its question into problem database. When the teacher responsible from interaction sees the left problem, he will answer student's question and add it into database.

3   Overall Design on Intelligent Interaction System

The overall architecture of intelligent interaction system can be divided into 5 modules:

(1) Student module. The student enters into interface after certification. Then he can input problem in line with human thinking and search answer in local database.

(2) Teacher module. Teachers enter into his interface after certification to check questions left by students. Then, they will input valuable questions and corresponding answers into database, as shown in Fig. 1.

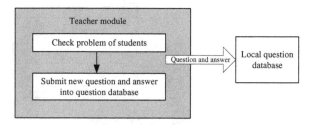

Fig. 1. Teacher working module

(3) Problem processing module. The first step processing on question is wording and obtains information from local database, including speech and property of words, so as to prepare for analysis next.

(4) Database searching module. Match obtained word information obtained from problem analysis module and exclude irrelevant questions according to certain rules and type, so as to find all candidate questions.

(5) Answer processing module. Compute similarity between candidate question and problem of users one by one and return maximum of similarity. Return 5 closest answers to avoid not getting satisfactory answer for inaccurate representation.

The information is transmitted among question processing module, student module, database searching module as well as answer processing module. The final result can be displayed on the student interface, as shown in Fig. 2.

4   Key Technologies Analysis

4.1   Automatic Word Segmentation

The system used word segmentation method based on string matching, which is also known as mechanical sub-word method. It matches to be analyzed character with word items in the dictionary according to certain strategy. If a string can be found in the dictionary, a match succeeds. According to different scanning direction, the method can be divided into forward match and reverse match. In accordance with priori matching of different length, it can be divided into maximize matching and minimize matching. In accordance with whether speech tagging combined with process of phase, it can be divided into pure sub-word method and the word mark combined with the integrated approach. Several commonly used methods include mechanical sub-word maximum matching forward, reverse maximum matching, minimum cut.

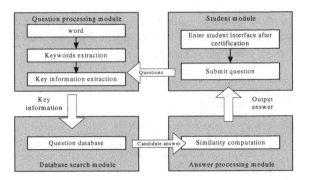

Fig. 2. Student query module

We can also combine above methods. For example, the forward maximum matching method and reverse maximum matching method combined to form two-way matching method. Because the characteristics of Chinese words into the word, forward and reverse minimum match the minimum match is rarely used. In general, the reverse match segmentation accuracy than a positive match, the phenomenon is also encountered less ambiguity. Results show that simply using the forward maximum matching error rate of 1/169, simply use the reverse maximum matching error rate of 1/245, but the accuracy is still far from meeting the actual needs. As the word is an intelligent decision-making process, mechanical methods can not solve the word segmentation phase of the two basic problems: segmentation ambiguity problem and the unknown word problem. Actual use of the word system all utilizes early mechanical sub-word as means to need a variety of other languages through information by increasing segmentation accuracy. For example, at the same time in the word syntactic, semantic analysis, use of syntactic information and semantic information to process ambiguous phenomenon has a complete thesaurus.

4.2   Semantic Analysis

In the interaction system, the descriptions on same question from different users are varying. For example:

Question 1: What is the concept of buffer? The keywords set after automatic segmentation are *operation system* and *concept*.

Question 2: What is the definition of buffer? The automatic segmentation set includes *operation system* and *definition*.

Question 3: What is buffer? The automatic segmentation set includes *buffer* and *what*.

Although these three questions have different description, the obtained keywords are also different, and the direct matching relation based on keywords are not high, but the objection meaning of expression is same. How to enable system understanding questions from users and obtain more accurate and comprehensive query, it needs to perform semantic analysis. Main word is to analyze on synonymous among words, so as to obtain more information from database. It is noteworthy that a word may have more than one part of speech. Therefore, a word may have many records in the database for its speech. If we can find word with same speech from analyzed result, the record information is assigned to this word. Otherwise, we may think that the first record information is that of keywords. Then the closeness of meaning between words can be used to determine whether they match.

4.3 Searching Technology

The similarity of sentence largely depends on similarity of nouns, verbs and adjectives in the sentence. In this system, the similarity of sentence is equivalent to superposition of different parts of speech. Similarity calculation is similar. Taking noun similarity computation as example, find sentence with least noun number and mark it as s_1, and another as s_2.

Definition: Set type of word a and b as $cno(a)$ and $cno(b)$, then the similarity between a and b is

$$sim(a,b) = 1 - \frac{|cno(a) - cno(b)|}{n}. \tag{1}$$

Where, the value of n is determined by number of synonym thesaurus in the database.

From (1), we can see that if a and b are synonymous terms, $|cno(a) - cno(b)| = 0$ and $sim(a, b)=1$, which means a and b are exactly matched. The relative position among words in the database is associated. Closer meaning means closer position and less $|cno(a) - cno(b)|$, larger $sim(a, b)$, which means two words are more matched.

The similarity of noun should firstly compare each noun a in s_1 with each b_j in s_2, and then take $\max\{sim(a,b_j)\}$. After a takes all nouns in s_1, the noun similarity between s_1 and s_2 is

$$Nsim(s_1,s_2) = \frac{\sum_{i=1}^{i=m} \max\{sim(a_i,b_j)\}}{r - m + 1} ; j = 0,1,\cdots,r . \tag{2}$$

Where, r is noun number in s_1 and m is noun number in s_2. If the difference between nouns number in s_1 and s_2, the noun similarity among sentences are smaller.

After pre-processing of user question, extract question record from database one by one and carry out pre-processing. As the noun insider determine subject of sentence to a large extent, if there is no same or similar nouns in two sentences, the sentence can not be list in candidate answers. In the candidate answer group, the similarity of user question and that in the database are ordered descending into list. The searching result is content in first node of list, which is the question with maximum similarity.

4.4 Data Table Design

The local database includes 5 tables, namely student information, teacher information, thesaurus, question bank and database. Among them, the student information and teacher information have three attributes, namely number, name and password to authenticate identities of students and teachers. The question database records problems submitted by students, including two attributes of *question* and *time*. The question database records large amount of common problems and confusing knowledge points, including three attributes as *No. question* and *answer*. The *No.* is the primary key to mark question. The design on question database is most important. The best approach is that different system has different aims. The integrity is the key to ensure system find accurate answer.

5   Conclusion

In the result verification process, we found that the proposed questions can find more accurate analysis and return closest answer. The biggest factor to affect accuracy is establishment of database. In the system, we mainly used relationship among words to perform matching, so design and implementation of segmentation system and thesaurus system are of most important. A complete segmentation system should be consistent with thesaurus system. As to specialized questions, we should build word database matching to it, so as to improve accuracy greatly. When common word database with 17809 synonyms used, the accuracy is 78% and consumes too many time. If specialized database with 1280 synonyms used, the accuracy improved to 95%. The speed also improves greatly. In addition, in the initially implementation of system, as the questions in the database are not sufficient, the returned answer may has some bias. So we should also divide sections on curriculum to implement simple indexing on directory-based indexing to speed up searching.

References

1. Xian, J., Mo, X.-L., Xi, J.-Q.: A question interaction system based on question pattern match. Journal of Shandong University (Natural Science) 9, 99–103 (2006)
2. Zhang, X.-H., He, P.-L., Liu, G.-R.: Study of Search Technology in Intelligent Question Interaction System. Microcomputer Applications 27, 261–263 (2006)
3. Wang, Q.: Design and Implementation of a Practical Intelligent Question Interaction System. Computer and Modernization 9, 110–113 (2007)
4. Wu, G.-Q., Zheng, F.: A method to build a super small but practically accurate language model for handheld devices. Journal of Computer Science & Technology 18, 747–755 (2003)
5. Cheng, J.-H.: Design or Words Module in Intelligent Q/A System Based on FAQ. Computer Technology and Development 18, 181–186 (2008)

An Improved Complex Network Community Detection Algorithm Based on K-Means

Yuqin Wang

Jilin Business and Technology College, Department of Information Engineering,
130062, Changchun, China
wyq_wyq@163.com

Abstract. In order to find community structure that exists in complex network structure, this paper introduced K-means approach to the complex network community structure of the research. This paper studies the complex network of community structure detection algorithm, through the existing algorithm learn and study, proposed an improved algorithm based on K-means. Not know the premise of community structure for the complex networks division, the algorithm is simple and easy to understand, and the algorithm was used in karate network, through experimental verification, experimental results show that the algorithm is effective.

Keywords: complex networks, community structure, k-means, detection algorithm, algorithm analysis.

1   Introduction

In recent years, with the complex network research, complex network of community structure as the basic properties of complex networks has become a hot research. People has been found between network nodes has same nature are often closely linked and together into societies, and community and between communities but no intensive links can be used to describe the sparse, this is the community structure [1]. Community structure is an important feature of many real complex networks. Search and analysis of community structure contribute to better understanding the internal structure of the network; analyze the nature of the network.. Currently many algorithms have been proposed for finding the existence of community structure in the complex networks. This paper studies the complex network of community structure detection algorithm, through the existing algorithm learn and study, proposed an improved algorithm based on K-means.

2   Community Structure Survey Algorithm

For more accurate seeks for in the network the community structure, achieves through the effective algorithm, the use as far as possible few information obtains as far as possible accurate cyber community structure goal. At present had already studied very

D. Jin and S. Lin (Eds.): Advances in FCCS, Vol. 2, AISC 160, pp. 243–248.

many algorithms, for example based on the side cluster coefficient's algorithm, based on the method which walks stochastically, based on q-Potts spinning method, based on resistance network method, Each kind the method which optimizes based on the modularity, each kind based on the spectrum analysis's method, based on dynamics synchronization's method, based on message center's method and so on, this article introduced the G-N algorithm.

2.1   G-N Algorithm

Girvan and Newman proposed that with on the one hand lies between the numbers to mark on the other hand each to the network connective influence. Some on the one hand on the other hand lies between the number to pass this side for the network in the most short-path's number. The most short-path to connect this apex to the variable least ways, Because community's side is between two spots the most short-path's road which must be taken, therefore the community on the one hand on the other hand lies between the number to be quite big continually, but the community interior on the one hand on the other hand lies between number quite few [3].

2.2   Algorithm Flows

The first step: compute network each on the one hand on the other hand lies between the number.

The second step: Discovers on the one hand lies between number biggest on the other hand, and its detachment. If on the one hand lies between number biggest on the other hand not to have one, may a stochastic choice detachment.

The third step: In the recompilation network surplus each on the one hand on the other hand lies between the numbers.

The fourth step: Is redundant second step, third step, until network in all side by detachment.

3   K-Means Algorithm

K-means clustering algorithm as a basis algorithms research of data mining and complexity network community division has a long history, proposed by MacQueen between the sample-based clustering algorithms similar to the indirect measure. K is the parameter of n objects into k community, so that communities internal with high similarity and community's similarity are very low [4]. K-means algorithm flow is as follows:

The first step: in the initial n nodes randomly selected k nodes, k nodes in each node as the first element of a community.

The second step: In the n nodes of set N remove the selected k nodes. Then select any node j from the collection of N, j and k-node compute the first element in the community from the d (G, in, j).

The third step: find the min (d (G, in, j)), in order to find with the node j representatives from the smallest point in. and in the community in a divided, while the node from the set N deleted.

Step four: cycle implementation of the second, three-step, until the set N is empty.

4   Improved Algorithms Introduced

On the one hand as k-means algorithm more dependent on the choice of the initial node, if k nodes randomly selected, the algorithm based on the choice of the initial node to get different results, in this algorithm can be based on node degree and network average degree of relationship to select the initial node, as long as the node is greater than the average degree of network put it as an initial node, rather than limit the initial number of nodes. The other hand, k-means algorithm requires prior knowledge of complex network is divided into the number of societies, in the actual network of community structure the division process is not applicable, Because the actual network of community structure are unknown, so this algorithm based on the nodes and the shortest path between the communities is smallest, the node joins the corresponding societies, Finally, according to association between the total number of neighbor nodes, set the threshold for associations to be merged, so there is no final definition of the number of societies, associations do not in advance have to know the number[5].

4.1   Network Average Degree $\bar{k}$

Complex network of nodes the strength of the lobbying ability and the size of the node's degree directly related. If a node of degree greater, indicating its lobby ability is greater, then it is easy to other nodes into their own in a community. In contrast, the degree of a node is smaller, then its lobby ability more weak, more easily persuaded by other nodes. Then we define the network average degree is of all nodes degrees in the network and all node the total of the ratio, to assess which nodes can be used as community centre node. Namely:

$$\bar{k} = \frac{1}{n}\sum_{i=1}^{n} k_i \tag{1}$$

4.2   Algorithm Description

1) Initialization. The complex network data into a simple graph format, the link situation of between the vertices stored in the adjacency matrix.

2) Calculate the degree of each node, to calculate the network average degree $\bar{k}$. If the node is greater than $\bar{k}$, put the node to remove from the initial node list, into the community list .

3) From the initial list the beginning of the first node s to start, calculate the s and the community each node in the list of the shortest path. From the results, find a shortest path to each node added to the first element of the list in the community, If there are two or more shortest paths, compare the degree of these communities the size of the first element, then the nodes join the maximum degree of that first element of the list community.

4) Repeat 3) until the initial list does not consider the node.

5) The various associations list to merge. Built a matrix B, where the elements B_{ij} for the two society first elements of the number of common neighbor nodes with the

first two elements of the degree of association of the ratio. The average of the B matrix as a threshold β, in B Each row to find a maximum (for the j-th column), if the maximum value is greater than β, put the two communities i, j merge, and update B matrix B $(i, j) = 0$; 6) Repeat 5) until the matrix B is not large compared with the value of β, associations merged to stop.

7) Finally, output community list of the various associations. Get the number of societies and associations the elements, the experimental results compared the actual results to calculated results and the accuracy of this algorithm.

5   Application and Analysis of Algorithms

5.1   Application of Algorithm

Karate club network is known as community structure, and online data disclosure, many researchers have been used to study the complex network community detection algorithm, Therefore, the division results for the of the network has credible, has been a validation tool as a complex network of community structure detected. Karate complex network is by observing the American karate club members between the relationships, according to among the members the usual contacts situation of the establishment of a network. The network includes 34 nodes, 78 edges, nodes represent the club members, while edges on behalf of the relationships between members. Figure 1 is the karate actual data abstraction karate network, which circle vertex on behalf of the manager members of club; square vertices represent the principal members of the teachers.

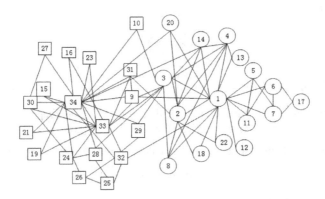

Fig. 1. Karate Network

Algorithm analysis of this complex network is as follows:

1) To calculate the network average degree is approximately equal to 4.6, the value of 4.

2) The initial lists and associations linked list as shown in Table 1:

Table 1. The Initialize chain

The initial list elements	Societies list element
5,6,7,8,10,11,12,13,15,16,17,18,19,20,	1,2,3,4,9,14,
21,22,23,25,26,27,28,29,30,31	24,32,33,34

3) Consider every element in the initial list, according to shortest path put the shortest node to join the linked list of the initial community, until the initial linked list is empty, the result in the following table:

Table 2. The community chain

Societies first element	The value of associations linked list
1	5,6,7,8,11,12,13,17,18,22
2,3,4,9,14,24	
32	25,26
33	24
34	10,15,16,19,20,21,23,27,28,29,30,31

According to the complex network of associations between Common neighbor the number of nodes and the first element of the associations between the total degree of the ratio, calculation of threshold β, according to threshold merged societies obtained the results in the following table:

Table 3. The result

Societies first element	Societies list element
1	2,3,4,5,6,7,8,11,12,13,14,17,18,22
34	9,24,25,26,10,15,16,19,20,21,23,27,28,29,30,31,32,33

5.2 Algorithm Analysis

We can see from the figure, the 20-node is a boundary node, the same links with the led by elements of 1 and 34 community, so there may be ambiguity. The karate network as a social network, when fully from the results of complex networks to study it did not take into account as a social network of external factors affect it. in different social conditions, The karate network division may be different . Therefore, as long as in a certain error range, you could say the algorithm is correct. Accuracy shown in Table 4

Table 4. The Accuracy of two algorithm

Algorithm	Accuracy
K-means clustering algorithm	94.12%
This paper algorithm	97.06%

Generally speaking, the accuracy of this algorithm compared to the unimproved algorithm has the superiority. However, the application of this improved algorithm is still the finding the method of shortest path. The improved algorithm of time complexity O (n3) ,time complexity of the algorithm is relatively large, it is still not suitable for large-scale community structure of complex networks division.

6   Summary

This paper uses the idea of k-means clustering algorithm, on the basis of improvement a similar algorithm. The advantage of this algorithm is on the one hand do not have to know in advance the number of societies, according to the node degree divided   and between nodes neighbor similarity to merge societies, came to the conclusion of the various communities. On the other hand this algorithm is relatively simple, the principle of the application is easy to understand. The node this algorithm to the karate network to be divided, to get good results. Verify this algorithm has feasibility.

References

1. Xie, W., Chen, H., Yang, X., Xiongyong: The study on degree distribution property of complex network based on cooperative communication. Journal of Electronics (China) 3(27), 224–229 (2010)
2. Du, N., Wang, B., Wu, B.: Community Detection in Complex Networks. Journal of Computer Science & Technology 4, 672–683 (2008)
3. Moody, J.: Race: School integration, and friendship segregation in America. Am. J. Social 107(3), 679–716 (2001)
4. Wang, X.: Improved k-means clustering algorithm in societies divided. Qinghai Normal University (Natural Science Edition) 2, 22–24 (2009)
5. Wang, X., Li, X., Chen, G.-R.: Complex network theory and its applications. Tsinghua University Press, Beijing (2006)

Diffusion-Driven Instability and Hopf Bifurcation in Spatial Homogeneous Brusselator Model

Bingfang Li[1] and Gaihui Guo[2]

[1] Department of Basic Course, Shaanxi Railway Institute, Weinan 714000, China
[2] College of Science, Shaanxi University of Science and Technology, Xi'an 710021, China

Abstract. This paper is concerned with the well known Brusselator system. For the spatial homogeneous model, the existence of Hopf bifurcation surrounding the interior equilibrium is obtained. Moreover, the direction of Hopf bifurcation and the stability of bifurcating periodic solutions are established. Our method is based on the center manifold theory. For the model with no-flux boundary conditions, the diffusion-driven instability of the interior equilibrium is studied. Finally, to verify our theoretical results, some examples of numerical simulations are included.

Keywords: Brusselator model, Hopf bifurcation, Diffusion-driven instability.

1 Introduction

This paper is concerned with the following Brusselator model

$$\begin{cases} u_t - d_1 u_{xx} = 1 - (b+1)u + bu^2 v, & x \in (0,\pi), \quad t > 0, \\ v_t - d_2 v_{xx} = a^2(u - u^2 v), & x \in (0,\pi), \quad t > 0, \\ u_x(x,t) = v_x(x,t) = 0, & x = 0, \pi, \quad t > 0, \end{cases} \quad (1)$$

where the unknown functions u, v represent the concentrations of two reactants; d_1, d_2 are diffusion coefficients of u, v respectively. The parameters a, b are fixed concentrations of other components.

The Brusselator model (1) was proposed by Prigogine and Lefever [1]. For this model, the existence and nonexistence of steady-state solutions were investigated in [2, 3]; the global dynamics of the Brusselator equations was studied in [4]. In what follows, we shall consider the existence and stability of Hopf bifurcation in (1). Here, we point out that Hopf bifurcation has been studied in [5]. But, in this paper, we mainly give the Hopf bifurcation analysis to (1) for the spatial homogeneous case. Moreover, the diffusion-driven instability region regarding the diffusion coefficients is established. Furthermore, some numerical simulations are also shown to support and supplement the analytical results. For the researches of Hopf bifurcation, we refer to [6,7] and references therein for details.

D. Jin and S. Lin (Eds.): Advances in FCCS, Vol. 2, AISC 160, pp. 249–255.
springerlink.com   © Springer-Verlag Berlin Heidelberg 2012

2   Hopf Bifurcation to the Spatial Homogeneous Model

The corresponding spatial homogeneous model takes the form

$$\begin{cases} u_t = 1 - (b+1)u + bu^2v, & t > 0, \\ v_t = a^2(u - u^2v), & t > 0. \end{cases} \tag{2}$$

Straightforward calculation shows that (2) has a unique equilibrium solution $U^* := (u^*, v^*) = (1,1)$. We first give the asymptotic stability of this positive constant solution. The Jacobian matrix of (2) at U^* is $A = \begin{pmatrix} b-1 & b \\ -a^2 & -a^2 \end{pmatrix}$. Clearly, $\mathrm{tr}A = b - 1 - a^2$, $\det A = a^2 > 0$. When $b > 1$, let $\tilde{a} = \sqrt{b-1}$.

Theorem 1. (see [5]) (i) U^* is locally asymptotically stable when $0 < b \leq 1$.
(ii) Assume $b > 1$. U^* is locally asymptotically stable when $a > \tilde{a}$ and unstable when $0 < a < \tilde{a}$.

Theorem 2. Assume $b > 1$. Then system (2) undergoes a Hopf bifurcation at U^* when $a = \tilde{a}$. The direction of Hopf bifurcation is subcritical and the bifurcating periodic solutions are orbitally asymptotically stable.

Proof. When $a = \tilde{a}$, A has a pair of imaginary eigenvalues $\pm i\sqrt{b-1}$. Let $\lambda = \alpha(a) \pm i\beta(a)$ be the roots of $\lambda^2 - \mathrm{tr}A\lambda + \det A = 0$, then $\alpha'(\tilde{a}) < 0$. By Poincaré-Andronov-Hopf Bifurcation Theorem, we know that system (2) undergoes a Hopf bifurcation at U^* when $a = \tilde{a}$.

To translate U^* to the origin, let $\tilde{u} = u - u^*$, $\tilde{v} = v - v^*$ and substitute them into (2). For convenience, we still denote $\tilde{u}$ and $\tilde{v}$ by u and v, respectively. Thus, the local system (2) becomes

$$\begin{pmatrix} u_t \\ v_t \end{pmatrix} = A \begin{pmatrix} u \\ v \end{pmatrix} + \begin{pmatrix} F^1(u,v,a) \\ F^2(u,v,a) \end{pmatrix}, \tag{3}$$

where $F^1(u,v,a) = bu^2 + 2buv + bu^2v$, $F^2(u,v,a) = -a^2u^2 - 2a^2uv - a^2u^2v$.

Let $\begin{pmatrix} u \\ v \end{pmatrix} = \begin{pmatrix} 1 & 0 \\ N & M \end{pmatrix} \begin{pmatrix} x \\ y \end{pmatrix}$, $M = -\dfrac{\sqrt{4a^2 - (b-1-a^2)^2}}{2b}$, $N = -\dfrac{a^2 + b - 1}{2b}$.

Then (3) becomes

$$\begin{pmatrix} x_t \\ y_t \end{pmatrix} = \begin{pmatrix} \alpha(a) & -\beta(a) \\ \beta(a) & \alpha(a) \end{pmatrix} \begin{pmatrix} x \\ y \end{pmatrix} + \begin{pmatrix} G^1(x,y,a) \\ G^2(x,y,a) \end{pmatrix}, \tag{4}$$

where $G^1(x, y, a) = F^1(x, Nx + My, a)$ and

$$G^2(x, y, a) = -\frac{N}{M}F^1(x, Nx + My, a) + \frac{1}{M}F^2(x, Nx + My, a).$$

In order to determine the stability and direction of periodic solutions, it follows from [6] that we only need to calculate the sign of $b(\tilde{a})$, which is given by

$$b(\tilde{a}) = \frac{1}{16}[G^1_{xxx} + G^1_{xyy} + G^2_{xxy} + G^2_{yyy}] +$$

$$\frac{1}{16\beta(\tilde{a})}[G^1_{xy}(G^1_{xx} + G^1_{yy}) - G^2_{xy}(G^2_{xx} + G^2_{yy}) - G^1_{xx}G^2_{xx} + G^1_{yy}G^2_{yy}],$$

where all the partial derivatives are evaluated at U^*. Note that when $a = \tilde{a}$,

$$\beta(\tilde{a}) = \sqrt{b-1}, \quad \tilde{M} = M\mid_{a=\tilde{a}} = -\frac{\sqrt{b-1}}{b}, \quad \tilde{N} = N\mid_{a=\tilde{a}} = -\frac{b-1}{b}.$$

By the expressions of $G^1(x, y, a)$ and $G^2(x, y, a)$, we have

$$b(\tilde{a}) = \frac{1}{16} \times 6b\tilde{N} + \frac{1}{16\sqrt{b-1}} \times 2b\tilde{M}(2b + 4b\tilde{N}) = \frac{-b-1}{8} < 0.$$

From $\alpha'(\tilde{a}) < 0$ and the above calculation of $b(\tilde{a})$, the conclusion is true. The proof is finished.

3   Diffusion-Driven Instability

From this section, we shall consider the effect of diffusion on the stability of the interior equilibrium U^* when $b > 1$.

It is well-known that the operator $u \to -u_{xx}$ with the above no-flux boundary conditions has eigenvalues and eigenfunctions $\mu_0 = 0$, $\mu_k = k^2$, $\varphi_0(x) = \sqrt{\frac{1}{\pi}}$, $\varphi_k(x) = \sqrt{\frac{2}{\pi}}\cos(kx)$ for $k = 1, 2, 3, \cdots$. The linearized system of (1) at U^* has the form

$$\begin{pmatrix} u_t \\ v_t \end{pmatrix} = L\begin{pmatrix} u \\ v \end{pmatrix} := \begin{pmatrix} d_1 & 0 \\ 0 & d_2 \end{pmatrix}\begin{pmatrix} u_{xx} \\ v_{xx} \end{pmatrix} + A\begin{pmatrix} u \\ v \end{pmatrix} \qquad (5)$$

Consider the characteristic equation $L(\varphi, \psi) = \lambda(\varphi, \psi)$ and let

$$(\varphi, \psi) = \sum_{k=0}^{\infty} (a_k, b_k)\cos(kx).$$

Then we obtain $\sum_{k=0}^{\infty} (A_k - \mu I)(a_k, b_k)^{\mathrm{T}} \cos(kx) = 0,$ where

$$A_k := A - k^2 D = \begin{pmatrix} b-1-d_1 k^2 & b \\ -a^2 & -a^2 - d_2 k^2 \end{pmatrix}.$$

It is clear to see that all the eigenvalues of L are given by the eigenvalues of A_k for $k = 0, 1, 2, \cdots$. Note that the characteristic equation of A_k is

$$\mu^2 - T_k \mu + D_k = 0, \quad k = 0, 1, 2, \cdots, \tag{6}$$

where

$$T_k = \mathrm{tr}A - (d_1 + d_2)k^2, \quad D_k = d_1 d_2 k^4 + [a^2 d_1 - (b-1)d_2]k^2 + \det A.$$

The interior equilibrium U^* of (1) is unstable when (6) has at least one root with positive real part. Note that $\mathrm{tr}A < 0$ when $a > \tilde{a}$. Then $T_k < 0$. Hence, (6) has no imaginary root with positive real parts. Therefore, U^* is unstable if and only if (6) has at least a positive real root. Define

$$f(k^2) := D_k = d_1 d_2 k^4 + [a^2 d_1 - (b-1)d_2]k^2 + \det A.$$

If $f(k^2) < 0$, then (6) has two real roots in which one is positive and the other is negative. When

$$a^2 d_1 - (b-1)d_2 < 0, \tag{7}$$

$f(k^2)$ will take the minimum value at $k^2 := k_{min}^2 = -[a^2 d_1 - (b-1)d_2]/2d_1 d_2$. In this case, $f(k^2) < 0$ is reduced to $a^4 d_1^2 - 2a^2(b+1)d_1 d_2 + (b-1)d_2^2 > 0$. From it we have

$$0 < \frac{d_1}{d_2} < \frac{(\sqrt{b}-1)^2}{a^2} \quad \text{or} \quad \frac{d_1}{d_2} > \frac{(\sqrt{b}+1)^2}{a^2}.$$

The second inequality should be deleted from (7). Hence, if d_1 / d_2 satisfies

$$\text{(H):} \quad 0 < \frac{d_1}{d_2} < \frac{(\sqrt{b}-1)^2}{a^2},$$

then $D_k < 0$ and the diffusion-driven instability occurs.

Theorem 3. Assume that $b>1$ and $a>\tilde{a}$, where $\tilde{a}=\sqrt{b-1}$, so that U^* is asymptotically stable for (2). Then U^* is an unstable equilibrium solution of (1) if (H) holds.

4   Numerical Simulations

In this section, we present some numerical simulations to illustrate and supplement our theoretical results above. Choose $b=5$ for all simulations. Then we have the critical value $\tilde{a}=2$.

For the spatial homogeneous model (2), by Theorem 2, we know that the equilibrium U^* is asymptotically stable when $a>\tilde{a}$, a Hopf bifurcation occurs when a passes through the critical value $\tilde{a}$, the direction of Hopf bifurcation is subcritical and the bifurcating periodic solutions are asymptotically stable. These facts are shown by the numerical simulations, see Fig. 1.

If the condition (H) holds, then the equilibrium U^* of (1) is unstable, that is, the diffusion-driven instability occurs. At this time, we can find spatial homogeneous periodic solutions to (1), see Fig.2. Moreover, plenty of numerical simulations suggest that there exist positive steady-state solutions to (1) for suitable parameters, see Fig.3.

We hope that these results could provide the corresponding theory basis for explaining and predicting some phenomena arising in autocatalysis.

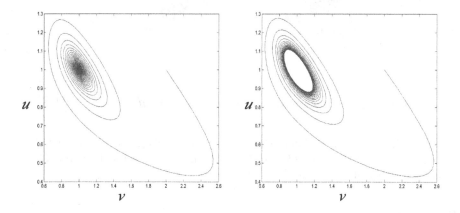

Fig. 1. Phase portraits of (2). Left: $a=2.01$, the interior equilibrium $U^*=(1,1)$ is asymptotically stable ; right: $a=1.99$, the interior equilibrium $U^*=(1,1)$ is unstable and (2) has nonconstant periodic solution surrounding it.

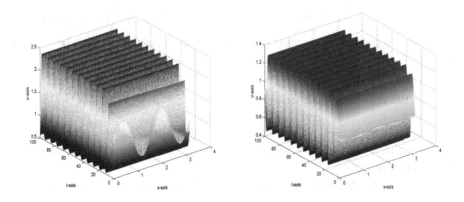

Fig. 2. Homogeneous periodic solution of system (1) with $d_1 = 1$, $d_2 = 10$, $a = 1.8$.

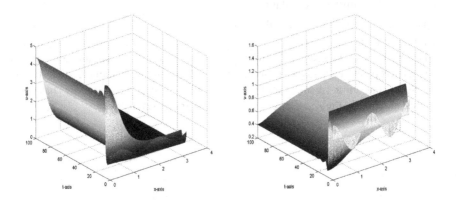

Fig. 3. Steady-state solution of system (1) with $d_1 = 0.1$, $d_2 = 2.5$, $a = 1.8$.

Acknowledgments. The authors thank the supports of the Fund Project of Shaanxi Railway Institute (No. 2011-43) and the Natural Science Basic Research Plan in Shaanxi Province of China (No. 2011JQ1015).

References

1. Prigogene, I., Lefever, R.: Symmetry breaking instabilities in dissipative systems II. J. Chem. Phys. 48, 1665–1700 (1968)
2. Brown, K.J., Davidson, F.A.: Global bifurcation in the Brusselator system. Nonlinear Anal. TMA 12, 1713–1725 (1995)
3. Peng, R., Wang, M.X.: Pattern formation in the Brusselator system. J. Math. Anal. Appl. 309, 151–166 (2005)

4. You, Y.: Global dynamics of the Brusselator equations. Dyn. Partial Diff. Eqns. 4, 167–196 (2007)
5. Li, B., Wang, M.X.: Diffusive driven instability and Hopf bifurcation in the Brusselator system. Appl. Math. Mech. 29, 825–832 (2008)
6. Yi, F.Q., Wei, J.J., Shi, J.P.: Diffusion-driven instability and bifurcation in the Lengyel-Epstein system. Nonlinear Anal. 9, 1038–1051 (2008)
7. Yi, F.Q., Wei, J.J., Shi, J.P.: Bifurcation and spatiotemporal patterns in a homogeneous diffusive predator-prey system. J. Differential Equations 246, 1944–1977 (2009)

Face Recognition Based on Adaptive Kernel Logistic Regression

Ziqiang Wang and Xia Sun

School of Information Science and Engineering, Henan University of Technology,
450001 Zhengzhou, China
{wzqagent,sunxiamail}@126.com

Abstract. Face recognition is a fundamental problem in many image processing tasks. Discriminating feature extraction and classifier selection are two key issues for face recognition approach. To effectively cope with the above two problems, a novel face recognition algorithm is proposed by using locality preserving projection and manifold adaptive kernel logistic regression. The experimental results on two face image data sets show that the proposed algorithm performs much better than the other popular face recognition algorithms in terms of recognition accuracy.

Keywords: face recognition, kernel logistic regression, locality preserving projection, manifold learning.

1 Introduction

Face recognition has attracted growing attention for real applications. Although numerous research efforts have been devoted to face recognition, the general face recognition problem is still a very challenging research issue due to the semantic gap between low-level visual features and high-level semantic concepts[1]. Since a major challenge of face recognition is that the captured face image data often lies in a high-dimensional feature space, two issues are central to all these face recognition algorithms: 1) discriminating feature extraction for face representation and 2) classification of a new face image based on the chosen feature representation. This work focuses on the two issues of feature extraction and classifier selection.

Linear subspace analysis has become a kind of popular feature extraction and representation method for face recognition due to its effectiveness and computational efficiency for feature extraction and representation. Principle component analysis (PCA) and linear discriminant analysis (LDA) are two classic algorithms for data reduction and feature extraction[2]. Numerous researches have shown that the LDA-based algorithms generally outperform PCA-based ones. However, both PCA and LDA algorithms see only the global Euclidean structure and cannot discover the nonlinear manifold structure hidden in the high-dimensional data. Recently, a number of research efforts have shown that the face images possibly reside on a nonlinear submanifold. To discover the underlying manifold structure for efficient face representation and recognition, a number of manifold learning algorithms have been

D. Jin and S. Lin (Eds.): Advances in FCCS, Vol. 2, AISC 160, pp. 257–262.
springerlink.com   © Springer-Verlag Berlin Heidelberg 2012

proposed and have been successfully applied to face recognition. Locality preserving projections (LPP) is the most representative manifold algorithm[3,4]. Therefore, we adopt the LPP algorithm for face image data reduction and feature extraction in this paper.

As for face recognition, classifier selection is another key issue after feature extraction. At present, the k-nearest neighbor (KNN) classifier is widely used in most face recognition systems. However, for large data sets, the computational demands for classifying patterns using KNN can be prohibitive. Support vector machine(SVM)[5] is a popular classifier algorithm used in recent years. However, the time complexity of SVM is cubic in the number of training points, it is thus computationally infeasible on very large data sets. Recently, logistic regression classifier[6] has attracted considerable attention for pattern classification due to its lower computational cost and higher classification accuracy. However, as a linear classifier, logistic regression is inadequate to describe the complexity of real face images because of illumination, facial expression, and pose variations. To deal with this limitation, nonlinear extensions of logistic regression through kernel trick are first proposed in this work, then the proposed manifold adaptive kernel logistic regression classifier is adopted as the face classification algorithm in the reduced lower-dimensional facial feature space in this paper.

2   Lower-Dimensional Facial Feature Extraction with LPP

In the appearance-based face recognition, the face image space is always of very high dimensionality. Due to the consideration of the curse of dimensionality, it is desirable to first project the face images into a lower-dimensional feature subspace in which the semantic structure of the image space becomes clear. LPP is a recently proposed manifold learning method for feature extraction and dimensionality reduction[3], it aims to find an embedding that preserves local information and best detects the essential manifold structure of the data set.

Given a set of face images $X = [x_1, x_2, \ldots, x_n] \in \mathbb{R}^{m \times n}$, where m is the dimension of each face image and n is the number of face images. LPP aims to find a linear transformation $V \in \mathbb{R}^{m \times d}$ that maps each image $x_i (i = 1, \cdots, n)$ in the m-dimensional space to a vector y_i in the lower d-dimensional space by $y_i = V^T x_i$ such that y_i represents x_i well in terms of the following optimal objective:

$$
\begin{aligned}
& \arg\min_V \sum_{i,j} (y_i - y_j)^2 S_{ij} \\
& = \arg\min_V \sum_{i,j} (V^T x_i - V^T x_j)^2 S_{ij} \\
& = \arg\min_V V^T XLX^T V
\end{aligned}
\tag{1}
$$

with the constraint

$$YDY^T = 1 \Rightarrow V^T XDX^T V = 1 \qquad (2)$$

where $Y = [y_1, y_2, \ldots, y_n] \in \mathbb{R}^{d \times n}$, S_{ij} denotes the similarity between the face images x_i and x_j, its definition is as follows:

$$S_{ij} = \begin{cases} 1, & \text{if } x_i \text{ is among the } k \text{ nearest neighbor of } x_j \\ & \text{or } x_j \text{ is among the } k \text{ nerest neighbor of } x_i. \\ 0, & \text{otherwise.} \end{cases} \qquad (3)$$

In addition, D is a diagonal matrix, its entries are column (or row) sums of S, $D_{ii} = \sum_j S_{ij}$, and $L = D - S$ is the graph Laplacian.

Then the transformation vector V that minimizes (1) under the constraint (2) is given by the minimum eigenvalue solution to the generalized eigenvalue problem:

$$XLX^T V = \lambda XDX^T V \qquad (4)$$

Thus, the problem of LPP is converted into finding the eigenvectors of the matrix $(XDX^T)^{-1} XLX$ associated with the smallest eigenvalues. For a new face image x, its lower-dimensional embedding can be calculated by

$$x \to y = V^T x. \qquad (5)$$

For face recognition, a problem arises that the matrix XDX^T cannot be guaranteed to be nonsingular. In this case, we can first apply PCA to remove the components corresponding to zero eigenvalues.

3   Adaptive Kernel Logistic Regression Classifier

Once facial feature are extracted with LPP algorithm, face recognition becomes a pattern classification task. Since kernel logistic regression is an effective and efficiency classifier for very large data sets, we use kernel logistic regression as face recognition algorithm in this paper.

Given the input face training samples $x = (x_1, \ldots, x_n)$ in the above LPP projected subspace, let $y \in \{+1, -1\}$ be the class label, logistic regression models the conditional probability of assigning a class label y to the input sample x according to the following equation

$$p(y|x) = \frac{1}{1 + \exp(-y(w^T x + b))} \qquad (6)$$

where w denotes the weights associated to input samples and b is the corresponding bias term.

As can be seen from (6), logistic regression is a linear classifier[6]. So it is inadequate to describe the complexity nonlinear facial structure features. To extend logistic regression to the nonlinear case, the adaptive kernel logistic regression(AKLR) classifier is proposed in the following.

The idea of AKLR is to solve the logistic regression in an implicit feature space constructed by the nonlinear mapping $\varphi : x \to \varphi(x) \in F$. In implementation, implicit feature vector does not need to be calculated explicitly, while it is just done by computing the inner product of two vectors in F with the following kernel function:

$$K\left(x_i, x_j\right) = \varphi\left(x_i\right) \cdot \varphi\left(x_j\right) \tag{7}$$

Consider a binary classification problem:

$$\left\{\left(x_i, y_i\right)\right\}_{i=1}^{n} \text{ and } y_i = \left\{+1, -1\right\} \tag{8}$$

where x_i is the training sample and y_i is the class label that x_i belongs to. Then, AKLR can be regarded as the below optimization problem:

$$\min_{f \in H_K} \frac{1}{n} \sum_{i=1}^{n} \ln\left(1 + e^{-y_i f(x_i)}\right) + \frac{\lambda}{2} \|f\|_{H_K}^2 \tag{9}$$

where H_K is a reproducing kernel Hilbert space(RKHS). The solution can be found through a Wolfe dual problem with the Lagrangian multipliers α_i . Thus for a given kernel function $K\left(,\right)$, the decision rule of AKLR classifier is given by

$$f\left(x\right) = \sum_{i=1}^{n} \alpha_i K\left(x_i, x\right) \tag{10}$$

According to the above derivation of AKLR, we can observe that different kernel function will produce different implicit kernel feature space. However, how to choose a suitable kernel function for a given application is still an open problem so far. In this research, motivated by the fact that manifold structure can enhance the discriminating power of the original kernel function[7], we employ the manifold adaptive kernel function

$$K\left(x, z\right) = \tilde{k}\left(x, z\right) - \tilde{k}_x^T \left(I + L\tilde{K}\right)^{-1} L\tilde{k}_z \tag{11}$$

where $\tilde{k}\left(,\right)$ is the original kernel function, $\tilde{k}_x = \left(\tilde{k}\left(x, x_1\right), \cdots, \tilde{k}\left(x, x_n\right)\right)$, and L is the graph Laplacian defined in Section 2. When an input kernel is deformed according to the manifold structure, the resulting kernel may be able to achieve better performance than the original input kernel.

4   Experimental Results

In this section, we investigate the performance of our proposed LPP plus adaptive kernel logistic regression(LPP+AKLR) algorithm for face recognition on two benchmark databases: Yale and ORL face databases. We compared LPP+AKLR algorithm with the PCA+KNN algorithm, the LDA+KNN algorithm, and LPP+KNN algorithm, three of the most popular face recognition algorithms.

The characteristics of the Yale and ORL databases are summarized as follows: The Yale data set[8] contains 165 gray scale images of 15 individuals. The ORL data set[9] contains 400 images of 40 individuals. All the face images are manually aligned, cropped and resized into 112×92, a standardized image commonly used in face recognition tasks. The number of nearest neighbors in our algorithm was set to be 6. Each data set was partitioned into randomly into a training set consisting of one half of the whole data set and a testing set consisting of the remainder one half of the whole data set. To reduce the variability, the splitting was repeated 10 times and the resulting accuracies were averaged.

The maximal average recognition accuracies as well as the corresponding optimal reduced dimensions obtained by all these algorithms on two face database are listed in Table 1 and Table 2, respectively. From the experimental results, we can make the following observations: Our proposed LPP+AKLR algorithm consistently outperforms the PCA+KNN, LDA+KNN and LPP+KNN algorithm. This is probably because our proposed algorithm not only extracts discriminating facial features with LPP algorithm, but also enhances the classification accuracy through manifold adaptive kernel logistic regression classifier. The PCA+KNN algorithm gives relative poor accuracy. This can be attributed to the fact that PCA is unsupervised and features extracted by PCA are not optimal for classification. Therefore, it is indeed necessary to simultaneously consider the feature extraction and classifier for designing practical face recognition algorithm.

Table 1. Performance comparison on the Yale database

Algorithm	Recognition accuracy	Dimension
PCA+KNN	76.4%	33
LDA+KNN	82.3%	14
LPP+KNN	90.8%	28
LPP+AKLR	95.2%	28

Table 2. Performance comparison on the ORL database

Algorithm	Recognition accuracy	Dimension
PCA+KNN	86.2%	199
LDA+KNN	93.5%	39
LPP+KNN	94.1%	40
LPP+AKLR	97.3%	40

5   Conclusion

In this paper, a novel face recognition algorithm by combination of LPP and AKLR is proposed. Experimental results demonstrate that the proposed algorithm achieves much better performance than other well-known face recognition algorithms.

Acknowledgment. This work is supported by the National Natural Science Foundation of China under Grant No.70701013, and the Natural Science Foundation of Henan Province under Grant No. 102300410020.

References

1. Liu, Q., Lu, H., Ma, S.: Improving Kernel Fisher Discriminant Analysis for Face Recognition. IEEE Transactions on Circuits and Systems for Video Technology 14, 42–49 (2004)
2. Duda, R.O., Hart, P.E., Stork, D.G.: Pattern Classification, 2nd edn. Wiley-Interscience, Hoboken (2000)
3. He, X., Niyogi, P.: Locality Preserving Projections. In: Advances in Neural Information Processing Systems, pp. 585–591. The MIT Press, Cambridge (2003)
4. He, X., Yan, S., Hu, Y., Niyogi, P., Zhang, H.-J.: Face Recognition Using Laplacianfaces. IEEE Transactions on Pattern Analysis and Machine Intelligence 27, 328–340 (2005)
5. Vapnik, V.N.: The Nature of Statistical Learning Theory. Springer, New York (1995)
6. Komarek, P., Moore, A.: Fast Robust Logistic Regression for Large Sparse Data Sets with Binary Outputs. In: Ninth International Workshop on Artificial Intelligence and Statistics, pp. 1–8. Oxford University Press, New York (2003)
7. Sindhwani, V., Niyogi, P., Belkin, M.: Beyond the Point Cloud: from Transductive to Semi-Supervised Learning. In: 22nd International Conference on Machine Learning, pp. 824–831. ACM Press, New York (2005)
8. The Yale database,
   http://cvc.yale.edu/projects/yalefaces/yalefaces.html
9. The Olivetti Research Laboratory (ORL) database,
   http://www.uk.research.att.com/facedatabase.html

The Uniqueness of Positive Solutions for a Predator-Prey Model with B-D Functional Response

Gaihui Guo[1] and Bingfang Li[2]

[1] College of Science, Shaanxi University of Science and Technology, Xi'an 710021, China
[2] Department of Basic Course, Shaanxi Railway Institute, Weinan 714000, China

Abstract. In this paper, we investigate a diffusive predator-prey model with Beddington-DeAngelis functional response, which is subject to homogeneous Dirichlet boundary conditions. Based on the fixed point index theory, a good understanding of the existence, uniqueness and stability of positive solutions is obtained when b is sufficiently small. We always reduce the proof of uniqueness and stability to the proof of the fact that any possible positive solution is non-degenerate and linearly stable.

Keywords: Predator-prey, Beddington-DeAngelis functional response, Index, Uniqueness.

1   Introduction

Let Ω be a bounded domain in R^N with smooth boundary $\partial\Omega$. Consider the following reaction-diffusion system

$$
\begin{cases}
-\Delta u = (a - u - \dfrac{bv}{1 + mu + kv})u, & x \in \Omega, \\[2mm]
-\Delta v = (c - v + \dfrac{du}{1 + mu + kv})v, & x \in \Omega, \\[2mm]
u = v = 0, & x \in \partial\Omega.
\end{cases}
\tag{1}
$$

where u and v represent the densities of prey and predator, respectively. The parameters a, b, c, d, m and k are constants with a, b, d positive and m, k non-negative; c may change sign. $u / (1 + mu + kv)$ is the so-called Beddington-DeAngelis (simply write as B-D) functional response. The corresponding ordinary differential system was introduced by Beddington [1] and DeAngelis [2].

Recently, we carried out qualitative analyses on (1) and obtained some existence, uniqueness and multiplicity results of positive solutions, see [3,4]. Motivated by the paper [5], where the uniqueness of positive solutions for a predator-prey model with predator saturation and competition was given when $b \lll 1$, in this paper, we shall

D. Jin and S. Lin (Eds.): Advances in FCCS, Vol. 2, AISC 160, pp. 263–269.
springerlink.com   © Springer-Verlag Berlin Heidelberg 2012

study the properties of positive solutions to (1) when $b \ll 1$, and further establish the uniqueness of positive solutions to (1) in this case. The proof of uniqueness is mainly based on the fact that any possible positive solution is non-degenerate and linearly stable. One can see [6] for detailed studies alone this line.

In order to present our main results, we need to give some notations and basic facts. For $q(x) \in C^{\alpha}(\overline{\Omega})$, let $\lambda_1(q) < \lambda_2(q) \leq \lambda_3(q) \leq \cdots$ be all eigenvalues of

$$-\Delta\phi + q(x)\phi = \lambda\phi \text{ in } \Omega, \quad \phi = 0 \text{ on } \partial\Omega.$$

As we know from [7], $\lambda_1(q)$ is simple and $\lambda_1(q)$ is strictly increasing in the sense that $q_1 \leq q_2$ and $q_1 \not\equiv q_2$ implies $\lambda_1(q_1) < \lambda_1(q_2)$. When $q(x) \equiv 0$, we denote $\lambda_1(0)$ by λ_1 for the sake of convenience. For any $a > \lambda_1$, it is well known from [7] that the problem $-\Delta u = (a - u)u$ in Ω, $u = 0$ on $\partial\Omega$ has a unique positive solution we denote by θ_a. It is also known that the mapping $a \to \theta_a$ is strictly increasing, continuously differentiable in (λ_1, ∞), and that $\theta_a \to 0$ uniformly on $\overline{\Omega}$ as $a \to \lambda_1$. Therefore, if $a > \lambda_1$, then (1) has a semi-trivial solution $(\theta_a, 0)$. Similar results hold with respect to another semi-trivial solution $(0, \theta_c)$ whenever $c > \lambda_1$. The main result of this work reads as follows.

Theorem 1. If $a > \lambda_1$, then (1) has at most one positive solution when $b \ll 1$.

2   Preliminary

In this section, we give some known results, which will be used later.

Lemma 2. (See [3]) Suppose that (u, v) is a positive solution of (1). Then (u, v) satisfies

$$u < \theta_a < a \text{ and } v < c + \frac{da}{1 + ma}.$$

Furthermore, $v > \theta_c$ if $c > \lambda_1$.

Let X be a real Banach space and W a closed convex set of X. W is called a total wedge if $\beta W \subset W$ for all $\beta > 0$ and $\overline{W - W} = X$. A wedge is said to be a cone if $W \cap (-W) = \{0\}$. For $y \in W$, define

$$W_y = \{x \in X : y + \gamma x \in W, \gamma > 0\} \text{ and } S_y = \{x \in \overline{W}_y : -x \in \overline{W}_y\}.$$

Let $F : W \to W$ be a compact operator with a fixed point $y \in W$ and denote by L the Fréchet derivative of F at y. Then L maps $\overline{W}_y$ into itself. We say that

L has property α on $\overline{W}_y$ if there exist $t \in (0,1)$ and $w \in \overline{W}_y \setminus S_y$ such that $w - tLw \in S_y$. We denote by $index_W(F, y)$ the fixed point index of F at y relative to W.

Lemma 3. (See [8])   Assume that $I - L$ is invertible on X.

(i) If L has property α on $\overline{W}_y$, then $index_W(F, y) = 0$.

(ii) If L does not have property α on $\overline{W}_y$, then $index_W(F, y) = (-1)^\sigma$, where σ is the sum of algebra multiplicities of the eigenvalues of L which are greater than 1.

Introduce the following notations: $X = C_0(\overline{\Omega}) \oplus C_0(\overline{\Omega})$; $W = P \oplus P$, where $P = \{ \varphi \in C_0(\overline{\Omega}) : \varphi(x) \geq 0, x \in \overline{\Omega} \}$, $D = \{(u,v) \in X : u < a, v < c + \frac{da}{1+ma} \}$; $D' = (\text{int}D) \cap W$. For any $t \in [0,1]$, define $A_t : D' \to W$ by

$$A_t(u,v) = (-\Delta + M)^{-1} \begin{pmatrix} tu(a - u - \dfrac{bv}{1+mu+kv}) + Mu \\ tv(c - v + \dfrac{du}{1+mu+kv}) + Mv \end{pmatrix},$$

where $M = \max\{ad, b(c+ad)\}$. Observe that (1) has a positive solution in W if and only if $A := A_1$ has a positive fixed point in D'. If $a, c > \lambda_1$, then $(0,0)$, $(\theta_a, 0)$, $(0, \theta_c)$ are the only non-negative fixed points of A which are not positive. The corresponding indices in W can be calculated in the following lemmas.

Lemma 4. (See [3])   Assume that $a > \lambda_1$.

(i) $Index_W(A, D') = 1$, $index_W(A, (0,0)) = 0$.

(ii) If $c > \lambda_1(-\frac{d\theta_a}{1+m\theta_a})$, then $index_W(A, (\theta_a, 0)) = 0$.

(iii) If $c < \lambda_1(-\frac{d\theta_a}{1+m\theta_a})$, then $index_W(A, (\theta_a, 0)) = 1$.

Lemma 5. (See [3])   Assume that $c > \lambda_1$.

(i) If $a > \lambda_1(\frac{b\theta_c}{1+k\theta_c})$, then $index_W(A, (0, \theta_c)) = 0$.

(ii) If $a < \lambda_1(\frac{b\theta_c}{1+k\theta_c})$, then $index_W(A, (0, \theta_c)) = 1$.

We state the existence of positive solutions to (1).

Theorem 6. (See [3]) (i) If $a > \lambda_1$ and $\lambda_1(-\frac{d\theta_a}{1+m\theta_a}) < c \le \lambda_1$, then (1) has at least a positive solution.

(ii) If $c > \lambda_1$ and $a > \lambda_1(\frac{b\theta_c}{1+k\theta_c})$, then (1) has at least a positive solution.

3   Proof of Theorem 1

We first consider the asymptomatic behavior of positive solutions as $b \to 0+$.

Theorem 7. If $a > \lambda_1$, then any positive solution of (1) (if exists) converges to $(\theta_a, \hat{v})$ as $b \to 0+$, where $\hat{v}$ is a unique positive solution of

$$-\Delta v = (c - v + \frac{d\theta_a}{1 + m\theta_a + kv})v, \quad x \in \Omega, \quad v = 0, \quad x \in \partial\Omega. \tag{2}$$

Proof. If (1) has positive solutions, then $a > \lambda_1$ and $c > \lambda_1(-\frac{d\theta_a}{1+m\theta_a})$. Note that the function $f(x, v) := c - v + \frac{d\theta_a(x)}{1+m\theta_a(x)+kv}$ is monotone decreasing for $v > 0$, and $f(x, v) < 0$ when $v \gg 1$. By Theorem 2.2 in [9], we see that (2) has a unique positive solution, denoted by $\hat{v}$, in view of $c > \lambda_1(-\frac{d\theta_a}{1+m\theta_a})$.

It is easy to see that the compact operator $A(u, v)$ converges

$$\hat{A}(u, v) = (-\Delta + M)^{-1} \begin{pmatrix} u(a - u) + Mu \\ v(c - v + \dfrac{du}{1 + mu + kv}) + Mv \end{pmatrix}$$

in D' as $b \to 0+$. So the fixed points of $A(u, v)$ converge to the fixed points of $\hat{A}(u, v)$ in D' as $b \to 0+$. Since $\hat{A}(u, v)$ in D' has a unique fixed point $(\theta_a, \hat{v})$, the result follows.

Now we discuss the stability of positive solutions to (1) when $b \ll 1$.

Theorem 8. If $a > \lambda_1$ and $c > \lambda_1(-\frac{d\theta_a}{1+m\theta_a})$, then any positive solution of (1) (if exists) is non-degenerate and linearly stable when $b \ll 1$.

Proof. It suffices to show that the linearized eigenvalue problem of (1) has no eigenvalue μ with $Re(\mu) \le 0$. To do this, a contradiction argument will be used by assuming that (1) with $b = b_i$ has a positive solution (u_i, v_i) which is either degenerate or linearly unstable for sequences $b_i \to 0$. Assume that there exists $(\xi_i, \zeta_i) \not\equiv (0,0)$ such that the following linearized problem

$$
\begin{cases}
-\Delta \xi_i - [a - 2u_i - \dfrac{b_i v_i (1 + k v_i)}{(1 + m u_i + k v_i)^2}]\xi_i + \dfrac{b_i u_i (1 + m u_i)}{(1 + m u_i + k v_i)^2}\zeta_i = \mu_i \xi_i, & x \in \Omega, \\
-\Delta \zeta_i - [c - 2v_i + \dfrac{d u_i (1 + m u_i)}{(1 + m u_i + k v_i)^2}]\zeta_i - \dfrac{d v_i (1 + k v_i)}{(1 + m u_i + k v_i)^2}\xi_i = \mu_i \zeta_i, & x \in \Omega,
\end{cases}
$$

$$(3)$$

has an eigenvalue μ_i with $Re(\mu_i) \leq 0$. We may assume $\| \xi_i \|_2^2 + \| \zeta_i \|_2^2 = 1$. By arrangements, we obtain

$$
\mu_i = \int_\Omega |\nabla \xi_i|^2 \, dx - \int_\Omega [a - 2u_i - \dfrac{b_i v_i (1 + k v_i)}{(1 + m u_i + k v_i)^2}] |\xi_i|^2 \, dx +
$$

$$
\int_\Omega \dfrac{b_i u_i (1 + m u_i)}{(1 + m u_i + k v_i)^2} \zeta_i \bar{\xi}_i dx + \int_\Omega |\nabla \zeta_i|^2 \, dx -
$$

$$
\int_\Omega [c - 2v_i + \dfrac{d u_i (1 + m u_i)}{(1 + m u_i + k v_i)^2}] |\zeta_i|^2 \, dx - \int_\Omega \dfrac{d v_i (1 + k v_i)}{(1 + m u_i + k v_i)^2} \xi_i \bar{\zeta}_i dx,
$$

where $\bar{\xi}_i$ and $\bar{\zeta}_i$ are the complex conjugates of ξ_i and ζ_i. By Lemma 2, we know that $0 < u_i < \theta_a$ and $\theta_c < v_i < c + da$. It is easy to see that $Im(\mu_i)$ is bounded and $Re(\mu_i)$ is bounded from below. Thus μ_i is bounded as we assume $Re(\mu_i) \leq 0$. By L^p estimate, we have $\| \xi_i \|_{W^{2,2}}$ and $\| \zeta_i \|_{W^{2,2}}$ are bounded. Hence, we can assume $\mu_i \to \mu$ with $Re(\mu) \leq 0$ and $(\xi_i, \zeta_i) \to (\xi, \zeta) \not\equiv (0,0)$ in H_0^1 strongly. From Theorem 7, it follows that $u_i \to \theta_a, v_i \to \hat{v}$, where $\hat{v}$ is a unique positive solution of (2). Letting $i \to \infty$ in (3), we see that (ξ, ζ) satisfies

$$
\begin{cases}
-\Delta \xi - (a - 2\theta_a)\xi = \mu \xi, & x \in \Omega, \\
-\Delta \zeta - [c - 2\hat{v} + \dfrac{d \theta_a (1 + m \theta_a)}{(1 + m \theta_a + k \hat{v})^2}]\zeta - \dfrac{d \hat{v}(1 + k \hat{v})}{(1 + m \theta_a + k \hat{v})^2}\xi = \mu \zeta, & x \in \Omega,
\end{cases}
$$

$$(4)$$

Obviously, $\mu \in R$. If $\xi \equiv 0$, then we have

$$
\mu \geq \lambda_1(-c + 2\hat{v} - \dfrac{d \theta_a (1 + m \theta_a)}{(1 + m \theta_a + k \hat{v})^2}) > \lambda_1(-c + \hat{v} - \dfrac{d \theta_a}{1 + m \theta_a + k \hat{v}}) = 0,
$$

a contradiction. Thus $\xi \not\equiv 0$ and from the first equality of (4), it follows $\mu > 0$ again. This shows that any positive solution of (1) is non-degenerate and linearly stable when $b \ll 1$.

Proof of Theorem 1.   We divide the proof into three cases.

(i) Assume that $a > \lambda_1$ and $c \leq \lambda_1 (-\frac{d\theta_a}{1+m\theta_a})$. It follows from Theorem 2.1 in [3] that (1) has no positive solution.

(ii) Assume that $a > \lambda_1$ and $\lambda_1 (-\frac{d\theta_a}{1+m\theta_a}) < c \leq \lambda_1$. The existence of positive solutions to (1) was given in Theorem 6. It suffices to show the uniqueness. From Theorem 7, we know that any positive solution of (1) is non-degenerate and linearly stable when $b \ll 1$. Hence, by compactness, A has at most finitely many positive fixed points in the region D'. Let us denote them by (u_i, v_i) for $i = 1, 2, \cdots, l$. Due to the non-degeneration and stability of any positive solution, $I - L$ is invertible in X and L has no eigenvalue greater than one, where L is the Fréchlet derivative of A at (u_i, v_i). Hence, L does not have the property α. From Lemma 3, it follows that $index_W (A, (u_i, v_i)) = 1$. By Lemmas 4, 5 and the additivity property of the index, we have

$$1 = \quad index_W (A, D') = index_W (A, (0,0)) + index_W (A, (\theta_a, 0)) +$$

$$\sum_{i=1}^{l} index_W (A, (u_i, v_i)) = 0 + 0 + l = l.$$

which shows that (1) has a unique positive solution for sufficiently small b.

(iii) Assume that $c > \lambda_1$. For any $\epsilon > 0$ small, there exists $b^* = \epsilon k$ small such that if $a \geq \lambda_1 + \epsilon$ and $b \leq b^*$, then we have $a \geq \lambda_1 + b/k > \lambda_1 (\frac{b\theta_c}{1+k\theta_c})$, and similarly, we get

$$1 = index_W (A, D') = index_W (A, (0,0)) + index_W (A, (\theta_a, 0)) +$$

$$index_W (A, (0, \theta_c)) + \sum_{i=1}^{l} index_W (A, (u_i, v_i)) = 0 + 0 + 0 + l = l.$$

Hence, (1) has a unique positive solution for $a, c > \lambda_1$ when $b \ll 1$.

Arguments in (i)-(iii) above imply that the conclusion in Theorem 1 holds true.

Acknowledgments. The first author thanks the support of the Natural Science Basic Research Plan in Shaanxi Province of China (No. 2011JQ1015).

References

1. Beddington, J.R.: Mutual interference between parasites or predators and its effect on searching efficiency. J. Animal Ecol. 44(1), 331–340 (1975)
2. DeAngelis, D.L., Goldstein, R.A., O'Neill, R.V.: A model for tropic interaction. Ecology 56(2), 881–892 (1975)
3. Guo, G.H., Wu, J.H.: Multiplicity and uniqueness of positive solutions for a predator-prey model with B-D functional response. Nonlinear Analy. TMA 72, 1632–1646 (2010)
4. Guo, G.H., Wu, J.H.: Existence and uniqueness of positive solutions for a predator-prey model with diffusion. Acta Mathematica Scientia 31, 196–205 (2011) (in Chinese)
5. Wang, M.X., Wu, Q.: Positive solutions of a prey-predator model with predator saturation and competition. J. Math. Anal. Appl. 345, 708–718 (2008)
6. Du, Y.H., Lou, Y.: Some uniqueness and exact multiplicity results for a predator-prey model. Trans. Amer. Math. Soc. 349(6), 2443–2475 (1997)
7. Ye, Q.X., Li, Z.Y.: Introduction to Reaction-Diffusion Equations. Science Press, Beijing (1990) (in Chinese)
8. Dancer, E.N.: On the indices of fixed points of mappings in cones and applications. J. Math. Anal. Appl. 91, 131–151 (1983)
9. Ko, W., Ryu, K.: Coexistence states of a predator-prey system with non-monotonic functional response. Nonlinear Anal. RWA 8, 769–786 (2007)

Improving Support Vector Data Description for Document Clustering

Ziqiang Wang and Xia Sun

School of Information Science and Engineering, Henan University of Technology,
450001 Zhengzhou, China
{wzqagent,sunxiamail}@126.com

Abstract. Document clustering has received a lot of attention due to its wide application in many fields. To effectively deal with this problem, a new document clustering algorithm is proposed by using marginal fisher analysis (MFA) and improved support vector data description (SVDD) algorithms in this paper. The high-dimensional document data are first mapped into lower-dimensional feature space with MFA, the improved SVDD is then applied to cluster the documents into different classes in the reduced feature space. Experimental results on two document databases demonstrate the effectiveness of the proposed algorithm.

Keywords: document clustering, support vector data description, marginal fisher analysis, dimensionality reduction.

1   Introduction

Document clustering is a key component for many practical applications. In the past decades, a large number of document clustering algorithms have been proposed to partition document data into clusters such that the document data within the same group are similar to each other, while the document data in different groups are dissimilar. With the explosive increase in document data on the Internet and the rapid advances of computer technology, clustering documents from a large-scale document database has become a hot research topic in the fields of pattern recognition and data mining[1]. However, a major challenge of document clustering is that the obtained document data often lies in a high-dimensional feature space. Thus it is often necessary to conduct dimensionality reduction to acquire an efficient and discriminative representation before formally conducting document clustering.

Latent semantic index(LSI)[2] is one of the most popular dimensionality reduction and feature extraction methods for document information process. However, LSI ignores the cluster structure while reducing the dimension. In addition, nonnegative matrix factorization (NMF) is also used to reduce the dimensions of document[3]. However, NMF is unsupervised and hence the derived coefficient matrix is unnecessary to be great at classification capability. Moreover, NMF algorithms are more computationally demanding due to its iterative update rule. While both LSI and NMF have attained reasonably good performance in document clustering, they may

D. Jin and S. Lin (Eds.): Advances in FCCS, Vol. 2, AISC 160, pp. 271–276.
springerlink.com   © Springer-Verlag Berlin Heidelberg 2012

fail to discover the underlying nonlinear manifold structure as they seek only a compact Euclidean subspace for document representation and clustering. To discover the intrinsic manifold structure of the document data, He et al. [4] developed the locality preserving indexing(LPI) method for document clustering. However, LPI algorithm is designed to best preserve data locality or similarity in the embedding space rather than good discriminating capability. In the recent research, Yan et al.[5] propose a new manifold learning algorithm termed marginal fisher analysis (MFA) for dimensionality reduction. While MFA has attained reasonably good performance in face recognition and gait recognition, the use of the MFA algorithm for document clustering in the context of data mining is still a research area where few people have tried to explore.

For document clustering, how to cluster different documents in the reduced feature space is another key issue. The support vector data description (SVDD)[6] algorithm is a recently emerged clustering method inspired by the support vector machine, it can generate cluster boundaries of arbitrary shape and deal with outliers. It was reported that SVDD could greatly enhance the clustering performance of NMF representation. Although the kernel function is at the heart of the SVDD algorithm, its construction has not been studied extensively. In this paper, we adopt a non-parametric adaptive kernel to further improve the clustering of the SVDD algorithm and apply it to the document clustering. The experimental results demonstrate the effectiveness of the proposed document clustering algorithm.

2   Document Dimensionality Reduction through MFA

MFA is a recently proposed linear dimensionality reduction method for manifold learning[5]. It is based on the graph embedding framework and jointly considers the local manifold structure and the class label information, as well as characterizing the separability of different classes with the margin criterion.

Given a set of document data $\{x_1, x_2, \ldots, x_n\} \subset \mathbb{R}^m$, let $X = [x_1, x_2, \ldots, x_n]$. Let $G = \{X, S\}$ be an undirected similarity graph with vertex set X and similarity matrix S. The corresponding diagonal matrix D and the Laplacian matrix L of the graph G are defined as follows:

$$L = D - S, \quad D_{ii} = \sum_j S_{ij} \tag{1}$$

MFA can be obtained by solving the following minimization problem:

$$W_{opt} = \arg\min_{W} \frac{W^T X (D - S) X^T W}{W^T X (D^p - S^p) X^T W} \tag{2}$$

where S is a similarity matrix defined on the data points in the intrinsic graph, and S_{ij}^p is a similarity matrix defined on the data points in the penalty graph. Their definitions are as follows:

$$S_{ij} = \begin{cases} 1, & \text{if } i \in N_{k_1}(j) \text{ or } j \in N_{k_1}(i) \\ 0, & \text{otherwise} \end{cases} \tag{3}$$

$$S_{ij}^p = \begin{cases} 1, & \text{if } (i,j) \in P_{k_2}\left(l(x_i)\right) \text{or} (i,j) \in P_{k_2}\left(l(x_j)\right) \\ 0, & \text{otherwise} \end{cases} \tag{4}$$

where $N_{k_1}(i)$ denotes the index set of the k_1 nearest neighbors of sample x_i that are in the same class, $l(x_i)$ is the class label of data point x_i, and $P_{k_2}\left(l(x_i)\right)$ is a set of data pairs that are the k_2 nearest pairs among the set $\left\{(i,j)\mid l(x_i) \neq l(x_j)\right\}$.

As can be seen from the objective function in (2), MFA aims to minimize the intraclass compactness and maximize the interclass separability simultaneously. Finally, with simple algebra steps, the optimal project vectors W can be regarded as the eigenvectors of the matrix $\left(X\left(D^p - S^p\right)X^T\right)^{-1} X\left(D - S\right)X^T$ associated with the smallest eigenvalues. Thus, the embedding is as follows:

$$x \rightarrow y = W^T x \tag{5}$$

where y is the lower-dimensional representation of the high-dimensional document data x, and W is the project matrix.

3   Document Clustering with Improved SVDD

Now, we obtain the lower-dimensional feature representation of high-dimensional data by using MFA algorithm. Then we adopt the support vector data description (SVDD)[6] as the document clustering algorithm since it can generate cluster boundaries of arbitrary shape and deal with outliers.

Given a set of document data points $\{x_i\}_{i=1}^n$ in the reduced feature space, and the nonlinear mapping function $\varphi(\ )$. SVDD aims to look for the smallest enclosing sphere of radius R by solving the following minimization problem:

$$\min\left(R^2 + C\sum_i \xi_i\right) \tag{6}$$

with the constraint

$$\left\|\varphi(x_i) - a\right\|^2 \leq R^2 + \xi_i, \ \xi_i \geq 0 \tag{7}$$

where a is the center of the sphere and ξ_i is the slack variables for soft constraint.

Using the Lagrange multipliers and the KKT complementarity conditions, the above primal optimization problem can be transformed into the following Wolfe dual problem:

$$\max\left(\sum_i K\left(x_i, x_i\right)\beta_i + \sum_{i,j}\beta_i\beta_j K\left(x_i, x_j\right)\right) \tag{8}$$

with the constraint

$$0 \leq \beta_i \leq C, i = 1,\ldots,n \tag{9}$$

where $K\left(x_i, x_j\right) = \varphi\left(x_i\right)\cdot\varphi\left(x_j\right)$ is the kernel function.

For each data point x, we can obtain the distance of its image in feature space from the sphere center in terms of

$$
\begin{aligned}
R^2\left(x\right) &= \left\|\varphi(x) - a\right\|^2 \\
&= K\left(x, x\right) - 2\sum_i K\left(x_i, x\right)\beta_i + \sum_{i,j}\beta_i\beta_j K\left(x_i, x_j\right)
\end{aligned}
\tag{10}
$$

Then the radius of the sphere is defined as

$$R = \left\{R\left(x_i\right) \middle| x_i \text{ is a support vector}\right\} \tag{11}$$

The contours that enclose the cluster in data space are defined by the set

$$R = \left\{x \middle| R\left(x\right) = R\right\} \tag{12}$$

From the above statement, we can observe that the kernel function selection is the heart of the SVDD algorithm since different kernel function will produce different constructions of implicit feature space. To further improve the clustering performance of SVDD algorithm, we adopt the following adaptive non-parametric kernel learning method.

Since the data-dependent kernel can be obtained via pairwise constraints[7], we can construct the following similarity matrix T to represent the pairwise constraints:

$$
T_{ij} = \begin{cases}
+1, & \left(x_i, x_j\right) \in S \\
-1, & \left(x_i, x_j\right) \in D \\
0, & otherwise
\end{cases}
\tag{13}
$$

where S denotes similar pairwise constraint(the data pairs share the same class), and D denotes dissimilar pairwise constraint(the data pairs have different classes). Let L be the normalized graph Laplacian matrix defined as

$$L = I - D^{-1/2}SD^{-1/2} \tag{14}$$

where I is the identity matrix. Then the manifold adaptive non-parametric kernel learning can be formulated the following minimization problem:

$$\min_{K \succ 0} \mathrm{Tr}\left(LK\right) + C\sum_{\left(x_i, x_j\right)\in S\cup D} l\left(T_{ij}K_{ij}\right) \tag{15}$$

where $\mathrm{Tr}(\ )$ is the matrix trace, C is the positive constant to control the tradeoff between the empirical loss $l(\ )$ and the intrinsic data manifold. Since the optimal problem in (15) belongs to typical semi-definite programming (SDP) problem, which can be easily computed using the standard SDP solver SeDuMi. By using the obtained optimal kernel function $K(\ )$, we can greatly improve the cluster performance of SVDD algorithm.

4   Experimental Results

In this section, we compare the proposed MFA plus SVDD(MFA-SVDD) clustering algorithm with LSI-based clustering(LSI) algorithm, NMF-based clustering(NMF) algorithm, and LPI-based clustering(LPI) algorithm. The numbers of the nearest neighbor k_1 and k_2 are set to 5 and 10, respectively. The experimental evaluation was performed on two well-known data sets in document clustering research: WebKB[8] and 20 Newsgroup (20NG)[9]. Each document was normalized into a TF-IDF representation, and the Euclidean distance was used as the distance measure.

The clustering performance is measured by two well-known metrics: the accuracy and the normalized mutual information(NMI)[3,4]. Each document data set was partitioned into randomly into a training set consisting of one half of the whole data set and a testing set consisting of the remainder one half of the whole data set.

The average accuracies as well as NMI values on the WebKB and 20NG are shown in Table 1 and Table 2, respectively. The main observations from the performance comparisons are summarized as follows: Our proposed MFA-SVDD algorithms achieves the best results in all the cases, which demonstrates the importance of utilizing both local manifold structure and class label information for dimensionality reduction, as well as clustering different classes with the improved SVDD algorithm.

Table 1. Performance comparison on the WebKB database

Algorithm	Accuracy	NMI
LSI	81.3	67.2
NMF	86.9	71.4
LPI	92.8	75.3
MFA-SVDD	97.4	82.6

Table 2. Performance comparison on the 20NG database

Algorithm	Accuracy	NMI
LSI	79.5	62.8
NMF	85.7	68.3
LPI	90.6	73.2
MFA-SVDD	96.8	78.4

5 Conclusion

In this paper, a new document clustering algorithm with improving SVDD is proposed. Experimental results on two document databases demonstrate that the proposed algorithm outperforms other popular document clustering algorithms.

Acknowledgment. This work is supported by the National Natural Science Foundation of China under Grant No.70701013, and the Natural Science Foundation of Henan Province under Grant No. 102300410020.

References

1. Berry, M.W.: Survey of Text Mining. Springer, New York (2003)
2. Deerwester, S.C., Dumais, S.T., Landauer, T.K., Furnas, G.W., Harshman, R.A.: Indexing by Latent Semantic Analysis. Journal of the American Society of Information Science 41, 391–407 (1990)
3. Xu, W., Liu, X., Gong, Y.: Document Clustering Based on Non-Negative Matrix Factorization. In: International Conference on Research and Development in Information Retrieval, pp. 267–273. ACM Press, New York (2003)
4. Cai, D., He, X., Han, J.: Document Clustering Using Locality Preserving Indexing. IEEE Transactions on Knowledge and Data Engineering 17, 1624–1637 (2005)
5. Yan, S., Xu, D., Zhang, B., Zhang, H.-J., Yang, Q., Lin, S.: Graph Embedding and Extensions: a General Framework for Dimensionality Reduction. IEEE Transactions on Pattern Analysis and Machine Intelligence 29, 40–51 (2007)
6. Tax, D.M.J., Duin, R.P.W.: Support Vector Domain Description. Pattern Recognition Letters 20, 1191–1199 (1999)
7. Zhuang, J., Tsang, I.W., Hoi, S.C.H.: SimpleNPKL: Simple Non-Parametric Kernel Learning. In: 26th International Conference on Machine Learning, pp. 1273–1280. ACM Press, New York (2009)
8. Craven, M., DiPasquo, D., Freitag, D., McCallum, A.K., Mitchell, T.M., Nigam, K., Slattery, S.: Learning to Extract Symbolic Knowledge From The World Wide Web. In: Fifteenth National Conference on Artificial Intelligence, pp. 509–516. MIT Press, Cambridge (1998)
9. 20 News Group,
   http://kdd.ics.uci.edu/databases/20newsgroups/20newsgroups.htm

Brain Tumors Classification Based on 3D Shape

Peng Wu[1], Kai Xie[1,2], Yunping Zheng[3], and Chao Wu[1]

[1] School of Electronics and Information, Yangtze University, Jingzhou, China
[2] Key Laboratory of Oil and Gas Resources and Exploration Technology of Ministry of Education, Yangtze University, Jingzhou, China
[3] School of Computer Science and Engineering, South China University of Technology, Guangzhou, China
peng_wu@189.cn, pami2009@163.com, zhengyp@scut.edu.cn

Abstract. Brain tumors are the second leading cause of cancer death in children under 15 years and young adults up to the age of 34, early detection and correct treatments based on accurate diagnosis are important steps to improve disease outcome. In this paper, a new method for brain tumors identification based on quantitative three-dimensional shape analysis is proposed. According to the character of magnetic resonance imaging (MRI) data and doctor's clinical experience, we defined three two-dimensional shape descriptors correlating with tumor type from different points of view; on this basis, four three-dimensional shape descriptors were defined to realize the automatic classification of brain tumors. The experiment result demonstrates that these shape descriptors can well represent the tumor character. The classification accuracy of regular/irregular tumor is 93.93% and that of benign/malignant tumor is 87.14%. To regular benign (RB)/regular malignant (RM)/irregular benign (IB)/irregular malignant (IM), the classification accuracy is 86.43%. This method can be used as a clinical image analysis tool for doctors or radiologists to tumor identification.

Keywords: Brain tumors identification, Three-dimensional shape analysis, MRI, Segmentation.

1   Introduction

Brain tumors are the second leading cause of cancer death in children under 15 years and young adults up to the age of 34. These tumors are also the second fastest growing cause of cancer death among humans older than 65 years [1]. Early detection and correct treatment based on accurate diagnosis are important steps to improve disease outcome.

Currently, magnetic resonance imaging (MRI) is an important tool to identify the location, size and type of brain tumor. Tissues usually become dense when diseased. Often, surrounding tissues are pulled toward the cancerous region, resulting in distortion. Masses are examined for location, shape, density, size, and definition of margins. Higher density is usually an indicator of malignancy, while lucent-centered lesions are usually benign. Cancerous lesions generally have a more irregular shape than benign lesions. Most benign masses are circumscribed, compact, and roughly elliptical. Malignant lesions usually have a blurred boundary and an irregular appearance [2].

In this article, we investigate region-based techniques to classify brain tumors into two groups as benign/malignant or regular/irregular, as well as into four groups as regular benign (RB), regular malignant (RM), irregular benign (IB), and irregular malignant (IM). The techniques are based upon measures of the morphology and appearance of masses. We use shape factors such as compactness, Fourier descriptors (FD's), and moments based upon contours of masses.

In Section 2, we present a review for analysis of abnormal brain tumor. Shape analysis procedures based upon compactness, FD's, and different types of moments are described in Section 3. Results of the classification experiments are presented and discussed in Section 4. Finally, conclusion is given in Section 5.

2   Analysis of Abnormal Brain Tumors

Brain tumors are very complicated because they have a wide range of appearance and effect on surrounding structures. Following are some of the general characteristics of brain tumors: (A) vary greatly in size and position, (B) vary greatly in the way they show up in MRI, (C) may have overlapping intensities with normal tissue, (D) may be space occupying (new tissue that moves normal structure) or infiltrating (changing properties of existing tissue), (E) may enhance fully, partially, or not at all, with contrast agent, and (F) may be accompanied by surrounding edema.

The basic characteristics of malignant tumors are (A) ragged boundaries, (B) initially only in white matter, possibly later spreading outside white matter, (C) margins enhance with contrast agent, inside does not, (D) accompanied by edema, and (E) infiltrating at first, possibly becoming space occupying when larger.

3   Materials and Methods

3.1   Image Acquisition

280 patients (164 males and 111 females) aged 16–67 with brain tumors: There were 169 tumors of low grade (I, II) and 111 tumors of high grade (III, IV). MR images of 280 brain tumors patients (before therapy) were acquired at Shang Hai Hua Shang Hospital from June 1999 to July 2002. All the patients were imaged on 1.5T MR scanner (GE Medical System) using a standard clinical imaging protocol. Fast spin echo sequence was used to obtain T1-weighted (T1) and contrast enhanced T1-weighted (CET1, gadolinium enhanced) images. In the axial plane, the 256×256 pixel images of the head were acquired with a 256 (8 bits). Three experienced radiologists manually mouse-traced and labeled (aided by computer) all tumor-contained slices for the classification of brain tumors.

3.2   Three Dimension Shape Description of Brain Tumors

It is well established that benign masses are usually round in appearance, and circumscribed with smooth contours. On the other hand, malignant tumors typically possess rough or irregular boundaries [3]. In computerized shape analysis, it is desirable to classify objects using robust features, which are independent of scaling, translation, and rotation [4], [5]. Compactness, FD's, and central invariant moments are the most commonly used methods for shape analysis.

Every tumor-contained slice includes two-dimensional shape of brain tumors. We can get three-dimensional shape [6-8] from all tumor-contained slices, and three dimension shape factors of brain tumors are described as following:

(1) Compactness Shape Factors C_{3D}

$$C_{3D} = \left[\sum_{i=1}^{N} C_i \right] / N \tag{1}$$

where C_i is compactness shape factor of nth slice. N is the number of all tumor-contained slices.

(2) Fourier Shape Factors F_{3D}

$$F_{3D} = \left[\sum_{i=1}^{N} (FF)_i \right] / N \tag{2}$$

where $(FF)_i$ is fourier shape factor of nth slice. N is the number of all tumor-contained slices.

(3) Moment-Based Shape Factors M_{3D}

$$M_{3D} = \left[\sum_{i=1}^{N} (F_3 - F_1)_i \right] / N \tag{3}$$

where $(F_3-F_1)_i$ is moment-based shape factor of nth slice. N is the number of all tumor-contained slices.

To describe the complexity of brain tumors, shape [9] is divided into two parts; XY direction and Z direction, These two parts describe the complexity of brain tumors along axis. Fig.1(a) includes three profiles that have the same radius circle; Fig.1(b) includes three profiles that have different radius circles. Compactness shape factors, fourier shape factors and moment-based shape factors of Fig.1(a) and Fig.1(b) is the same, but shape of two is different along Z direction.

We define a new factor irregular degree Z_{3D} to describe the difference along Z direction:

$$Z_{3D} = \frac{\left[\sum_{i=1}^{N-1} (Area_i \cap Area_{i+1}) \right]}{\left[\sum_{i=1}^{N-1} (Area_i) \right]} \tag{4}$$

where $Area_i$ is area of brain tumors in nth slice, $Area_i \cap Area_{i+1}$ is area of brain tumors intersection between neighbor slices; Higher Z_{3D} describe large change along axial direction while lower Z_{3D} describe small change along axial direction.

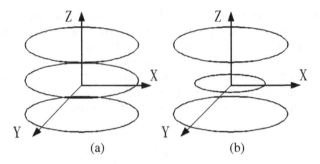

Fig. 1. Illustration of irregular degree along axis direction

4   Result and Discussion

4.1   Shape Extraction

A 3×3 median filter was employed to reduce possible noise within the MRI images. In some cases, e.g. the background is black, using only the global threshold technique is not enough to segment brain tumors from the background because the contrast varies within the image. The automatic threshold technique developed by Panigrahi, Misra, Bern, and Marley can be applied to segment the background from the images of brain tumors. The boundaries of brain tumors are obvious in the segmented images and can be efficiently detected. The segmentation step is necessary to obtain a closed and continuous boundary of brain tumors, which is difficult to be obtained directly by the traditional edge detection technology. Then the Canny edge detector is used to detect the shape of brain tumors with low and high thresholds. The edges in row 3 of Fig. 2 are closed and continuous, which is very important for further processing.

4.2   Classification

Numbers of different types of tumors in the databases used in this study are shown in Table 1.

The percent accuracy rates of benign/malignant classification using all of the features are given in Table 2. The shape measurements that depend on boundary information gave benign/malignant classification rates of 93.93% for the databases.

The regular/irregular classification [10] rates of 87.14% for the database were obtained using the combination of C3D, F3D, M3D and Z3D, as shown in Tables 3.

The four-group (RB, RM, IB, and IM) classification result of 86.43% for the database was obtained using the combination of C3D, F3D, M3D and Z3D, as given in Table 4.

The identification results of different shape features are shown in Table 5.The results indicate that the shape measurements of all features provides high accuracy in four groups classification; also give similar results. It is important to note that the shape factors (without Z3D) do not provide four-group classification accuracy beyond 51.38% Thus, the results establish the complementary roles of Z3D for shape factors in distinguishing between the four types of tumors.

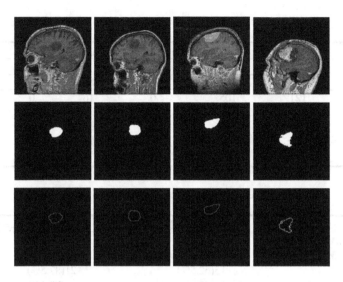

Fig. 2. Results of the shape extraction algorithms, row 1: original images of brain tumors; row 2: segmentation results; row 3: results of edge detection. Column I: regular benign (RB); column II: regular malignant (RM); column III: irregular benign (IB); column IV: irregular malignant (IM).

Table 1. Numbers of different types of tumors in the databases used in this study

Database	RB	RM	IB	IM	Total
-	121	23	48	88	280

Table 2. Details of the best regular/ irregular classifier for the database

Classification using C_{3D}, F_{3D}, M_{3D} and Z_{3D}				
Database	#cases	Classified as regular	Classified as irregular	%Correct
Regular	144	129	15	89.58
Irregular	136	2	134	98.53
Total	280	-	-	93.93

Table 3. Details of the best benign/malignant classifier for brain tumors

Classification using C_{3D}, F_{3D}, M_{3D} and Z_{3D}				
Database	#cases	Classified as Begin	Classified as Malignant	%Correct
Begin	169	151	18	89.35
Malignant	111	18	93	83.78
Total	280	-	-	87.14

Table 4. Details of the best four-group classifier for the database

| Database | #cases | Classification using C_{3D}, F_{3D}, M_{3D} and Z_{3D} | | | | %Correct |
| | | Classified as | | | | |
		RB	RM	IB	IM	
RB	121	105	0	16	0	86.78
RM	23	0	19	0	4	82.60
IB	48	0	6	42	0	87.50
IM	88	0	12	0	76	86.36
Total	280	-	-	-	-	86.43

Table 5. The identification results of different shape features

Using feature	Begin($N_b=24$)	Malignant($N_m=72$)
$C_{3D}, F_{3D}, M_{3D}, Z_{3D}$	21(87.5%)	50(69.44%)
C_{3D}, F_{3D}, M_{3D}	20(83.33%)	37(51.38%)
F_{3D}, M_{3D}, Z_{3D}	15(62.5%)	21(87.5%)
C_{3D}, M_{3D}, Z_{3D}	19(79.16%)	45(72.22%)
F_{3D}, M_{3D}, Z_{3D}	20(83.33%)	46(70.83%)
M_{3D}, Z_{3D}	14(58.33%)	41(56.94%)
C_{3D}, Z_{3D}	18(75%)	41(56.94%)
F_{3D}, Z_{3D}	16(66.67%)	42(58.33%)

5   Conclusion

The computer-aided diagnosis system of brain tumors has great clinical significance for alleviating doctor's pressure and reducing misdiagnosis rate. In this paper, a new method of brain tumors identification based on quantitative three-dimensional shape analysis is proposed. According to the character of magnetic resonance imaging data and doctor's clinical experience, we defined three two-dimensional shape descriptors correlating with tumor type from different points of view; on this basis, four three-dimensional shape descriptors were defined to realize the automatic classification of brain tumors. The experiment result demonstrates that these shape descriptors can well represent the tumor character. The classification accuracy of regular/irregular tumor is 93.93% and that of benign/malignant tumor is 87.14%. To benign (RB)/regular malignant (RM)/irregular benign (IB)/irregular malignant (IM), the classification accuracy is 86.43%.

This paper has, therefore, established the usefulness of the proposed shape measurements in tumor classification. An innovation of this work is that we intend to develop methods for automatic detection of tumor areas as well as for the derivation of their boundaries.

Acknowledgements. This work has been partially supported by CNPC Innovation Foundation (2010D-5006-0304), Specialized Research Fund for the Doctoral Program of Higher Education of China (20070532077), Natural Science Foundation of Hubei

Province of China (2009CDB308), Educational Fund of Hubei Province of China (Q20091211, B20111307).

References

[1] The Brain Tumor Society, http://www.tbts.org
[2] Chen, Y., Gunawan, E., Low, K.S., Wang, S.-C., Soh, C.B., Putti, T.C.: Effect of lesion morphology on microwave signature in 2-D ultra-wideband breast imaging. IEEE Transactions on Biomedical Engineering 55, 2011–2021 (2008)
[3] Rangayyan, R.M., El-Faramawy, N.M., Desautels, J.E., Alim, O.A.: Measures of acutance and shape for classification of breast tumors. IEEE Transactions on Medical Imaging 16, 799–810 (1997)
[4] Levine, M.D.: Feature extraction: A survey. Proceedings of the IEEE 57, 1391–1407 (1969)
[5] Pavlidis, T.: Algorithms for shape analysis of contours and waveforms. IEEE Transactions on Pattern Analysis and Machine Intelligence 2, 301–312 (1980)
[6] Shen, L., Rangayyan, R.M., Desautels, J.E.L.: Calcifications, Detection And Classification of Mammographic. In: Bowyer, K.W., Astley, S. (eds.) State of the Art in Digital Mammographic Image Analysis, pp. 198–212. World Scientific Publishing Co., Inc. (1994)
[7] Persoon, E., Fu, K.S.: Shape Discrimination Using Fourier Descriptors. IEEE Transactions On Systems Man And Cybernetics 7, 170–179 (1977)
[8] Lin, C.C., Chellappa, R.: Classification of partial 2-d shapes using fourier descriptors. IEEE Transactions on Pattern Analysis and Machine Intelligence 9, 686–690 (1987)
[9] Shen, L., Rangayyan, R.M., Desautels, J.L.: Application of shape analysis to mammographic calcifications. IEEE Transactions on Medical Imaging 13, 263–274 (1994)
[10] Gupta, L., Srinath, M.D.: Contour sequence moments for the classification of closed planar shapes. Pattern Recognition 20, 267–272 (1987)

Study of Genetic Neural Network Model in Water Evaluation

Xuemei Meng

Jilin Business and Technology College, Department of Information Engineering,
130062, Changchun, China
mxmyymh@163.com

Abstract. Due to the defects of traditional water assessment method, this paper constructs a model based on genetic neural network model. This paper analyzes the BP neural network and the basic principles and characteristics of GA, Discusses the theoretical foundation of optimizing the weights and thresholds of BP neural network based on GA, This model is used to actual water evaluation, which proved the optimized neural network based on GA has been improved in convergence rate and convergence property. The results of the evaluation with the actual water quality results and fuzzy evaluation method, evaluation results of the index evaluation are compared, the results show that the genetic neural network model is suitable to water Evaluation.

Keywords: genetic algorithm, BP neural network, GA-BP, evaluation.

1 Introduction

The water quality assessment is the basis of the water environment management, which directly affects the formulation of water-use planning; it plays a very important position in the water environment management. Since the 1970s, scholars have proposed dozens of quality evaluation methods, but in general can be divided into three categories: the index evaluation method, hierarchical clustering evaluation method, fuzzy comprehensive evaluation method. However, the above methods are applicable conditions and limitations; there is a way or another drawback. In this paper, genetic neural network used in water quality evaluation this area, as an example to introduce the genetic algorithm with neural networks is how to combine the neural network training. The paper discusses in detail the operation of genetic algorithm to optimize the BP network, and experiments show that the GA optimized BP network convergence speed and convergence of the performance has been improved.

2 Genetic Algorithms

The genetic algorithm is based on Darwinian evolution, simulation of biological evolution process of the calculation model, embodies the struggle for existence, survival

D. Jin and S. Lin (Eds.): Advances in FCCS, Vol. 2, AISC 160, pp. 285–290.
springerlink.com © Springer-Verlag Berlin Heidelberg 2012

of the fittest, survival of the fittest "competition mechanism. The main feature of the genetic algorithm is groups the information exchange between individuals and groups search strategy. The genetic algorithm starting from any initial population, which form groups of each individual are to a solution of the posed problem search space, through selection, crossover and mutation operation making the population from generation to generation into getting in the search space area better and better until the optimal solution [1]. Therefore, the genetic algorithm is a new global optimization search algorithm, its robustness, simple and versatile, suitable for parallel processing of these significant characteristics.

For the study of the genetic algorithm, there are still some controversial, some very different design principles and academic point of view and sometimes difficult to unify. The theoretical basis links of the genetic algorithm are still relatively weak, but many applications and examples to fully simulate the natural evolution of the search process can often produce very simple, universal and strong robustness of the calculation method [2].

3   BP Neural Network

BP neural network based on supervised learning, the use of non-linear differentiable function as a transfer function of the feed forward neural network. It was founded in 1986 by Rumelhart and McClelland proposed multilayer before back-propagation to the network algorithm or back-propagation based on Error Back Propagation (EBP) algorithm [3]. Figure 1 shows typical BP network topology.

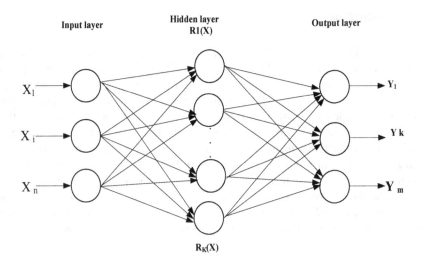

Fig. 1. Typical BP network topology

4   Water Quality Evaluation Model Based on Genetic Neural Network

In this paper, BP neural network as a modeling tool to build a water quality evaluation model, to reflect the water quality assessment corresponding relationship between the complex input and output.

4.1   Genetic Neural Network Model

BP neural network has fully distributed storage structure, capable of massively parallel information processing, in particular its nonlinear systems has highly simulate ability, But at the same time, it also has natural shortcomings, mainly in the slow rate of convergence or can not converge easily fall into local minima of the error function, and vulnerable to the impact of overtraining. Local minima to a large extent by adjusting the initial weights to solve, and the convergence slow or unable to convergence problems is often as network training late into the local minimum lead. Therefore, if in the early stage of network training can take effective way to optimize the initial weights and thresholds, and discard the original kind of random method to determine the right value, you can better solve the above problems [4].

The genetic algorithm has macro-search capability and good performance of global optimization, as well as the characteristics of its strong robustness and implicit parallelism, etc., So use it to optimize BP neural network training before the weights and threshold values, able to overcome the shortcomings of the BP neural network prediction. The genetic algorithm optimization neural network mainly includes three aspects: the optimization of connection weights; network structure optimization; learning rule optimization. In this paper, the genetic algorithm is application in neural network connection weights.

4.2   Genetic Algorithm Optimization BP Neural Network

BP neural network is to modify the weights and threshold values based on gradient descent method, to the network initial weights and thresholds is extraordinarily sensitive, different initial values are likely to lead to different results for training, or into a local minimum or not at all convergence. The global search ability of genetic algorithm of to find optimal solution or performance good suboptimal solutions can reduce the sensitivity of the BP network initial weights [5].

The specific procedure is: First, the genetic algorithm to generate an initial population, this group, each individual corresponds to a set of weights and thresholds of the neural network; Secondly, each individual in accordance with the average error of the network to determine its "merits" to decide behind the evolution of the probability of the next generation. Special note is that each individual's fitness unlike function optimization so intuitive, the network error is the calculation of the network's actual output and expected output. After several iterations to get one individual the last, this individual to represent the right value of a genetic algorithm can find the optimum weights in the number of iterations, mean network error is minimum. Use this weight as the initial weights of BP neural network training, in theory, than the initial weight training has certain advantages. So that the Genetic algorithm combination of neural networks to realize the complementary advantages [6].

5   Model Simulation Experiment Results

Procedures for water quality assessment process is build up to using the existing mature on the basis of J2EE open source framework Struts, Spring, Hibernate, the Tomcat server as running SQL Server 2005 as database server, training sample 4000 data, the test sample data 15 amended.

ID	PH	DO	COD	KH3	TYPE
1	7.22	5.7	5	0.11	3
2	7.02	7.16	3.6	0.44	2
3	7.02	8.53	5.6	0.29	3
4	6.53	4.08	4.5	0.07	4
5	6.92	0.5	8.4	0.04	6
6	7.26	7.23	3.2	0.02	2
7	6.83	7.85	1.3	0.05	1
8	8.33	7.88	4	0.49	2
9	8.54	8.05	3.4	1.95	5
10	7.41	4.74	2.4	0.11	4
11	7.47	4.3	4.1	0.17	4
12	7.45	6.12	0.8	0.09	2
13	7.52	2.37	5	0.41	5
14	7.53	0.6	5.2	1.08	6
15	8.38	5.6	7.1	8.62	6

Fig. 2. Data set of test samples

5.1   GA-BP Evaluation Results Analysis

The test data set Shown in Figure 2 of 15, numbered from 1 to 15. Number of hidden units is set to 12, maximum iteration number is 20, the target error of 0.02, learning rate is 0.8, population size is 20, crossover rate 0.8, mutation rate of 0.1, the maximum number of iterations of 4, for a given run the network in 20 times the number of iterations, the network error to 0.027, the water level of evaluation out of the water quality test samples shown in Table 1:

Table 1. Comparing to Actual Water Level

Number	1	2	3	4	5	6	7	8	9	10	11	12	13	14	15
GA-BP	II	I	III	III	V	I	I	I	V	II	III	I	VI	VI	III
Actual	III	II	III	IV	VI	II	I	II	V	IV	IV	II	V	VI	VI

This table can be seen that most of the water quality evaluation results is quite consistent with the actual water level, only very few of the predictions with the actual level difference 2 levels, from the overall forecast results is quite satisfactory.

5.2   Compare to GA-BP Method and Fuzzy Comprehensive Evaluation (FCM) Method

Using Fuzzy comprehensive evaluation method on the same test sample data carry quality assessment, results are shown in Table 2.

Table 2. Comparing to Result of Fuzzy Comprehensive Method

Number	1	2	3	4	5	6	7	8	9	10	11	12	13	14	15
GA-BP	II	I	III	III	V	I	I	I	V	II	III	I	VI	VI	III
FCM	III	I	IV	IV	II VI	II	II	II	VI	V	IV	I	VI	VI	VI

Through the analysis of the results of two evaluation methods in the table, the FCM evaluation on the part of the sample forecast is consistent with the GA-BP method, Evaluation results grade more accurate the water quality level of VI in water samples predict , the water quality level of V in water samples predict predicted to VI. This may be due to the evaluation factors standard in these water samples, fuzzy comprehensive evaluation has played a larger effect, thereby increasing the grade of its evaluation results.

5.3   Compare Analysis to GA-BP Method and the Index Evaluation Method

Using Index evaluation method on the same test sample data quality assessment, the results shown in Table 3.

Table 3. Comparing to Result of Index Method

Number	1	2	3	4	5	6	7	8	9	10	11	12	13	14	15
GA-BP	II	I	III	III	V	I	I	I	V	II	III	I	VI	VI	III
Index	I	II	III	II	VI	II	I	II	V	IV	IV	II	V	V	VI

From the analysis results of data in the table, the index evaluation method, in general, the results of more "optimistic". In the actual water quality level of the water samples, the level of evaluation results is significantly lower, which may be index evaluation method using a weighted superposition of the Water Quality Index ignored individual small number of excessive pollution factor on the water environment more characteristics, the Individual exceeded factor "submerged" to the final water quality index, and thus is more optimistic about the evaluation results.

6   Conclusion

From traditional water quality evaluation method exists the problem of insufficient, analysis and described on the basis of the BP neural network and GA algorithm. Discusses the theoretical basis Based on GA optimization BP neural network weights and thresholds of the Water Quality Assessment Model and confirmed the reliability and accuracy of the models from theoretical and practical applications.

References

1. Zhou, M., Sun, S.: Principle and application of genetic algorithms. National Defense Industry Press, Beijing (2001)
2. Li, L., Bing, Z.L.: The genetic algorithm a combination of evaluation model based on the genetic algorithm. Hefei University of Technology (Natural Science Edition) 22 (2004)
3. Ni, S., Bai, Y.: BP neural network model in the underground water quality assessment. Systems Engineering Theory and Practice 20(8), 124–127 (2000)
4. Guo, X., Zhu, Y.: Evolutionary neural network based on genetic algorithm. Tsinghua University 40(10), 116–118 (2000)
5. Liu, T.: Arable land grading evaluation based on neural networks and genetic algorithms. Shandong Normal University (2009)
6. Wu, Q.: Intelligent weather forecasting system based on genetic neural network. Fudan University (2003)

Study on the Recognition Algorithms
for TDR Reflected Wave

Jianhui Song[1,*], Guohua Zhang[2], Peixin Sun[2], Naili Feng[2], and Shaolin Xu[2]

[1] School of Information Science and Engineering, Shenyang Ligong University,
Shenyang, 110159, P.R. China
hitsong@126.com
[2] Shenyang Artillery Academy of PLA, Shenyang, 110162, P.R. China

Abstract. The measuring accuracy of the TDR cable length measurement system relies on the recognition of the reflected wave. The reflected wave is susceptible to be affected by noise and the reflected wave detecting error is comparatively large. In order to correctly identifying and detecting the reflected wave, this dissertation studies four reflected wave recognition algorithms, which are established according to the threshold method, centroid method, polynomial fitting method and the wavelet modulus maxima method respectively. The advantages and disadvantages of these algorithms are analyzed. The repeatability and accuracies of the reflected wave arrival time for these algorithms are compared through practical applications.

Keywords: TDR, reflected wave, recognition algorithms, measuring accuracy.

1 Introduction

Compared with the traditional measurement methods, time domain reflectometry (TDR) technology has an advantage of non-destruction, portability and high-precision, which is an ideal cable length measurement method[1,2]. The detection signal is unavoidable to be affected by various kinds of noises. The noises have a huge interference on the useful signal extraction and detection. In order to correctly identifying and detecting the reflected wave, this dissertation studies four reflected wave recognition algorithms, which are established according to the threshold method, centroid method, polynomial fitting method and the wavelet modulus maxima method respectively. The recognition accuracies of the reflected wave arrival time for these algorithms are compared through practical applications.

2 TDR Cable Length Measurement Theory

TDR is a very useful measuring technology based on high-speed pulse technology [3,4]. The cable length measuring principle is very simple. The test voltage pulse is injected into one end of the cable, and the pulse will be reflected at the end of the

---

[*] Corresponding author.

D. Jin and S. Lin (Eds.): Advances in FCCS, Vol. 2, AISC 160, pp. 291–296.
springerlink.com © Springer-Verlag Berlin Heidelberg 2012

cable. By measuring the time interval between the injection pulse and the reflection pulse, the cable length can be obtained by assuming the velocity as constant. The formula for length measurement is

$$l = \frac{v \cdot \Delta t}{2} \tag{1}$$

Where, l —cable length,

$\quad\quad v$ —signal propagating velocity,

$\quad\quad \Delta t$ —time interval between the injection pulse and the reflection pulse.

The time interval between the injection pulse and the reflection pulse relies on the recognition of the reflected wave.

3 The Reflected Wave Recognition Algorithms

3.1 The Threshold Method

The simplest method for determining the arriving time of the reflected wave is to use the time at which the received signal crosses some predetermined amplitude threshold[5]. The threshold is chosen to be some small level near the zero baseline of the signal. A timer is started when the direct arriving pulse crosses the threshold and is then stopped when the reflected pulse crosses the same threshold. The time difference is then translated into a cable length. The advantage of this technique is its simplicity; it is very cost-effective to implement. Furthermore, this method has the advantage that results can be accumulated in real time without the need for time-consuming intervention by a computer to perform calculations.

This method has limited application because of the difficulty in determining the appropriate threshold for distorted, noisy, or unknown waveforms. If the signal is distorted with excessive ringing, the oscillations will cause false triggering of the timing circuits. If noise is present, it can cause false triggering of the timer, especially in cases where low level reflected wave is measured. If two pulses are closely spaced and overlapping, the timer will begin counting but will not have a trigger to stop. The cable is dispersive and consequently neither the front edge of the pulse or the pulse shape is preserved causing further errors. In typical power cables, attenuation is heavy and thus the reflected pulse is often much lower in amplitude than the direct pulse. These nonideal propagation characteristics contribute to the difficulty of obtaining delay estimates using a predetermined amplitude threshold as a trigger reference for the two pulses.

The basic difficulty associated with this method is the lack of a reference point on the pulse waveform that can be used as a time delay measurement reference. The chosen empirical formula of the reference point is

$$x_{\text{thres}} = m \cdot \sqrt{P_n} \tag{2}$$

Where x_{thres} is the reference point, m is parameters selected by the user, P_n is noise power.

3.2 The Centroid Method

The centroid method takes the centroid of the reflected wave as the location of feature points, and using the centroid algorithm to find the centroid of the reflected wave.

Without considering the noise, the time centroid of the estimated value is

$$\tilde{t} = \frac{\sum_i t_i x_i}{\sum_i x_i} \tag{3}$$

Where t_i is time coordinate; x_i is signal amplitude.

The centroid method takes the centroid of the reflected wave as the location of feature points. It can significantly reduce the influence of each of the signal measurement data. Therefore, it can eliminate the system errors, reduce the random error, and improve the measurement stability and repeatability. The measurement accuracy is not limited by the oscilloscope sampling rate.

3.3 The Polynomial Fitting Method

Polynomial fitting method is a mathematical method using certain fitting function and discrete sample data to reconstruction a continuous signal. For a system whose sampling rate has been identified, using the interpolation method of segmentation can improve the measurement accuracy limited by the sampling rate of the sampling oscilloscope.

For the time domain reflected signal, the sample data can be expressed as (x_i, y_i) $(i = 1, 2, ..., N)$, $y_i = y(x_i)$, that is:

$$(x_1, y_1), (x_2, y_2), ..., (x_i, y_i), ..., (x_N, y_N) \tag{4}$$

Where N is the length of the sample data, (x_i, y_i) is the number i sampling point of the measured signal sample data, x_i is the number i sampling point of time coordinate, y_i is the number i sampling point of signal amplitude

The front of the reflected wave can be fitted by a quadratic polynomial as:

$$p(x) = a_0 + a_1 x + a_2 x^2 \tag{5}$$

3.4 The Wavelet Modulus Maxima Method

Singularities and irregular structures often carry the most important information in signals, and the interesting information is given by transient phenomena such as peaks. By decomposing signals into elementary building blocks that are well localized both in space and frequency, the wavelet transform can characterize the local regularity of signals[6].

If the point (a_0, b_0) meet the follow formula:

$$\frac{\partial W_f(a_0, b_0)}{\partial t}\Big|_{t=t_0} = 0 \tag{6}$$

(a_0, b_0) is the local maximum point. For $\forall t \in (t_0, \delta)$, if:

$$|W_f(a_0, t)| \le |W_f(a_0, t_0)| \tag{7}$$

(a_0, b_0) is the modulus maxima. As the TDR reflected wave is analyzed by wavelet modulus maxima method, the reflected wave arrival time is determined as the singularities, that is the location of the modulus maxima.

4 Experimental Results and Analysis

Take the RVV 300/300V PVC insulated cable as example, the cable length of 30.00m is measured 51 times with the sampling rate of 2.5GHz. The measurement results are firstly pre-processed by wavelet de-noising technology. Then the threshold, centroid, polynomial fitting and wavelet modulus maxima method are used to identify the reflected wave front. Table 1 is the average and standard deviation of time interval measurement results.

Table 1. Average and standard deviation of time interval measurement results with 30.00m cable (ns)

	Threshold		Centroid		Polynomial fitting		Modulus maxima	
	de-noised before	de-noised after	de-noised before	de-noised after	de-noised before	de-noised after	de-noised before	de-noised after
average	316.43	316.28	355.08	355.06	316.77	316.22	317.69	316.44
standard deviation	0.68	0.16	0.49	0.33	0.25	0.16	1.38	0.23
average standard deviation	0.09	0.02	0.07	0.05	0.03	0.02	0.19	0.03

It can be seen from table 1 that the measuring accuracy of the centroid and polynomial fitting is high because of less affected by noise. The algorithms have function of eliminating noise, therefore the measuring accuracy can only be little improved by using noise cancellation technology to pre-process the measurement results. Both algorithms can break through the accuracy limited by sampling rate. As the centroid method uses the signal threshold of the entire reflected waves and the different thresholds are correspond to different measurement time, thus its measuring accuracy is comparatively large. The threshold method and the wavelet modulus maxima method are susceptible to be affected by noise. The measuring accuracy can be significantly improved by de-noising pre-processing the measurement results. The measuring accuracy of both algorithms is limited by the sampling rate.

The cable length of 86.75m is measured 55 times with the sampling rate of 1.0 GHz. The measurement results are firstly pre-processed by wavelet de-noising technology. Then the threshold, centroid, polynomial fitting and wavelet modulus maxima method are used to identify the reflected wave front. Table 2 is the average and standard deviation of time interval measurement results.

Table 2. Average and standard deviation of time interval measurement results with 86.75m cable (ns)

	Threshold		Centroid		Polynomial fitting		Modulus maxima	
	de-noised before	de-noised after	de-noised before	de-noised after	de-noised before	de-noised after	de-noised before	de-noised after
average	922.14	924.47	980.95	980.68	923.63	924.37	918.40	916.27
standard deviation	2.26	0.40	2.38	1.10	0.83	0.38	3.64	1.96
average standard deviation	0.31	0.05	0.32	0.15	0.11	0.05	0.49	0.27

It can be seen from table 2 that the measurement signal attenuation increases and the signal to noise ratio decreases along with the measurement cable length increases, leading to the increasing of the measurement error. At this point, the measuring accuracy can be significantly improved by de-noising pre-processing the measurement results.

Note that, the time interval detection points of these algorithms are different, therefore, the time interval measurement results obtained by these algorithms are not the same. If the theoretical value of the time interval needs to be determined, the velocity must be known. In fact, the propagation velocity is a complex process which is affected by many factors, and it can not be directly determined[5,7]. In order to obtain the propagation velocity, the measurement conditions must be defined. That is in the compare of the time interval measurement algorithms, only the length of the cable under test is surely known. Therefore, this article gives the theoretical value of the cable length under test, and only the repeatability and accuracy of these time interval measurement algorithm are discussed.

5   Conclusions

The measuring accuracy of the TDR cable length measurement system relies on the recognition of the reflected wave. In order to correctly identifying and detecting the reflected wave, this dissertation studies four reflected wave recognition algorithms, which are established according to the threshold method, centroid method, polynomial fitting method and the wavelet modulus maxima method respectively. The experimental results show that the centroid and polynomial fitting is high because of less affected by noise. The algorithms have function of eliminating noise. The measuring accuracy can only be little improved by using noise cancellation technology to pre-process the measurement results. Both algorithms can break through the accuracy limited by sampling rate. The threshold method and the wavelet modulus maxima method are susceptible to be affected by noise. The measuring accuracy can be significantly improved by de-noising pre-processing the measurement results.

Acknowledgement. The study is funded by the 2011 Education Department of Liaoning Province (Project No.:L2011036).

References

1. Dodds, D.E., Shafique, M., Celaya, B.: TDR and FDR Identification of Bad Splices in Telephone Cables. In: 2006 Canadian Conference on Electrical and Computer Engineering, pp. 838–841 (2007)
2. Cohen, L.: Time-frequency Distributions-A Review. Proceedings of IEEE 77(7), 941–981
3. Dodds, D.E., Shafique, M., Celaya, B.: TDR and FDR Identification of Bad Splices in Telephone Cables. In: 2006 Canadian Conference on Electrical and Computer Engineering, pp. 838–841 (2007)
4. Du, Z.F.: Performance Limits of PD Location Based on Time-Domain Reflectometry. IEEE Transactions on Dielectrics and Electrical Insulation 4(2), 182–188 (1997)
5. Steiner, J.P., Reynolds, P.H.: Estimating the Location of Partial Discharges in Cables. IEEE Transactions on Electrical Insulation 27(1), 44–59 (1992)
6. Gilany, M., Ibrahim, D.K., Eldin, E.S.T.: Traveling-Wave-Based Fault-Location Scheme for Multiend-Aged Underground Cable System. IEEE Transactions on Power Delivery 22(1), 82–89 (2007)

Velocity Frequency Characteristics of Time Domain Reflectometry Cable Length Measurement System

Jianhui Song[1,*], Zhiyong Tao[2], Guohua Zhang[2], Xuchen Lv[2], and Shaolin Xu[2]

[1] School of Information Science and Engineering, Shenyang Ligong University,
Shenyang, 110159, P.R. China
hitsong@126.com
[2] Shenyang Artillery Academy of PLA, Shenyang, 110162, P.R. China

Abstract. Traveling wave velocity is the key to the measuring accuracy of the time-domain reflectometry cable length measurement system. The velocity value is affected by many factors. In order to reduce the velocity frequency characteristics impact on time-domain reflectometry cable length measurement system, the effect of frequency on traveling wave velocity is studied, the velocity curve of traveling wave as a function of frequency is established. The traveling wave velocity in coaxial cable as a function of cable length is measured. The experimental results show that the signal frequency is different in different cable lengths because of attenuation and different frequency corresponds to different velocity.

Keywords: Velocity, Frequency characteristics, Time domain reflectometry, Nonlinear.

1   Introduction

Compared with the traditional measurement methods, time domain reflectometry (TDR) technology has an advantage of non-destruction, portability and high-precision, which is an ideal cable length measurement method[1,2]. Traveling wave propagation velocity is the key to the measuring accuracy of the time-domain reflectometry cable length measurement system. In fact, the propagation velocity is a complex process which is affected by many factors[3,4]. The accurate velocity value is difficult to be decided. The reference velocity is only an estimate one. In order to reduce the velocity frequency characteristics impact on time-domain reflectometry cable length measurement system, the effect of frequency on traveling wave velocity is studied in this paper.

2   The Traveling Wave Velocity Expression

Considering the frequency-dependent transmission line parameters, the general formula of the traveling wave velocity is[5]:

D. Jin and S. Lin (Eds.): Advances in FCCS, Vol. 2, AISC 160, pp. 297–302.
springerlink.com   © Springer-Verlag Berlin Heidelberg 2012

$$v = \frac{\omega}{\beta} = \frac{\omega}{\sqrt{\frac{1}{2}[\omega^2 L_0 C_0 - R_0 G_0 + \sqrt{(R_0^2 + \omega^2 L_0^2)(G_0^2 + \omega^2 C_0^2)}]}} \tag{1}$$

Where, R_0 is the resistance per unit cable length ; L_0 is the inductance per unit cable length ; G_0 is the conductance per unit cable length ; C_0 is the capacitance per unit cable length ; ω is angular frequency.

Take coaxial cable as example, for RG59, the capacitance per unit cable length is 100pF/m, the conductance per unit cable length is 200 μS/m . The propagation constant can be approximately expressed as

$$\gamma \approx \sqrt{(R_0 + j\omega L_0) j\omega C_0} = \sqrt{-\omega^2 L_0 C_0 + j\omega R_0 C_0} \tag{2}$$

The capacitance per unit cable length can be expressed as

$$C_0 = \frac{2\pi\varepsilon}{\ln(b/a)} \tag{3}$$

Where, ε is the insulation dielectric constant between the inner and outer conductance layer ; a is the inner conductance radius ; b is the outer conductance radius.

The dielectric constant ε of the insulated medium can be expressed as:

$$\varepsilon = \varepsilon_r \varepsilon_0 \tag{4}$$

Where, $\varepsilon_0 = 8.854 \times 10^{-12}$; ε_r is the relative dielectric constant.

Ideally, the inductance per unit cable length can be expressed as

$$L_0' = \frac{\mu}{2\pi} \ln\frac{b}{a} \tag{5}$$

Where, μ is the media penetration in the coaxial cable between the inner and outer conductance.

The inner inductance is

$$L_a = \frac{\mu\delta_a}{4\pi a} \tag{6}$$

The outer inductance is

$$L_b = \frac{\mu\delta_b}{4\pi b} \tag{7}$$

The inductance is the sum of L_0', L_a and L_b

$$L_0 = L_0' + L_{s0} f^{-1/2} \tag{8}$$

Where,

$$L_{s0} = \frac{\mu^{1/2}}{4\pi^{3/2}}(\frac{\sigma_a^{-1/2}}{a} + \frac{\sigma_b^{-1/2}}{b}) \tag{9}$$

The resistance per unit cable length can be expressed as

$$R_0 = 2\pi L_{s0} f^{1/2} \tag{10}$$

Formula 2 can be expressed as:

$$\gamma = j\beta_0\sqrt{1 + (1-j)\frac{L_{s0}}{L_0'}f^{-1/2}} \tag{11}$$

Where,

$$\beta_0 = \omega\sqrt{L_0'C} \tag{12}$$

From formula 5 and 9, the follow formula can be obtained

$$\frac{L_{s0}}{L_0'} = \frac{1}{2\sqrt{\pi\mu}}(\frac{\sigma_a^{-1/2}}{a} + \frac{\sigma_b^{-1/2}}{b})(\ln\frac{b}{a})^{-1} \tag{13}$$

Doing song algebraic transformation, the propagation constant can be approximately expressed as

$$\gamma = \beta_0\frac{1}{2}\frac{L_{s0}}{L_0'}f^{-1/2} + j\beta_0(1+\frac{1}{2}\frac{L_{s0}}{L_0'}f^{-1/2}) \tag{14}$$

The phase shift constant can be expressed as

$$\beta = \mathrm{Im}\{\gamma\} = \beta_0(1 + \frac{1}{2}\frac{L_{s0}}{L_0'}f^{-1/2}) \tag{15}$$

According to the relationship between the traveling wave velocity and the phase shift constant, the traveling wave velocity can be derived as

$$v_p = \frac{\omega}{\beta} = \frac{\omega}{\beta_0(1+\frac{1}{2}\frac{L_{s0}}{L_0'}f^{-1/2})} = \frac{\omega}{\omega\sqrt{L_0'C_0}(1+\frac{1}{2}\frac{L_{s0}}{L_0'}f^{-1/2})} = \frac{1}{\sqrt{L_0'C_0}(1+\frac{1}{2}\frac{L_{s0}}{L_0'}f^{-1/2})} \tag{16}$$

3   Experimental Results and Analysis

Figure 1 is the velocity curve of traveling wave as a function of frequency derived from formula 16. It can be seen from figure 1 that the traveling wave velocity is close to be a constant when the frequency is high. That is, at high frequencies, the velocity is only affected by cable insulation medium. However, when at low frequencies, the frequencies have great impact on velocity.

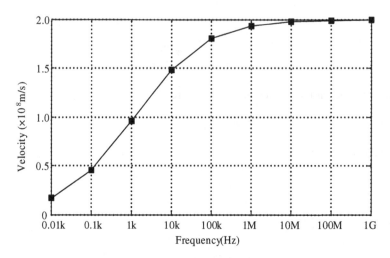

Fig. 1. Velocity curve of traveling wave as a function of frequency

As the signal frequency is different in different cable lengths because of attenuation and different frequency corresponds to different velocity. Therefore, the traveling wave of different cable length under test is different. Figure 2 is the actual measuring variation diagram of velocity with cable length.

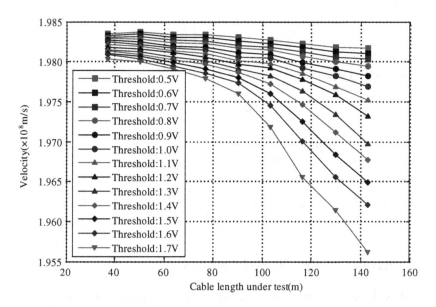

Fig. 2. Actual measuring variation diagram of velocity with cable length

The following conclusions can be drawn from Figure 2:

(1) With the increase of the cable length under test, the high frequency components gradually reduced because of the attenuation, that is the traveling wave velocity measured values changes with different cable length.

(2) The higher the threshold, the smaller the traveling wave velocity, the more obvious the traveling wave velocity measured value changes with cable length.

In fact, the insulating dielectric constant is not a constant value and it has frequency dependent characteristics, its expression is:

$$\varepsilon_r(\omega) = \varepsilon_r'(\omega) - j\varepsilon_r''(\omega) \tag{17}$$

The following expression is the empirical formula of the low-voltage paper-oil cable insulation dielectric constant at 20°C[6]:

$$\varepsilon_r = 2.5 + \frac{0.94}{1 + (j\omega \cdot 6 \times 10^{-9})^{0.315}} \tag{18}$$

The dielectric constant curve of paper-oil insulation as a function of frequency derived from formula 18 is shown in figure 3.

It can be seen from figure 3 that the dielectric constant of the insulating materials is affected by frequency. Therefore, when considering the frequency characteristics of the traveling wave velocity, the frequency-dependent characteristics of the distribution parameters must also be studied.

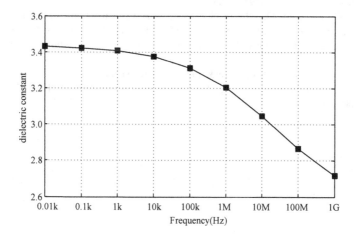

Fig. 3. Dielectric constant curve of paper-oil insulation as a function of frequency

4   Conclusions

The effect of frequency on traveling wave velocity is studied, the velocity curve of traveling wave as a function of frequency is established. The effect of frequency on the dielectric constant is studied, the dielectric constant curve of paper-oil insulation as a function of frequency is established. The traveling wave velocity in coaxial cable as a function of cable length is measured. The experimental results show that with the increase of the cable length under test, the high frequency components gradually reduced because of the attenuation, that is the traveling wave velocity measured values changes with different cable length. The higher the threshold, the smaller the traveling wave velocity, the more obvious the traveling wave velocity measured value changes with cable length. Therefore, the cable length measurement should be nonlinear compensated.

Acknowledgement. The study is funded by the 2011 Education Department of Liaoning Province (Project No.:L2011036).

References

1. Dodds, D.E., Shafique, M., Celaya, B.: TDR and FDR Identification of Bad Splices in Telephone Cables. In: 2006 Canadian Conference on Electrical and Computer Engineering, pp. 838–841 (2007)
2. Du, Z.F.: Performance Limits of PD Location Based on Time-Domain Reflectometry. IEEE Transactions on Dielectrics and Electrical Insulation 4(2), 182–188 (1997)
3. Shutt-Aine, J.E.: High-Frequency Characterization of Twisted-Pair Cables. IEEE Transactions on Communications 49(4), 598–601 (2001)
4. Ei Din, E.S.T., Gilany, M., Aziz, M.M.A., Ibrahim, D.K.: A Wavelet-Based Fault Location Technique for Aged Power Cables. In: 2005 IEEE Power Engineering Society General Meeting, vol. 3, pp. 2485–2491 (2005)
5. Buccella, C., Feliziani, M., Manzi, G.: Detection and Localization of Defects in Shielded Cables by Time-Domain Measurements with UWB Pulse Injection and Clean Algorithm Postprocessing. IEEE Transactions on Electromagnetic Compatibility 46(4), 597–605 (2004)
6. Gustavsen, B., Martinez, J.A., Durbak, D.: Parameter Determination for Modeling System Transients-Part II: Insulated Cables. IEEE Transactions on Power Delivery 20(3), 2045–2050 (2005)

Natural Computing and Intelligent Algorithms in Materials Image Processing Technology

Jing Zhang

Basic Courses Department, Beijing Union University,
Beijing, China
zhang1jing4@sina.com

Abstract. Intelligent algorithm is simulated biological theory of intelligent systems constitute a new type of information processing technology, has been widely used in industrial engineering, information processing and other fields. In recent years, in the field of image processing and analysis, intelligent algorithms are also widely used technology. In this paper, intelligent algorithm technique in medical image segmentation, image registration and the application of computer-aided technology and research are reviewed, representative described techniques and algorithms.

Keywords: materials image processing, intelligent algorithms, information technology.

1 Introduction

Intelligent calculation was also known as "soft computing", is subject to their natural (biological) law of enlightenment, according to its principle, mimic the problem solving algorithm. Get inspiration from nature, imitating the structure of the invention creation, this is the bionics. This is one aspect of the nature of learning. On the other hand, we can also use the bionic principle of design (including the design of algorithms); this is the idea of intelligent computing. This many aspects, such as artificial neural network technology, simulated annealing, tabu search algorithm, ant colony model.

Image processing is a series of computer image processing to achieve the desired result. Common handling of digital images, image coding, image enhancement, image restoration, image segmentation and image analysis and so on. Image processing generally refers to digital image processing. Although some treatment methods you can use the optical or analog technology, but they are nowhere near as flexible digital image processing and convenience, and thus into the digital image processing the main aspects of image processing. [1]

2 Artificial Neural Network

Logical thinking is based on logical rules of reasoning process; it first information into concepts, and using symbols, then, according to symbolic computation in serial mode logic; this process can be written as a serial command, the computer implementation. However, the intuitive mind is a distributed storage of information together; the result is a sudden arises between ideas or solutions to problems.

Artificial neural network simulation of the human way of thinking is a non-linear dynamical system, which feature information distributed storage and parallel co-processing. However, a large number of neurons in a network system can realize the behavior is very colorful. Artificial neural network applications in image processing cases abound, such as image compression, image segmentation, edge detection, image enhancement, image recognition and so on.

Discrete categories of data (such as ground truth data), in many cases do not have statistical significance. For the high-dimensional data, Bayes standards required by the covariance matrix will be difficult. In addition, this approach with people's visual interpretation of image classification methods are very different, so people who want to find a visual interpretation and classification more similar to the computer classification.[2]

Neural network classification of remote sensing image processing applications, the most extensive and in-depth, with the theory of artificial neural network system development, neural network classification of remote sensing image processing is increasingly becoming an effective means, and gradually replaces the trend of the maximum likelihood method. Artificial neural network has self-organizing, nonlinear, self-learning ability in many areas, it played an important role. I believe that with artificial neural network theory and practice of development in the near future, it will play in the field of image processing, a greater role.

Remote sensing image classification is the use of computers through various types of remote sensing image features in the spectral and spatial information analysis, feature selection, and use some means to feature space is divided into non-overlapping sub-space, and then all image naturalization to each pixel to sub-space. Traditional remote sensing image classification based on remote sensing data in computer statistical value characteristic of the sample data and training to carry out a statistical relationship between the classification of surface features, and more generally for Bayes statistical theory based on the maximum likelihood method, but with the dimension of the space remote sensing data expanding the number, the method began to reveal weaknesses. Multi-source, multi-dimensional remote sensing data may not have the normal characteristics. [3]

3   Simulated Annealing Algorithm

Simulated annealing algorithm was first proposed by Kirkpatrick and other areas used in combinatorial optimization, which is based on the solution strategy Mente-Carlo iterations of a random optimization algorithm, the starting point is based on the physical annealing process of solid material and the general combinatorial optimization problems the similarity between. Simulated annealing starting from a higher initial temperature, with decreasing temperature parameters, with the probability of jumps in the solution space characteristics of the objective function of a random search for global optimal solution, i.e. the probability of a local optimal solution can jump out and ultimately tends to the global optimum. Simulated annealing algorithm is a general optimization algorithm, probability theory; global optimization algorithm with the performance, the project has been widely used, such as control engineering, production scheduling, machine learning, signal processing and other fields. To improve the success rate of the image mosaic, is proposed based on adaptive simulated annealing and multi-resolution search strategy of the new method of stitching images automatically.

Due to the improved image mosaic algorithm combining simulated annealing search method precision and multi-resolution high-efficiency, multi-resolution than the traditional direct search algorithm for image stitching higher success rate, especially for the less than ideal image quality of medical ultrasound images showed good robustness.[4] The new algorithm first adaptively select the registration area, then the mutual information as similarity evaluation standard, combined with adaptive simulated annealing and the idea of multi-resolution search strategy for image translation and rotation, respectively, parameters of global optimization and local search, and finally of image stitching. Through the noisy digital images and medical ultrasound images of the simulations splicing experiments showed that the new algorithm of multi-resolution than traditional direct search method has high accuracy, speed and noise immunity advantages.

Traditional simulated annealing algorithm clear thinking, simple in principle, can be used to solve many practical problems. But there are many drawbacks, scholars and their corresponding improvement has been improved simulated annealing algorithm, including: thermal annealing method, a memory of the simulated annealing algorithm, with a return to the search of the simulated annealing algorithm, several optimization law, tempering annealing algorithm.

In the traditional simulated annealing process, the algorithm terminates at a predetermined stopping criteria for S, such as: control parameter t is less than a sufficiently small positive number; succession of several Markov chains in solution have not been any improvement; two a Markov chain have been derived from the difference of the absolute value of the solution is less than a sufficiently small positive number and so on. However, simulated annealing search process is random, and when t is large can accept some deterioration in the solution, and as the t-value decay, deterioration of the solution decreases the probability of being accepted up to 0. On the other hand, some of the current solution to achieve the optimal solution must be a temporary deterioration of the "ridge" and therefore, these guidelines can not guarantee that the final solution to stop just once throughout the search process to reach the optimal solution.

Therefore, the increase in the traditional algorithm in a memory device so that it can remember encountered during the search the best solution, when annealed at the end of the final solution obtained with the memory device in the solution as more and those who take final results.

4   Tabu Search

Tabu search algorithm by introducing a flexible storage structure and the corresponding guidelines to avoid circuitous taboo search, and to pardon some of the guidelines through the contempt fine was taboo status, thereby ensuring a variety of effective explore and ultimately global optimization.[5] Tabu search (Tabu Search or Taboo Search, referred to as TS) the idea was first proposed in 1986 by the Glover, it is an extension of local search area, is a global optimization algorithm step by step, is a process of human intelligence simulation.

Proposed a tabu search framework based on intelligent memory, the actual implementation process in the nature of the problem can be targeted to do the design, this paper gives the basic tabu search process, based on the algorithm of how to design the key steps a useful summary and analysis.

Tabu search in order to continue to move to the neighborhood would rely on expanding the search space; the neighborhood movement in the current solution, based on the specific mobile strategy in accordance with a certain number of new solutions, these new solutions is called neighborhood solution, the number of new solutions as the size of neighborhood solutions. Neighborhood movement, the literature also called neighborhood operations, neighborhood structure, and neighborhood transformation.

When adding a taboo tabu object table, set the term length for the taboo, the search process for each iteration, the taboo tabu table automatically by a term of the object, when the object of a taboo term of 0, from taboo table deleted. Each length is taboo tabu table object in the survival time, also known as tabu object term.

Greedy local search is based on the field of sustainable thinking in the field of the current solution to search, although the algorithm is generic and easy to implement, and easy to understand, but its search performance depends entirely on the structure and the early solution of the field, especially glimpse into local minimum and can not guarantee of global optimization. Search for local areas, in order to achieve global optimization, you can try through the following: the probability of acceptance of inferior solutions controllability to escape local minima, such as simulated annealing; to expand the search area of the structure, such as TSP's 2-opt extended to k-opt; multi-point parallel search, such as evolutionary computation; variable structure field search. In addition, TS is used to avoid the taboo circuitous search strategy, it is a deterministic local minimum sudden jump strategy.

Tabu search involves the area, taboo table, the length of taboo, candidate solutions, contempt criteria and other concepts, we first use an example to understand the tabu search and its important concept, then give the general algorithm flow. Tabu search is a reflection of artificial intelligence, is an extension of the local area search. Tabu search is to mark the most important ideas have been searching the corresponding local optimal solution of some of the objects, and the iterative search further tries to avoid these objects (rather than an absolute prohibition cycle) to ensure an effective search for different ways of exploration.

Solution evaluation function, the relevant literature is also known as the fitness function, fit or function of the value of fitness function. Directly to the optimization objectives as an evaluation function is a simple, intuitive way, and any equivalent transformation and optimization target function can be used as evaluation function. Tabu search for the solution space, through the evaluation function to calculate the corresponding evaluation function value, the evaluation function value representative of the degree to understand the pros and cons. [6] Sometimes, the calculation of the objective function are more difficult or time consuming, you can take the values reflect the characteristics of the problem instead of the target calculation, but must ensure that the target eigenvalue problem of a consistent and optimal. According to the characteristics of the problem may be the better evaluation function value, and vice versa may be as small as possible. Based on mathematical methods, both goals can be equivalent conversion.

5   Ant Model

Comprehensive model of the different ant types of combinatorial optimization problems on the application, the establishment of a colony model of generalized model of a typical ant colony model to achieve the implementation process and key elements of the analysis, pointing out the different ant colony algorithm for the essential difference between models with the traveling salesman problem, the second assignment problem, and network routing problems typical combinatorial optimization problem, an overview of the ant colony optimization model in static and dynamic portfolio optimization problem in the application and finally discuss the ant colony model in the modeling, implementation, and theoretical research in the field future direction. Ant colony system has a distributed organizational model for solving complex combinatorial optimization problems, distributed control to provide a good idea, so the ant colony system behavior and self-organizing capacity of the research has attracted the interest of many researchers. Ant's group optimization is the ant colony system is an important research area, new models, new methods, new applications continue to emerge. Bionics paper the mechanism of ant colony model is described. [7]

Ant colony clustering algorithm used in image-based cluster analysis of the characteristics of emotion, and the original ant colony clustering algorithm stopped improving. With the launching of the image retrieval system, rational organization and management of image database, image retrieval has become the crucial point. After computing the Euclidean distance between samples to determine the starting ants, simulated ants pick up and drop food behavior, based on the primary color of the extracted image features, image clustering emotional stop. Experiments indicate that the clustering algorithm can achieve better results and high retrieval efficiency.

Disposal of other areas such as image color clustering, image contraction, image retrieval, image fusion, and so also have the use of color clustering is a basic computer vision and graphics in the disposition effect. Experiments indicate that the algorithm stops with a robust color quantization and color distortion.

Ant colony algorithm is proposed to extend the application of fractal image coding time crunch improved methods. Related to basic fractal image contraction algorithm, experiments indicate that the basic protection method robust built in image quality, expedite the speed of fractal image crunch.

6   Conclusion

Image processing from a large number of optimization problems, the optimization algorithm within the computer field has always been a hot issue. Optimization algorithm consists of heuristic algorithms and smart random algorithm. Heuristic depends on the nature of the problem of understanding, is a local optimization algorithm. Intelligent algorithms do not rely on random nature of the problem, search the solution space according to certain rules, until the search to the approximate optimal solution or optimal solution, are global optimization algorithms, represented by a genetic algorithm, simulated annealing, tabu search algorithm, ant colony models.

Acknowledgments. The work is supported by Funding Project for Academic Human Resources Development in Institutions of Higher Learning under the Jurisdiction of Beijing Municipality (PHR (IHLB)) (PHR201108407). Thanks for the help.

References

1. Bianchini, M., Frasconi, P.: Learning Without Local Minima in Radial Basis Function Networks. IEEE Transaction on Neural Networks 6(3), 749–756 (1995)
2. Li, S.-R., Ebong, I.E.: Tunneling-Based Cellular Nonlinear Network Architectures for Image Processing. IEEE (2009)
3. Suzuki, K., Li, F., Sone, S., et al.: Computer-aided diagnostic scheme for distinction between benign and malignant nodules in thoracic low-dose CT by use of massive training artificial neural network. IEEE Trans. Med. Image 24(9), 1138–1150 (2005)
4. Babaguchi, N., et al.: Connectionist model binarization. In: Proc. 10th ICPR, pp. 51–56 (1990)
5. Joo, S., Moon, W.K., Kim, H.C.: 26th Annual International Conference of the IEEE EMBS (2004)
6. Satirapod, C., Trisirisatayawong, I., Homniam, P.: Establishing Ground Control Points for High-resolution Satellite Imagery Using GPS Precise Point Positioning. In: Proceedings of 2003 IEEE International Geo Sciences and Remote Sensing Symposium, IGARSS 2003, July 21-25, vol. 7, pp. 4486–4488 (2003)
7. Cascio, D., Fauci, F., Magro, R., et al.: Mammogram segmentation by contour searching and mass lesions classification with neural network. IEEE Transactions on Nuclear Science 53(5), 2827–2833 (2006)
8. Papadopoulos, A., Fotiadisb, D.I., Likas, A.: Anautomatic microcal-cification detection system based on a hybrid neural network classifier. Artif. Intell. Med. 25(2), 149–167 (2002)
9. Dunstone, E., Andrew, J.: Super-high, scale invariant image compression using a surface learning neural network. In: International Symposium on Speech (1994)
10. Cortes, C., et al.: A network system for image segmentation. In: Proc. Intl. Joint Conf. on Neural Network, vol. 1, pp. 121–125 (1989)

The Study on Information Technology Patent Value Evaluation Index Based on Real Option Theory[*,**]

Jie Chen[1] and Zeming Yuan[2]

[1] School of Management, Tianjin Radio &TV University, 300191 Tianjin, P.R. China
[2] School of Business, Tianjin University of Finance and Economic, 300222 Tianjin, P.R. China
jie5068@126.com

Abstract. The establishment of a set of information technology patent value evaluation system can not only solve the core problem of patent transaction, but also effectively promote the patent as the embodiment of the value of intangible property. This research firstly studies The application model of option pricing, including Pricing method and A general paradigm; Then, it analyzes The influence factors of Information technology patent value, including Economic factors and Profit factor. Information technology patent value evaluation index based on the option pricing, including the design of quantitative index and the design of qualitative index. Finally, it makes suggestions about Future prospect.

Keywords: information technology patent, real option theory, value evaluation index.

1 Introduction

Research on the Evaluation of Information technology patent is under the background of promoting the implementation of the national intellectual property strategy. The aims include that we will promote the information technology independent innovation achievements in intellectual property right, commercialization and industrialization, develop intellectual property results in the exchange and the use value of intellectual property rights transactions, and we will provide the necessary theoretical basis.

In general, information technology patent has some features, such as the meaning more microscopic abstract, higher technological content and so on. More importantly, the information technology high speed update and market development of the rapid change of information technology patent value evaluation effectiveness put forward higher requirements. Therefore the establishment of a set of information technology patent value evaluation system can not only solve the core problem of patent transaction, but also effectively promote the patent as the embodiment of the value of intangible property.

[*] This research is supported by National Social Science Foundation of China (No:10BJY103).

[**] This research is supported by Tianjin Science and Technology Project (No:11ZLZLZT02100).

2   The Application Model of Option Pricing

2.1   Pricing Method

Option is a financial derivative, its pricing is mainly depends on the original asset price change. As a kind of risk assets, original asset price changes randomly. If the original asset prices are determined, as a derivative of the option or the securities price is determined.

The Pricing Mechanism of Financial Options based on two important premise: Non-Arbitrage Equilibrium Analysis Risk-Neutral valuation method. The cost of capital depends on the use of capital, rather than the source of capital. Financial market transactions are zero NPV behavior. Financial market transactions are zero NPV behavior by the no arbitrage equilibrium analysis, Once the chance of arbitrage, arbitrager's action will promote market equilibrium, so that it make the same capital cost equal. Therefore it can be considered two assets which produce the same cash flow are equal, or two assets can be copied. each other. The two assets which can be copied each other, if they transact in the market, they must have the same equilibrium price.

The essence of the option :Option buyer can further use new information in a period of time, they can reduce the expected future uncertain extent, so as to make more reasonable judgment and decision , thus it can bring gains or losses reduction.

On the basis of no arbitrage equilibrium, the core idea of the real option pricing, is: As determining the value of investment opportunity and investment strategy, investors should not simply use the subjective probability method or utility function, investors should seek the method which is established in the market of the maximum value of the project.

The key of real option pricing is to use the dynamic replication technology: in the capital market we seek an actual assets or items which have the same risk characteristics of tradable securities with the evaluate actual assets, it is known as the" twin securities", and we build portfolio with the securities and risk free bond , get the corresponding real option characteristics of income.

2.2   A General Paradigm

The nature of the evaluation assets directly affects the properties of real option. The frame provides identification and structure of real option pricing process in the analysis of evaluation assets under the premise of progressive nature, the specific processes are as follows:

According to the uncertainty and risk factors of the evaluation assets we discuss investment problems, what is the value and forms? It mainly due to the influence factors, how to quantify the performance.

According to the flexibility of investment itself in decision-making, or properly constructing the clause which is contain the option value , or obtaining   preliminary information by the initial investment , so we can obtain one or more choice and determine the basic types of real options. As the same as designing and applicating the real options, we can consider solve the problem of the real options.

Establishing the mode of the solving process, which is the technology strongest portion in the whole process, which is need a relatively profound mathematical knowledge and financial knowledge, the specific solution is often needed computer, at the same time, we should analysis various factors on the impact of the conclusion.

Analyzing the results, considering the problem of parameter sensitivity, which must be considered when we solve the issues. Because the market is uncertain, we must take into account the influence of variety factors, Analyzing the changes and its influences corresponding to the value of real options when we make the investment decision.

According to the actual situation of the project, we should redesign it. After completing the above steps, we can consider the influence of the real option and l analyze the initial assumptions, according to the results of calculation and analysis, at the same time considering the actual situation, we should adjust the real option design details, and even find new sources of uncertainty, we find new option characteristics and corresponding to the type of real options, we make decisions repeatly.

3 The Influence Factors of Information Technology Patent Value

3.1 Economic Factors

The economic influence factors of information technology patents included financial cost aspect and financial revenue aspect.

Financial cost aspect

Information technology patent belongs to the technical commodity, it is mainly intellective labor results, it is a complex mental labor creation, therefore the patent need a lot of basic research , software and hardware configuration ,before formal R & D. After the R & D success, it need to be put into production, these will affect the direct financial costs of the patent. Therefore, financial cost including initial investment, operation cost improvement cost.

Financial income aspect

Information technology patent value is influenced by direct financial gains, investors when investors purchase the patent, they pay more attention to the patent if it can bring the excess profit, when the owner price the information technology patent ,whose direct financial income is an important factor under considering.

3.2 Profit Factor

Information technology patent value will be affected by their profit ability, because it has greater profitability in the future, it will face the smaller market risk, it may bring more and more potential earnings. Information technology patent profitability depends on many aspects, therefore the evaluation of information technology patent profitability is very complex. After large amount of investigation, information technology patents, technical, market and other factors have close contact, therefore, this article will sum up in the following four aspects in Profit factor: Technical factors, Market factors, Management factors, Legal factors.

4   Information Technology Patent Value Evaluation Index Based on the Option Pricing

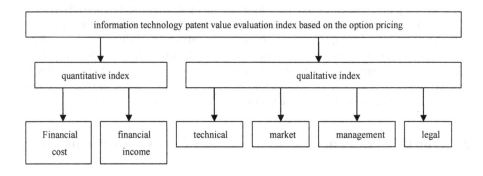

Fig. 1. Information technology patent value evaluation index based on the option pricing

4.1   The Design of Quantitative Index

Quantitative indexes reflect the economic factors of patent value, it refers to invest financial costs and obtain financial income in the field of information technology patent.

The financial cost

Financial costs included the plan of patent, development of patent, testing of patent and implementation stages of patent, mainly including the following aspects:

Basic research cost. Information technology patent is a system project in high-tech fields, such as the SAP large ERP software, which is   needed a large number of basic research, perhaps these research did not bring immediate profit, but it is essential premise in the subsequent development.

Software and hardware purchase cost. Hardware purchase cost mainly refers to the purchase of the necessary infrastructure, such as the purchase of the server, computer, printer and laying network cost. The software purchase cost mainly refers to the purchase of software inputs, such as the purchase of system software, database software and antivirus software costs etc.

The introduction of high-tech talent cost. Information technology patent right is a kind of wisdom and creation achievement, which need hire professionals by considerable costs .

Conversion cost. Profits generated from the product after the patent research is successful it need product, it convert into purchasing for the enterprise, which is conversion cost in the process of conversion.

Maintenance cost. Maintenance cost mainly refers to the daily maintenance cost after the patent products put into production. For example, hardware and software upgrades, system data backup and troubleshooting.

Recurrent expenditure. Recurrent expenditure mainly refers to the using some of the consumables spending during the process of patent production, such as paper, storage and so on.

Financial income

Financial revenue is mainly used to measure direct benefits by using the information technology patent, mainly including the following contents: Income increase, cost reduction, various cost savings.

4.2   The Design of Qualitative Index

Qualitative index reflects the influence of patent value profit ability factors. After the research, we discover information technology patent profitability mainly by the technology, market, business management and law development .Therefore the design of the qualitative indexes will cover the four aspects.

Technical index

Technology innovation is a technology patent core competitiveness, therefore, the profit which is brought by technology innovation is the important part of patent income, it mainly includes the following content: technology innovation, technology content and scope of application, technology maturity, technical substitution level.

Market index

Information technology patent acquires gain after the commodity during the market competition. Market indicators can include the following aspects. market capacity, market demand lever, market competitive ability, remaining economic life

Management index

Information technology patent put into using, like ERP software, the application of enterprise business process, will produce major effect, while the internal process is the key of improving the performance of enterprise, in addition, information technology has a certain role in promoting the development of enterprises, these belong to the management profit ability, specific content as follows: business process optimization, knowledge capital accumulation, organizational structure reform.

Law index

Information technology patent is approved by the law, the law gives the right person value rights for the exclusive right, including the following contents: Independence, Implementation status, the legal status of stability degree.

5 Future Prospect

The evaluation index system includes quantitative index and qualitative index, quantitative indicators evaluate direct economic benefits, the qualitative indexes evaluate the profitability, the overall goal is assessment of the value of information technology patent. It is noted that the established index system of relatively complete, but in response to specific information technology patents may have different priorities, during the process of evaluation according to different evaluation objects, we should select different combinations of indicators, we do not need use all.

References

1. Cobb, B.R., Charnes, J.M.: Real Options Valuation. In: Proceedings of the 2007 Winter Simulation Conference (2009)
2. Hausman, J.: Real Alternatives and Incentitives Options and Patent Damageto Innovate. Journal Comps: The Legal Treatment of Non-inferingilation (2006)
3. Berman, B., Woods, J.D.: Positioning 1P for Shareholder Value Managing Intellectual Property (March 2004) (accessed on October 11, 2004)
4. Yi, W.: The research on the evaluation of computer and automation technology patent based on option pricing theory. Beijing Jiaotong University (2011)

The Design and Implementation of an Ordering System for Restaurants Based on 3G Platform

Ming Xia[1], XiaoMin Zhao[1], KeJi Mao[1], Yi Fang[2], and QingZhang Chen[1,*]

[1] College of Computer Science and Technology, Zhejiang University of Technology,
Hangzhou, 310023, China
{xiaming,zxm,maokeji,qzchen}@zjut.edu.cn
[2] Beijing Yinxin Changyuan Science & Technology Co., Ltd, Beijing, 100080, China
myfy@163.com

Abstract. An ordering system can help restaurants to increase the efficiency in ordering process. In this paper, we designed and implemented an ordering system for restaurants based on 3G platform. The system provides both customers and waiters a customized handheld 3G terminal to view up-to-date menu and issue ordering commands, and the system will quickly forward the order to kitchen to notify chefs to prepare dishes. With this ordering system, restaurants can greatly reduce operation cost, and improve their customer satisfaction.

Keywords: Ordering System, Restaurant, 3G, Android.

1 Introduction

Catering industry in China is developing rapidly in these years. However, the ordering process, in which customers read a printed menu and then tell waiter their order, remains almost unchanged. This traditional process is more and more limiting the developing potential of catering companies. For instance, dishes served by a restaurant typically change very frequently, and once the dishes served changed, the menu must be reprinted. At the other hand, it is very difficult to record the favorites of each customer and provide recommendation service to customers.

There are several currently available digital ordering systems for restaurants, such as BOLI wireless ordering system. But most of these systems are only designed for waiters to give ordering commands, and a customer still needs to read printed menu and tell waiter his/her order. At the same time, these systems are mostly developed based on ISM 433Mhz or IEEE 802.11 wireless communication technology, which limits the working range.

This paper introduces a digital ordering system based on 3G platform [1]. In this system, ordering process is completed using a customized handheld 3G terminal [2], which retrieves the up-to-date dishes information from a server, generates and displays the menu on its screen to waiter/customer, and at last transmits the ordering

---

[*] Corresponding author.

D. Jin and S. Lin (Eds.): Advances in FCCS, Vol. 2, AISC 160, pp. 315–321.
springerlink.com © Springer-Verlag Berlin Heidelberg 2012

results back to data server. It can also be extended to be able to remember the ordering results to provide recommendation service to customers.

2   System Design

The system adopts three-layer architecture, as shown in Fig. 1.

Fig. 1. Three-layer architecture of the system.

The system can be generally divided into two parts: the part for enterprise users and the part for individual users (such as waiters and customers). In the part for enterprise users, registered users can log in to manage the system, e.g., upload the dish lists to the data server and check all available orders. In the part for individual users, registered users can log in to view the dish list and issue ordering commands.

We will then elaborate on the details of the system design through working process analysis.

(1) Enterprise users

Business process of the enterprise users are shown in Fig. 2. One of the major tasks of the enterprise user is to upload the dish information, including the dish's catalog, style, name, price, discounts, and etc..

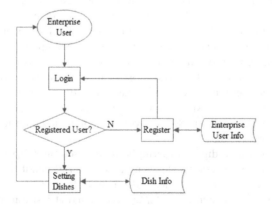

Fig. 2. Enterprise user business flow.

(2) Individual users

Customer or waiter can log in to check the dishes served by the restaurant and issue ordering commands. The data server will then send the ordering results to a printer located at the kitchen, and chefs will prepare food according to the customers' order. At last, customers will be required to pay for their order through ZHIFUBAO, and the whole business process finishes, as shown in Fig. 3.

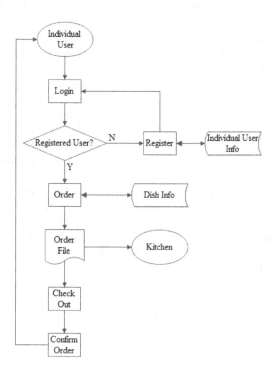

Fig. 3. Individual user business flow.

2.1   The Design of Weighted Traversal Algorithm Based on Priority

The orders issued by customers and waiters must be waiting in a queue before processing. In order to improve the system performance, we designed a weighted traversal algorithm based on the weighted priority scheduling algorithm [3], as shown in Fig. 4.

The algorithm works as follows:

(1) Put tasks of all clients into a "Standby queue" for scheduling.

(2) Give each client a preset service priority, and calculate its weight, as shown in Eq. 1 (here we assume that there are n clients):

$$Weight[i] = \frac{Priority[i]}{\sum\limits_{i=1}^{n} Priority[i]} \cdot \qquad (1)$$

Take a 4-client queue as an example, if we set each client's priority to: Priority[1]=5, Priority[2]=3, Priority[3]=1 and Priority[4]=1, Then Client[1]'s weight can be calculated as: Weight[1]=5/(5+3+1+1)=50%. Similarly, Weight[2]=30%, Weight[3]=10%, and Weight[4]=10%.

(3) Pick up a fixed number of tasks from the standby queue, and put these tasks in the initial client task queues, in which tasks that belong to different clients are separated into different queues.

(4) Setup an array to store the weight of each client, and arrange the array by the descending order of weight.

(5) Pick up tasks of the client who has the highest weight from the initial client task queues, and put them into the final client task queue. Then the algorithm picks up the tasks of the client who has the second highest weight. This process repeats until the algorithm reaches the client who has the lowest weight, and the number of tasks of each client that can be moved into the final client task queue is: count(FinalClientTaskQueue[i])=floor(count(InitialClientTaskQueues[i])*weight[i]), in which FinalClientTaskQueue[i] is the tasks of Client[i] in the final client task queue, IntitialClientTaskQueues[i] is the tasks of Client[i] in the initial client task queues, the function count(x) is the number of a given data x, and the function floor(x) is the value rounded to the next lowest integer of x.

(6) Return to Step (3).

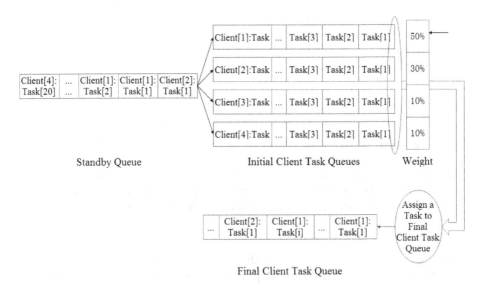

Fig. 4. Weighted traversal algorithm based on priority.

3   System Implementation

We implemented the system based on Java, and employed Eclipse as the development environment because it includes a standard plug-in components set such as JDT (Java Development Tools) [4], and can dynamically load these plug-in components [5]. Eclipse also provides a dedicated Plug-in Development Environment (PDE) [6] for efficient plug-in development.

The implemented three-layer architecture of the system is given in Fig. 5. For foreground users, they mainly access the system to issue order through 3G handheld terminals, including both phones and tablet PCs. These terminals run android 2.2 as their operating systems. The ordering service system contains foreground and background application server and a dedicated database server which stores information of the dishes and users. For background users, they typically access the system for maintenance tasks, including uploading the dish information, managing individual users, and etc..

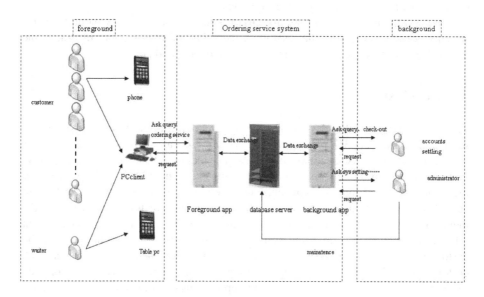

Fig. 5. Implemented three-layer architecture of the system.

The ordering service system contains four major modules: system settings, dish settings, ordering and reporting, as shown in Fig. 6.

The system settings module enables enterprise users to manage the account information of the system, including adding and deleting user, changing user's password, updating user's personal information and managing the access control list. It also allows the enterprise user to maintain and backup system data.

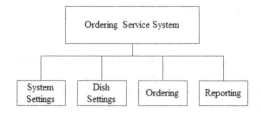

Fig. 6. Modules of the ordering service system.

The dish settings module is available for enterprise users, and it allows enterprise users to input the list of dishes served by the restaurant, and manage the category, name, pictures, price and discount information of each dish.

Ordering module is mainly designed for individual users. Individual users can use this module to view the menu, and issue ordering commands, i.e., input the name and quantity of dish, room and table number, note, and etc.. It also provides secured checkout and payment functions after ordering.

Reporting module provides enterprise users a detailed daily/weekly/monthly/yearly summary of the ordering in the restaurant, including the total number of orders, the total value of orders, the most frequently ordered dishes, and etc..

4   Conclusions

This paper presents the design and implementation of an ordering system for restaurants based on 3G platform. The system uses a 3G handheld terminal as interacting device, which displays up-to-date menu on its screen, and allows customer/waiter to directly issue ordering commands. The system will then forward the ordering commands to kitchen, and chefs can prepare food according to customers' requirements. With this system, restaurants will not have to reprint the menu once the list of dishes changed, thus greatly increase the management efficiency. At the same time, the system can provide richer information about the customers. For instance, the system can be simply extended to remember each customer's ordering information. Through data mining, we can provide recommendation service to customers, and improve customer experience.

Acknowledgments. This work is supported by the Natural Science foundation of Zhejiang Province (Y1110649).

References

1. Gao, P., Zhao, P., Chen, Q.T.: 3G Technology Q & A. The People's Posts and Telecommunications Press, Beijing (2009)
2. Fu, C.D., Huang, C.M., Tang, X.S.: 3G Terminal Technology and Application. The People's Posts and Telecommunications Press, Beijing (2007)

3. Jiang, Z.H., Qi, W.J.: The Analysis and Evaluation of Common Job Scheduling Algorithms. Journal of Leshan Teachers College 12, 57–59 (2008)
4. Chinese Eclipse Community, http://www.ceclipse.org/
5. Li, Z.X., Yang, R.L.: Java EE Web Programming: Eclipse Platform. China Machine Press, Beijing (2008)
6. Developer Works, IBM,
   http://www.ibm.com/developerworks/cn/linux/opensource/os-ecov/

An Empirical Research on University Students' Intention to Use Online Courses

Hanyang Luo[1,2], Sai Wang[2], and Xiaoling Li[3]

[1] Shenzhen Graduate School, Harbin Institute of Technology, Shenzhen 518055, China
[2] College of Management, Shenzhen University, Shenzhen 518060, China
[3] Office of Psychology, Overseas Chinese Town Middle School, Shenzhen 518053, China
hanyang@szu.edu.cn

Abstract. Based on technology acceptance model and innovation diffusion theory, this research proposes a model to examine how university students intend to use online courses. The model is empirically tested using an online questionnaire survey. The research results indicate that perceived usefulness, perceived ease of use and teachers' induction are significant antecedents of university students' attitude to online courses, which further prominently positively affect their intention to use. This research also finds that the positive effect of perceived innovation compatibility on attitude has been almost completely mediated by perceived usefulness. But computer self-efficacy doesn't have remarkable positive effect on the attitude to online courses.

Keywords: perceived innovation compatibility, perceived usefulness, perceived ease of use, computer self-efficacy, online courses, attitude, intention.

1   Introduction

In recent years, many researchers have engaged in plenty of studies on the development and application of online courses. But it seemed that some of them overlooked a basic problem: Would students like to use such online courses? Which factors determine students' intention to use online courses? The analysis of these determinants is of great importance. Without students' usage intention, the modern educational technology, advanced educational thought and innovative teaching mode integrated into these online courses will not be realized at all.

2   Literature Review

In 1986, Davis proposed the famous Technology Acceptance Model (TAM), which introduced two new constructs, perceived usefulness (PU: the belief that using an application will increase one's performance) and perceived ease of use (PEOU: the belief that one's use of an application will be free of effort). According to TAM, both PU and PEOU could predict an individual's attitude concerning the use of an application [1].

In 1995, Rogers used well-established theories in sociology, psychology, and communications to develop an approach to study the diffusion of innovations, which

D. Jin and S. Lin (Eds.): Advances in FCCS, Vol. 2, AISC 160, pp. 323–328.
springerlink.com   © Springer-Verlag Berlin Heidelberg 2012

was known as Innovation Diffusion Theory (IDT). According to IDT, an innovation generates uncertainty, and uncertainty motivates an individual to seek more information about alternatives. Diffusion is then the process by which an innovation is communicated through certain channels over time among the members of a social system [2].

As for online courses, based on constructivism learning theory, Hanyang Luo and Xiaoling Li presented some design principles such as guidance and initiative, real scene and collaborative learning, and then discussed the overall structural design of online courses [3]. Andria Young and Chari Norgard found that the following areas are important for student satisfaction with online instruction: interaction between student and professor, interaction among students, consistent course design across courses, technical support availability, and flexibility of online courses [4]. Hassan M. Selim used a structural equation model to fit and validate the Course Website Acceptance Model (CWAM) and the results indicated that course website usefulness and ease of use was proved to be key determinants of the acceptance and usage of course website [5].

3   Conceptual Model and Theoretical Hypotheses

3.1   Conceptual Model

Based on Technology Acceptance Model and Innovation Diffusion Theory, we proposed an integrated conceptual model of university students' behavioral intention to use online courses (see Fig. 1).

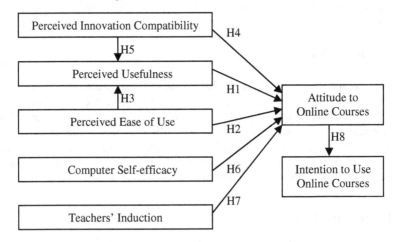

Fig. 1. An integrated model of intention to use online courses

3.2   Theoretical Hypotheses

PU and PEOU. In our study, perceived usefulness is the degree to which a university student believes that using an online course will enhance his or her learning performance. Information system researchers have discovered that perceived usefulness has a positive effect on the attitude to information systems [1]. Therefore, we hypothesize that:

H1: Perceived usefulness will have a positive effect on the attitude to online courses.

In our research, perceived ease of use is the degree to which a student believes that using an online course will be free of effort. Information system researchers have found that perceived ease of use has a positive effect on the attitude to and perceived usefulness of information systems [1]. Thus, we hypothesize that:

H2: Perceived ease of use will have a positive effect on the perceived usefulness of online courses.

H3: Perceived ease of use will have a positive effect on the attitude to online courses.

Perceived Innovation Compatibility. Perceived innovation compatibility is the degree to which the innovation (such as an online course) is perceived to be consistent with the potential users' (such as university students') existing values, previous experiences and needs. Some researchers have found that perceived innovation compatibility has a direct effect on perceived usefulness of and attitude to information system. Thus we hypothesize that:

H4: Perceived innovation compatibility will have a positive effect on the attitude to online courses.

H5: Perceived innovation compatibility will have a positive effect on perceived usefulness.

Computer Self-efficacy. Computer self-efficacy is defined as an individual's perceptions of his or her ability to use computers in the accomplishment of a task. The more experience a university student acquires online, the more important are concerns of control over personal information, which implies that computer self-efficacy will have a positive effect on his or her attitude to information systems such as online courses. Thus, we hypothesize that:

H6: Computer self-efficacy will have a positive effect on attitude to online courses.

Teachers' Induction. In fact, in our study, teachers' induction plays a role as social norm, which is defined as the degree to which university students perceived that others (especially their friends, classmates and teachers) approved of or encouraged their using online courses. Some researchers have confirmed that social norm played an important role in a user's attitude to information system. Therefore, we hypothesize that:

H7: Teachers' Induction will have a positive effect on students' attitude to online courses.

Attitude. According to Theory of Reasoned Action, favorable attitudes toward an act or event would lead to a positive intention to perform the act or adopt the event. In the online context, Hanyang Luo et al. found that consumers' favorable attitude toward a website will positively affect their willingness to use it [6]. Therefore, we suppose:

H8: Attitude to online courses will have a positive effect on intention to use them.

4   Research Methodology

In order to test our model, we collected data via an online questionnaire. We developed multi-item scales for each variable, most used five-point Likert-type interval scales ranging from strongly disagree (1) to strongly agree (5). Then, we asked some students in Shenzhen University to answer a series of questions regarding their experience on some online courses. At last, due to some reasons such as incomplete data, 362 observations could be used in data analysis.

5   Result

5.1   Measurement Model

The reliability and validity of measurement for each construct was tested by using exploratory and confirmatory factor analysis based on the 362 samples.

Exploratory factor analysis. A principal component analysis with varimax rotation was used to examine measures. Factors with eigenvalue above 1.0 were extracted in

Table 1. The results of principal component analysis

Items	Perceived Ease of Use	Perceived Usefulness	Computer Self-efficacy	Attitude	Perceived Innovation Compatibility	Intention	Teachers' Induction
IC1	.064	.040	.148	.058	**.815**	.150	.208
IC2	.312	.387	.153	.197	**.603**	.052	-.122
IC3	.160	.418	.154	.192	**.637**	.189	-.107
PU1	.281	**.573**	.020	.233	.289	.125	.200
PU2	.029	**.848**	.094	.141	.062	-.004	.139
PU3	.187	**.640**	.159	.154	.259	.370	.099
PE1	**.704**	.016	.162	.191	.355	.190	.244
PE2	**.787**	.147	.134	.195	.222	.027	.077
PE3	**.776**	.166	.086	.047	-.063	.258	-.035
SE1	.167	-.005	**.864**	-.030	.123	-.059	.135
SE2	.052	.134	**.820**	-.026	.069	.126	.089
SE3	.094	.113	**.711**	.220	.157	.208	-.212
TI1	.175	.316	.069	.212	.073	.160	**.736**
TI2	-.075	.084	.034	.375	.162	.150	**.734**
AT1	.306	.197	.052	**.796**	.147	.079	.031
AT2	.267	.304	.020	**.659**	.028	.331	-.064
INT1	.322	.088	.161	.358	.271	**.615**	.123
INT2	.179	.144	.119	.170	.143	**.844**	.109

Components

Table 2. The results of reliability using Cronbach's alphas

Variables	Perceived Ease of Use	Perceived Usefulness	Computer Self-efficacy	Attitude	Perceived Innovation Compatibility	Intention	Teachers' Induction
alphas	0.775	0.754	0.767	0.762	0.755	0.768	0.662

each construct; these cumulatively explained over 73.84% of total variance. Items with low loadings on the intended factor or high cross-loadings on other factors were removed. Table 1 reports the results of principal component analysis. The resulting scales were then evaluated for reliability using Cronbach's alphas. As shown in Table 2, all but one had acceptable reliability (alphas > 0.70).

Confirmatory factor analysis. Confirmatory factor analysis was performed with LISREL 8.70. The fit of the overall measurement model was estimated by various indices. RMSEA showed the discrepancy between the proposed model and the population covariance matrix, to be 0.062, which was lower than the recommended cut-off of 0.08. All other indices (CFI, NFI, IFI, GFI) exceeded the commonly acceptance levels (0.90), demonstrating that the measurement model exhibited a fairly good fit with the data.

5.2   Structural Model

We examine the path coefficients of the structural model. This involved estimating the path coefficients and R^2 value. Path coefficients indicated the strengths of the relationships between the independent and dependent variables, whereas the R^2 value was a measure of the predictive power of a model for the dependent variables. Fig. 2 shows the structural model.

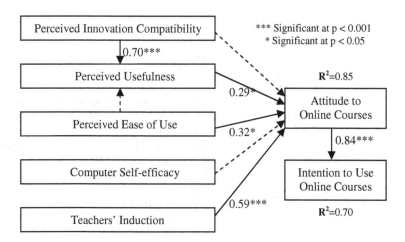

Fig. 2. Structural model

328   H. Luo, S. Wang, and X. Li

As shown in Fig.2, the path coefficients respectively from perceived usefulness, perceived ease of use, to attitude, are significant at the $P < 0.05$ level, the path coefficient from teachers' induction to attitude is very significant at the $P < 0.001$ level. Therefore, the data indicate support for H1, H2 and H7. But the path coefficients respectively from perceived innovation compatibility, computer self-efficacy, to attitude, are not significant, and the path coefficient from perceived ease of use to perceived usefulness is not significant, either. Therefore, H3, H4 and H6 aren't supported. The path coefficients respectively from perceived innovation compatibility to perceived usefulness, and from attitude to intention are the most significant at the $P < 0.001$ level, which strongly support our hypotheses H5 and H8.

6 Discussion

The most interesting result in our study is that, perceived innovation compatibility has a very significantly positive effect on perceived usefulness, but it doesn't have a remarkably positive effect on the attitude to online courses, which meant that perceived innovation compatibility has no significant direct effect on the attitude. In fact, the positive effect of perceived innovation compatibility on attitude has been almost completely mediated by perceived usefulness.

As we supposed, perceived usefulness, perceived ease of use have significantly positive effect on university students' attitude to online courses, which further has an extraordinarily positive effect on their intention to use online courses.

Acknowledgments. This research is supported by the Planned Project for Philosophy and Social Science in Guangdong Province (No. GD11CGL05), Shenzhen University Teaching Innovation Research Foundation (No.JG2011023) and Shenzhen University Laboratory and Apparatus Management Research Foundation (No. 2011028).

References

1. Davis, F.D.: Perceived Usefulness, Perceived Ease of Use, and User Acceptance of Information Technology. MIS Quarterly 13(3), 319–340 (1989)
2. Rogers, E.: Diffusion of Innovations. Free Press, New York (1995)
3. Luo, H., Li, X.: Research on the Design of Network Courses Based on Constructivism. In: Proceedings of the 1st International Workshop on Education Technology and Computer Science, pp. 283–287. IEEE Press, Piscataway (2009)
4. Young, A., Norgard, C.: Assessing the quality of online courses from the students' perspective. The Internet and Higher Education 9(2), 107–115 (2006)
5. Selim, H.M.: An empirical investigation of student acceptance of course websites. Computers & Education 40(4), 343–360 (2003)
6. Luo, H.: Research on initial trust in a B2C E-vendor. In: Lin, S., Huang, X. (eds.) CESM 2011, Part II. CCIS, vol. 176, pp. 438–444. Springer, Heidelberg (2011)

Research on Antecedents and Gender Differences of University Students' Computer Self-efficacy

Hanyang Luo[1,2], Zhini Li[2], and Xiaoling Li[3]

[1] Shenzhen Graduate School, Harbin Institute of Technology, Shenzhen 518055, China
[2] College of Management, Shenzhen University, Shenzhen 518060, China
[3] Office of Psychology, Overseas Chinese Town Middle School, Shenzhen 518053, China
hanyang@szu.edu.cn

Abstract. Computer self-efficacy plays an important role in the acceptance and use of new information technology. However, little is known about the antecedents of computer self-efficacy. This research proposes a model to probe into which factors affect university students' computer self-efficacy. The model is empirically tested using an online questionnaire survey. The research results indicate that encouragement, anxiety, openness and agreeableness are significant antecedents of computer self-efficacy. Taking gender into account, our research results show that agreeableness is significantly related to computer self-efficacy for female but not for male, and that openness is significantly related to computer self-efficacy for male but not for female. Interestingly, encouragement has a more significantly positive effect on computer self-efficacy for male students than for female ones.

Keywords: computer self-efficacy, encouragement, anxiety, openness, agreeableness, gender.

1   Introduction

Computer self-efficacy plays an important role in the acceptance and use of new information technology, many researchers have engaged in plenty of studies on its effect on the application of information system and its evaluation index systems. However, little is known about the antecedents of computer self-efficacy. The present study therefore aims at answering the following question: What is the relationship between some stable personality traits (such as openness and agreeableness) and computer self-efficacy? How do some teaching tactics (such as encouragement and demonstration) affect computer self-efficacy? Does gender moderate this relationship?

2   Literature Review

Defined as people's judgment of their capabilities to organize and execute courses of action required to attain designated types of performances, self-efficacy is concerned not with the skills one has but with judgments of what one can do with whatever skills one possesses [1]. Computer self-efficacy (CSE) is defined as an individual's perceptions of his or her ability to use computers in the accomplishment of a task.

D. Jin and S. Lin (Eds.): Advances in FCCS, Vol. 2, AISC 160, pp. 329–334.

Empirical evidence indicates that individuals with a high CSE are more likely to form positive perceptions of IT and use IT more frequently [2]. CSE has also been showed to be positively related to playfulness and negatively related to computer anxiety.

On the other hand, as defined as an enduring pattern of reactions and behaviors across similar situations, personality includes five factors, most commonly called neuroticism, extraversion, openness, conscientiousness, and agreeableness. Huma Saleem et al.'s research results indicate that four of the above five stable personality traits contribute to explain computer self-efficacy. Taking gender into account, their results show that the traits of neuroticism, extraversion, and agreeableness are significantly related to computer self-efficacy for women but not for men [4].

3   Conceptual Model and Theoretical Hypotheses

3.1   Conceptual Model

Based on theories on computer self-efficacy and personality trait, we proposed an integrated conceptual model of university students' computer self-efficacy (see Fig. 1).

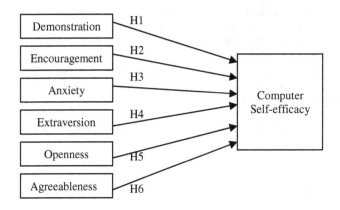

Fig. 1. An integrated model of computer self-efficacy

3.2   Theoretical Hypotheses

Demonstration. Some research results indicate that experiences with computer (including vicarious experience acquired by teachers' operation demonstration) have strong and significant effects on computer self-efficacy beliefs. Therefore, we hypothesize that:

H1: Demonstration will have a positive effect on university students' computer self-efficacy.

Encouragement. Students' computer self-efficacy can be enhanced by verbal persuasion which is a word of encouragement or a positive reinforcement. Such words could strengthen their abilities to carry out tasks at hand, using information technology. Thus we hypothesize that:

H2: Encouragement will have a positive effect on university students' computer self-efficacy.

Anxiety. Researchers have found the relationships between higher computer self efficacy, lower computer anxiety, more positive attitudes towards the Internet and longer reported use of the Internet [4]. Thus, we hypothesize that:

H3: Anxiety will have a negative effect on university students' computer self-efficacy.

Extraversion. Extraversion is characterized by terms such as gregarious, energetic, and self-dramatizing. Empirical research results indicate that extraversion is positively related to self-efficacy, positive attitude and enjoyment of IT use [5]. Therefore, we hypothesize that:

H4: Extraversion will have a positive effect on university students' computer self-efficacy.

Openness. Openness is characterized as imaginative, curious, open-mindedness, and original. Individuals who score high on openness are more likely to have positive attitudes to learning experiences in general and therefore may also be more willing to learn new information technologies [4]. Therefore, we suppose:

H5: Openness will have a positive effect on university students' computer self-efficacy.

Agreeableness. Agreeableness is marked by descriptors such as warm, tactful, and considerate. Empirical evidence indicates that low agreeableness is associated with higher scores on computer anxiety and lower CSE [6]. Therefore, we hypothesize that:

H6: Agreeableness will have a positive effect on university students' computer self-efficacy.

4   Research Methodology

In order to test our model, we collected data via an online questionnaire. We developed multi-item scales for each variable, most used five-point Likert-type interval scales ranging from strongly disagree (1) to strongly agree (5). Then, we asked some students in Shenzhen University to answer a series of questions regarding their experience on some online courses. At last, due to some reasons such as incomplete data, 406 observations could be used in data analysis.

5   Result

5.1   Measurement Model

The reliability and validity of measurement for each construct was tested by using exploratory and confirmatory factor analysis based on the 406 samples.

Exploratory factor analysis. A principal component analysis with varimax rotation was used to examine measures. Factors with eigenvalue above 1.0 were extracted in each construct; these cumulatively explained over 75.46% of total variance. Items with low loadings on the intended factor or high cross-loadings on other factors were removed. Table 1 reports the results of principal component analysis. The resulting scales were then evaluated for reliability using Cronbach's alphas. All had acceptable reliability (alpha > 0.70), as shown in Table 2.

Table 1. The results of principal component analysis

Items	Components					
	Anxiety	Encouragement	Agreeableness	Openness	Extraversion	Demonstration
DEM1	-.032	.128	-.045	-.021	.045	**. 868**
DEM2	-.015	.163	.143	.035	-.026	**. 842**
ENC1	-.043	**. 855**	.129	.118	.041	.208
ENC2	-.069	**. 940**	.056	.094	.088	.069
ENC3	-.103	**. 925**	.029	.077	.043	.081
ANX1	**. 805**	-.103	.007	-.072	-.051	.009
ANX2	**. 701**	-.021	.125	-.050	.002	.109
ANX3	**. 899**	-.055	-.084	-.003	-.013	-.121
ANX4	**. 873**	-.033	-.098	.019	-.007	-.080
EXT1	-.002	.073	.137	.127	**. 856**	-.001
EXT2	-.100	.053	.311	.334	**. 658**	.035
EXT3	.022	.103	.584	.201	**. 588**	.001
OPN1	-.098	.094	.140	**. 827**	.126	.067
OPN2	.010	.100	.214	**. 855**	.167	.010
OPN3	-.018	.098	.227	**. 844**	.157	-.066
AGR1	.002	.041	**. 770**	.100	.239	.048
AGR2	.023	.073	**. 798**	.216	.055	-.001
AGR3	-.058	.068	**. 829**	.206	.137	.078

Table 2. The results of reliability using Cronbach's alphas

Variables	Anxiety	Encouragement	Agreeableness	Openness	Extraversion	Demonstration
alphas	0.835	0.918	0.794	0.863	0.750	0.719

Confirmatory factor analysis. Confirmatory factor analysis was performed with LISREL 8.70. The fit of the overall measurement model was estimated by various indices. RMSEA showed the discrepancy between the proposed model and the population covariance matrix, to be 0.053, which was lower than the recommended cut-off of 0.08. All other indices (CFI, NFI, IFI, GFI) exceeded the commonly acceptance levels (0.90), demonstrating that the measurement model exhibited a fairly good fit with the data.

5.2   Structural Model

We examine the path coefficients of the structural model. This involved estimating the path coefficients and R^2 value. Path coefficients indicated the strengths of the relationships between the independent and dependent variables, whereas the R^2 value was a measure of the predictive power of a model for the dependent variables. Fig. 2 shows the structural model.

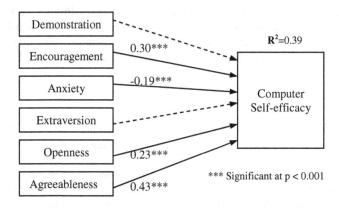

Fig. 2. Structural model for all samples

As shown in Fig.2, the path coefficients respectively from encouragement, anxiety, openness and agreeableness to computer self-efficacy, are significant at the $P < 0.001$ level. Therefore, the data indicate support for H2, H3, H5 and H6. But the path coefficient from demonstration to computer self-efficacy is not significant. Thus, H1 isn't supported. To our surprise, contrary to our anticipation, extraversion has a negative effect on computer self-efficacy, which runs counter to our hypothesis H4.

6   Discussion

As shown in Fig.3 and Fig.4, taking gender into account, our research results show that the trait of agreeableness is significantly related to computer self-efficacy for female students but not for male ones, and that the trait of openness is significantly related to computer self-efficacy for male students but not for female ones. The most interesting result in our study is that, encouragement has a more significantly positive effect on computer self-efficacy for male students than for female ones.

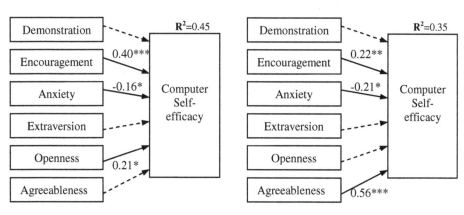

Fig. 3. Structural model for male samples   **Fig. 4.** Structural model for female samples

Acknowledgments. This research is supported by the Planned Project for Philosophy and Social Science in Guangdong Province (No. GD11CGL05), Shenzhen University Laboratory and Apparatus Management Research Foundation (No. 2011028) and Shenzhen University Teaching Innovation Research Foundation (No. JG2011023).

References

1. Imhof, M., Vollmeyer, R., Beierlein, C.: Computer use and the gender gap: The issue of access, use, motivation, and performance. Computers in Human Behavior 23, 2823–2837 (2007)
2. Compeau, D.R., Higgins, C.A., Huff, S.: Social cognitive theory and individual reactions to computing technology: A longitudinal study. MIS Quarterly 23(2), 145–158 (1999)
3. Hackbarth, G., Grover, V., Yi, M.Y.: Computer playfulness and anxiety: Positive and negative mediators of the system experience effect on perceived ease of use. Information & Management 40(3), 221–232 (2003)
4. Saleem, H., Beaudry, A., Crotea, A.: Antecedents of computer self-efficacy: A study of the role of personality traits and gender. Computers in Human Behavior 27, 1922–1936 (2011)
5. Hunsinger, M., Poirier, C.R., Feldman, R.S.: The role of personality and class size in student attitudes toward individual response technology. Computers in Human Behavior 24, 2792–2798 (2008)
6. Korukonda, A.R.: Differences that do matter: A dialectic analysis of individual characteristics and personality dimensions contributing to computer anxiety. Computers in Human Behavior 23, 1921–1942 (2007)

Research on Remote Sensing Image Feature Classification Based on Improved ACO

Chong Cheng[1] and Yong Zhang[2]

[1] Huazhong University of Science and Technology, 430074 Wuhan, China
[2] Wuhan University of Science and Technology, 430081 Wuhan, China
86887048@qq.com, appserver@yahoo.cn

Abstract. Based on researches of traditional ant colony optimization algorithm and its various applications, the optimization efficiency of algorithm was improved by optimizing movement rules of single ant and parameter selection. The algorithm was then applied to remote sensing image classification. Combining with characters of remote sensing image, an ant colony clustering algorithm based on entropy and pheromone was presented, which uses defined ant pheromone and short-term memory to correct its random movement. The information entropy was used to compare and replace group similarity and probability conversion function in traditional algorithm. The feature selection was optimized continuously to achieve better algorithm and efficiency improvement. Simulation experiment results also verify the effectiveness and feasibility of algorithm.

Keywords: remote sensing image, classification, ant colony, pheromone, entropy.

1 Introduction

As to the complexity of terrain and remote sensing process, same terrain on the image shows various spectral features, namely the synonym spectrum phenomenon. The images has same spectral feature are often different terrain feature, namely same spectrum different object phenomenon, which result in complex classification. It also has serious wrong and missing classification phenomenon. The classification accuracy is also low. Obviously, as to complex remote sensing imaging process, the spectral characteristics can not accurately correspond to same class. It is not sufficient to perform factor classification based on brightness [1].

Ant Colony Optimization (ACO) Algorithm is a new heuristic search algorithm. Remote sensing image classification is a clustering and combined optimization problem. Preliminary research and experiments results have proved that ACO can achieve better classification effect. The texture information of image can also be utilized, which has better robustness. Artificial ant colony can reduce interference and effectively use neighborhood information of pixel as artificial ant can effectively communicate via pheromone on the pixel, which is a robust operation environment. Based on researches of traditional ant colony optimization algorithm and its various applications, an ant colony clustering algorithm based on entropy and pheromone

D. Jin and S. Lin (Eds.): Advances in FCCS, Vol. 2, AISC 160, pp. 335–340.
springerlink.com   © Springer-Verlag Berlin Heidelberg 2012

was presented and applied into remote sensing image classification. The paper is organized is as follows: section 2 introduces application status of ant colony optimization algorithm in image classification; section 3 presented improved ant colony clustering algorithm; section 4 analyzes performance of improved algorithm with experiments and section 5 concludes our work.

2 Application of ACO in Remote Sensing Image Classification

To address on multi-source and multi-dimension remote sensing data, the traditional remote sensing image classification method based on probability and statistics theory has some drawbacks, for example, when remote sensing data does not meet the normal prior probability distribution, maximum likelihood classification categories will appear offset and classification accuracy can not be guaranteed. The applications of ACO in remote sensing image process mainly include two aspects. One is clustering combining ant and pixel and the other is feature optimization with ACO.

2.1 Clustering Combining Ant and Pixel

Remote sensing image recognition is the technique process to divide all pixels in the image into several types according to its features. Taking single ant as basic unit, firstly analyze pixel according to its own recognition rules, and then search pheromone path in 8 neighbor fields. Along the path, it reaches next pixel and put pheromone on the path. For specific feature of ant, they always follow along the path with high pheromone. The pheromone along this type of pixel is always high, so ants tend to search on this kind of pixel. As time increases, the following behaviors of large amount of ants also enhance the pheromone density of this pixel. When the density exceeds some value, we can determine that the pixel belongs to this type [2, 3].

2.2 Feature Optimization with ACO

To improve application effect of ACO in remote sensing image classification, we also brought out classification with combined feature of images. In the image classification process, it is a vital step to extract the most effectively representation classification feature. In case of classification with ACO, the processing of ant is based on single pixel. Therefore, we should also extract feature of single pixel. In a remote sensing image, the gray information is considered to be the most basic and important information. On the other hand, texture feature describes attributes of pixel and its neighbor pixels, so the texture feature shows the neighbor information of image. With texture information, we can better distinguish these images with similar visual features. To perform image classification using grey feature and texture feature can make up for lack of single feature classification and improve the efficiency of classification. In the experiment, the grey information of image was utilized. The grey co-occurrence matrix was also used to compute texture feature. With these parameters, we can describe the mean and variance of image [4-5].

3 Ant Colony Clustering Algorithm Based on Pheromone and Entropy

3.1 Algorithm Description

Ordinary ACO and its improvement algorithms need to set large amount parameters. Improper parameters setting will result in not ideal clustering result, and even failure. Therefore, researchers are constantly looking for new breakthrough, so that it tends to improve. The paper combines pheromone and entropy to present ant colony clustering algorithm based on entropy and pheromone, which uses defined ant pheromone and short-term memory to correct its random movement. The information entropy was used to compare and replace group similarity and probability conversion function in traditional algorithm. The feature selection was optimized continuously to achieve better algorithm and efficiency improvement. It also changes determination rules to pick and drop data object.

Entropy is the best measure for uncertainty, which is also an important concept in modern dynamical systems and through theory. The definition of entropy is as follows: assume x is a random variable and X is the set of all possible values, $p(x)$ is possibility function whose value is x, so the entropy $E(x)$ is defined as

$$E(x) = -\sum_{x \in X} p(x) \log p(x) . \tag{1}$$

As to a vector $x = \{x_1, x_2, \cdots, x_n\}$ with multiple variables, the computation of entropy is

$$E(x) = -\sum_{x_1 \in X_1} \cdots \sum_{x_n \in X_n} p(x_1, x_2, \cdots, x_n) \log p(x_1, x_2, \cdots, x_n) . \tag{2}$$

Where, $p(x_1, x_2, \cdots, x_n)$ is multivariate distribution function and $X_1, X_2, \cdots, X_n$ is the possible value set of appropriate vector items.

Referring to the character that information entropy contained in the subspace clustering entropy than the entropy does not contain a small cluster, the entropy can be introduced into traditional ACO to change its determination rules to pick and drop data object, so as to reduce parameter number and achieve the target of accelerating clustering speed. Around the data object o_i, less is the entropy, the similarity of it with the neighbor data object is larger. On the contrary, larger entropy means less similarity. This strategy can be described as: an ant load object o_i moves to another grid without data object, compute object information entropy in the surrounding $s \times s$. Assume the entropy before not place object o_i is larger than that after the object was placed, drop the object o_i. An empty load ant moves to object o_i, compute object entropy in the surrounding $s \times s$. Assume the entropy before not pick object o_i is larger than that after the object was picked, pick up the object o_i.

Assume each data object include n independent attributes $A_1, A_2, \cdots, A_n$, the set of possible value is $X_1, X_2, \cdots, X_n$ and object entropy $E(s^2)$ in the area $s \times s$ is

$$E(s^2) = -\sum_{i=1}^{n} \sum_{x \in X_n} p(x) \log p(x) . \tag{3}$$

Where, $p(x) = \dfrac{numOf\_x}{numCase}$ and $numOf\_x$ is data object number that meet condition $A_i = x$ in area $s \times s$; the $numCase$ is total number of data object in $s \times s$.

3.2   Algorithm Flow

According to the above description, the flow of presented improved algorithm is shown in Fig. 1.

In the improved algorithm, only the factor that neighbor s parameter can affect ant movement status. So, each time of place data object can reduce entropy of block and each time of pick up can increase entropy. It can also accelerate clustering process, so that data object with greater similarity will quickly gather together.

4   Experiment and Results Analysis

4.1   Experiment Steps

The paper used Landsat MMS images and aviation SAR images matching to its space in North China as target to obtain characteristics and to perform classification experiment. The experiment area is plain in North China, and the coverings on it can be divided into 4 types as water body, vegetation, residential area and others.

Based on ACO theory and above mentioned method, we designed the following experiment steps:

Step 1: Input remote sensing image $Y = [X_1, X_2, \cdots, X_n]$, where $X_i (i = 1, 2, \cdots, n)$ is the eigenvector of pixel i and N is pixel number of inputted image.

Step 2: Define structure of ant, using variable x and y to represent location of ant in the image. The variable *direction* is defined to help ant conduct next pixel selection.

Step 3: Randomly distribute ants to the image.

Step 4: Set $n=0$, where n is iteration number.

Step 5: If the pixel meet classification rule, ant release some amount information pheromone. Then change to next pixel by computing transmission probability. If the condition can not been satisfied, randomly transmit to next pixel.

4.2   Experiment Result Analysis

The experiment was carried on computer with 3.0GHz CPU and 1GB memory. The operation environment is MATLAB 6.5. In the experiment, all pixels in image were conducted 2-classification. Table 1 show the average classification accuracy and classification speed of Landsat MMS images and aviation SAR images with single grey feature and ACBEP algorithms. The classification accuracy is obtained from random sampling of classification and the confusion matrix to statistic degree with actual types.

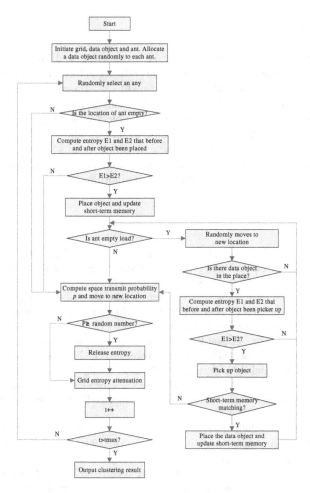

Fig. 1. Improved algorithm flow

From Table 1 we can know that the classification accuracy is improved with the proposed improvement algorithm. The experimental results also show that the running time decrease with the ACBEP algorithm. For single image, the running time is 479s, which is less than that of grey feature. The average classification has improved from 74.8% to 86.9%.

Table 1. Classification result comparison

Methods	Grey-feature	ACBEP algorithm
Average classification accuracy/%	74.8	86.9
Running time/s	821	479

5   Conclusion

Classification of remote sensing image is a clustering and combined optimization problem. On the basis of basic ACO algorithm, an ant colony clustering algorithm based on pheromone and entropy was brought out. Based on algorithm based on pheromone, the method modified determination rule to pick up and place data object. Local area information entropy of current data object was computed and compared to form status transition strategy. The parameter setting number was also reduced and stability of algorithm was improved. The experiment results show that this improved algorithm can improve classification accuracy and decrease operation time. In the future, we will focus on improve robustness of algorithm.

References

1. Zhang, H., Huang, F., Guo, J., Liu, M.: Automatic Classification of Remote Sensing Image Using Ant Colony Clustering Algorithm. In: Proceedings of 2nd International Congress on Image and Signal Processing (CISP 2009), pp. 1–4 (2009)
2. Li, R., Meia, X., Liu, J.: New method of remote sensing image classification based on Ant Colony. In: Proceedings of 2011 6th IEEE Joint International Information Technology and Artificial Intelligence Conference (ITAIC), pp. 375–379 (2011)
3. Song, Q., Guo, P., Jia, Y.: Ant Colony Optimization algorithm for remote sensing image classification using combined features. In: Proceedings of 2008 International Conference on Machine Learning and Cybernetics, pp. 3478–3483 (2008)
4. Wang, S.-G., Yang, Y., Lin, Y.: Automatic Classification of Remotely Sensed Images Based on Artificial Ant Colony Algorithm. Computer Engineering and Applications 29, 77–80 (2005)
5. Yin, Q., Guo, P.: Multi-spectral Remote Sensing Image Classification with Multiple Features. In: Proceedings of the Sixth International Conference on Machine Learning and Cybernetics, pp. 360–365 (2007)
6. Tian, Y.-Q., Guo, P., Lyu, M.R.: Comparative Studies on Feature Extraction Methods for Multispectral Remote Sensing Image Classification. In: Proceedings of 2005 IEEE International Conference on Systems, Man and Cybernetics, pp. 1275–1279 (2005)

Research and Realization of the Measurement and Control for Manipulation Signals of Vehicle Driving Simulator

Zhai Yuwen[1], Yang Xiao[1], and Ai Xuezhong[2]

[1] College of Mechanical & Electrical Engineering, Jiaxing University,
Jiaxing 314001, Zhejiang, China
[2] College of Information & Control Engineering, Jilin Institute of Chemical Technology,
Jilin City 132022, China

Abstract. The scheme of measurement and control for manipulation signals of vehicle driving simulator is introduced in this paper. The signal processing circuits for steering angle, throttle opening, clutch travel and brake travel are given. The synthesis methods of engine rotation rate sinusoidal signal, speed/mileage pulse signal and steering force PWM driving signal are analyzed. The interface circuit of RS485 communication is presented.

Keywords: Driving simulation, Measurement and control, Data acquisition, Single chip computer, Signal detection.

1   Introduction

The vehicle driving simulation system generates the real-time driving environment with virtual visual and audio effect and movement emulation through the simulative driving cabin and computer, which can make trainee feel driving a real vehicle with telepresence. It is convenient to train driving operation instead of real vehicle and helpful for standardization and normalization of driving training. Moreover, with the advantages of energy-saving, safety, low-cost, weatherproof operation and high-efficiency, the system is of great significance for training the driver of heavy truck or military vehicle. So it is necessary to develop vehicle driving simulator applicable to national situation of traffic and road [1][2][3].

2   Principle of the Vehicle Driving Simulation System

The vehicle driving simulation system consists mainly of driving cabin, 6-degree-of-freedom (6-DOF) movement platform, control unit, hydraulic station, servo valve, visual computer, control computer and linear-and-angular shift sensors. It provides the driving environment with real driving cabin of a car or truck in an enclosed simulation room in front of which LED screen is hung on to display the dynamic traffic situation generated by the visual computer [4].

D. Jin and S. Lin (Eds.): Advances in FCCS, Vol. 2, AISC 160, pp. 341–346.
springerlink.com   © Springer-Verlag Berlin Heidelberg 2012

The driver, sitting in the driving cabin, steers the wheel and other mechanism to output manipulating instructions which is sent to the visual computer through the data acquisition system. Meanwhile, the visual computer transmits the informations of vehicle speed and road situation to the control computer via peer to peer network in accordance with UDP protocol. Then the control computer implements the servo control of 6-DOF movement platform, calculates the real-time data through the mathematical model of reverse rotating torque of steering gear, and outputs the corresponding force feeling through the reverse torque motor so that the nearly true moving feeling can be presented [5].

3   System Framework

The framework of the measurement and control system for the manipulation signals of vehicle driving simulator is shown in Fig.1. The single chip computer C8051F060 is taken as the core of the system. Four shift signals within -5V to +5V, representing steering, throttle, brake and clutch travel, are respectively acquired by four shift sensors, then sent to the pre-amplifier and converted to the voltage signals within the range of 0 to 2.4V that the A/D converter in C8051F060 can accept. The Hall Sensor is employed to acquire 10 binary signals representing switching position, such as low/high beam of headlamps, steering turn right/left, and so on. All above input signals are sent to C8051F060. After processed by C8051F060, the engine rotation rate sinusoidal signal and speed/mileage pulse signal are synthesized respectively. The communication between visual computer and C8051F060 is implemented by RS485 interface circuit.

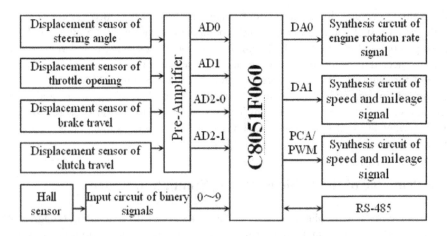

Fig. 1. Framework of the measurement and control system for the manipulation signals of vehicle driving simulator

4 Hardware Circuit

4.1 Pre-amplifier Circuit

Four displacement signals, representing steering, throttle, brake and clutch travel, are respectively acquired by four displacement sensors which outputs is within the range of -5V to +5V, then sent to the pre-amplifier and converted to the voltage signals within the range of 0 to 2.43V that the A/D converter in C8051F060 can accept.

Fig. 2 illustrates the Pre-Amplifier Circuit in which the low-drift operational amplifier OPA2335 is selected.

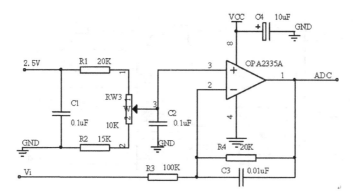

Fig. 2. Pre-Amplifier Circuit

Vi is the output voltage of the shift sensor (also is the input voltage of the Pre-Amplifier); Let $R1$=20K, $R2$=20K, $R3$=100KΩ, $R4$=20KΩ, when set V_+ =1V by adjusting the rheostat, then the output voltage can be calculated as (1):

$$ADC = (1+\frac{R4}{R3})V_+ - \frac{R4}{R3}Vi = 1.2 - 0.2Vi \qquad (1)$$

When Vi varies from -5V to +5V, the output voltage ADC is within the range of 2.2 to 0.2V, which can meet the demand of A/D converter in C8051F060.

4.2 Input Circuit for Binary Signals

The input circuit for binary signals, representing switching position, such as low/high beam of headlamps, steering turn right/left, and so on, is shown in Fig. 3. The Photoelectric Coupler TLP521-4 is used to convert the input binary signals into standard voltage signals and send them to C8051F060, meanwhile isolate the input signals from C8051F060 electrically.

4.3 Signal Synthesis Circuits for Engine Rotation Rate and Mileage Meter Signals

The function signal generator MAX038 is used to generate different signals to simulate the engine rotation rate signal and mileage signal and send them to motor after processing by C8051F060 so as to control the vehicle speed. The engine rotation rate signal is controlled by sinusoidal signal, and the mileage signal is controlled by square wave signal.

Tachometer, whose input pulse signal is from the breaker contact of ignition system, is a device to measure the engine rotation rate. In the normal operation of engine, the breaker contact switch on and off frequently forming the pulse signal. The frequency of the pulse signal is in direct proportion to the engine rotation rate. The vehicle running speed can be calculated based on its engine rotation rate; thereby the reduction or increase of vehicle's gears can be determined.

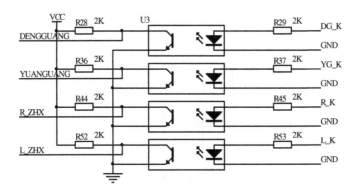

Fig. 3. Input Circuit for Binary Signals

The engine rotation rate, controlled by a sinusoidal signal, is in fact the rotation rate of crankshaft. A sinusoidal signal with adjustable amplitude and adjustable frequency is generated by MAX038. The frequency of the sinusoidal signal is controlled by C8051F060 so that the engine rotation rate can be controlled. Fig. 4 illustrates the sinusoidal signal generation circuit.

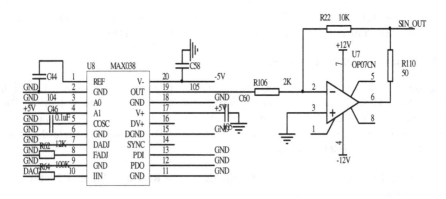

Fig. 4. Sinusoidal Signal Generation Circuit

The value of mileage can be calculated through the real-time measurement value of the frequency of the pulse signal inputting to C8051F060, as the frequency of the pulse signal is in direct proportion to the engine rotation rate. Then the mileage value calculated is saved in serial port data register and sent to the visual computer to display, so the values of vehicle running speed and mileage can be shown to the driver intuitively. The pulse signal is generated by MAX038 and sent to C8051F060.

4.4   High Power PWM Motor Driving Circuit

The DC Motor is driven by MOSFET combining with PWM square wave from C8051F060. The DC Motor Driving Circuit is shown in Fig. 5.

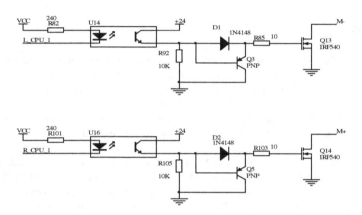

Fig. 5. Motor Driving Circuit

4.5   RS-485 Communication Interface Circuit

The communication between C8051F060 and visual computer is implemented based on RS485 protocol. Fig.6 illustrates the RS-485 Interface Circuit [6].

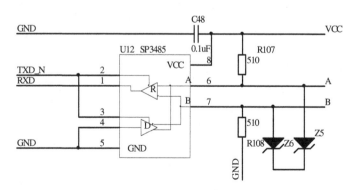

Fig. 6. RS-485 Interface Circuit

5   Software Design

The system software is programmed with modules by C51. The initialization is undertaken by the main program. Subroutines of data acquisition and processing, signal output and display, serial communication are designed [7].

6   Conclusion

The measurement and control system for manipulation signals designed in this paper has been used in the developed driving simulator. And the application inspection shows that the system can realize the signal detection accurately.

References

1. Cai, Z.-F., Zhang, A.-Y.: Study of automobile emulated driving model and simulation. Journal of Zhejiang University (Engineering Science) 36(3), 327–330 (2002)
2. Du, F., Li, W.-W., Zhong, Y.-J., Tang, X.-B.: The Application of Virtual Reality in driving simulator. Control & Automation 22(29), 292–295 (2006)
3. Liu, D.-P., Miu, X.-D., Wang, C.-J., Li, S.-M., Gao, Y.: Development State of Automobile Driving Simulator and Its Key Technology. Highways & Automotive Application (5), 53–59 (2010)
4. Yao, Q., You, F., Xu, J.-G., Wang, S.-M.: Design and Realization of Automobile Driving Simulation System Based on Real Scene. Machinery & Electronics (11), 7–10 (2010)
5. Ai, X., Zhao, D., Tang, X., Feng, S.: Realization of the Special Purpose Vehicle's Driving Simulation System with Steering Force Tele-presence. Journal of Wuhan University of Technology (Transportation Science & Engineering) 32(2), 222–224 (2008)
6. Zhao, X.: Design of Extend MODBUS Protocol for RS485 Bus Module of Measurement and Control. Automation & Instrumentation 22(2), 37–40 (2007)
7. Qiu, X., Li, A., Yin, N.: The Design of Control System of the Vehicle Simulator. Journal of Huangshi Institute of Technology 24(3), 32–35 (2008)

Fast 3D Modeling and Deformation in Virtual Reality Display of Special Necessities for Ethnic Minorities

Min Li[1], Yong Tang[2], JinLai Guo[3], LingZhi Wu[4], XiuChun Zhao[3], ZhiJie Song[4], LiXin Shi[1], and RuiRui Zheng[1]

[1] School of Information and Communication Engineering,
Dalian Nationalities University, Dalian 116605, China
[2] School of Information Science and Engineering, Yanshan University
Qinhuangdao 066004, China
[3] School of Electromechanical and Information Engineering,
Dalian Nationalities University, Dalian 116605, China
[4] Graduate School of Yanshan University,
Qinhuangdao 066004, China

Abstract. A way of fast 3D modeling and deformation driven by 2D template is presented to realize fast conceptual design and avoid some defects of reconstructing 3D models from 2D sketches, such as irregular sketched contour and immature sketch recognition techniques. The fish entity is taken as an example for experiments. According to the shape characters of various fishes, a 2D fish contour was drawn with Bessel curves and some control points were set in 2D template to modulate various parts of the size, shape and location of the fish. The experiment results show that this method can create various 3D fish models rapidly and improve the digital display of special necessities for ethnic minorities significantly.

Keywords: Special necessities for ethnic minorities, Virtual reality, 3D fish models, 2D contour.

1 Introduction

Fast 3D modeling is an important research direction of computer graphics. The traditional 3D modeling tool is so complicated that most people are difficult to master it in the short term, so the method of 3D modeling based on 2D sketches becomes one of research focus. But there exist some problems such as irregular sketched contour and immature sketch recognition techniques. Another method of 3D modeling based on 3D template deformation is also an important research direction. With this method, various 3D models can be got without reconstruction, but there exist some problems. For example, topology is not simple to create complex models, the template is too fine to generate models quickly, and especially 3D models cannot be used for modeling moving objects because identifying the mouse accurately requires an object at rest.

D. Jin and S. Lin (Eds.): Advances in FCCS, Vol. 2, AISC 160, pp. 347–355.
springerlink.com © Springer-Verlag Berlin Heidelberg 2012

Investigating the contour interpretation rules shows that, there are a lot of hand-drawn process requirements in order to accurately identify 2D hand-drawn outline such as the maximum number of intersecting lines at same point, the clarity and continuity of stroke and so on [1]. Unclear and discontinuous 2D hand-drawn outline will limit the applications of model projection technology and model reconstruction techniques [2]. Weidong Wang et al spent a lot of work to guide the user on the operation and the elimination of arbitrary drawing [3]. Olga A. karpenko et al described the outline and cross-section of objects with lines, which supports the elimination, bending and extrusion operations, can quickly generate a simple model of circular objects, but simple elements is not suitable for creating complex models [4]. Ryan Schmidt and others used a simple interface to build 3D model, but it is needed to rotate the camera frequently and pay attention to the view problem, so it is rather difficult to operate for novice [5]. Alec Rivers and others used two orthogonal 2D hand-drawn outlines on the canvas to reconstruct 3D models, which can generate complex models without paying attention to the view problems, but it is time-consuming to hand-draw two images [6]. As for 3D modeling method based on 3D template deformation, Jun Mitani et al adopted cube topology, but a simple model can only be generated for simple structures [7]. Ramesh Raskar et al adopted sophisticated templates to produce high-quality 3D model, but a separate editing each curve is required and the production rate of 3D model is affected [8]. Andrew Nealen and others thought the hand-drawn contours of objects as the characteristic line of a 3D model [9]. Ellen Dekkers and others used contours as the characteristic line of an object, which supports flexible operations to adjust the shape and size of the object, but requires an object at rest for the accurate identification of mouse operation, so it is only suitable for industrial design not moving objects modeling [10].

For the above, 2D hand-drawn method and 3D model deformation method are combined here, and 3D modeling method driven by 2D template is presented. In this method, firstly the contour of 3D template deformation techniques is extracted to replace hand-drawn graphics on the canvas as a template and several types of control points are set to adjust the shape of the object, then control points are adjusted according to the corresponding relations between 2D templates and 3D models to generate various 3D models.

2   Design of 2D Template Fish

2.1   Drawing 2D Template Fish

A fish entity can be divided into head, mouth, body, fins and tail, of which the head, mouth, body and tail can be drawn with a Bezier curve, and fins with a straight line. The 2D template fish is shown in Fig. 1.

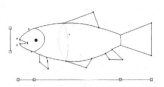

Fig. 1. 2D template fish

In order to adjust a fish entity, three kinds of control points: horizontal control points, vertical control points and the control points on fish are set. Dragging the vertical control points up or down can adjust the vertical height of the fish; dragging the horizontal control points left or right can adjust the length of the fish head, body and tail; dragging the control points on fish can adjust the size, shape and position of the fish mouth and fins. Adjusted 2D template fish is shown in Fig. 2. Dragging the curve of the fish thickness can adjust the thickness of a fish.

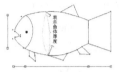

Fig. 2. Adjusted 2D template fish

To ensure the authenticity of the fish shape, some treatments should be done on characteristics of the various parts. The control points of fish body and fins are set to control fins conveniently and not to affect the fish shape; the common control point, which is the intersection of the fish head, body and tail, is adopted to avoid various parts of the fish entity out of touch.

2.2   2D Contour Adjustment

For two vertical control points, if their effects were only upper half and lower half of the fish respectively, there will be distortion on fish body when moving either one of the control points. Fig. 3 is the effect diagram of very large mouth and dislocated eye after dragging control point 0 upwards.

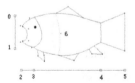

Fig. 3. 2D template fish after dragging control point 0 upwards

To avoid the above, the highest point must be corresponding to the control point 0 and the lowest point to the control point 1. Set the impact factor of 1 for the control point 0 to the highest point and the impact factor of 0 to the lowest point, the impact factor to the control point i on fish entity between them is:

$$k_{0i} = (y_l - y_i)/(y_l - y_h) \tag{1}$$

where y_i is the y value of the control point i, y_l the lowest point and y_h the highest point.

On the contrary, set the impact factor of 0 for the control point 1 to the highest point and the impact factor of 1 to the lowest point, the impact factor to the control point i on fish entity between them is

$$k_{1i} = (y_i - y_h)/(y_l - y_h) \qquad (2)$$

$$k_{0i} + k_{1i} = 1 \qquad (3)$$

Suppose that Δt_0 is the amount of movement up and down for the control point 0 and Δt_1 for the control point 1, the amount of movement up and down for the control points on the fish entity influenced by them is as follow respectively:

$$\Delta t_{0i} = \Delta t_0 * k_{0i} \qquad (4)$$

$$\Delta t_{1i} = \Delta t_1 * k_{1i} \qquad (5)$$

Similarly, the impact factors for the horizontal control points to the x value of the control point j on fish entity are also different. Set one point at fish mouth front end corresponding to the control point 2, one point at intersection of fish head and fish body corresponding to the control point 3, one point at intersection of fish tail and fish body corresponding to the control point 4 and one point at fish tail back end corresponding to the control point 5. Suppose that the control point j is only influenced by adjacent control points i and $i+1$, Δt_i is the amount of movement in the x direction for the point i and Δt_{i+1} the amount of movement in the x direction for the point $i+1$, the amount of movement in the x direction for the point j is:

$$\Delta t_j = \Delta t_i * k_{i,j} + \Delta t_{i+1} * k_{i+1,j} \qquad (6)$$

where $k_{i,j}$ and $k_{i+1,j}$ are the impact factors for horizontal control points i and $i+1$ to the control point j on the fish entity.

$$k_{i,j} = (x_{i+1} - x_j)/(x_{i+1} - x_i) \qquad (7)$$

$$k_{i+1,j} = (x_j - x_i)/(x_{i+1} - x_i) \qquad (8)$$

where x_i is the x-axis coordinate value of the control point i, x_{i+1} the control point $i+1$ and x_j the control point j on the fish entity.

To ensure the true shape of the fish entity, adjustable range of control points has the following requirements: the order of the control points in vertical, horizontal and the tail can not be reversed; the fish entity can not be set to zero thickness.

3   Generating 3D Fish Model

3.1   Drawing 3D Fish Model

According to the contour data of 2D template fish, the 3D fish body, head and tail are drawn with a Bezier surface and fish fins with a polygon. In order to reflect the authenticity of 3D fish, the thickness of the second curve on the fish body should be the most and their thickness on both sides should be increased smoothly. The 3D fish model is displayed in Fig. 4.

Fig. 4. The 3D fish model

A Bezier surface is formed by four group curves through interpolation and each group curve has four control points. The drawing structure of 3D fish model is shown in Fig. 5. The solid lines represent the four group curves forming a Bezier surface and the points on lines are the control points. No.5 and No.6 are the curves in the screen, No.2 is the common curve of fish head and body and No.3 is the common curve of fish body and tail.

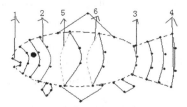

Fig. 5. The drawing structure of 3D fish model

The data of curve 1, 2, 3 and 4 are provided by the corresponding 2D contour and the curve data among them are obtained by linear interpolation. The curve 1 corresponds with 2D fish mouth, the curve 2 intersection of fish head and body, the curve 3 intersection of fish body and tail, and the curve 4 the end of fish tail.

3.2   3D Fish Model Formula

3D coordinates of control points are determined by 2D control point coordinates. Various impact factors are set for every control point and the coordinates of all control points on the 3D fish model can be obtained, the formulas are as follows.

Suppose that w_i is the impact factor of control point i to the control points on fish model, Δf_i the displacement of control point i and f the 3D fish model before deformation, 3D fish model after deformation f' is:

$$f' = f + \sum_{i=n}^{m} w_i \Delta f_i \tag{9}$$

$$f' = (X', Y', Z') \tag{10}$$

$$f = (X, Y, Z) \tag{11}$$

where X', Y', Z' represent the x, y, z values of various control points after deformation respectively and X, Y, Z before deformation. Formula (9) can give following the formulas of x, y and z axis.

$$f_x' = f_x \pm w_i \Delta f_i \quad i \in 6 \tag{12}$$

$$f_y' = f_y - \sum_{i=n}^{m} w_i \Delta f_i \quad i \in (0,1) \tag{13}$$

$$f_z' = f_z - \sum_{i=n}^{m} w_i \Delta f_i \quad i \in (2,3,4,5) \tag{14}$$

Set d the thickness of the fish at its widest point, the impact factor of the control point 6 to the x value of the control point j on the 3D fish entity is:

$$w_{6j} = 2 * |x_j| / d \tag{15}$$

For the formula (13), the impact factors of the control points 0 and 1 to the y value of the control point j on fish entity is as follows respectively.

$$w_{0j} = (y_j - y_l) / (y_h - y_l) \tag{16}$$

$$w_{1j} = (y_h - y_j) / (y_h - y_l) \tag{17}$$

$$w_{0j} + w_{1j} = 1 \tag{18}$$

For the z value, because the control points on fish entity are only influenced by two adjacent 2D horizontal control points, if z_r is the z value of the control point on 3D fish entity corresponding to 2D control point $n+1$ and z_l the z value of the control point on 3D fish entity corresponding to 2D control point n, the impact factor of the point j between the two control points is:

$$w_{n,j} = (z_r - z_j) / (z_r - z_l) \tag{19}$$

$$w_{n+1,j} = (z_j - z_l)/(z_r - z_l) \tag{20}$$

$$w_{n,j} + w_{N=1,j} = 1 \tag{21}$$

The control point data of 3D fish fins are provided by the corresponding control points on 2D fish template. When the coordinates of each control point on 3D fish entity are obtained from the above formulas, the 3D fish model can be drawn.

3.3 Movement of 3D Fish

The movement of a fish tail is the main aspect of fish entity movements, including not only moving around but also bending. In order to control the fish entity conveniently, it is supposed that the deformed shape of fish tail is in line with the parabolic equation, the formula (22) is the vertex expression, in which a, h, k are constants and a is not zero. The greater a is, the greater the degree of bending of the tail is. The Bezier surface on the tail is segmented in detail to control the shape of the tail flexibly and accurately. Suppose that the Bezier surface on the tail is formed by 4*4 control points, the coordinates of any point on the Bezier surface can be get through formula (23), in which $\omega_{ij}(u,v) = \omega_i(u)\omega_j(v)$, V_{ij} is the displacement between V_i and V_j, which are the control points on the Bezier surface, u and v are any value from 0 to 1, $\omega_i(u)$ and $\omega_j(v)$ can be calculated from the formula (24).

To ensure that the reproduction of your illustrations is of a reasonable quality, we advise against the use of shading. The contrast should be as pronounced as possible.

$$y = a(x-h)^2 + k \tag{22}$$

$$P(u,v) = \sum_{i=0}^{M-1}\sum_{j=0}^{N-1} \omega_{ij}(u,v)V_{ij} \tag{23}$$

$$\omega_i^N(t) = \frac{N!}{i!(N-i-1)!}(1-t)^{N-i-1}t^i \tag{24}$$

$$\arg\min\left\| a(X-h)^2 + k - Y \right\| \tag{25}$$

where X and Y is the set of coordinates of points on the surface, adjusting the coordinates to meet the optimization equation (25) can make the tail swing more vivid.

4   Experiment Results

3D fish simulation models are built based on VS2005 and OpenGL software platform and 2D template fish contours are drawn with MFC. The hardware environment is mainly the PC with Intel Pentium Dual E2160, dual-core 1.8GHz, 1G RAM and

NVIDIA GeFroce 8500T graphics. Fig. 6 and Fig. 7 are the effect pictures of 3D fish models after adjusting control points and control lines.

The experiment results show that the size and shape of the 3D fish model can be changed through changing the size and shape of 2D template fish by adjusting control points. Eventually the various 3D fish models with different size and shape can be generated.

Fig. 6. Various 3D fish models after adjusting control points

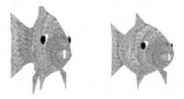

Fig. 7. 3D fish models of different thickness of a fish entity

5   Conclusion

A way of fast 3D modeling driven by 2D template is presented. The experiment results show that: this method does not need painting base, simply adjust the control points in the template to get the 2D template contours, the 3D fish model can be generated. It is a simple and fast modeling method. It can avoid appearing discontinuous and unclear lines during hand-drawn process, remove the steps of selecting point of view, solve the problem of requiring drawing people to have a certain professional level and eliminate the non-realistic caused by requiring objects stationary when adjusting moving object models on the 3D. Meanwhile as long as the template is fine enough, the method can produce any shape of 3D fish model. The biggest advantage of this method is in line with people's daily habits, representing a fish with the side view profile of fish But as the real fish's body shape changing, a 2D template is difficult to apply to all types of fish. This is inadequacy of the method and also the content of follow-up work.

Acknowledgments. This paper is supported by "National Science and Technology Support Program of China" under Grant No.2009BAH41B05.

References

1. Varley, P., Company, P.: Sketch input of 3D models: Current directions. In: Proceedings of 2nd International Conference on Computer Vision Theory and Applications, pp. 85–91 (2007)
2. Prados, E.: Application of the theory of the viscosity solutions to the Shape From Shading problem. University of Nice-Sophia Antipolis (2004)
3. Wang, W.D., Georges, G.G.: A Survey of 3D Solid Reconstruction from 2D Projection Line Drawings. Computer Graph. 12(2), 137–158 (1993)
4. Karpenko, O.A., Hughes, J.F.: Smoothsketch: 3d free-form shapes from complex sketches. ACM Transactions on Graphics 25(3), 589–598 (2006)
5. Schmidt, R., Khan, A., Singh, K., et al.: Analytic drawing of 3d scaffolds. ACM Transactions on Graphics 28(5), 1–10 (2009)
6. Rivers, A., Durand, F., Igarashi, T.: 3D Modeling with Silhouettes. ACM Transactions on Graphics 29(4), 1–8 (2010)
7. Mitani, J., Suzuki, H., Kimura, F.: 3D sketch: sketch-based model reconstruction and rendering. Kluwer Academic Publishers press, USA (2002)
8. Raskar, R., Kara, L.B., Shimada, K.: Sketch-based 3dshape creation for industrial styling design. IEEE Computer Graphics and Applications 27(1), 60–71 (2007)
9. Nealen, A., Igarashi, T., Sorkine, P., et al.: Fibermesh: designing freeform surfaces with 3d curves. ACM Transactions on Graphics 26(3), 41–49 (2007)
10. Dekkers, E., Kobbelt, L., Pawlicki, R., et al.: A Sketching Interface for Feature Curve Recovery of Free-Form Surfaces. In: Proceedings of the SIAM/ACM Joint Conference on Geometric and Physical Modeling, pp. 235–245 (2009)

A Novel PCA-SVM Flow Pattern Identification Algorithm for Electrical Resistance Tomography System

Yanjun Zhang[1] and Yu Chen[2,*]

[1] Heilongjiang University Harbin, China, 150080
[2] Northeast Forestry University Harbin, China, 150040
lg_chenyu@yahoo.com.cn, zyj716@hotmail.com

Abstract. The two-phase flow measurement plays an increasingly important role in the real-time, on-line control of industrial processes including fault detection and system malfunction. Many experimental and theoretical researches have done in the field of tomography image reconstruction. However, the reconstruction process cost quite long time so that there are number of challenges in the real applications. An alternative approach to monitor two-phase flow inside a pipe/vessel is to take advantage of identification of flow regimes. This paper proposes a new identification method for common two phase flow using PCA feature extraction and SVM classification based on electrical resistance system measurement. Simulation was carried out for typical flow regimes using the method. The results show its feasibility, and the results indicate that this method is fast in speed and can identify these flow regimes correctly.

Keywords: Electrical Resistance Tomography, Flow regime identification, Principal component analysis, SVM.

1   Introduction

Two-phase flow is a mixed-flow pattern widely found in nature, especially in the chemical, petroleum, electricity, nuclear power and metallurgical industries[1]. Two-phase flow regime identification plays an increasingly important role in the automation process of energy industry. It can provide valuable information for a rapid and dynamic response which facilitates the real-time, on-line control of processes including fault detection and system malfunction, many experimental and theoretical researches have done in the field of image reconstruction. However, the visualization process cost quite long time so that there are number of challenges in the real applications. An alternative approach to monitor two-phase flow inside a pipe/vessel is to take advantage of identification of flow regimes instead of image reconstruction, especially for those stable and simple Two-phase flows. Two-phase flow regimes not only affect the two-phase flow characteristics and mass transfer, heat transfer performance, but also affect the system operation efficiency and reliability, while the other parameters of the two-phase flow measurements have a great impact. Therefore,

---

* Corresponding author.

two-phase flow regimes on-line identification for oil mixed transportation systems such as two-phase process of analysis, testing and operation are important[2,3,4].

In the horizontal pipe and vertical pipe flow pattern is different, mainly because of the horizontal pipe by gravity due to the role of media in the pipeline at the bottom of a sedimentation trend. For different media, the classification of flow patterns are also differences in the actual should be divided according to specific circumstances. Present a common flow patterns are: the level of flow, core flow, circulation and trickle and so on[5].

In this paper, a principal component analysis of SVM algorithm for flow pattern identification[6,7], algorithm to meet the convergence conditions and to simplify the complex pre-processing steps, greatly reducing the computational Experimental results show that the algorithm is effective, complexity, improve the speed of the identification. Experimental results show that the algorithm can obtain a higher recognition rate, and provide a new and effective method for flow pattern identification algorithm in electrical resistant tomography.

2   The Basic Principles of Principal Component Analysis

Principal component analysis (PCA) is a statistical analysis of data in an effective way[8-9]. With the aim of space in the data as much as possible to find a set of data variance explained by a special matrix, the original projection of high dimensional data to lower dimensional data space, and retains the main information data in order to deal with data information easily.

Principal component analysis is a feature selection and feature extraction process, its main goal is to enter a large search space characteristics of a suitable vector, and the characteristics of all the main features extracted. Characteristics of the selection process is to achieve the characteristics of the input space from the space map, the key to this process is to select feature vectors and input at all the features on the projector, making these projectors feature extraction can meet both the requirements of the smallest error variance. For a given M-dimensional random vector $X = [x_1, x_2, \ldots x_m]^T$, For its mean E[X]=0, the covariance C_x can be expressed as follows:

$$C_x = E[(X - E[X])(X - E[X])^T] \qquad (1)$$

Because of E[X]=0, covariance matrix is therefore auto-correlation matrix as follow:

$$C_x = E[XX^T] \qquad (2)$$

Calculate eigen values of C_x $\lambda_1, \lambda_2, \ldots, \lambda_m$ and the corresponding normalized eigenvector $\omega_1, \omega_2, \ldots \omega_m$ as the following equation:

$$C_x \omega_j = \lambda_i \omega_j \qquad i = 1, 2, \ldots m \qquad (3)$$

express the matrix as follows:

$$Y = \omega^T X \qquad (4)$$

With a linear combination of eigenvectors can be reconfigurable X , The following formula:

$$X = \omega Y = \sum_{i=1}^{m} \omega_i y_i \tag{5}$$

Characteristics obtained through the selection of all the principal components, and in the feature extraction process, then select the main features to achieve the purpose of dimensionality reduction.The mean of the vector analysis can be written as:

$$E[Y] = E[\omega^T X] = \omega^T E[X] = 0 \tag{6}$$

The covariance matrix C_y is the autocorrelation matrix of Y , it can be written as:

$$C_Y = E[YY^T] = E[\omega^T XX^T \omega] = \omega^T E[XX^T \omega] \tag{7}$$

In the truncated Y, it is necessary to ensure the cut-off in the sense of mean square deviation is the optimal. $\lambda_1, \lambda_2, ..., \lambda_m$ can only consider the first L largest eigenvalue, with these characteristics for reconstruction of X , the estimated value of reconstruction is as follows:

$$\hat{X} = \sum_{i=1}^{L} \omega_i y_i \tag{8}$$

Its variances are met as follows:

$$e_L = E[(X - \hat{X})^2] = \sum_{i=L+1}^{m} \lambda_i \tag{9}$$

According to the formula (9), the current characteristic value L is larger, the minimum mean square error can be achieved. Also the formula as follows:

$$\sum_{i=1}^{m} \lambda_i = \sum_{i=1}^{m} q_{ii} \tag{10}$$

Where q_{ii} is the diagonal matrix element of C_X , the contribution rate of variance as follows:

$$\varphi(L) = \frac{\sum_{i=1}^{L} \lambda_i}{\sum_{i=1}^{m} \lambda_i} = \frac{\sum_{i=1}^{L} \lambda_i}{\sum_{i=1}^{m} q_{ii}} \tag{11}$$

When $\varphi(L)$ is large enough, you can pre-L constitute a feature vector space $\omega_1, \omega_2, ... \omega_L$ as a low-dimensional projection space, thus completing the deal with dimensionality reduction.

3   Flow Pattern Identification Based on SVM

A support vector machine (SVM) is a general-purpose learning algorithm in statistics and computer science for a set of related supervised learning methods that analyze data and recognize patterns, used for classification and regression analysis[10]. It stems from the frame work of statistical learning theory or Vapnik-Chervonenkis (VC) theory and was originally developed for pattern recognition problem. The standard SVM takes a set of input data and predicts, for each given input, which of two possible classes forms the input, making the SVM a non-probabilistic binary linear classifier[11]. Given a set of training examples, each marked as belonging to one of two categories, an SVM training algorithm builds a model that assigns new examples into one category or the other[12]. An SVM model is a representation of the examples as points in space, mapped so that the examples of the separate categories are divided by a clear gap that is as wide as possible. New examples are then mapped into that same space and predicted to belong to a category based on which side of the gap they fall on[13].

Assume that the given train set is $\{(x_1, y_1), (x_2, y_2), \cdots, (x_l, y_l)\}$, satisfying

$$y_i[(\omega X_i) + b] - 1 \geq 0 \qquad i = 1, 2, \cdots l \tag{12}$$

And these given data points each belong to one of two classes which is defined as 1 and -1.In the equation 12, l stands for the index of samples set, the y_i is either 1 or -1, indicating the class to which the point belongs.

The goal of learning is to find the maximum-margin hyperplane that divides the points having $y_i = 1$ from those having $y_i = -1$. By using geometry, we find the distance between these two hyperplanes is $2/\|\omega\|$. To get the maximum the distance equals to get minimum of $\|\omega\|^2$, The hyperplane which satisfies the equation 12 and makes $\|\omega\|^2$ minimum is defined the optimization hyperplane.

Finding the optimal classification hyperplane is transformed into an optimization problem. Usually, the problem is discussed under three cases: linearly separable, linearly inseparable and non-linear. Linear separable and linearly inseparable can now be solved by standard quadratic programming techniques and programs as the formula, and there is only a minimum point, it can be solved by standard Lagrange multiplier method.

$$Q(a) = \sum_{i=1}^{l} a_i - \frac{1}{2} \sum_{i,j=1}^{l} a_i a_j y_i y_j (X_i \bullet X_j) \tag{13}$$

The constraint condition is:

$$\sum_{i=1}^{l} y_i a_i = 0 \qquad a_i \geq 0, \qquad i = 1, 2, \cdots, n \tag{14}$$

Where, a_i is Lagrange multiplier.

In nonlinear case, the train data are mapped into a high dimensional linear feature space through nonlinear tranformation $\phi(\cdot)$. Since only the dot product operation is performed between the vectors, using the kernel function can avoid complex calculations in higher dimensional space. The kernel function K satisfies:

$$K(X_i, X_j) = \phi(X_i) \cdot \phi(X_j) \tag{15}$$

The goal function of quadratic programming problem is changed into:

$$Q(a) = \sum_{i=1}^{l} a_i - \frac{1}{2} \sum_{i,j=1}^{l} a_i a_j y_i y_j K(X_i \bullet X_j) \tag{16}$$

4   Experiments Results

This paper implemented experiments for several common flow regimes: bubbly flow, core flow, annular flow, laminar flow, empty pipe and full pipe. The simulations were implemented by using MATLAB on a computer running at the AMD Phenom Triple-Core 2.31Ghz with 2G of RAM.

The approach based on PCA and SVM identifier is tested on laminar flow, empty pipe, trickle flow, full pipe, the core flow and circulation flow. 72 samples were collected for all flow regimes in the experiments and 68 sets of data can be correctly identified.

svm type is c_svc, training is RBF kernel function, the RBF kernel parameter γ is 0.125, the number of categories is 6.

The identification accuracy rate are computed to verify the confidence of the system. They are listed in Table 1. The average identification accuracy is 94.444%. This proved that PCA feature extraction method combined with SVM classifier has good classification ability for flow regime identification. It can be observed that the simpler flow regime, the higher accuracy rate. The accuracy rate of identification of trickle flow is significantly lower than other flow regimes. It can be explained by the complexity of trickle flow.

Table 1. The Identification Accuracy Rate

Flow regime	samples	Accuracy Rate
Empty pipe	10	100%
Full pipe	10	100%
Core	12	91.667%
Circulation	13	92.3076%
laminar	12	100%
Trickle	15	86.6667%

5   Conclusions

This paper presents a principal component analysis feature extraction of the symmetric subspace flow pattern identification algorithm and SVM classifier in ERT system. This method simplifies the complex pre-processing steps, greatly reduces the computational complexity and improves the recognition speed. The experimental results show that the algorithm is effective, the approach presented in this paper to obtain a higher recognition rate in the ERT. Future research should focus on improving the recognition rate while ensuring at the same time, as much as possible to reduce noise or changes in media distribution factors, the impact of flow pattern identification. In practical industry process, however, the flow regimes will be more complex and volatile. To improve the identification accuracy, more flow regimes should be collected and trained.

Acknowledgement. The Corresponding author is Yu Chen for this paper, and this work is partially funded by the Science and Technology Foundation of Education Department of Heilongjiang Province (11551344, 12513016), Natural Science Foundation of Heilongjiang Province (F201019), Postdoctoral Fund of Heilongjiang Province.

References

1. Yang, X.D., Ning, X.B., Yin, Y.L.: Fingerprint image preprocessing technique and feature extraction algorithm. Journal of Nanjing University Natural Sciences 4(42), 351–361 (2006)
2. Areekul, V., Watchareeruetai, U.: Separable gabor filter realization for fast fingerprint enhancement. In: IEEE International Conference, vol. 3(1), pp. 5–10 (2005)
3. Karhumen, J., Joutsensalo, J.: Represention and Separation of Signals Using Nonlinear PCA Type Learning. Helsinki University of Technology 8, 1083–1091 (2001)
4. Ham, F.M., Kim, I.: Extension of the Generalized Hebbian Algorithm for Principal Component Extraction. Appliction and Science of Neural Networks 3455, 85–274 (1998)
5. da Rocha Gesualdi, A., Manoel de Seixas, J.: Character recognition in car license plates based on principal components and neural processing. Neural Networks 14, 206–211 (2002)
6. Wang, L., Wang, X., Feng, J.: In: On image matrix based feature extraction algorithms. IEEE Transactions on Systems, Man, and Cybernetics-Part B: Cybernetics: Accepted for Future Publication 99, 194–197 (2005)
7. Tao, J., Jiang, W.: Improved two-dimensional principal component analysis based on the palmprint identification. Optical Technology 33(2), 283–286 (2005)
8. Xing, X., Luo, W.: State estimation in power system based on GRNN algorithm. Heilongjiang Electric Power 33(1), 50–54 (2011)
9. Aizerman, M., Braverman, E., Rozonoer, L.: Theoretical foundations of the potential function method in pattern recognition learning. Automation and Remote Control 25, 821–837 (1964)
10. Teukolsky, S.A., Vetterling, W.T., Flannery, B.P.: Numerical Recipes: The Art of Scientific Computing. Cambridge University Press, New York (2007)
11. Zhang, L.F., Wang, H.X.: Identification of two-phase flow regime based on support vectormachine and electrical capacitance tomography technique. Chinese Journal of Scientific Instrument 30(4), 812–916 (2009)
12. Liu, H.Y., Chen, D.Y., Li, M.Z.: Image Reconstruction Algorithms of Electrical Capacitance Tomography Based on Support Vector Machine 11(4), 1–5 (2006)

Personalized Information Recommendation of E-Commerce

Bo He

School of Computer Science and Engineering, ChongQing University of Technology,
400054 ChongQing, China
heboswnu@sina.com

Abstract. The traditional information recommendation of E-commerce could not satisfy current E-commerce enterprises and users' needs. This paper divided E-commerce users into new users and other users. It proposed the recommendation strategy based on user segmentation. On the base of these, it designed a personalized information recommendation system of E-commerce, namely, PIRSE. The experimental results indicate that the recommendation strategy is feasible.

Keywords: E-commerce, Personalized information recommendation, User profiles, User segmentation.

1 Introduction

Personalized information recommendation service of E-commerce[1] has becoming an important research task. The traditional recommendation strategy could not satisfy current E-commerce enterprises and users' needs. This paper establishes user profiles and puts forward a recommendation strategy based on user segmentation[2], designs a personalized information recommendation system of E-commerce, namely, PIRSE.

2 User Profiles

Establishment of user profiles[3] is the base of realizing personalized information recommendation of E-commerce.

2.1 Definition of User Profiles

Definition 1. User profiles can be described as an aggregation, namely, $U_i=\{(d_{i,1},w_{i,1}), (d_{i,2},w_{i,2}),\ldots,(d_{i,n},w_{i,n})\}$, and $d_{i,j}\in T$. $T=\{t_1, t_2,\ldots,t_m\}$ is an aggregation of user interest subjects, $w_{i,j}\in [0,1]$ is the weight of user interest subjects, and $w_{i,1} + w_{i,2}+\ldots+w_{i,n}=1$.

D. Jin and S. Lin (Eds.): Advances in FCCS, Vol. 2, AISC 160, pp. 363–367.
springerink.com   © Springer-Verlag Berlin Heidelberg 2012

2.2   Construction of User Profiles

User profiles are constructed by three ways:

(1)   According to the information of registration, PIRSE gets elementary user profiles.
(2)   PIRSE picks up user profiles according to mining request of user.
(3)   PIRSE adjusts user profiles according to the estimation of mining results.

When a user registers, PIRSE makes the user answer a series of questions. The system produces a user interest tree according to the answers. The interest subjects is more and more specific from the root to the leaf of the interest tree. The process from root to leaf is the process of discovering user interest subjects. The more deep the layer of user interest tree is, the bigger of the weight of the user interest subjects is. According to human-computer interaction, PIRSE takes out primary user interest subjects and establishes user profiles.

When a user inputs mining request each time, the system picks up user interest subjects from it and renews user profiles. The algorithm is described as follows.

Algorithm 1. Renewal algorithm of the user profiles

Input: mining request of user is I, constant c, and $0 \leq c \leq 1/n$, the previous user profiles $\{(d_{i,1}, w_{i,1}), (d_{i,2}, w_{i,2}), \ldots (d_{i,n}, w_{i,n})\}$.
Output: the renewed user profiles $\{(d_{i,1}, w_{i,1}), (d_{i,2}, w_{i,2}), \ldots (d_{i,n}, w_{i,n})\}$.
Methods: According to the following steps.

(1)   Picking up effective vocabulary L from I.
(2)   If $L \in \{d_{i,1}, d_{i,2}, \ldots d_{i,n}\}$, suppose $L = d_{i,p}$, then transfer to (3), or else (4).
(3) $(d_{i,p},\ w_{i,p}+c) \longrightarrow (d_{i,p},\ w_{i,p})$, then transfer to (6).
(4)   Add L into $\{d_{i,1}, d_{i,2}, \ldots d_{i,n}\}$, suppose $d_{i,q} = L$, then $w_{i,q} = min\{w_{i,j}|1 \leq j \leq n\}$. If $c > w_{i,q}$, then transfer to(5), or else (6).
(5) $(L,\ c) \longrightarrow (d_{i,q}, w_{i,q})$.
(6)   If there is other effective vocabulary form I, then transfer to (1).

(7) $w_{i,j} = w_{i,j} / \sum_{k=1}^{n} w_{i,k}$.

3   The Recommendation Strategy Based on User Segmentation

The user of E-commerce is described as follow.

$User = \langle NewUser, OtherUser \rangle$

$NewUser = \{User \mid User\_current - User\_start \leq \sigma\}$

$OtherUser = \{User \mid User\_current - User\_start > \sigma\}$

$User\_start$ is the first time of visit.

$User\_current$ is the current time.

σ is threshold.

The recommendation strategy based on user segmentation provides two kinds of recommendation algorithms. One is the intelligent information recommendation algorithm based on user model clustering, namely IRUMC algorithm[4]. The other is Web intelligent information recommendation method based on collaborative filtering, namely WIIRM algorithm[5]. IRUMC algorithm fits new users. WIIRM algorithm fits other users.

The recommendation strategy based on user segmentation is described as follows.

if (the user belongs to new users)

PIRSE adopts IRUMC algorithm;

else

PIRSE adopts WIIRM algorithm;

4   Personalized Information Recommendation System of E-Commerce

On the base of user profiles and the recommendation strategy based on user segmentation, this paper designs a personalized information recommendation system of E-commerce, namely, PIRSE. PIRSE includes the interface module, the preprocessing module, the mining module and the on-line recommendation module and. PIRSE is described as fig 1.

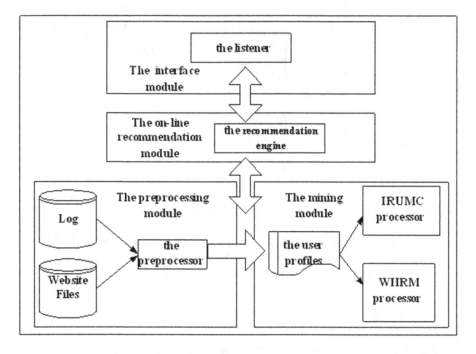

Fig. 1. FIRSE

The interface module includes the listener. The interface module is the interface of user and system.

The preprocessing module includes the preprocessor. The preprocessing mainly includes data cleaning, user recognition, session identification, path supplementation and user pattern recognition, etc.

The mining module includes IRUMC processor and WIIRM processor. IRUMC processor treats data by IRUMC algorithm. WIIRM processor treats data by WIIRM algorithm.

The on-line recommendation module includes the recommendation engine. The recommendation engine can gain recommendation sets by recommendation algorithms.

5   Related Experiments

The experimental data comes from the log of a website in July 2010. PIRSE adopts the IRUMC algorithm and WIIRM algorithm, as fig 2.

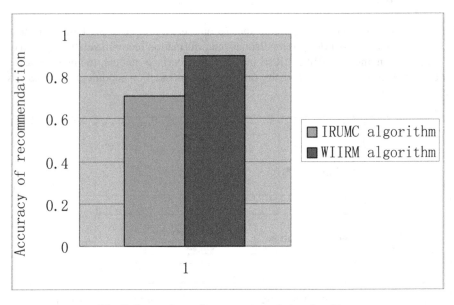

Fig. 2. Comparison of two recommendation algorithms

The accuracy of two recommendation algorithm is 0.71 and 0.9. Hence, IRUMC fits new users. WIIRM algorithm fits other users. The experimental results indicate that the recommendation strategy based on user segmentation is feasible.

6   Conclusion

The paper designs a personalized information recommendation system of E-commerce, namely, PIRSE. The experimental results indicate that the recommendation strategy is feasible. It is believed that personalized information recommendation of E-commerce will have a magnificent future.

Acknowledgment. The paper is supported by the ministry of education humanity and social scientific research items grant No. 09yjc870032.

References

1. Marko, B.: An Adapative Web Page Recommendation Service. In: 1st International Conference on Autonomous Agents, Marina del Rey (Feburuary 1997)
2. Xu, M., Qiu, Y.H., Qiu, J., Wang, X.F.: Segmenting your users in recommender system. In: Proceeding of the IASTED International Conference: Modeling and Simulation 2003, pp. 584–588. ACTA Press, Palm Springs (2003)
3. Zeng, C., Xing, C.X., Zhou, L.Z.: A survey of personalization. Jounal of Software 13(10), 1952–1961 (2002) (in Chinese with English abstract)
4. Bo, H., Yang, W., Zhang, J.X., Wang, Y.: Intelligent information recommendation algorithm based on user model clustering. Computer Engineering and Design 27(13), 2360–2361 (2006) (in Chinese with English abstract)
5. Bo, H.: Web Intelligent Information Recommendation Method Based on Collaborative Filtering. Library and Information Service 54(19), 115–118 (2010) (in Chinese with English abstract)

Chaos Synchronization of Coupled Neurons via H-Infinity Control with Cooperative Weights Neural Network

Yuliang Liu[1], Ruixue Li[1], Yanqiu Che[1,2,*], and Chunxiao Han[1]

[1] Tianjin Key Laboratory of Information Sensing & Intelligent Control, Tianjin University of Technology and Education, Tianjin 300222, P.R. China
yqche@tju.edu.cn
[2] Institute of Semiconductors, Chinese Academy of Sciences, Beijing 100083, P.R. China

Abstract. In this paper, an H-infinity control with a cooperative weights neural network is proposed to realize the synchronization of two gap junction coupled chaotic FitzHugh-Nagumo (FHN) neurons. We first use a cooperative weights neural network to approximate the unknown nonlinear function. Then we employ the H-infinity control technique to attenuate the effects caused by unmodelled dynamics, disturbances and approximate errors. Finally, by Lyapunov method, the overall closed-loop system is shown to be stable and chaos synchronization is obtained. The control scheme is robust to the uncertainties such as unmodelled dynamics, ionic channel noises and external disturbances. The simulation results demonstrate the effectiveness of the proposed control method.

Keywords: Chaos synchronization, FitzHugh-Nagumo (FHN) model, Synchronization control, H-infinity control, Cooperative weights neural networks.

1 Introduction

Chaotic systems have complex dynamical behaviors that depend sensitively on tiny variance of initial conditions with bounded trajectories. Chaotic oscillations of individual neurons may be responsible for many regular regimes of operation [1]. Many neuronal models have been developed to simulate chaos of real neurons [2]. On the other hand, synchronization of chaotic dynamical systems has attracted a great deal of attraction due to their applications in various fields [3, 4]. Without control, N identical coupled neurons can eventually synchronize only when the coupling strength is above certain critical value [5] which may be beyond the physiological condition. Thus, many control methods have been proposed to achieve chaos synchronization [6, 7].

In this paper, we investigate synchronization control of two coupled chaotic FitzHugh-Nagumo (FHN) neurons. Under external electrical stimulation, the

---

[*] Corresponding author.

D. Jin and S. Lin (Eds.): Advances in FCCS, Vol. 2, AISC 160, pp. 369–374.
springerlink.com   © Springer-Verlag Berlin Heidelberg 2012

individual FHN model may show chaotic behaviors [5]. To achieve chaos synchronization, we present an H_∞ controller based on a neural network with cooperative weights. We first use a neural network to approximate the uncertain nonlinear function of the dynamical system by on-line adjusting the cooperative factor. Then we use the H_∞ tracking technique to attenuate the effects caused by unmodeled dynamics, disturbances and approximate errors. The proposed controller not only guarantees closed-loop stability, but also assures an H_∞ tracking performance for the coupled system.

The rest of the paper is organized as follows: In Sec. 2, we present the dynamics of gap junction coupled neuronal system of two FHN neurons under external electrical stimulations without control. In Sec. 3, we design the adaptive controller with H_∞ tracking performance for chaos synchronization of the coupled FHN neurons. The simulations given in Sec. 4 demonstrate the effectiveness of the control. Conclusion of this paper is given in Sec. 5.

2   Model of Two Coupled FHN Neurons

The model of gap junction coupled FHN neurons is described by:

$$
\begin{aligned}
\dot{X}_1 &= X_1(X_1 - 1)(1 - rX_1) - Y_1 + g(X_2 - X_1) + I + d_1 \\
\dot{Y}_1 &= bX_1 \\
\dot{X}_2 &= X_2(X_2 - 1)(1 - rX_2) - Y_2 + g(X_1 - X_2) + I + d_2 + u \\
\dot{Y}_2 &= bX_2
\end{aligned}
\tag{1}
$$

where X_i and Y_i $(i = 1, 2)$ are rescaled membrane voltage and recovery variable of two FHN neurons, respectively. g is the coupling strength of gap junction, and I is the external electrical stimulation in form of $I = A\cos(\omega t)/\omega$ with A, $\omega = 2\pi f$ the amplitude and frequency, respectively. Throughout this paper, we fix parameters b, r, A, f at values 1, 10, 0.1 and 127.1, respectively. d_1 and d_2 are added to simulate disturbances or noise in ionic channels. According to the result of [5], when $d_1 = d_2 = 0$, if the individual neurons are chaotic, then the synchronization occur only when the coupling strength beyond 0.5. As shown in

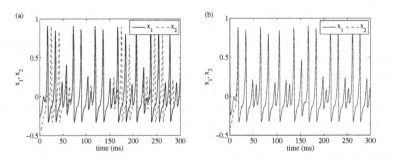

Fig. 1. System responses X_1, X_2 with coupling strengths (a) $g = 0.02$, and (b) $g = 0.7$

Fig. 1, for examples, if the coupling strength of the gap junction $g = 0.02 < 0.5$ (Fig. 1a), the synchronization cannot occur; the synchronization occurs when $g = 0.7 > 0.5$ (Fig. 1b). The term u is the added control force (synchronization command) such that the dynamical behaviors of the two coupled FHN neurons are synchronous no matter how large the coupling strength is.

3   Chaos Synchronization Using Adaptive H_∞ Control

3.1   Description of the Error Dynamics under Control

Let $e_X = X_2 - X_1$ and $e_Y = Y_2 - Y_1$, then the error dynamical system of the coupled neurons can be expressed as

$$\dot{e} = Ae + B\left[f(x) + u + d\right] \tag{2}$$

where $d = d_2 - d_1$, $e = [e_X, e_Y]^T$, $A = [-2g, -1; b, 0]$, $B = [1, 0]^T$, and $f(x) = X_2(X_2 - 1)(1 - rX_2) - X_1(X_1 - 1)(1 - rX_1)$ is the nonlinear part of the error dynamics with $x = [X_1, X_2]^T$. The problem of synchronization between the coupled systems can be translated into a problem of how to realize the asymptotical stabilization of the system Eq. (2) at origin. The goal is to design the controller $u(t)$ such that $\lim_{t \to \infty} \|e(t)\| = 0$.

3.2   Adaptive H_∞ Controller Design

The traditional radial basis function (RBF) neural network (NN) consists of one hidden layer and one linear output layer, expressing as $f_{tnn} = \theta^T \phi(x)$ [8], where the input vector $x \in \Omega_x \subset R^2$ with Ω_x being a compact set, the weight vector $\theta \in \Omega_\theta \subset R^m$ with m being the NN node number, and basis functions $\phi(x)$ chosen as the commonly used Gaussian functions with fixed centers and widths. In our design procedure, unknown continuous function $f(x) : R^2 \to R$ is to be approximated by the RBFNN with cooperative weights $f_{tnn} = \alpha \theta^T \phi(x) = \alpha \eta(x)$, where $\eta(x) = \theta^T \phi(x)$ and α is the cooperative factor. The neural network with cooperative weights can harmonize the change of all the weights by on-line adjusting the cooperative factor. Thus the hidden nodes dynamically increase or decrease so as to prevent the neural network from over-fitting or under-fitting, which can accelerate the fitting speed and improve the control precision.

We approximate $f(x)$ in the following form

$$f(x) = \alpha^* \eta(x) + \epsilon \tag{3}$$

where α^* is the ideal constant cooperative factor, and ϵ is the approximate error. Let $\hat{\alpha}$ be the estimate of α^*, and denote $\tilde{\alpha} = \hat{\alpha} - \alpha^*$ as the cooperative factor estimate error. Then the error dynamics Eq. (2) becomes

$$\dot{e} = Ae + B[\alpha^* \eta(x) + u + \omega] \tag{4}$$

where $\omega = \epsilon + d$ is the lumped uncertainty.

Theorem 1: Consider the error dynamics Eq. (2) with an uncertain nonlinear function $f(x)$, choose the control input as

$$u = -k^T e - \hat{\alpha}\eta(x) - \frac{1}{2\rho^2}B^T Pe \qquad (5)$$

where the feedback gain vector $k = [k_1, k_2]^T$ is chosen such that $A_1 = A - Bk^T$ is Hurwitz. $\rho > 0$ is an attenuate level, and the matrix $P = P^T > 0$ is chosen as the solution of Lyapunov matrix equation: $A_1^T P + PA_1 = -Q$, where $Q = Q^T > 0$ is a given matrix. And choose the adaptive update laws for $\hat{\alpha}$ as:

$$\dot{\hat{\alpha}} = \eta(x)B^T Pe, \qquad (6)$$

then the H_∞ tracking performance [9] for overall system satisfies the following relationship:

$$\frac{1}{2}\int_0^T e^T(t)Qe(t)dt \leq \frac{1}{2}e^T(0)Pe(0) + \frac{1}{2}\tilde{\alpha}^2(0) + \frac{1}{2}\rho^2\int_0^T \omega^2(t)dt \qquad (7)$$

Proof: Consider the Lyapunov function

$$V = \frac{1}{2}e^T Pe + \frac{1}{2}\tilde{\alpha}^2 \qquad (8)$$

Differentiating it and noting relevant equations , we have

$$\begin{aligned}
\dot{V} &= \frac{1}{2}\left[e^T P\dot{e} + \dot{e}^T Pe\right] + \tilde{\alpha}\dot{\hat{\alpha}} \\
&= -\frac{1}{2}e^T Qe - \frac{1}{2\rho^2}\left(B^T Pe\right)^2 + \omega B^T Pe \qquad (9) \\
&\leq -\frac{1}{2}e^T Qe + \frac{1}{2}\rho^2\omega^2
\end{aligned}$$

Integrating both sides of Eq. (7) from $t = 0$ to $t = T$, and noting $V(T) \geq 0$, we obtain

$$\begin{aligned}
\frac{1}{2}\int_0^T e^T Qedt &\leq V(0) + \frac{1}{2}\rho^2\int_0^T \omega^2 dt \\
&\leq \frac{1}{2}e^T(0)Pe(0) + \frac{1}{2}\tilde{\alpha}^2(0) + \frac{1}{2}\rho^2\int_0^T \omega^2 dt
\end{aligned} \qquad (10)$$

This completes the proof. Thus the tracking performance [9] is achieved for a prescribed attenuation level , and the synchronization of coupled system (1) can be obtained.

4   Simulation Results

In this section, numerical simulations are carried out for the synchronization of the coupled FHN neuron systems via the proposed adaptive H_∞ NN control.

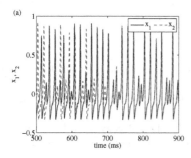
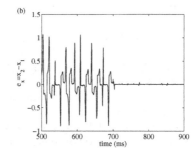

Fig. 2. System responses before and after control: (a) waveforms of X_1 and X_2, and (b) the corresponding error $e_X = X_2 - X_1$

We choose the design parameters in the simulations as $k = [3 - 2g, 1]^T$ and $Q = [1, 0; 0, 1]$ then $P = [0.25, 0.25; 0.25, 1.25]$ and $\rho = 0.1$. The initial states are $[X_1(0), Y_1(0)] = [0.1, 0.1]$ and $[X_2(0), Y_2(0)] = [0.11, 0.11]$, and the initial cooperative factor $\hat{\alpha}(0) = 0$.

We illustrate synchronization of two coupled chaotic FHN neurons with ionic channel noise. Specifically, d_1 and d_2 are Gaussian random noise with mean zero and variance 0.2 which are added to the system at the beginning of the simulation. We switch on the controller at time $t = 700$ms. The corresponding responses of the system are given in Fig. 2. Before the control is implemented, the two neurons exhibit their own chaotic dynamical behaviors and are not synchronized. After the controller is applied, the errors converge to round of zero rapidly and the synchronization is obtained.

5   Conclusion

In this paper, synchronization of two gap junction coupled chaotic FHN neurons in external electrical stimulations via the adaptive H_∞ control has been investigated. We use a neural network with cooperative factor to approximate the uncertain nonlinear part of the synchronization error system, and an H_∞ control to eliminate the approximate errors, ionic channel noise and disturbances. Based on the Lyapunov stability theorem, stability of close loop error system can be guaranteed by proper choice of the control parameters, which means synchronization of the coupled chaotic neurons. The simulation results have demonstrated the efficiency and robustness of the proposed control method.

Acknowledgment. This work is supported by The NSF of China (Grants No. 50907044, No. 60901035, No. 61072012, No. 61172009, and No. 61104032) and the Science and Technology Development Program of Tianjin Higher Education (Grant No. 20100819). We would also acknowledge the support of Tianjin University of Technology and Education (Grants No. KYQD10009, No. KJ10-01 and No. KJ11-04).

References

1. Rabinovich, M.I., Abarbanel, H.D.I.: The Role of Chaos in Neural Systems. Neuroscience 87, 5–14 (1998)
2. Chay, T.R.: Chaos in a Three-Variable Model of an Excitable Cell. Physica D 16, 233–242 (1985)
3. Pecora, L.M., Carroll, T.L.: Synchronization in Chaotic Systems. Phys. Rev. Lett. 64, 821–824 (1990)
4. Pecora, L.M., Carroll, T.L., Johnson, G.A., Mar, D.J., Heagy, J.F.: Fundamentals of Synchronization in Chaotic Systems, Concepts, and Applications. Chaos 7, 520–543 (1997)
5. Wang, J., Deng, B., Tsang, K.M.: Chaotic Synchronization of Neurons Coupled with Gap Junction under External Electrical Stimulation. Chaos, Solitons & Fractals 22, 469–476 (2004)
6. Lian, K.Y., Liu, P., Chiang, T.S., Chiu, C.S.: Adaptive Synchronization Design for Chaotic Systems via a Scalar Driving Signal. IEEE Trans. Circuits Systems I. 49, 17–27 (2002)
7. Wang, C., Ge, S.S.: Synchronization of Two Uncertain Chaotic System via Adaptive Backstepping. Int. J. Bifur. Chaos 11, 1743–1751 (2001)
8. Gupta, M.M., Rao, D.H.: Neuro-Control Systems: Theory and Applications. IEEE Press, New York (1994)
9. Chen, B.S., Lee, C.H., Chang, Y.C.: H_∞ Tracking Design of Uncertain Nonlinear SISO Systems: Adaptive Fuzzy Approach. IEEE Trans. Fuzzy Syst. 4, 32–43 (1996)

Information Service Architecture of Virtual-Machine-Based Grid System

Yan Junhao and Lv Aili

College of Computer Science and Technology, Henan Polytechnic University,
No.2001, Century Avenue, Jiaozuo City, Henan Province, China
{yanjh,lvaili0210}@hpu.edu.cn

Abstract. To ease the management of virtualized distributed grid systems, it is important to develop an information service to deliver the resource information and system status. This paper focuses on target system model and high level information requirement. By studying the information's collecting, translating, providing and indexing, we build an information service architecture for the grid infrastructure that consists of distributed virtual machines.

Keywords: Grid computing, information service, virtual machine.

1 Introduction

Grid computing technology [1] offers exciting solutions for parallel and distributed computing. It can provide reliable, collaborative, and secure access to remote computational resources, as well as distributed data and scientific instruments. Although great advances have been made in the field of grid computing, users are still expected to meet some difficulties in employing grid resources. Grid computing's most important research issue is QoS provision. Resources shared on computational grids, in general, cannot guarantee the QoS of resource provision.

Another important research focus of grid computing is to provide a customized runtime environment for grid applications. In general, grid applications demand customized execution environments. But, sometimes it is difficult for grid resources to provide different runtime environments for multiple grid applications.

Virtualization [2] is the process of presenting a logical grouping or subset of computing resources so that they can be accessed in abstract ways that give benefits over the original configuration. Virtualization technology can bring various advantages such as resource customization, performance guarantee and isolation, and easy resource management. Employing virtual machines as computing environments for grid applications might therefore be a promising solution to address these challenges.

1) Virtual machines are allocated with specified resources required by grid applications, for example, CPU bandwidth, memory size, and storage capacity. Thus, QoS is guaranteed since virtual machine resources are dedicated to grid applications during the execution.

D. Jin and S. Lin (Eds.): Advances in FCCS, Vol. 2, AISC 160, pp. 375–380.
springerlink.com © Springer-Verlag Berlin Heidelberg 2012

2) Various preconfigured virtual machine templates can be prepared for different grid applications. When a grid application arrives at a compute center, a virtual machine can be created from a certain virtual machine template and started on demand. Therefore, grid applications can always find the desired execution environments if customized virtual machines are prepared. Furthermore, grid users can also configure the allocated virtual machines to build the desired computing environments.

Therefore, virtualized distributed grid infrastructures have been adopted to help a user to build an advanced computing environment.

2   Information Service of Virtual-Machine-Based Grid System

To ease the management of virtualized distributed grid systems, it is important to develop an information service to deliver the resource information and system status. In this paper, we present our work on an information service for grid infrastructures that consists of distributed virtual machines. The system is hierarchical and includes the following components [3].

— **Information collector:** The information collector runs in the virtual machines and functions to get the virtual machines' resource information.

— **Information translator** and provider: The information provider collects resource information from multiple information collectors inside one computing center or manage domain and provides the information to high-level grid services of various virtual organizations (VOs).

— **High-level grid information service (GIS):** In the VO level, the GIS or other services such as Globus WebMDS [4] can demand virtual machine resource information, which is provided by multiple information providers. Our implementation reuses various existing technologies as building blocks, for example, distributed software agents and Globus monitoring and discovery system. Our contribution is to design and implement an integrated solution for a virtual-machine-based GIS.

3   Related Works

3.1   Virtual Machine and Distributed Virtualized Infrastructure

For several years, the grid computing research community has shown interest in virtual machines and virtual environments. The typical virtual machine monitor (VMM) or hypervisor setup includes the Xen VMM [5], VMware server/ESX server [6], and user-mode Linux (UML) [7]. The Globus alliance recently implemented the concept of a virtual workspace that allows a grid client to define an environment in terms of its requirements, manage it, and then deploy the environment on the grid. Some distributed virtualized infrastructures have been developed with virtual machines. The Xen Grid Engine follows an approach to create dynamic virtual cluster partitions using para-virtualization techniques. Amazon Elastic Compute Cloud (EC2) is a Web service that provides resizable compute capacity in the cloud, which is enabled by Xen virtual machines. The aforementioned research provides a background and context for our research on distributed virtualization infrastructure. Our work is aimed toward distributed infrastructures, which are composed of multiple compute sites, and provides a number of distributed virtual machines. This paper presents an innovative implementation of distributed information services for distributed virtualized systems.

3.2   Information Service for Grid Computing

The Grid Information Service (GIS) is a vital part of any grid software infrastructure, providing a fundamental mechanism for discovery and monitoring and, hence, for planning and adapting the application behavior [8]. In a grid environment, the description, discovery, and monitoring of resources is very challenging due to the diversity, large numbers, transient membership, dynamic behavior, and geographical distribution of the entities in which a user might be interested. An information collector is an agent or sensor for collecting resource information and provides it to an information service. Information from multiple information services are aggregated to an aggregated information service to provide a snapshot of a number of resources. Information services also register themselves to a directory service. Users can query the directory service to discover information services and thereafter query the information services or aggregated information services.

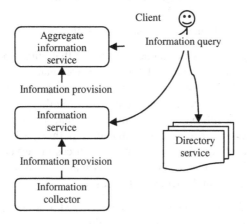

Fig. 1. GIS model

Current information services for grid computing cannot be applied on distributed virtual machine infrastructures due to various reasons, for example, heterogeneous and hierarchical resource management and virtual machine information organization. This paper devotes itself into the field of GIS research. Compared with other implementations of the GIS, our implementation of the information service presented in this paper develops a novel framework and a set of software implementations to provide distributed virtual machine information for grid computing. The grid-level information service of our implementation uses the GIS architecture model shown in Fig. 1 as it is widely employed by the grid community.

4   System Architecture of the Information Service

As discussed above, virtual machines are widely used in grid computing. A virtual-machine-based grid computing system is hierarchical and distributed, thus requiring a model to represent it. Furthermore, management services, for example, an information service, are demanded o develop the virtual-machine-based grid computing system.

We propose a system model to describe a distributed, hierarchical, and heterogeneous virtual-machine-based grid system, which contains distributed Computer Sites interconnected by networking (see also Fig. 2).

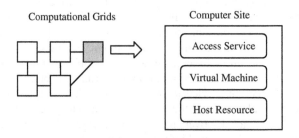

Fig. 2. Target system model

Each Computer Site logically consists of the following levels.

1) The compute site provides an access service that allows remote users to access resources of the computer center. The access service can be offered by existing grid middleware, a portal, Web services, or any functionality that support remote steering. Our information service is developed and integrated in this level.

2) In the middle level exists virtual machines that are backed by host resources. The information service operates virtual machines in this level.

3) The fabric level contains various host resources or servers, which are installed with VMMs. Host resources offer multiple virtual machines. The back ends of the information service are implemented in this level.

Virtual-machine-based grid systems are characterized by some special features, which bring research challenges for deploying, monitoring, and operating the system, such as site autonomy, hierarchy, heterogeneity, large-scale distribution. So, we design an architecture suit for Virtual-machine-based grid systems shown as follow.

As Fig.3 shown, the information agent resides inside a virtual machine and collects resource information. Information agents communicate with the information collector to provide resource information to the higher level. We implement the CIM [9] Object Manager (CIMOM) agent for the VMware ESX server with the VMM support. It also can implement application-level lightweight Java agent for other type of virtual machine, including Xen and the VMware ESX server.

A site information service works on the access point of a compute site and collects all virtual machine resource information by communicating with the information agents. The site information service consists of several components like the information collector, the information translator, and the information provider for the Globus index service:

— **Information collector:** The information collector runs in the virtual machines and functions to get the virtual machines' resource information. It is deployed in the access point at the compute site level and communicates with multiple information agents. An information collector contains the following components: query sever, information engine, information cache, and agent registry.

— **Information translator:** The information translator translate CIM Information to GLUE [11] Information. We use CIM Schema for Virtual Machine Information. Itt provides a common definition of management information for systems, networks, applications, and services. The VMware CIM SDK provides a CIM interface for developers to build management applications. The GLUE schema represents an abstract model for grid resources and mappings to concrete schemas that can be used by information services within grids. Various types of resources that are shared on computational grids should be described in a precise and systematic manner.

— **Information provider:** The information provider collects resource information from multiple information collectors inside one computing center or manage domain and provides the information to high-level grid services of various virtual organizations (VOs). We build an information provider from virtual machines and provide information to the Aggregator Framework.

— **Globus index service:** Globus index service is used for the grid-level information service. The grid index service can collect information and publish the information to clients. The grid index service can also register to each other in a hierarchical fashion in order to aggregate data at several levels.

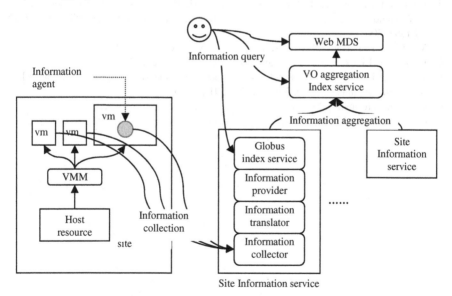

Fig. 3. Software architecture of the information service [3]

5   Conclusion

This paper has implemented information service architecture of virtual machine pools in a computer center for grid computing. The information service will can monitor virtual machines backed by popular VMMs such as the VMware server and the VMware ESX server. The architecture of information service for monitoring distributed virtual machines and applications on production Grids is extremely useful for scientific applications.

Acknowledgment. The authors of this paper wish to show many thanks to the support from Youth Foundation of Henan Polytechnic University (Q2010-54), the Natural Science Foundation of Henan Provincial Department of Education under Grant No. 2011A520014 and Soft Science Study of Henan Province (102400450064).

References

1. Foster, I., Kesselman, C., Tuecke, S.: The physiology of the grid: An open grid services architecture for distributed systems integration. Open Grid Service Infrastructure Work Group, Global Grid Forum, Tech. Rep. (June 2002)
2. Seawright, L., MacKinnon, R.: VM/370—A study of multiplicity and usefulness. IBM Syst. J. 18(1), 4–17 (1979)
3. Wang, L., von Laszewski, G., Chen, D., Tao, J., Kunze, M.: Provide Virtual Machine Information for Grid Computing. IEEE Transactions on Systems, Man, and Cybernetics—Part A: Systems and Humans 40(6), 1362–1374 (2010)
4. WebMDS, http://www.globus.org/toolkit/docs/4.0/info/webmds/
5. Barham, P., Dragovic, B., Fraser, K., Hand, S., Harris, T., Ho, A., Neugebauer, R., Pratt, I., Warfield, A.: Xen and the art of virtualization. In: Proc. 19th ACM SOSP, pp. 164–177 (October 2003)
6. VMware Virtualization Products, http://www.vmware.com/
7. User Mode Linux, http://user-mode-linux.sourceforge.net/
8. Jie, W., Cai, W., Wang, L., Procter, R.: A secure information service for monitoring large scale grids. Parallel Comput. 33(7/8), 572–591 (2007)
9. Common Information Model, http://www.dmtf.org/standards/cim/
10. GLUE Schema v1.3, http://glueschema.forge.cnaf.infn.it/Spec/V13

Construction of Public Information Network Platform in Urban Intelligent Communities with Outdoor Activities

Binxia Xue[*], Zhiqing Zhao, and Weiguang Li

Architectural School, Harbin Institute of Technology, No. 66 Xidazhi Str.,
Nangang Dist., Harbin, China
binxia68@126.com, Zhaozq88@126.com, lwg0605@163.com

Abstract. Based on the main technical characteristics and functional classification of contemporary intelligent community as well as the profound interpretation to the management system of public information, the construction of public information network platform in intelligent communities with outdoor activities and its technical roadmap are proposed. Moreover, with the advances of modern multimedia and network communication technology, the virtual and realistic public activities are connected to improve the neighborhood relations, dispel social differentiation and promote the social integration.

Keywords: urban intelligent community, virtual-reality, interaction, platform of public information network, construction.

1 Introduction

Depending on the development of modern science and technology, intelligent community, as a new emerging type, integrates the information transmission, network technology, information integration and intellectual residential system. It also indentifies the arrival of intellectual society, which regards the digitalization and networking as the platform. For the intelligent communities in China, the superior related to technology and management is pursued to bring more convenience on the function and life for residents as well as create the atmosphere embracing the nature. The quality of safety, comfort, convenience and humanization are embodied in the intelligent community with beautiful environment and perfect function, which is conducive to the sustainable operation of public facility system.

From the aspect of constitutive relations, the intelligent community derives from the network on the basis of intelligent architecture and it integrates the 4C, namely computer, communication & network, control, communication and intelligent card. The system of information integration network, which consists of service management center, access network and home intelligent system, is established with the aim to receive, transmit, collect and monitor the integration of multi-information, service and management, property management and security and residential intellectual system effectively and in real time. Hence the high-tech intelligent means are provided for the

---

[*] Corresponding author.

D. Jin and S. Lin (Eds.): Advances in FCCS, Vol. 2, AISC 160, pp. 381–387.
springerlink.com   © Springer-Verlag Berlin Heidelberg 2012

service and management of residential quarters in order to achieve the efficient and fast value-added service and management as well as offer the safety and comfortable home environment [1][2]. At present, the main function covers living environment, information service environment, safety environment, property management and service environment [3]. The security system includes alarm system, video intercom system and monitoring system; the property management system refers to remote meter reading system of water, electricity, gas and hot water, automatic toll collection system and background music system; network information service system includes computer network and comprehensive information service system, which provides online shopping, electronic communities, online education and cable television services. In addition, automatic management system of electrical and mechanical equipment plays as the technical support to manage and control the normal operation of pipe network and public facilities and energy-saving and emission reduction in communities. It is concluded that these elements mostly have satisfied the material function, which is reflected in the equipment upgrading and technical advancement. However the potential value and function haven't be exerted in the exploitation and utilization of promoting resident interaction, enhancing cohesion of community and easing social differentiation for the virtual community.

Today a prominent phenomenon lies in that social differentiation and gated communities are becoming increasingly serious with the urbanization development and progress of social material. Meanwhile according to the various economic income, social status, cultural orientation and values, urban residential space is divided into, hence resulting in the isolation and exclusion of social space and separation of the living areas. For the current urban communities, residents exist in the more independent and individual form, destroying the integrity and continuity of the urban space. Unfortunately it causes the gap between the people and the lost of the public space in the community. Therefore a general used method or effective way is urgently needed to remedy the declining relationship in the community. The public information network platform can be improved in the intelligent community with the outdoor activities, which can serve to construct the multi-leveled contact to further get rid of the illusion and dependence brought by the virtual technology. Consequently this could promote community flow, improve the utilization of public space and enrich the quality of the community life.

2   General Construction

The shared characteristic of intelligent communities is the high ability to interact the information among residents, communities and societies, which is able to offer the more perfect security environment, property management and excellent service of community public information for the residents. Along with the development of a new generation of network integration technology, the integrated solution, which consists of network video, control, networking, transmission, storage and extended application, has greatly strengthened the ability of the network system in remote surveillance, management and operation and database accessibility. In addition, the exchanging ability of information and database among automatic system of residents, service platform of community public information, disaster prevention system and property management system is also enhanced for different users. Also the seamless connection with internet is achieved through the firewall [4] [5]. It is possible to enjoy the multi information of visual-realistic multi-media for the residents and ultimately the public activities and communication in real life is promoted as well as the social integration and vitality of community is upgraded.

The platform of public information network with outdoor activities, mainly taking the information management system as the core, realizes its construction through the integration and expansion of other intelligent system, including communication, smart home, implementation and application of public facilities and security application. This new kind of information management system should highlight the primary content of virtual community, which covers e-commerce, book selling, comprehensive information, and video on demand. At the same time, to satisfy the requirement of functional expansion and resource sharing, it is necessary to integrate and absorb the functions of other systems according to its own conditions and needs. Such as the digital television subsystem under the project of communication application system, subsystem of community broadband local network and telephone communication subsystem, which could offer TV shopping, information search, voice, fax, e-mail, video for conference, video telephone, video text and communication with multimedia; as for the subsystem of home information release and message, belonging to the project of smart home system, provides the possibility to send and submit message and group discussion in the form of point to point or point to multi points between the property centers and residents through the virtual community net.

In this context, the residents in the community will establish a platform of real time release of information and resource and communication community with multi function and multi points through the public information platform with outdoor activities, realizing the interaction between the individual intention and public environment (see in Figure 1). For instance, if one propose some outdoor activities in the central Greenland or the square through the virtual community net page, the real-time virtual community will collect the feedback from other residents in s short period of time (if it is concerned) and display the related suggestion or response. Even the residents can receive the video of the related public places through the network integration technology. When the call is a kind of collective recreational activities, such as balls, cards, ballroom dancing or chatting, the full range information service of the virtual community will give out the valuable reference for the intended people as well as promote the occurrence of the public events. Another significant possibility lies in the face-to-face discussion or causal gatherings of the common topics among the residents, which will also greatly improve the utilization rate of the halls and cultural centers in the communities.

It is worth to stress that the content of the public information electronic display subsystem under the project of system of public equipment implementation and application will be displayed in phase through a column, displaying the specified text and images, TV, video, DVD, video cameras and other video programs. However, the remote login service of the subsystem of the community broadband local network make it possible to extract information and database, participate in the electronic community communication and monitor security state and electrical equipment through logging the home computer in different places under the cooperation of the smart home system.

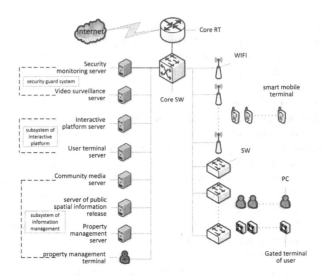

Fig. 1. Network topological graph of Intelligent Community

Based on the function of traditional electronic community, this public information network platform constructs the real-time communication and information feedback model among the users through the new generation of network video integration technology. Due to this, the virtual indoor activities are linked with the real outdoor public recreational activities, unlike the traditional intelligent communities, which is only limited to the online shopping with man-machine dialogue, e-mail, online education, online entertainment and remote medical consultation.

3   Technical Support

Comparing with the traditional virtual community, the public information platform includes more complicated functional demand and target, which thus put forwards higher requirement for the operation pattern of network and the performance and synergy of its three primary levels, information level, controlling level and equipment level (sensing/executive level).

First of all, various factors should be taken into account for the construction of intelligent community and starting from the reliability, openness and sustainable development of intelligent system, It is meaningful to converge three platforms of the information network system in the community, which is composed by the physical platforms of intelligent system (PDS system), technical platform (computer network) and operation platform (field controlling bus). Both the Ethernet or cable HFC network is possible to support of the network operation and the combination also can be used [7].

Then the system integration and connection should be applied to achieve the compatibility and integration of different technologies and equipments. System integration is defined in the national standard as intelligent subsystems of various functions are connected physically, logically and functionally in intelligent buildings with the aim to realize information comprehension and resource sharing. The intelligent system of the property management includes subsystem of security prevention, subsystem of interactive platform and subsystem of information management (Figure 2). Gating terminals, computers and smart mobile phones can access these subsystems through the network and read the information to give the response.

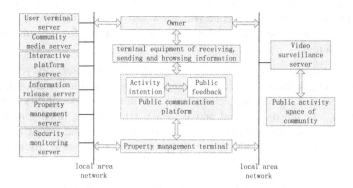

Fig. 2. Operation procedure of the public information network platform

At last, the optimization of technology integration should be focus on. Systems are organized with the scientific procedure to work or control dynamically in the optimal way. The controllability of the intelligent community depends on the each system and the controlling strategies of computer integration controlling system station [3].

Based on the above principles, the following supporting technologies are applied in the construction of public information network platform combining with the state of the art development of network communication technology.

Connection of the user terminal and subsystems: 1.Intelligent mobile terminals. User could connect the subsystem server through the WIFI wireless network.2. User's computer could access the subsystem server though LAN of the community. 3. Reading and response of the integrated information for the gating terminal. The user's terminal server is connected through the LAN of the community.

Network integration approaches: 1. The subsystem server is connected with the core switch though six unshielded twisted –pair (CAT6) to achieve 1Gbps data transfer speed. 2. Switched of each unit and the core switch are connected with optical fiber through photo electricity converter. User's computer is linked with unit switch via CAT5e.3.Network surveillance camera with NET is selected as the video surveillance equipment and for the gated terminal, the one who support TCP / IP protocol is priority, which is connected with video surveillance service through LAN.4.Video cable, twisted pair and fiber are applied respectively according to the position for the video display in the public area.

4   Design Points

Technically, four problems should be noted in the construction of public information network with outdoor activities: integration, openness appropriateness and response.

Integration. Central integration management system, as the key of the intelligent community construction, is able to conserve resource and optimize the performance. Subsystems of the integrated systems in the intelligent residential areas are all produced by varied factories following by different standards, such as data transfer bus, hardware interfaces, software interfaces and control algorithms. The integrated controlling system needs to cooperate with these subsystems to reduce the complexity of the systems. The same standard and protocol are applied for the subsystems and integrated systems.

Openness. Considering the operating cycles of the computer, 50 years and the renovation and upgrading cycles of the computer system and information facilities, a high intelligent community with closed and less upgrading possibility definitely is attached with less reasonable and active than a intermediate intelligent residential areas with openness and upgrading possibility. Therefore the construction of the intelligent communities should obey the principle of openness to maintain the upgrading and expanding space in the life cycle of the residence [6].

Appropriateness. For the intelligent community, the communication network serves as the physical platform for the comprehensive information service, the connection with external WAN and the intelligent property management. The ability of the network to provide comprehensive information and information service should be considered for its appropriateness as well as the advanced nature,, scalability, cost performance and the affordability of the investment cost for the developers(users) [7]. The priority solutions need to be selected relying on the network system, equipment and cabling of the construction during different periods to conserve resource.

Response. The network platform of the virtual community, which should reflect the true value, needs to connect the communication among people in virtual world with the realistic life environment to convert it into the public activities in real life. The intelligent community is traditionally defined as the comprehensive site for residence, work, study and entertainment. Due to the network, work, study, entertainment, shopping, medical, tourism and social communication all can be realized without leaving the homes. However the platform of public information network with outdoor activities will reverse this understanding and realize the realistic across in the aspect of sociology through the technical integration and creative planning of public community.

From the social level, the urban space accommodates the orderly collection and activities of the people and thing as well as reflects the social relationship features among people and the cultural context hidden. Under the urban environment with high density, the relationship of the individual residents is increasingly split, transient, superficial and utilitarian [8]. In response, the interactive community with outdoor activities intends to incorporate the urban life into the growing and interdependent general relationship to realize its development. And the integration of the culture and society will be achieved through the promotion of the interpersonal relationship in the community and the allocation and utilization of the facilities and the public space.

5   Conclusion

The development of modern society increasingly leads to the fragile and lost of the interpersonal contact. Under this context, the intelligent community not only offers the convenient life, but also aggravates the material neighborhood, depending on the virtual world. Combining with technology and society, the construction of the public information network is conducive to get rid of the addiction to the cyber world to realize the interaction with the real life and strengthen the structure of the public realm in the community. Consequently the virtual community, which becomes a useful mean and media, will find out a perfect fit for the residents between material and void, independence and share and privacy and communication.

References

1. Zhang, S.: The Network Planning Design of the Intelligent Community. Industry and Planning Forum 9(6), 121–123 (2010) (in Chinese)
2. Housing Industry Office of Construction Ministry. Demonstration Projects Construction and Technical Guidelines of National Resident Intelligent System (trial version) (April 1999) (in Chinese)
3. Yu, Z.: Controllability and Integrated Control of Intelligent Community. New Architecture 5(5-7) (1999) (in Chinese)
4. Li, X.: Analysis of the Video Surveillance System in Intelligent Community. Intelligent Building and Information 132(11), 100–102 (2007) (in Chinese)
5. Hu, Z.: Network Integration of Intelligent Building. Intelligent Building and Information 81(8), 20–21 (2003) (in Chinese)
6. Wei, H.: Analysis of Intelligent Solution for the Residential Community. Building Energy Efficiency 37(8), 41–43 (2009) (in Chinese)
7. Jiao, F.: Communication Network Design of Intelligent Community. Modern Electronic Technology 158(15), 10–12 (2003) (in Chinese)
8. Grafmeyer, Y.: Urban Sociology. Tianjin People's Publishing House (2005) (in Chinese); Xu, W. (Translation)

Blind Image Restoration Based on Wavelet Transform and Wiener Filtering

FengQing Qin

Department of Computer Science and Technology, Yibin University, Yibin, 644007, China

Abstract. A blind image restoration method is proposed to estimate the high resolution image when the point spread function (PSF) of the imaging system is unknown or partially known. Firstly, the Gaussian PSF of the degraded image is estimated through Wavelet transform method. Then, utilizing the estimated PSF, high resolution image is estimated through Wiener filtering image restoration method. In order to justify the influence of PSF estimation on the quality of the restored image, image restoration is performed on different PSF. Experimental results show that the deviation of the Gaussian PSF is estimated with high accuracy, as well as that the PSNR and quality of the estimated image increase around the real PSF.

Keywords: Blind Image Restoration, Point Spread Function, Wavelet Transform, Wiener Filtering.

1   Introduction

Image restoration refers to restoring a high resolution image from the observed degraded image with low resolution. In recent years, image restoration has been applied to many areas such as astronomy, remote sensing imaging and medical image restoration etc. In each area, although the source of the image or the cause of degradation may be different, the essence are the same that the degraded image can be expressed as the convolution of the original high resolution image with the point spread function of the imaging system.

In some cases, the PSF is known, and such image restoration method is called linear restoration, which is research by most of the current image restoration algorithms. However, in many practical applications, the PSF of the imaging model is most likely unknown or is known only within a set of parameters, and such image restoration is called blind restoration. In most algorithms, the PSF is assumed to be a known PSF with given parameters, which does not meet the real imaging model of optical devices and the performance of the restoration algorithm will decrease. Thus, blind image restoration problem [1-6] arises naturally and is expressed as estimating a high-resolution image and the PSF simultaneously, which is one advanced issue and challenge in image processing.

The foremost difficulty of blind image restoration is rooted in the fact that the observed image is an incomplete convolution [7]. The convolution relationship around the boundary is destroyed by the cut-off frequency, which makes it much

D. Jin and S. Lin (Eds.): Advances in FCCS, Vol. 2, AISC 160, pp. 389–395.

more difficult to identify the PSF of the imaging system. Gaussian PSF is the most common blur function in many imaging system, it is mainly considered in many algorithms. The Gaussian PSF estimation based on Wavelet transform [6] is adopted to estimate the standard deviation here.

In this paper, a blind image restoration method is proposed. For and given degraded image, Gaussian PSF is estimated through Wavelet transform method. Image restoration is performed through Wiener filtering algorithm. Experimental results justify the fact that PSF estimation plays an important part in blind image restoration.

2   PSF Estimation

2.1   Gaussian Point Spread Function (PSF)

Gaussian point spread function (PSF) is the most common blur function of many optical measurements and imaging systems. Generally, the Gaussian PSF may be expressed as follows:

$$h(m,n) = \begin{cases} \frac{1}{\sqrt{2\pi}\sigma}\exp\{-\frac{1}{2\sigma^2}(m^2+n^2)\}(m,n)\in R \\ 0 \qquad\qquad\qquad others \end{cases} \qquad (1)$$

Where, σ is the standard deviation; R is a supporting region. Commonly, R is much smaller than the size of image, and often is denoted by a matrix with size of $K \times K$. K is often an odd number, and given by experience. From the mathematical description of Gaussian PSF, we can see that the standard deviation (σ) needs to be estimated.

2.2   Lipschitz Exponent

The singularity of signal can be characterized by Lipschitz exponent accurately. For a signal $f(t)$, the character around a given point t_0 is expressed as follows:

$$\left| f(t_0+\Delta t) - p_n(t_0+\Delta t) \right| \le A\left| h \right|^\alpha \qquad (n < \alpha < n+1) \qquad (2)$$

Where, Δt is a sufficiently small value; $p_n(t)$ is a n-order polynomial, and α is the Lipschitz exponent of $f(t)$ at the point t_0.

Generally, the Lipschitz exponent represents the singularity of a function at a given point. The bigger the α is, the smoother the function at this point will be; the smaller the α will be, the more singular the function at this point will be.

2.3   Wavelet Modulation Maxima

Taking the assumption that the wavelet function $\varphi(t)$ is continuously differentiable, and the attenuation rate at the infinite point is $O(1/(1+t^2))$. If the wavelet transformation of f(t) satisfies the following condition:

$$\left|W_s f(t)\right| \le \left|K s^\alpha\right| \tag{3}$$

Namely,

$$\log\left|W_s f(t)\right| \le \log K + \alpha\left|K s\right| \tag{4}$$

Then, the Lipaschitz exponent of $f(t)$ at interval $[a,b]$ is α. Where, K is a constant; s is the scale of wavelet transformation; $|W_s f(t)|$ is the wavelet modulation of $f(t)$ at scale s.

From (4),we can see that, if $\alpha>0$, the wavelet modulation maxima increase with the scale s; if $\alpha<0$, the wavelet modulation maxima decrease with the scale s; if $\alpha=0$, the wavelet modulation maxima doesn't change with the scale s.

2.4   Gaussian PSF Estimation

If $f(x,y)$ is the high resolution image, and $g(x,y)$ is the blurred image, the wavelet modulation between them is expressed as follows:

$$\left|W_s g(x,y)\right| = \frac{s}{s_0}\sqrt{\left|W_{s_0}^{(1)} f(x,y)\right|^2 + \left|W_{s_0}^{(2)} f(x,y)\right|^2} = \frac{s}{s_0}\left|W_{s_0} g(x,y)\right| \tag{5}$$

Where,

$$s_0 = \sqrt{s^2 + \sigma^2} \tag{6}$$

Thus, for the blurred edge point, the relationship of the wavelet modulation, the scale, and the standard deviation of Gaussian PSF can be expressed as follows:

$$\left|W_s(g(x,y)\right| \le K\frac{s}{s_0} s_0^\alpha \tag{7}$$

Wavelet transformation at multiple scale (more than 3) is performed on the blurred image edge, and the corresponding wavelet modulation maxima at different scales will be calculated. By solving equations, the constant (K), the Lipschitz exponent (α), and the standard deviation of the Gaussian PSF (σ) will be gained respectively.

3   Image Restoration Method

Wiener filtering is broadly applied in signal and image processing. Practical experience shows that it is a deconvolution technique with good restoration effect and small amount of calculation.

In Wiener filtering image restoration algorithm, both the blur function and the statistical character of system noise are considered. The noise is assumed to be a random process. The aim is to make the mean square error between the original image

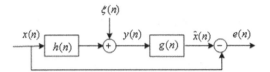

Fig. 1. The sketch map of Wiener filtering

and the estimated image to be the least. According to this idea, he sketch map of Wiener filtering may be expressed in Fig.1.

The observed low resolution image may be expressed as follows:

$$y(n) = \sum_{k=-\infty}^{\infty} x(n-k)h(k) + \xi(n) = x(n)*h(n) + \xi(n) \tag{8}$$

Where, * is the convolution operation; $x(n)$ is the original high resolution image; $\xi(n)$ is Gaussian white noise with zero mean; $h(n)$ is the blur function, which can be presented by point spread function.

It's expected to find a filter with input $y(n)$, and the output is the estimated image expressed as follows:

$$\hat{x}(n) = \sum_{k=-\infty}^{\infty} y(n-k)g(k) = y(n)*g(n) \tag{9}$$

The mean square error between the original image and the estimated image is expressed as follows:

$$e^2 = E\{|x(n) - \hat{x}(n)|^2\} \tag{10}$$

If the image and the noise are irrelevant, the minimum of the error function may be calculated in frequency domain. When the discrete Fourier transform (DFT) method is used to estimate the restored image, the Wiener filter may be expressed as follows:

$$X = \frac{H^*Y}{|H|^2 + S_{nn}/S_{xx}} \tag{11}$$

where, X, Y and H are the DFT of the real image (x), the blurred image (y) and the blur function (h) respectively; * denotes the conjugate operation; S_{nn} and S_{xx} denote the power spectrum of the noise and the real image. As it is usually very difficult to estimate S_{nn} and S_{xx}. the Wiener filter is usually approximated by the following formula:

$$X = \frac{H^*Y}{|H|^2 + \Gamma} \tag{12}$$

where, Γ is a positive constant, which is often taken as a experience value.

4   Experiments

In order to test the effectiveness of the proposed method objectively and subjectively, experiments are performed on simulated image. The original image 'camera.bmp' of size 256×256 is shown in Fig.2. The HR image is passed through the LR imaging model as shown in Fig.3. Firstly, the orignal image is convolved with a Gaussian PSF to simulate a blurred image. Here, the size of Gaussian PSF is taken as 7, and the original standard deviation (σ_0) is taken as 1.2 respectively.

Fig. 2. The original image   **Fig. 3.** The blurred image

Utilizing the Gaussian PSF estimation based on Wavelet transform, the estimated standard deviation ($\hat{\sigma}$) is 1.254. The estimated Gaussian PSF is shown in Fig.4. The relative estimation error of the standard deviation is: $|\sigma_0 - \hat{\sigma}| / \sigma_0 = 0.045$.

In order to justify the importance of PSF estimation in blind image restoration, in the case of the size of Gaussian PSF is 7, Wiener image restoration is performed at different estimated standard deviation taken as the range of [0.1, 3], with an increment

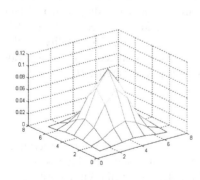
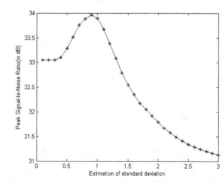

Fig. 4. The estimated PSF.   **Fig. 5.** The PSNR at different standard deviation

(a) The restored image ($\hat{\sigma}$ =0.1, PSNR=33.0480)   (b) The restored image ($\hat{\sigma}$ =1.254, PSNR=33.2194)

(a) The restored image ($\hat{\sigma}$ =2, PSNR=31.7917)   (b) The restored image ($\hat{\sigma}$ =3, PSNR=31.1313)

Fig. 6. The Restored images at different estimation standard deviations.

of 0.1, and the corresponding PSNR at different estimated standard deviation is shown in Fig.5. The PSNR of the restored image around the real PSF is higher. When the estimated standard deviation is 0.1, 1.254, 2, and 3 respectively, the SR reconstructed images are shown in Fig.6 (a)-(d) respectively. The quality of the restored image around the real PSF is better. When the estimated PSF is far away from the real value, the restored image is illegible and appears ringing effect.

5   Conclusions

A blind image restoration method based on Wavelet transform and Wiener filtering is proposed. The standard deviation of Gaussian PSF is estimated with high accuracy. Utilizing the estimated PSF, image is restored through Wiener filtering algorithm. In order to test the role of PSF estimation on the quality of restored image, blind image restoration is acted at different stand deviations. Experimental results justify the great importance of PSF estimation in blind image restoration. When the PSF is around the real value, the PSNR and the quality of the restored image will increase, and vice versa.

Acknowledgements. This paper is supported by the project supported by Scientific Research Found of Sichuan Provincial Education Department (11ZA174), the project supported by Application Fundamental Research Found of Sichuan Provincial Scientific and Technology Department (2011JY0139), the project supported by the Science and Technology Bureau of Yibin City (2011SF016), the project supported by the Doctor Scientific Research Starting fund of Yibin University (2009B02).

References

1. Seghouane, A.-K.: A Kullback-Leibler Divergence Approach to Blind Image Restoration. IEEE Transactions on Image Processing 20(7), 2078–2083 (2011)
2. Luo, Y.H., Fu, C.Y.: Midfrequency-based real-time blind image restoration via independent component analysis and genetic algorithms. Optical Engineering 50(4) (2011), doi:10.1117/1.3567072
3. Wang, Z.F., Tang, Y.D.: Semi-blind image restoration based on Chan-Vese denoising model. Chinese Optics Letters 6(6), 405–407 (2008)
4. EI-sallam, A.A., Boussaid, F.: Spectral-based blind image restoration method for thin TOMBO imagers. Sensors 8(9), 6108–6124 (2008)
5. Jirik, R., Taxt, T.: Two-dimentional blind Bayessian deconvolution of medical ultrasound images. IEEE Transactions on Ultrasonics Ferroelectrics and Frequency Control 55(10), 2140–2153 (2008)
6. Tao, Q.-C., Deng, H.-B.: Wavelet transform based Gaussian point spread function estimation. Optical Technique 30(3), 284–288 (2004)
7. Zou, M.Y.: Deconvolution and Signal Recovery. Defence Industry Publishing (2004)

3D Environment Modeling with Hybrid of Laser Range Finder and Range Camera

Chenglu Wen[1], Ling Qin[1], Siyuan Lin[1], and Qingyuan Zhu[2,*]

[1] Cognitive Science Department, Xiamen University, 361005 Fujian, China
Fujian Key Laboratory of the Brain-like Intelligent Systems, 361005 Fujian, China
clwen@xmu.edu.cn, aql.yi5637@163.com, prolins@foxmail.com
[2] Department of Mechanical and Electrical Engineering, Xiamen University,
361005 Fujian, China
zhuqy@xmu.edu.cn

Abstract. 3D environment modeling achievement gives important reference for mobile robot navigation and dangerous situation searching. In this paper, we present a method that acquire 3D model for indoor environment with hybrid of laser range finder and range camera. 3D laser point cloud can be obtained by the coordinate work between the 2D laser range finder and a pan-tilt. RGB along with depth information can be provided by range camera for every pixel, which provides fast matching for the registration between the laser and pixel point cloud, and finally achieves accurate 3D model for the environment with RGB and exact depth information. We evaluated our method on indoor scenes with different contents, and the results shown the effectiveness of the method to provide fast and accurate model for indoor environment with vision and depth information.

Keywords: 3D modeling, laser range finder, range camera, registration.

1 Introduction

Building of 3D model for environment provides exact description and obtains better understanding of the environment for mobile robot application, like object recognition, obstacle defining during the robot navigation in serve situation, and robot simultaneous localization and mapping problem in unknown environment. Many sensors have been applied for 3D modeling of environment, like monocular camera [1], laser finder [2-3], range camera [4], and pictures [5], etc. However, using of only one sensor cannot give complete and accurate description of the environment.

Hybrid of different sensors can achieve fusion and much more information about environment at one time [6]. For example, 3D model of environment by laser range finder can present environment with accurate 3D point cloud providing geometric structure but without textural or color information, and 3D model of environment by regular camera could not provide depth information. There are certain applications of hybrid of camera and laser range finder, but they are computational costly and complicated. With some consumer range cameras emerging recently, we present a method of integration of laser range finder and consumer range camera, which obtains a fast model of environment with both visualization and appearance. The 3D fusion model can be used for better scene understanding and environment mapping,

2   System Overview

3D laser ranging system consists of a 2D laser range finder LMS100 and a pan-lilt MV-5959. The 2D Laser range finder is mounted on the pan-tilt platform, and 3D point cloud can be achieved by laser finder sampling with pan-tilt pitching.

For the placement of laser range finder and depth camera, suppose geometric parameters of laser finder establishment are $(h, m)^T$, where h is the distance between the rotation axis of pan-tilt and the ground, m is the distance between the rotation axis of pan-tilt and laser ranger finder center. All these parameters should be calibrated before the test. For a spatial point detected by laser finder, the 3d polar coordinate is presented as $P(w, \beta, \alpha)^T$, where w is distance between the laser range finder center and the spatial point, β is horizontal scan angle of laser finder, and α is the vertical rotation angle of laser finder.

Laser scan point in Euclidean coordinate $P(x, y, z)^T$ can be calculated by:

$$\begin{bmatrix} x \\ y \\ z \end{bmatrix} = \begin{bmatrix} \sin\alpha\cos\beta & \cos\alpha \\ \sin\beta & 0 \\ -\cos\alpha\cos\beta & \sin\alpha \end{bmatrix} \begin{bmatrix} w \\ m \end{bmatrix} + \begin{bmatrix} 0 \\ 0 \\ h \end{bmatrix}. \tag{1}$$

3   Sensor Calibration

Calibration for multi-sensor is required for sensor data fusion. We conduct calibration of the laser range finder and depth camera separately. A checkerboard placing with different positions and angles against the ground were used as calibration scenes, and laser range finder and camera acquire data for every scene. Follow with the camera calibration and registration together of camera and laser finder (As shown in Fig.2).

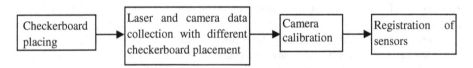

Fig. 1. Flow chart of registration of two sensors

3.1   Camera Calibration

A traditional camera calibration method is adopted in this paper [7]. Firstly, detect all the corners in calibration images, then five of cameras' internal and external parameters are obtained by solving a linear equation using orthogonal rotation matrix with ignoring the radial distortion of camera. Then, camera's radial distortion coefficient is estimated by least squares method. Internal and external parameters optimization is conducted by minimum re-projection error criteria. The equation of camera calibration model is represented as:

$$z_c \begin{bmatrix} u \\ v \\ 1 \end{bmatrix} = M_1 M_2 X_w = \begin{bmatrix} a_x & 0 & u_0 & 0 \\ 0 & a_y & v_0 & 0 \\ 0 & 0 & 1 & 0 \end{bmatrix} \begin{bmatrix} r_c & t_c \\ 0^T & 1 \end{bmatrix} \begin{bmatrix} X_w \\ Y_w \\ Z_w \\ 1 \end{bmatrix}. \tag{2}$$

Where M_1 is determined by a_x, a_y, u_0, v_0, which are internal camera parameters only related to internal construction of camera. M_2 is determined by transformation relationship between the camera and world coordinate system, which are external camera parameter.

3.2   Plane Parameter of Laser

For laser range finder frame, random sample consensus (RANSAC) algorithm is used to fit a plane to the laser 3D point cloud, where two plane parameters were estimated in one laser frame by finding the plane with maximal number of points [8]. The algorithm is shown as follows:

(1) Randomly select three points in the initial point cloud, and calculate the plane equation as: $ax + by + cz = d$, then calculate the distance between every point in point cloud to the plane i by $D_i = |xa_i + yb_i + zc_i - d_i|$.

(2) Accumulate the number of inliers to N when $D_i \leq t$, where t is threshold.

(3) Repeat until maximal inliers reaches for calibration plane.

(4) Plane fitting with these inliers points to get plane equation and then plane parameters (a, b, c, d) are estimated.

An example of calibration checkerboard for camera and laser was shown in Fig.2. Where image for camera calibration is shown in Fig.2(a), 3D point cloud by laser range finder is shown in Fig.2(b), and the checkerboard plane fitted is shown in Fig.2(c).

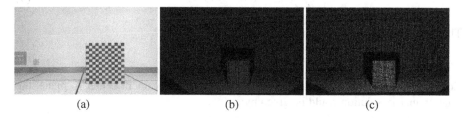

(a)   (b)   (c)

Fig. 2. Calibration checkerboard image used. (a) Camera image. (b) 3D point cloud by laser range finder. (c) Calibration plane fitted.

3.3   Registration Together with Laser and Camera

Registration of laser and camera together means to estimate the rigid transformation (rotation R and translation t) between the laser and camera frame. Suppose a point P_l in laser coordinate can register to one only corresponding point P_c in camera coordinate by:

$$P_c = R \cdot P_l + t. \tag{3}$$

Suppose the orientation of calibration image plane i in camera coordinate is $\theta_{c,i}$, and the distance between the plane and the camera is $S_{c,i}$, they are given by:

$$\theta_{c,i} = r_{c,i} \begin{bmatrix} 0 \\ 0 \\ -1 \end{bmatrix} = -r_{3,c,i}. \tag{4}$$

$$S_{c,i} = \theta_{c,i}^T \cdot t_{c,i}. \tag{5}$$

Where $R_{3,c,i}$ is the third line of rotation matrix of image frame i in camera coordinate system $r_{c,i}$, $t_{c,i}$ is translation vector of image frame i in camera coordinate. Suppose the orientation of calibration plane i in laser coordinate system is $\theta_{l,i}$, and the distance between the plane and the laser is $S_{l,i}$, they are given by:

$$\theta_{l,i} = \begin{bmatrix} a_i \\ b_i \\ c_i \end{bmatrix} \tag{6}$$

$$z_c S_{l,i} = d_i. \tag{7}$$

Then set laser and camera coordinate systems presented as:

$$\theta_c = [\theta_{c,1} \quad \cdots \quad \theta_{c,n}]^T, \alpha_c = [\alpha_{c,1} \quad \cdots \quad \alpha_{c,n}]^T$$
$$\theta_l = [\theta_{l,1} \quad \cdots \quad \theta_{l,n}]^T, \alpha_l = [\alpha_{l,1} \quad \cdots \quad \alpha_{l,n}]^T \tag{8}$$

Where, n is the number of calibration frames. Registration together with laser and camera is realized by minimizing the difference between the laser distance and camera transformation distance, which given by:

$$\min_t \sum_t [S_{l,i} - (S_{c,i} - \theta_{c,i}^T t)]^2 \tag{9}$$

Then we can estimate a solution for translation t_o by:

$$t_o = (\theta_c \theta_c^T)^{-1} \theta_c (S_c - S_l) \tag{10}$$

Considering minimizing the angle between the laser and camera center to plane, the rotation matrix solution could be given by:

$$R_o = \arg\max_R \sum_i \theta_{c,i}^T (R\theta_{l,i}) \tag{11}$$

For the initial transformation (R_o, t_o) we got above, iterative optimization is adopted to minimize the distance from the points selected by RANSAC to calibrated plane. Supposed $P_{l,i}$ presents the 3D point cloud of calibration plane in i[th] laser frame given by:

$$P_{l,i} = \begin{bmatrix} P_{l,i}^{(0)} & \cdots & P_{l,i}^{(m)} \end{bmatrix} \tag{12}$$

Where $P_{l,i} \subseteq R^{3*1}$, m $= m_i$ means the number of laser points in this laser frame. The optimization is given by minimizing the difference between the camera distance and laser transformation distance iteratively shown as:

$$\arg\min_{R,t} \sum_{i=1}^{n} \frac{1}{m_i} \sum_{i=1}^{n} [\theta_{c,i}^T (R x_{l,i}^{(j)} + t) - S_{c,i}]^2 \tag{13}$$

The depth information of each point in range camera finally was used to applied registration one more time with the transformation result above.

4   Fusion of Camera and Laser Data

For an environment scene, we obtain both camera image and laser frame using camera and 3D laser ranging system. For every point P_l in 3D laser point cloud, we project it to camera coordinate system by $P_c = R \cdot P_l + t$ using transformation parameters (R, t) achieved by registration above. Only the projected points that match to the pixel point in camera image will give color information defining as RGB. Then a 3D model of the scene can be formed by these points with RGB information, which visualize the scene by colored 3D model. An example of laser image and fusion image of a scene is shown in Fig.3.

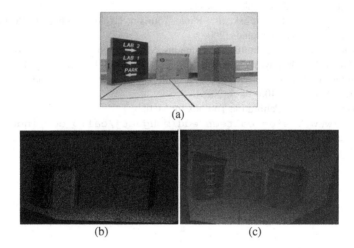

Fig. 3. Laser image and fusion image. (a) Color image from camera. (b) 3D laser point cloud. (4) Fusion image of color image and laser frame.

5   Conclusions

A method for 3D environment modeling with hybrid of laser range finder and range camera was proposed in this paper, which achieve 3D environment model with accurate depth information and color information for mobile robot application online. The combination of laser range finder and range camera was achieved by registration together of two sensors, and the depth point cloud from range camera was registered to laser range finder depth point cloud to optimize the transformation between the two sensor coordinate systems. The fusion of laser range finder and range camera was performed by projecting laser point cloud to range camera point cloud by rigid transformation calculated by registration of two sensors. Furthers work will work on alignment of different scenes.

Acknowledgments. Thanks to Foundation of Fujian Province of China (No. 2011J05159) for financial support.

References

1. Clemente, L., Davision, A., Reid, I., Neira, J., Tardos, J.: Mapping Large Loops with a Single-held Camera. In: Proceedings of Robotics: Science and Systems (2007)
2. Thrun, S., Burgard, W., Fox, D.: A Real-time Algorithm for Mobile Robot Mapping with Applications to Multi-robot and 3D Mapping. In: Proceedings of the IEEE International Conference on Robotics & Automation (2000)
3. Triebel, R., Burgard, W.: Improving Simultaneous Mapping and Localization in 3d Using Global Constraints. In: Proceedings of National Conference on Artificial Intelligence (2005)
4. Henry, P., Krainin, M., Herbst, E., Ren, X., Fox, D.: RGB-D Mapping: Using Depth Cameras for Dense 3D Modeling of Indoor Environment. In: Proceedings of International Symposium on Experimental Robotics (2010)
5. Snavely, N., Seitz, S., Szeliski, R.: Photo Tourism: Exploring Photo Collections in 3D. ACM Transactions on Graphics (2006)
6. Newman, P., Sibley, G., Smith, M., Cummins, M., Harrison, A., Mei, C., Posner, I., Shade, R., Schroter, D., Murphy, L., Churchill, W., Cole, D., Reid, I.: Navigating, Recognizing and Describing Urban Spaces With Vision and Laser. International Journal of Robotics Research 28(11-12), 1406–1433 (2009)
7. Bouguet, J.: Camera Calibration Toolbox for Matlab,
   http://www.vision.caltech.edu/bouguetj/calib_doc/index.html
8. Konolige, K., Agrawal, M.: FrameSLAM: From Bundle Adjustment to Real-time Visual Mapping. IEEE Transactions on Robotics 25(5), 1066–1077 (2008)

Advanced Configuration Design for Industry Control Based on OPC Technology

Hui Zhao

School of Information Engineering, Zhengzhou University, Zhengzhou 450001, China

Abstract. With the enlarging of the production scale, the top of the monitoring software need connect more the lower local equipment to achieve process monitoring and control, this needs more software support of device driver, and the change of the hardware's features need to write another new driver, which undoubtedly increase the burden of the software developers, at this time, OPC standard arises in this market condition. OPC provides a standard, open, seamless connection and unified interface. In this article, they developed a configuration program based on OPC technology. That is, through the definition of OPC equipment, which is done in the Kingview, people can put the site signal and monitoring and control software conveniently to link up, which can realize data exchange and easily sharing.

Keywords: OPC, configuration software, Kingview, monitoring and control.

1   Introduction

In the production site, people need to monitor and control data of site equipment and running situation on time ,which is able to master the accurate production status ,quickly find the equipment failure and improve labor productivity .However, our ultimate goal is to maximize industrial efficiency. The application of supervision and configuration software can realize the monitoring and control of production on time. The configuration software gets real-time data through the I/O drivers from the I/O devices, then after necessary processing to the data, on the one hand, it directly displays graphics on PC screen, on the other hand, according to configuration requirements and operation instruction, it sends data to the I/O device and then it adjusts the control parameters. With the enlarging of the production scale, the top of the monitoring software need connect more the lower local equipment to achieve process monitoring and control, this needs more device driver software support, software developers need research a different driver according to distinct hardware equipment characteristics, and the change of the hardware's features need to write another new driver, which undoubtedly increase the burden of the software developers, making them unable to devote themselves to their core product development ,at this time ,OPC standard arises in this market condition.

In the field of industrial control, data exchange is the core of the system operation, between different manufacturer monitoring software and site equipment, OPC provides a standard, open, seamless connection unified interface. Through

D. Jin and S. Lin (Eds.): Advances in FCCS, Vol. 2, AISC 160, pp. 403–407.
springerlink.com   © Springer-Verlag Berlin Heidelberg 2012

standard OPC interface, they will conveniently make the signal and monitoring software, man-machine interface software link and realize data exchange and easily sharing. After Using OPC standard, instead of software developers, hardware developers research unified interface program for their product, which avoiding lots of problems such as repetitive developments, various drivers among different developers, not supporting hardware features change, access conflict and so on.

2   OPC Technology Theory and Its Application

2.1   The Core of the OPC Technology

OPC (OLE for Process Control) is formulated and advocated a nonprofit technology standard by the OPC Foundation Organization .OPC is based on Microsoft's OLE (now Active X), COM and DCOM technology, owning the flexibility and efficiency, and supporting all of the programming language. It defines a set of common, high performance standard interface. This specification greatly increases interoperability among automation system, the equipment system and commercial office system.

The realization of the OPC technology includes two parts, OPC server and client application of OPC. For hardware manufacturers, they generally need develop OPC server which is suitable for the hardware communication, for configuration software, it requires to develop OPC customer program. OPC customer application which meets the OPC standard can access any OPC server programs from any manufacturers.

OPC server realizes its object and interface. OPC server is composed of Server, Group and Item. Triadic relation is contained in turn. OPC server object is used to provide relevant information of server object itself, and as the container of OPC group. OPC group offers relevant information about group object itself, and gives mechanism of organizing and administrating item. Data item is the most small logic unit of reading date and writing date. OPC Item is a representative of physical connection from OPC service to data source. OPC term cannot be directly visit by OPC client program, because there is not COM interface corresponding to item in the OPC rules, all items visit can be achieved through OPC group which contains OPC item. Each data item includes three variables: Value, Quality and Time Stamp. Data value is in VARIANT forms. There are two kinds of interfaces in the OPC standard: Custom Interface which is used by the client and server programmers and Automation Interface which supports

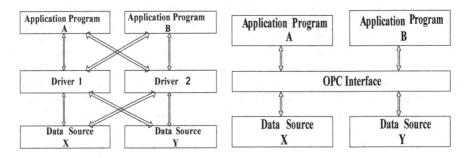

Fig. 1. Traditional data exchange   **Fig. 2.** Data exchange under OPC standard

the development of commercial application. The former is provided by OPC server, the latter may be not offered. OPC customer application realizes the communication with equipment through so two kinds of OPC interface in industrial control network.

2.2 Application of OPC Technology in the Configuration Software

In September 1997, the OPC standard version 1.0 has first released, it broke the traditional way of DDE and occupies the dominant status with its peculiar openness, interconnectivity and high efficiency. And OPC server accessing to process data can overcome heterogeneous network structure and the difference between the network protocols. The rapid development of OPC technology not only makes configuration software own a wider application domain expansion and permeability, and promote the configuration software more standardization. In practical applications, most configuration software have their own proprietary real-time database and historical database, the databases are exploited by developers, if people want to visit databases, they have to write different codes, and only through calling the API functions provided by the developers or some other special way, which no doubt is very complicated. If the developers who design proprietary database of the monitoring software also can provide a OPC server accessing the database at the same time, the client will not need know interface characteristics, it is ok that as long as people write client service program following OPC standard, which greatly simplifies the development process.

The communication between OPC server and configuration software is a process that the client accesses OPC server, in fact, it is the process which a typical customer visit out-of-process COM components. It can be seen that the customer program and COM components program communicates with each other through the interface.

At present, domestic popular configuration software companies are Rockwell's Rsview32, Intellulion's IHX, the Simon's Wincc, Beijing Kunlun's Mcgs, respectively, etc. The configuration software support all OPC technologies, they not only can respectively be OPC server and also provide man-machine interface and software test which are both as the OPC clients , because of the openness of the OPC technology itself, the client of any system can visit any other system's data server, which have realized remote data transmission.

Obviously, the continuous increasing of OPC international foundation members and the sustaining improvement of OPC communication performance are important guarantee that whether it will can keep a foothold in communication protocols of configuration software at a long term or not.

3   Monitoring and Configuration Program Development Based on OPC Technology

Taking Kingview 6.53 as an example, this paper develops a monitoring and configuration program based on OPC technology. Kingview 6.53 is software platform and development environment which both belong to the monitoring layer of automatic control system, provide users with general level of software tools which can rapidly construct monitoring function of industrial automatic control system by using flexible configuration mode. Kingview 6.53 has the standard OPC interfaces , also it can be used as both the OPC client and OPC server to provide services to other OPC client, the system sees it as OPC client according to need in this paper.

Kingview 6.53 can also link any more the OPC Servers, each server is taken as an external device. And people can define, add or delete it, just like a PLC or instrument equipment as well. After simple configuration to DCOM of the computer, Kingview can also act as OPC server. In this paper, two computers, which are both loaded Kingview to be treated as the server and the client respectively, connect with each other by cable, in order to simulate to send and receive data between Kingview and OPC server.

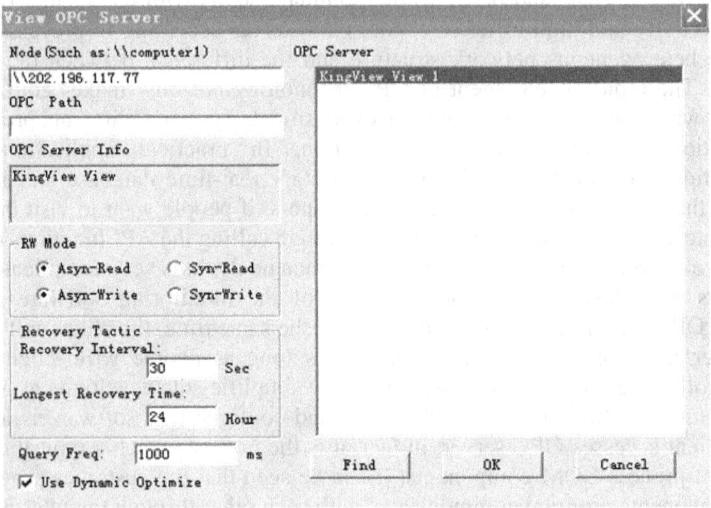

Fig. 3. Setting of the OPC server

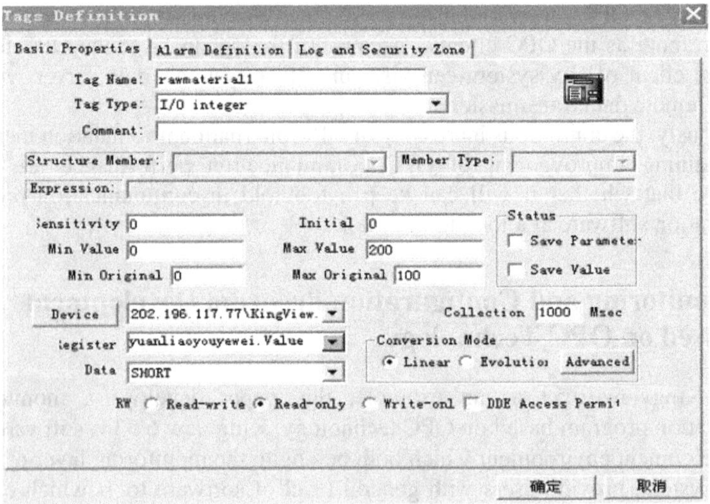

Fig. 4. Definition of variables

Generally speaking, the engineering person need firstly define physical parameters of communication in the OPC server, then, they need the definition of configuration software variables corresponding to the lower level machine variables (data item). Here, first of all, they define OPC equipment.

Then, people open Kingview 6.53, which means the start of OPC client. After that, it is time to define the variables and lower level machine variables.

In the operation system, Kingview connects the corresponding OPC server in order to automatically complete data exchange with OPC server.

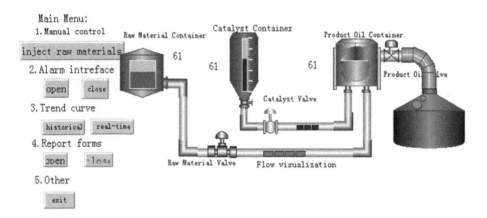

Fig. 5. The main monitoring interface

4   Conclusion

This paper briefly introduces the OPC technology's kernel and application of configuration software, and it gives an example of monitoring the configuration program, after many trials , the system has high refresh rate of data, good real-time performance and fine stability, which has achieved the expected purpose.

References

1. Zhang, M.: Research and Development on OPC Client and Server Based on New DCS. Shandong University Master's Thesis (2010)
2. Qi, J., Tang, W.: Software design for industry control based on OPC specification. Journal of Shandong University of Architecture and Engineering 18(4), 68–70 (2003)
3. Wang, J.: The Configuration Software Development Based on OPC and Its Application on Telecom Power Supervisory Control System. Guangdong University of Technology Master's Thesis (2008)

Nonlinear Synchronization of New Hyperchaotic System

Yan Yan and Longge Zhang

Department of Mathematics and Physics, North China Electric Power University,
Baoding 071003, P.R. China
longgexd@163.com

Abstract. In this paper, a nonlinear synchronization scheme for a new hyperchaotic system is presented. Based on the Lyapunov stability theory, the error system is stable under the proposed method, and the error states tend to zeros rapidly. Finally numerical simulations are provided to show the effectiveness and feasibility of the developed method.

Keywords: synchronization, hyperchaotic system, nonlinear.

1   Introduction

Chaos is an interesting phenomenon of nonlinear systems. A deterministic chaotic system is remarkable for its character of sensitive to the change in initial conditions[1]. Chaos synchronization has attracted a great deal of attention ever since Pecora and Carroll [2] established a chaos synchronization scheme for two identical chaotic systems with different initial conditions. And then it has been extensively studied due to its potential application in technological applications [3–6]. On the other hand, hyperchaos system characterized with more than one positive Lyapunov exponent [7], and its adoption of higher dimensional chaotic systems enhances the security of communication scheme so it has attracted more and more attention. Various effective methods have been presented to synchronize various chaotic and hyperchaotic systems, for example, adaptive control [8–9], linear and nonlinear feedback control [10–12], active control [13], adaptive feedback control [14] and so on.

In this paper, the nonlinear synchronization of a new hyperchaotic system is invested. In section 2, the new hyperchaotic system is introduced. In section 3, the nonlinear synchronization of the system is invested, and the simulation results of the method is included to verify the method in section 4. Section 5 gives the conclusion.

2   New Hyperchaotic Systems

In paper [15], a new four-scroll hyperchaotic system is evolved from a novel 4D nonlinear smooth autonomous system

D. Jin and S. Lin (Eds.): Advances in FCCS, Vol. 2, AISC 160, pp. 409–413.
springerlink.com   © Springer-Verlag Berlin Heidelberg 2012

$$\begin{cases} \dot{x} = ax - yz \\ \dot{y} = xz - by \\ \dot{z} = cxy - dz + gxw \\ \dot{w} = kw - hy \end{cases} \tag{1}$$

Where $x, y, z, w \in R$ are the state variables, and a, b, c, d, g, h and k are systems' positive constant parameters. When $a = 8, b = 40, c = 2, g = 5, h = 0.2, k = 0.05$ and $d = 14$, the system displays a single four-scroll hyperchaotic attractor. The four-scroll hyperchaotic attractor is figured in Fig.1(a)-(d).

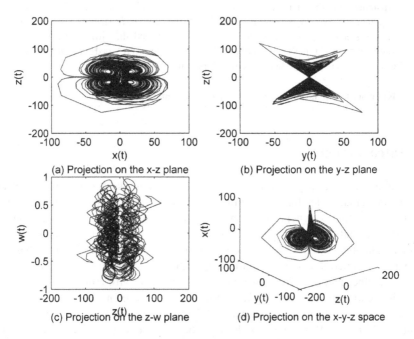

Fig. 1. Phase portraits of the four-scroll hyperchaotic attractor for a=8,b=40,c=2,d=14, g=5,h=0.2,k=0.05

3   Nonlinear Synchronization Controllers

The synchronization of the two identical systems is investigated. The objective of synchronization is to control the error between the slave system and master system tend to zero. For system (1), the master and slave systems can be defined as followed:

$$\begin{cases} \dot{x}_1 = ax_1 - y_1 z_1 \\ \dot{y}_1 = x_1 z_1 - by_1 \\ \dot{z}_1 = cx_1 y_1 - dz_1 + gx_1 w_1 \\ \dot{w}_1 = kw_1 - hy_1 \end{cases} \tag{2}$$

and

$$\begin{cases} \dot{x}_2 = ax_2 - y_2 z_2 + u_1 \\ \dot{y}_2 = x_2 z_2 - by_2 + u_2 \\ \dot{z}_2 = cx_2 y_2 - dz_2 + gx_2 w_2 + u_3 \\ \dot{w}_2 = kw_2 - hy_2 + u_4 \end{cases} \tag{3}$$

where the scripts 1 and 2 stand for the master and slave system respectively, and u_1, u_2 u_3 and u_4 are the controllers that are designed to be synchronized of the two hyperchaotic systems.

Define the systems' errors as $e_1 = x_2 - x_1, e_2 = y_2 - y_1, e_3 = z_2 - z_1, e_4 = w_2 - w_1$. Then the error system

$$\begin{cases} \dot{e}_1 = ae_1 - y_1 e_3 - z_2 e_2 + u_1 \\ \dot{e}_2 = -be_2 + x_2 e_3 + z_1 e_1 + u_2 \\ \dot{e}_3 = -de_3 + cx_2 e_2 + cy_1 e_1 + gx_2 e_4 + ge_1 w_1 + u_3 \\ \dot{e}_4 = ke_4 - he_2 + u_4 \end{cases} \tag{4}$$

The controller $U = [u_1, u_2, u_3, u_4]^T$ can be selected as follows:

$$\begin{aligned} u_1 &= -(a+1)e_1 + y_1 e_3 + z_2 e_2 \\ u_2 &= -x_2 e_3 - z_1 e_1 \\ u_3 &= -cx_2 e_2 - cy_1 e_1 - gx_2 e_4 - ge_1 w_1 \\ u_4 &= -(k+1)e_4 + he_2 \end{aligned} \tag{5}$$

Using equ.(4), (5) and (6), the error system is

$$\begin{cases} \dot{e}_1 = -e_1 \\ \dot{e}_2 = -be_2 \\ \dot{e}_3 = -de_3 \\ \dot{e}_4 = -e_4 \end{cases} \tag{6}$$

Select the candidate Lyapunov function of system (8) as

$$V = (e_1^2 + e_2^2 + e_3^2 + e_4^2)/2 \tag{7}$$

Then its derivative is

$$\dot{V} = e_1 \dot{e}_1 + e_2 \dot{e}_2 + e_3 \dot{e}_3 + e_4 \dot{e}_4 = -e_1^2 - be_2^2 - de_3^2 - e_4^2 < 0 \tag{8}$$

Based on the Lyapunov stability theory, the error system is stable. Then the synchronization of the slave and master system is released.

4   Numerical Simulation

In this section, to verify and demonstrate the effectiveness of the method in section 3, we will display the numerical results for the master and slave systems. In the simulations, we select the parameters of hyperchaotic system as $a = 8, b = 40$, $c = 2$, $g = 5$ $h = 0.2, k = 0.05, d = 14$. The initial values of the drive and response systems are $[10,1,10,1,12,8,0,9]$. Fig.2 displays the effectiveness of the proposed method.

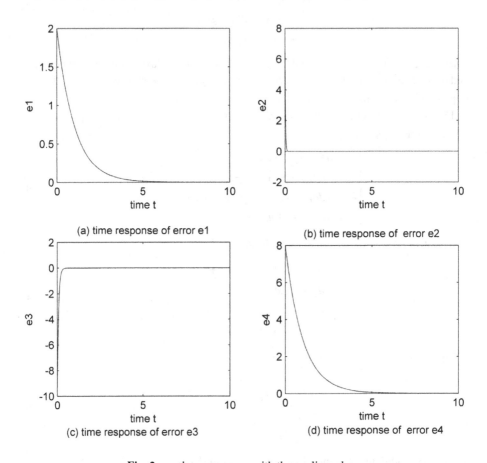

Fig. 2. synchronous errors with the nonlinear law

References

1. Chen, G., Dong, X.: From chaos to order. World Scientific (1998)
2. Pecora, L.M., Carroll, T.L.: Synchronization in chaotic systems. Phys. Rev. Lett. 64, 821–824 (1990)
3. Xiao, M., Cao, J.: Synchronization of a chaotic electronic circuit system with cubic term via adaptive feedback control. Commun. Nonlinear Sci. Numer. Simul. 14, 3379–3388 (2009)
4. Posadas-Castillo, C., Cruz-Hernández, C., López-Gutiérrez, R.M.: Synchronization of chaotic neural networks with delay in irregular networks. Appl. Math. Comput. 205, 487–496 (2008)

5. Boukabou, A.: On the control and synchronization design for autonomous chaotic systems. Nonlinear Dynamics and Systems Theory 8(2), 151–167 (2008)
6. Guo, H.-M., Zhong, S.-M., Gao, F.-Y.: Design of PD controller for master–slave synchronization of Lur'e systems with time-delay. Appl. Math. Comput. 212(1), 86–93 (2009)
7. Murali, K., Tamasevicius, A., Mykolaitis, G., Namajunas, A.: Hyperchaotic System with Unstable Oscillators. Nonlinear Phenomena in Complex Systems 3(1), 7–10 (2000)
8. Park, J.: Adaptive synchronization of hyperchaotic Chen system with uncertain parameters. Chaos, Solitons Fractals 26, 959–964 (2005)
9. Tang, Y., Fang, J.: Adaptive synchronization in an array of chaotic neural networks with mixed delays and jumping stochastically hybrid coupling. Commun. Nonlinear Sci. Numer. Simul. 14, 3615–3628 (2009)

10. Zhang, Q., Lu, J.: Chaos synchronization of a new chaotic system via nonlinear control. Chaos, Solitons Fractals 37, 175–179 (2008)
11. Tian, L.X., Xu, J., Sun, M., et al.: On a new time-delayed feedback control of chaotic systems. Chaos, Solitons and Fractals 39, 831–839 (2009)
12. Yau, H.-T., Yan, J.-J.: Chaos synchronization of different chaotic systems subjected to input nonlinearity. Appl. Math. Comput. 197(2), 775–788 (2008)
13. Naseh, M.R., Haeri, M.: Robustness and robust stability of the active sliding mode synchronization. Chaos, Solitons Fractals 39, 196–203 (2009)
14. Zhou, X.B., Wu, Y., Li, Y., et al.: Adaptive control and synchronization of a novel hyperchaotic system with uncertain parameters. Appl. Math. Comput. 203, 80–85 (2008)

A Design of Data Integration Using Cloud Computing

Yushui Geng[*] and Jisong Kou

School of Management, Tianjin University, Tianjin, China
gys@spu.edu.cn

Abstract. To better solve the complexity and diversity in data integration, we propose a new idea of data integration platform based on SaaS service. Firstly, we analyze the situation of today' data integration. Secondly, We give the model architecture of data integration ant its work principles. Thirdly, we analyze the model's characters. This model of platform is more intelligent, efficient and personalized in solving complicate data integration.

Keywords: data integration, cloud computing, SaaS.

1 Introduction

Cloud computing in some form or another can be traced back to the early sixties. Cloud computing has been in existence in some form or the other since the beginning of computing. Cloud computing is an "all encompassing" term that includes any service that is delivered over the network, says Tucker [1]. However, the advent of vastly improved software, hardware and communication technologies has given special meaning to the term cloud computing and opened up a world of possibilities. It is possible today to start an ecommerce or related company without investing in datacenters. This has turned out to be very beneficial to startups and smaller companies that want to test the efficacy of their idea before making any investment in expensive hardware. Corporations like Amazon, SalesForce.com, Google, IBM, Sun Microsystems, and many more are offering or planning to offer these infrastructure services in one form or another. An ecosystem has already been created and going by the investment and enthusiasm in this space the ecosystem is bound to grow [2].

This thesis tries to define and explain a model of data integration platform that can provide SaaS service about data in cloud or enterprise internal databases. This platform tries to include a heavy dose of data synchronization and data replication, which enables both a multi-step integration process between internal and external resources, and that archives data across service provider applications and internal databases.

---

[*] Biographies: Yushui Geng, born in 1965, male, Jinan people of Shandong Province, professor, Doctor's degree graduate, research field in information management and information system; Jisong Kou, born in 1947, male, Tianjin people, professor, Doctor's degree, doctoral supervisor.

D. Jin and S. Lin (Eds.): Advances in FCCS, Vol. 2, AISC 160, pp. 415–419.

2   The Situation of Today's Data Integration

Recommendation weight Data integration involves combining data residing in different sources and providing users with a unified view of these data [3]. This process becomes significant in a variety of situations, which include both commercial (when two similar companies need to merge their databases) and scientific (combining research results from different bioinformatics repositories, for example) domains. Data integration appears with increasing frequency as the volume and the need to share existing data explodes [4]. It has become the focus of extensive theoretical work, and numerous open problems remain unsolved.

Demand for the cloud is certainly on the upswing, but could easily be derailed if it turns out that it can't function with legacy infrastructure without putting data through a lot of complicated processes. Data integration across internal and cloud platforms is crucial if the technology is to live up to its promise as a global resource pool capable of extreme scalability and long-distance load-balancing prowess. But now that more organizations are gaining experience on the cloud, the industry seems poised to take the next step- integrating cloud services with legacy enterprise infrastructure. After all, few organizations will simply junk systems and technology that they've spent millions to acquire just to port everything over to someone else's data center.

3   Design of Data Integration Service Platform

Data stored in the cloud is not easily accessible for integration with applications *not* residing in the cloud, which can definitely be a roadblock to adopting public cloud computing. The availability of and ease of access to data stored in the public cloud for integration, data mining, business intelligence, and reporting - all common enterprise application use of data - will certainly affect adoption of cloud computing in general. The benefits of saving dollars on infrastructure (management, acquisition, maintenance) aren't nearly as compelling a reason to use the cloud when those savings would quickly be eaten up by the extra effort necessary to access and integrate data stored in the cloud.

You can also just leave that data out there in the cloud, implement, or take advantage of if they exist, service-based access to the data and integrate it with business processes and applications inside the data center. You're relying on the availability of the cloud, the Internet, and all the infrastructure in between, but like the solution for integrating with salesforce.com and other SaaS offerings, this is likely the best of a set of "will have to do" options.

The issue of data and its integration has not yet raised its ugly head, mostly because very few folks are moving critical business applications into the cloud and admittedly, cloud computing is still in its infancy. But even non-critical applications are going to use or create data, and that data will, invariably, become important or need to be accessed by folks in the organization, which means access to that data will - probably sooner rather than later - become a monkey on the backs of IT.

3.1   Software as a Service (SaaS)

This is defined as the delivery of an application across the internet. The application is hosted as a web application or service and is made available to users either through a browser interface or via well known web service interfaces. Some of the services offered are free. The distinction between SaaS and earlier applications delivered over the Internet is that the latter were developed specifically to leverage web technologies such as the browser and provide an alternate user interface to the user. The browser solved the problem of thick clients allowing applications to write a single front end that worked on all browsers. The users did not need to install any client code. With SaaS the applications are now services that can be accessed from a browser as well as via APIs from other services. According to Pete Koomen, a product manager at Google, advances in SaaS were possible because of the inroads made by rich web applications brought about by advances in web2.0 technologies. This increased the use of web-based applications like mail, calendar, maps etc. that eventually led to these applications being exposed as services [5, 6].

In this model of data integration platform, we use SaaS technology. All the functions are provided as the services by the data integration.

3.2   The Model of Data Integration Platform

The new design model of data integration platform is shown as Fig. 1.

Resource Data and Result Data are two interface of the platform. These two interfaces are in the form of service. Among them, the interface of Resource Data is used to get the data from the enterprise internal systems or the data in the cloud to the data integration platform. The interface of Result Data is used to put the data to user in the form of service. The interface of Resource Data is similar to the adapter that can visit various kinds of data resources.

In the data integration platform, data are filtered and combined. Finally the data are produced by the users' requirements. Users call SaaS service to get the data that they wanted.

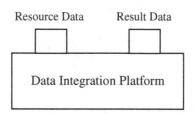

Fig. 1. The module architecture of data integration platform

3.3   The Work Principle of Data Integration Platform

By the Resource Data service, the platform gets the needed data. Every data resource is described and recorded as one Data Node. The Data Node includes data resource address, kinds and raw data object. The raw data maybe are from the enterprise internal system database. Maybe they are from a private or public cloud. So this part would be achieved very difficultly.

Data Table consists of part raw data in Data Node and statistical information for example average values. The procedure is according to the data filter and processing principles that are made by users.

Targets are the data for the Result Data service. In another words, targets are some data that users needs. These data are red by calling Result Data service. This procedure is according to the Data Joint principles that are also made by users.

For data integration users, the platform is interactive and open. This platform is presented to users in the form of SaaS software. Users do not store the raw data, middle data and target data in their machines.

In addition, graphical interface would be presented when this platform is running. Users select and set Data Node's name, address, kinds and data content. Then users can define the Data Table's architecture and the corresponding principles. Later users can make the Data Joint principles. The target data are produced automatically.

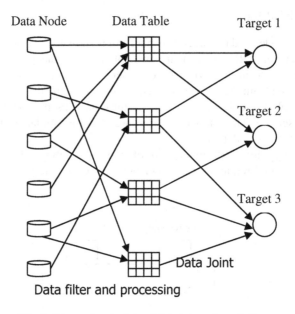

Fig. 2. The work principle of data integration platform

4   Analysis of Data Integration Service Platform

This model of data integration service platform solves the problem how to read data in the cloud at enterprise internal nodes and how to combine the enterprise internal system data with the data in software as service (SaaS) software. This data integration platform is designed based on service concept. So this data integration platform is smart, extensible and open.

5   Conclusions

In this paper, we designed a novel and efficient platform of data integration using cloud computing. In the following research, we would further explore the details of the Resource Data service. The efficiency when the service of Result Data is running is also an important problem.

References

1. Mathew, R., Spraetz, R.: Test Automation on a SaaS Platform. In: 2009 International Conference on Software Testing Verification and Validation, ICST, pp. 317–325 (2009)
2. Godse, M., Mulik, S.: An Approach for Selecting Software-as-a-Service (SaaS) Product. In: 2009 IEEE International Conference on Cloud Computing, CLOUD, pp. 155–158 (2009)
3. Lenzerini, M.: Data Integration: A Theoretical Perspective. In: PODS 2002, pp. 233–246 (2002)
4. Ullman, J.D.: Information Integration Using Logical Views. In: Afrati, F.N., Kolaitis, P.G. (eds.) ICDT 1997. LNCS, vol. 1186, pp. 19–40. Springer, Heidelberg (1997); Hai, H., Sakoda, S.: SaaS and integration best practices. Fujitsu Scientific and Technical Journal, 45(3), pp. 257-264 (July 2009)
5. Cao, Y., Zhang, M.: Integration of Enterprise Application Based on SOA. In: World Congress on Software Engineering, WCSE, vol. 3, pp. 227–231 (2009)
6. Delin, Q.: Design of Medical Insurance Supervision System Based on Active Data Warehouse and SOA. In: 2009 WRI World Congress on Computer Science and Information Engineering, CSIE, vol. 3, pp. 45–49 (2009)
7. Moore, B., Mahmoud, Q.H.: A service broker and business model for saas applications. In: 2009 IEEE/ACS International Conference on Computer Systems and Applications, AICCSA, pp. 322–329 (2009)
8. Erl, T.: SOA Design Patterns. Prentice Hall PTR, Upper Saddle River (2009)
9. Atkinson, M.P., van Hemert, J.I., Han, L., Hume, A., Liew, C.S.: A distributed architecture for data mining and integration. In: Proceedings of the Second International Workshop on Data-aware Distributed Computing, Garching, Germany, June 09-10, pp. 11–20 (2009)
10. Gu, Q., van Vliet, H.: SOA decision making - what do we need to know. In: Proceedings of the 2009 ICSE Workshop on Sharing and Reusing Architectural Knowledge, May 16, pp. 25–32 (2009)
11. Blanton, J., Leski, S., Nicks, B., Tirzaman, T.: Making SOA work in a healthcare company. In: Proceeding of the 24th ACM SIGPLAN Conference Companion on Object Oriented Programming Systems Languages and Applications, Orlando, Florida, USA, October 25-29 (2009)

Towards a Method for Management Information System Based SOA

Yushui Geng[*] and Jisong Kou

School of Management, Tianjin University, Tianjin, China
gys@spu.edu.cn

Abstract. Recently SOA is a research hot because it is open and flexible. But all people only concern its application in a certain field. Almost no one puts forward a general method for different field systems. Also there are no service-oriented developing software and tools. This paper proposes a method for the development of management information systems based on SOA. Business process can be designed by interactive interface. Users can complete the system design by simply mouse dragging action.

Keywords: SOA, MIS, Web Service.

1 Introduction

Service-orientation is the new architectural paradigm that answers the challenge by centering on XML and Web services standards that revolutionize how developers compose systems and integrate them over distributed networks. This paper proposes a Service orientation solution to the management information system based on SOA (Service-oriented architecture).

It is obvious that the world is now a global village enabled by improved information technology and further increases in the society's dependence on computing appear inevitable. Consequently upon this fact, the study of trends and developments in the technology industry cannot be overemphasized. The application of technology to business processes has to a great extent changed the structure and mode of operations in organizations. However the critical path to success is not the technology itself, but its effective application to various elements of the business, as the failure of a new technology is not in the elements of the technology itself but its application. Several technologies which would have benefited organizations by improving their business processes failed as a result of the way those technologies were applied at that point in time. Why are new technologies been developed in the Information Technology circle everyday despite the ample of them that we have around? Over the past decades, the IT industry has been battling with the maintenance cost associated with mainframe managed systems which are a considerable drain on

---

[*] Biographies: Yushui Geng, born in 1965, male, Jinan people of Shandong Province, professor, Doctor's degree graduate, research field in information management and information system; Jisong Kou, born in 1947, male, Tianjin people, professor, Doctor's degree, doctoral supervisor.

D. Jin and S. Lin (Eds.): Advances in FCCS, Vol. 2, AISC 160, pp. 421–426.
springerlink.com   © Springer-Verlag Berlin Heidelberg 2012

IT department budgets in any business organization. These high costs are results of interactions between the outdated platforms and non-mainstream languages on which these legacy systems are built. Despite this weakness, fewer organizations are ready to take the step towards change while some don't want to even think about it. The main problem is that these systems had always been considered too valuable and costly to risk modernized or changed.

This paper proposes a method for the development of service-oriented management information systems.

2   Service-Oriented Software Architecture Theory

Service-oriented architecture is about the evolution of business processes, applications and services from today's legacy-ridden and smooth integration of disparate applications to a world of connected businesses, accommodating rapid response to change and utilizing vast degrees of business automation. SOA can be viewed as a computing methodology or approach to building IT systems in which business services i.e. services provided by an organization to clients are the key organizing principles used to align IT systems with the needs of the business. It is a set of general design principles that enables organizations to change business processes on the wing and respond to the shifting demands of the business in a manner that would be impractical or cost-prohibitive using conventional application development and resources allocation [1].

2.1   The Overview of SOA

SOA is not a specific language or technology, but an architecture pattern and software design thought.

SOA is an attempt to provide a set of principles or governing concepts that are used during the phases of systems development and integration. Such architecture attempts to package functionality as interoperable services within the context of the various business domains that use it. Several departments within a company or different organizations may integrate or use such services - software modules provided as a service -even if their respective client systems are substantially different. SOA is an attempt to develop yet another means for software module integration. Rather than defining an API, SOA defines the interface in terms of protocols and functionality. An endpoint is the entry point for such an SOA implementation [2].

Service-orientation requires loose coupling of services with operating systems, and other technologies that underlie applications. SOA separates functions into distinct units, or services [3], which developers make accessible over a network in order to allow users to combine and reuse them in the production of applications. These services communicate with each other by passing data from one service to another, or by coordinating an activity between two or more services.

Services generally adhere to the following principles of service-orientation [4]:

1) abstraction
2) autonomy
3) composability
4) discoverability

5) **formal contract**
6) **loose coupling**
7) **reusability**
8) **statelessness**

2.2 Web Services and SOA

Architectures can operate independently of specific technologies. Designers can implement SOA using a wide range of technologies [5, 6, 7], including:

1) **SOAP, RPC**
2) **REST**
3) **DCOM**
4) **CORBA**
5) **Web Services**
6) **WCF (Microsoft's implementation of Web service forms a part of WCF).**

Earlier SOA used COM or ORB based on CORBA specifications and recent SOA stresses on web services using standard description (WSDL), discovery (UDDI) and messaging (SOAP). Service oriented architecture may or may not use web services. Many implementers of SOA have begun to adopt an evolution of SOA concepts into a more advanced architecture called SOA 2.0 [8].

Web services can implement a service-oriented architecture. Web services make functional building-blocks accessible over standard Internet protocols independent of platforms and programming languages. These services can represent either new applications or just wrappers around existing legacy systems to make them network-enabled.

Each SOA building block can play one or both of two roles:

1) Service Provider
The service provider creates a web service and possibly publishes its interface and access information to the service registry.

2) Service consumer
The service consumer or web service client locates entries in the broker registry using various find operations and then binds to the service provider in order to invoke one of its web services.

The features of Web Service are very suitable for the realization of SOA architecture [9].

2.3 Benefits of SOA

SOA provides a layer of abstraction that enables an organization to continue leveraging its investment in IT by wrapping these existing assets as services that provides business functions.

1) **Easy Integration and Complexity Management.** The integration point in Service Oriented Architecture is the service specification and not the implementation. This provides implementation transparency and minimizes the impact when infrastructure and implementation changes occur [10].

2) **Responsively and Faster Time to Market.** The ability to compose new services out of existing ones provides a distinct advantage to an organization that has to be agile to respond to demanding business needs. This leads to rapid development of

business services and allows an organization to respond quickly to changes and reduce its time-to-market.

3) Cost and Increased Reuse. With core business services exposed in a loosely coupled manner, they can be more easily used and combined based on business needs. This means less duplication of resources, more potential for reuse, and lower costs [11].

4) Improved Flexibility. Flexibility concerns the ability of solutions to be altered in the face of changing business and technology requirements. SOA provides the flexibility and responsiveness that is critical to businesses to survive and thrive.

5) Division of Responsibility. Service-oriented development aims to separate business logic from the data thereby gives the ability to more easily allow business people to concentrate on business issues, technical people to concentrate on technical issues, and for both to collaborate using the service contract.

The existing mainframe systems also called legacy systems are not capable of some of the mentioned benefits. For example, the cost of maintenance a legacy system is very expensive and does not support critical business application because business logic and data are not separated. Organizations potentially can continue getting value out of existing resources instead of having to rebuild a new system from the scratch if they could employ an effective migration path from the legacy systems to a service based system.

3   The New Method of Service-B

The Service-B refers to the block of service. It is Similar to the component in the object oriented technology. The main idea of the method of Service-B is to provide a way to allow users to assemble existing service resources and new service resources according to their individual needs promptly. The method Service-B includes two levels: interface level and library level shown as Fig. 1.

Drag and set parameter: The business users run Service-B system and drag interface elements from logical components lists to design application system. The interface elements include the front program of existing system, the affairs components of existing (developed by CORBA/RMI/DCOM) and the new subsystem. The business users could set parameters of interface elements according to specific application.

Logical express: In order to facilitate the users' operation, it is needed that all kinds of services are expressed logically as the dynamic icons.

Business process: This method is based on business process-driven. The system users need to analyze and design the system by business process mode. The future system likely includes new function modules in addition to the functions in existing system.

Service description: New subsystems or new function modules are described by WSDL (Web Services Description Language) language. The existing system also are described by WSDL whether they were developed by CORBA/RMI/DCOM or not. In service-oriented architecture there is only the concept of service.

Service issued: All of web services are registered in UDDI(Universal Description, Discovery and Integration) centre.

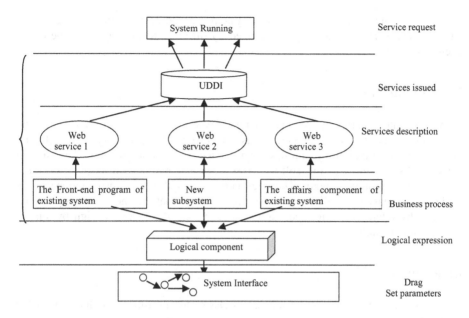

Fig. 1. The Method of Service-B

Service request: When the application system is running, interface elements can send service requests to UDDI so that to get the needed service.

The service library contains various services. The services can be general ones for all fields. The services also can be professional for the field bank, rails, Gis and, etc.

4   Analysis of the New Method

The outstanding advantages of this method lie in the following two aspects:

Firstly, this development method greatly lessens the requirements for developers and greatly shortens the development time. The more important thing is to allow users to make their efforts to research the system business instead of the new technology and the implementation details. The user's main work is analyzing business process, rather than learning the development of language and studying development tool.

Secondly, users can describe the business requirements simply and can construct the application system at high-speed and low-cost by dragging. This method breaks the barrier of collaboration between the computer experts, domain experts, business designers and business executives. By business-driven mode we can build the required services and their associations of various services.

The method of Service-B has the following characteristics:

First: service as the center. Service is taken as the centre of service registration, search, description and calls.

Second: on-demand service. Users can dynamically look for and combinate one or several services to complete a specific function.

Third: openness. All kinds of legacy systems can be easily encapsulated as web services. The various subsystems developed by different technologies can be easily integrated.

5   Conclusions

SOA is the latest methodology to solve the software complexity and relevance in the long-standing evolutionary process. This paper presents a method of developing management information system based on service-oriented architecture. This system model can reuse the existing software resources and reduces the cost of system's design and development again. The intelligence and agility of information system is evidently increased. Finally, the most important is that business process can be designed by interactive interface. Users can complete the system design by simply mouse dragging action.

References

1. Davis, Z.: Strategies for SOA Success. Ziff Davis Media Custom Publishing (December 2005)
2. Laue, P.: Consulting Architect on SOA Project Automating the Claims Handling Process, Interview (2007)
3. Walker, L.: IBM Business Transformation Enabled by Service Oriented Architecture. IBM System Journal 46(4) (2007)
4. Newcomer, E., Lomow, G.: Understanding SOA with Web Services. Addison-Wesley (2005)
5. Tao, L.: Lixin Tao, Shifting Paradigms with the Application Service Provider Model. Computer 34(10), 32–39 (2001)
6. Scheibler, T., Leymann, F.: A Framework for Executable Enterprise Application Integration Patterns. In: I-ESA (2008)
7. Weerawarana, S., Curbera, F., Leymann, F., Storey, T., Ferguson, D.F.: Web Services Platform Architecture: SOAP, WSDL, WS-Policy, WS-Addressing, WS-BPEL, WS-Reliable Messaging and More. Prentice Hall PTR, Upper Saddle River (2005)
8. Decker, G., Kopp, O., Leymann, F., Weske, M.: Interacting services: From specification to execution. Data & Knowledge Engineering 68(10), 946–972 (2009)
9. Cao, Y., Zhang, M.: Integration of Enterprise Application Based on SOA. In: World Congress on Software Engineering, WCSE, vol. 3, pp. 227–231 (2009)
10. Delin, Q.: Design of Medical Insurance Supervision System Based on Active Data Warehouse and SOA. In: 2009 WRI World Congress on Computer Science and Information Engineering, CSIE, vol. 3, pp. 45–49 (2009)
11. Mathew, R., Spraetz, R.: Test Automation on a SaaS Platform. In: 2009 International Conference on Software Testing Verification and Validation, ICST, pp. 317–325 (2009)
12. Godse, M., Mulik, S.: An Approach for Selecting Software-as-a-Service (SaaS) Product. In: 2009 IEEE International Conference on Cloud Computing, CLOUD, pp. 155–158 (2009)

Research on Comprehensive Evaluation System of Forestry Circular Economy

Wen Qi and Xiaomei Zhang

Economics and Management College of Northeast Agricultural University,
Mucai Street 59,1 50030 Harbin, China
qiwen2277@163.com, zhangxiaomei0451@yahoo.com.cn

Abstract. This study constructs a primary evaluation system of forestry circular economy and also determines a scientific and objective evaluation method. This paper indicated the principles of constructing the evaluation system and the characteristics of forestry circular economy, used qualitative and quantitative analysis methods, and then worked out the evaluation system which including one target, three kinds of guidelines indices and seventeen specific indices and the evaluation method of principal component analysis.

Keywords: forestry circular economy, evaluation, indices system.

1 Introduction

Recent years, there is a rapid growth of China's economy. The result is the problems such as high consumption, pollution and restraints of resources and environment, become increasingly severe. To radically solve resources and environment problems, the government timely made a significant decision to develop circular economic. After several years' practice, it's clear to see that to develop circular economy in China is an effective way not only to ensure the rapid economy growth, but also to relieve resources and environment presses. For forestry production facing such severe conditions of resource shortages and environmental pressures, it is particularly exigent and indispensable to develop circular economy actively and effectively.

Forest is very important resources for human's survival and development. It constructs ecological environment and provide forestry production. Facing pressures brought by the lack of forestry resources and forestry enterprises are managing difficulty, the urgent matter for forestry industry is to extensively and effectively develop circular economy, to fully, reasonably and circularly make use of limited forestry resources, and farthest improve utilization ratio of resources. To scientifically evaluate the development level of forestry circular economy is the basis of making decision for advancing further development of forestry circular economy. Therefore, discussing scientific evaluation system of forestry circular economy is tremendously significant for developing forestry circular economy actively and effectively.

D. Jin and S. Lin (Eds.): Advances in FCCS, Vol. 2, AISC 160, pp. 427–431.
springerlink.com © Springer-Verlag Berlin Heidelberg 2012

2   The Principles of Constructing Evaluation Indices System

Whether the indices system is scientific and reasonable directly impacts on the evaluation results. Therefore, the indices must be able to scientifically, objectively, rationally and as comprehensively as possible to reflect the all aspects which impact development situation of forestry circular economy. However, it is very difficult to establish a set of both scientific and rational evaluation system. To this end, some certain principles are necessary for analysis and judgement.

2.1   Systemic and Hierarchical

To develop forestry circular economy is a complex project and its goal is fully develop ecological, economic and social benefits. So, the evaluation indices system must be able to comprehensively reflect every aspect of circular economy. Thus, it must be able to synthetically reflect above-mentioned three functions. Meanwhile, the normal operation of forestry circular economy system depends on functional groups from different levels. So, indices system should also have a hierarchical structure. This means the higher indices are the integration of lower ones and the lower indices.

2.2   Scientific and Practical

Establishment of the evaluation indices system should be based on scientific theories. To construct the indices system depends on accurately comprehending and grasping the essence of forestry circular economy. The method used should be scientific. The evaluation indices system also must be operational. This means that the data should be obtainable or facile and the operation of the system is viable.

2.3   Dynamic and Comparable

The development of forestry circular economy is a long-term and complex process, which determines the evaluation indices system should have dynamic feature and be able to synchronously reflect the present situation and the developmental tendency of forestry circular economy. In addition, the system should remain relatively stable, only in which way, the evaluation results can be compared horizontally and vertically.

2.4   Representative and Concise

The representative indices can reflect the development situation and essence of forestry circular economy should be chosen. The specific indices under every category should be able to explain the matter from different angles. The indices have same or approximate meaning should not be chosen. Indices should be structured and concise and each index should have clear meaning and be relative independent.

3   The Thinking of Cnstructing Evaluation Indices System

First, at present in china the problem of constructing the evaluation indices system of forestry circular economy is still in research stage and there still not has a set of comprehensive evaluation indices system. Therefore, when constructing the indices system must draw lessons from the already relatively mature indices system such as the indices system of sustainable development and the experiences of studying the evaluation indices system of circular economy. But the characteristics of forestry circular economy must be considered.

In addition, when constructing the indices system, the main thing to do is selecting indices and determining the structural relation among them usually by qualitative and quantitative analysis methods. The construction process can be divided into two stages, namely process of primarily selecting indices and process of perfecting the indices. When primarily selecting the indices, can divide the several indices have been chosen into several parts according to evaluation objects and measurement targets. It means gradually subdividing the indices so that each category of indices can be described by specific statistical indicators. The indices system after primary selection may not be scientific, so it must take scientific tests to further test the feasibility and accuracy of each index. Should check whether the data of the evaluation indices can be attained, and whether the method of calculation, the calculation range and the calculation content is correct and in accordance with its goal.

4 Constructing the Evaluation Indices System

The evaluation indices system of forestry circular economy is composed by tree layers of the target layer, the guidelines layer and the indices layer. Forestry circular economy runs in a form of closed forestry production system, in which people use circular economics method in forestry production and put the terminal materials and energies back into the input to achieve cyclic utilization of materials and energies. Its goal is, crossing forestry production process and forest products' life cycle, to minimize the input of forestry resources and the creation and emission of wastes, to improve the cyclic utilization and the productive efficiency of forestry industry and accomplish the unification of the economic, social and ecological benefits.

Thus, the target layer should fully illuminate the development situation of forestry circular economy to identify the total effect and development tendency and the certain extent that has reached after implementing, regulating and controlling forestry circular economy. The guidelines layer should reflect the characteristics of forestry circular economy that it pays attention to the ecological benefits, in forestry production use circular economics method of low input, consumption and emission and its ultimate goal of the harmonious development of economy, society and ecology. Indices layer is composed by each factor impacts the guidelines layer. They are some specific indices including restriction indices and supporting indices and reflecting guidelines indices from different aspects. This study, based on above-mentioned idea and principles, constructs a primary evaluation system of forestry circular economy which including one target, three kinds of guidelines indices and seventeen specific indices.

Table 1. The evaluation indices system of forestry circular economy

Target layer	Guidelines layer	Indices layer
		GDP of forestry industry (hundred million yuan)
	Development of economy and society	Productivity(%)
		Average income of forestry practioners (yuan/person)
		Increment of employment rate caused by developing forestry circular economy(%)
Development situation of forestry circular economy	Utilization of resources	Energy(standard coal) consumption every ten thousand yuan GDP (t/ten thousands yuan)
		Water consumption every ten thousands yuan GDP(m^3/ten thousand yuan)
		Input amount of woodlands every ten thousands yuan(ha/ten thousands yuan)
		Timber consumption amount every ten thousands yuan(m^3/ten thousands yuan)
		Proportion of disposable wooden products(%)
		Utilization efficiency of woodlands(%)
		Utilization efficiency of timber(%)
		Comprehensive utilization rate of wooden products wastes(%)
		Comprehensive utilization rate of three industrial wastes(%)
	Environmental protection	Forest coverage(%)
		Forest growing stock(ten thousands m^3)
		Discharges of wooden products wastes every ten thousands yuan (m^3/ten thousands yuan)
		Discharges of industrial wastes every ten thousands yuan(m^3/ten thousands yuan)

5   Determining the Evaluation Method

Before apply the evaluation system has been constructed, the evaluation method that can scientifically and objectively analyze the object of evaluation must be determined. In this study, the method of principal component analysis (PCA) is adopted.

PCA is a multivariate statistical analysis method which changes numerous indices into several comprehensive indices according to the way of reducing dimensionality. Its principle is, to change a set of given relevant variables into another group of irrelevant variables by mathematical transformation and rank the new variables in order of decreasing variance and then select the representative principal components. The weight of each principal components determined by above-mentioned method is determined by the magnitude of the contribution rate of the comprehensive factor. This helps overcome the shortcoming that the weight human-making exist subjective assumption and make the evaluation results is objective, reasonable and unique.

PCA is a method widely applied in comprehensive evaluation. PCA can eliminate the correlativity among the index samples, extract a few representative indices on the premise of remaining the main sample information and objectively obtain the reasonable weights of main indices. The principle components have been obtained express a linear combination of original variables. If the final comprehensive indices include all components, the result will be accurately retain 100% of the information provided by the original variables. Even give up several components, it will assure that 85% of the information will be reflected in the comprehensive scores so that the comprehensive evaluation results will be more credible. Consequently, PCA can well give expression to the characteristics of the comprehensive evaluation system of forestry circular economy in this study and then achieve the comprehensive evaluation of forestry circular economy.

References

1. Xie, Y., Zhang, Z.: Connotation and Hierarchy of Recycling Economy on Forestry. Issues of Forestry Economics, 11–14 (2009)
2. Chi, C., Jiang, J.: Research Contents and Construction ideas of the Evaluation Indices System of Circular Economy. Technology Economics, 5–7 (2006)
3. LuSmith, Y.: Review of Research on China's Evaluation Indices System of Circular Economy. Environmental Protection and Circular Economy, 15–17 (2010)

Research on the Data Warehouse Technology and Aided Decision Support in Telecommunication Marketing System

Wenfeng Jiang

School of Information, Shandong Polytechnic University, Jinan 250353, China

Abstract. As social market changes constantly and competition among corporations becomes more and more serious, there is an urgent need for the telecommunication companies to take full advantage of detailed business data of various subsystems of telecommunication marketing, through the comprehensive analysis of a large number of detailed authentic historical data to grasp the telecommunication marketing conditions of the whole company timely and accurately, and scientifically predict the trend of the telecom marketing development. The plan focuses on how to build a data warehouse and how to use data analyzing tools for decision support. In the plan, the aided decision support system consists of three parts: data warehouse system, data extraction, transformation and load tools (ETL) and aided decision support tools. The system provides an integrated solution to making efficient and high-quality aided decision support of telecom marketing system.

Keywords: Decision Support, Data Warehouse, OLAP.

1 Introduction

In this part, I'd like to make clear the definitions of two key terms—data warehouse and decision support system and the significance of th research.The founder of the Data Warehouse W. H. Inmon defined data warehouse as "a subject-oriented, integrated, time-variant, and nonvolatile collection of data in support of management's decision making process"[1]. Decision support system (DSS) is a system offering information that helps people to make decisions. It doesn't make decisions for human beings, but help them to make decisions [2]. According to Power, "Information Systems researchers and technologists have built and investigated Decision Support Systems (DSS) for approximately 40 years"[3]. George Dantzig, Douglas Engelbart, Jay Forrester did the pioneering work in this field. J.C.R. Licklider, T.P. Gerrity and Gordon Davis all contributed greatly to the system. It was brought to a great height of devotement by Peter Keen and Charles Stabell, whose textbook "provided the first broad behavioral orientation to decision support system analysis, design, implementation, evaluation and development[2]. Sprague and Carlson (1982) defined DSS as "a class of information system that draws on transaction processing systems and interacts with the other parts of the overall information system to support the decision-making activities of managers and other knowledge workers in organizations (p. 9) [4].

D. Jin and S. Lin (Eds.): Advances in FCCS, Vol. 2, AISC 160, pp. 433–438.
springerlink.com   © Springer-Verlag Berlin Heidelberg 2012

(1) Significance of data warehouse and aided decision support in the tele-communication marketing system

It is of great significance for the telecom companies to make good use of the business data of telecommunication marketing and analyze valuable materials in the business data.

(2) Problems existing in the construction of the telecom marketing data warehouse and aided decision support system. Nowadays the big companies of our country have realized the importance of the computer-aided decision support. They have invested a lot to construct their own aided decision support systems.

2   Overall Plan

Our plan focuses on the designation of data warehouse and data model, combining the current mature tools, designing data loading tools and decision support tools for the plan. The entire plan as shown below:

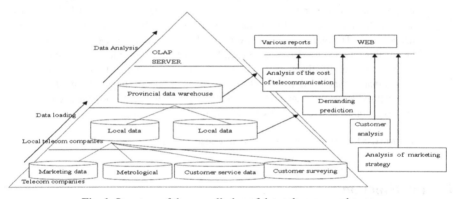

Fig. 1. Structure of the overall plan of data telecom warehouse

Data warehouse system: data warehouse is used to store, manage the data of the warehouse. This is the base of the whole telecom marketing decision support system.

Extract, transform and load (ETL) tools and aided decision support tools: The role of ETL tool is to load the source data in the database into the data warehouse. The quality of the loading will greatly influence the rationale of the subsequent data analysis and decision-making, thereby influence the accuracy of decision-making.

3   Construction of Telecom Marketing Data Warehouse and Aided Decision Support System

As noted above, the construction of telecom marketing data warehouse and aided decision support system is a systematic project. The development of data warehouse is a "spiral" way. Therefore, the data warehouse environment is set up in accordance with iterative development approach, which first establishes a small part of the system, and then builds another part. The establishment of a data warehouse should follow four steps:

First, designing the structure of data warehouse according to the method of data warehouse. Second, fully understanding the needs of decision-makers of the companies; making clear specific interests of the companies, and dividing in accordance with the principle of priority; Third, realizing the migration of the business operational systems and external data source to the data warehouse (ETL).Forth, providing the information to users at all levels of the company by means of visualization with the information displaying tools.

3.1   Designation of the Structure of the Overall Data Warehouse

We designed the structure of data warehouse according to the features of the telecom. Since the telecom centers of provinces and various cities both have the needs of decision analysis, while their analysis are not the same. We, therefore, designed a hierarchical structure of data warehouse. Coarse granularity is suitable for macro decision support, to make strategic decisions; Fine granularity is suitable for micro-decision support, to make timely and rapid market responses. Fine-grained data warehouse is designed for different areas of address, to form data mart. The entire structure of data warehouse is shown in the following. In which, local date warehouse includes not only business databases for manufacturing, but also includes Web databases based on customers' feedbacks.

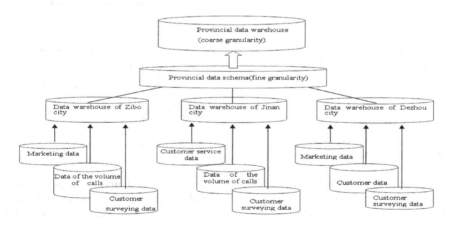

Fig. 2. Hierarchy chart of data telecom warehouse

This warehouse system supports users define different granularities to reflect the features of the data under different granularities. If the user appoints the synthetic granularity of the data, the system can synthesize the data according to the new granularity automatically, and form a multi-dimensional model under the new granularity.

3.2   Data Warehouse Modeling

The designation of the data model to store data for the data warehouse is also needed after the construction of the structure of the entire data warehouse. In order to offer the users a convenient way to make complex analysis as well as a direct display of the data, the data of the data warehouse are often organized into multi-dimensional model. How to choose the facts and dimensions needs the corporation of telecom experts and software development experts. In the case, the volume of calls is the fact. The perspectives to analyze the volume of calls are dimensions. The dimensions are mainly business fields, regions, time and the categories of the calls. Dimensions can be represented as levels. Time can be organized as year, quarter, month; categories of calls can be organized as main types and detailed subsections. By this structure, the data warehouse model to the analysis of the volume of calls can be displayed as follows:

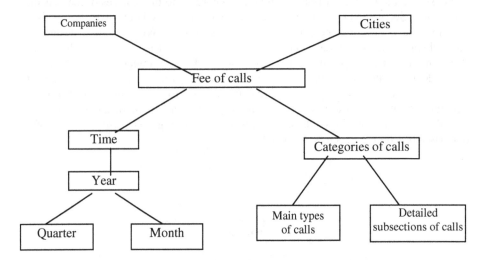

Fig. 3. Analysis model of fee of calls

3.3   Data Loading

To extract data from business database into the data warehouse, data processing, transformation, and aggregation are firstly needed. After that, the data can be loaded into the data warehouse. The quality of the data loading will greatly influence the subsequent data analysis and the rationale for decision-making, thereby influence the accuracy of decision-making.

The entire ETL process is shown as above. First extracting the suitable data from the business database, and putting them into the data warehouse buffer, then processing the data in the data warehouse buffer according to the rules of data cleansing and transformation. The final step is to load the processed data from the data warehouse buffer into the warehouse.

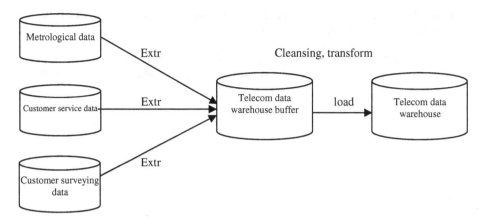

Fig. 4. Process of ETL

3.4   Analyzing Data, Displaying Analysis Result, and Supporting

Decision-Making

The aim of the construction of the data warehouse and data loading is to get the rationale for decision-making. With the combination of the operation of different data analysis, the result of the aided decision making support can be acquired. When the digging and the analyzing end, the aided decision making support system should vividly display the result of the analysis to the decision makers so that they can understand it easily and make correct decisions.

4   Conclusion

In summary, our data warehouse and aided decision support system provide an integrate solution to the efficient and high-quality aided decision support. A suitable data warehouse is designed for the telecom companies. The whole structure supports flexible user-defined data granularity. A flexible data model is designed, enabling users adjust data model according to their own needs. Taking the features of the telecom industry into consideration, we designed a system that can support a variety of ETL tools to load data, support a variety of decision analyzing tools, data mining tools, and can display the results in several ways.

References

1. Inmon, W.H.: Building the Data Warehouse. QEDWiley (1991)
2. Keen, P.G.W., Scott-Morton, M.S.: Decision Support Systems: An Organizational Perspective. Addison-Wesley, Reading (1978)

3. Power, D.J.: A Brief History of Decision Support Systems. DSSResources.COM, World Wide Web, version 4.0 (March 10, 2007),
   http://DSSResources.COM/history/dsshistory.html
4. Engelbart, D.C.: Augmenting Human Intellect: A Conceptual Framework (October 1962)
5. Sprague Jr., R.H., Carlson, E.D.: Building Effective Decision Support Systems, Inc. (1982)

NC Feeding Servo System and Self-adaptive Multi-rule Factor Controller

YuHu Zuo

Linyi University, Linyi Shandong, 276005, China
zuotiger@163.com

Abstract. The principle of self-adaptive rule factor fuzzy controller is introduced and self-adaptive multi-rule factor fuzzy controller is designed. The fuzzy controller is simulated with fuzzy toolboxes and Simulink environment in Matalab. It is applied to the NC feeding servo system and the results show that the dynamic performance and stable deviation of control system are improved.

Keywords: self-adaptive multi-rule factor, fuzzy controller, NC feeding servo system.

1 Introduction

With the development of science and manufacturing technology, product quality and species diversity are becoming better and better. People need high accuracy NC equipment, but manufacture accuracy of machine tool parts has reached a high level, and it is very difficult to further improve system accuracy by improving the precision of machine body. Meanwhile, as electronic technology progress, especially the computing speed of computer, making the advanced control algorithm can be applied in practice, which realize real-time adjustment controller of compensation and the parameters of system in order to improve the performance of system. Therefore, research and development high speed high precision NC servo system is the key problems and the effective way in CNC system development.

2 Design of Self-adaptive Rule Factor Fuzzy Controller

2.1 Structure of Self-adaptive Rule Factor Fuzzy Controller

Self-adaptive rule factor fuzzy controller is based on the common fuzzy controller and introduction of a rule factor, which analytical expressions of model structure is shown as follows type:

$$U = \alpha E + (1 - \alpha)EC$$

$\alpha \in [0,1]$ is rule factor; $E = Y - R$ is deviation , $EC(k) = E(k) - E(k-1)$ $(K = 0,1,2,\cdots)$ is variation of deviation , which is discrete time variable. Its structure is shown in figure 1.

D. Jin and S. Lin (Eds.): Advances in FCCS, Vol. 2, AISC 160, pp. 439–442.
springerlink.com   © Springer-Verlag Berlin Heidelberg 2012

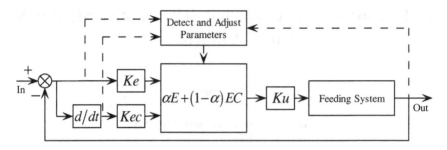

Fig. 1. The structure of self-adaptive rule factor fuzzy controller

By adjusting the value of the rule factor α, different characteristic factor of the fuzzy model can be got. α is the weighting coefficient of deviation and variation of deviation, which has the distinct physical significance. The fuzzy model with rule factor not only easily adjusts the rules, and overcomes defects of determining the control rules by the experience, which is beneficial to analyse and design the fuzzy controller by analytical methods.

2.2   Principle of Self-adaptive Multi-rule Factor Fuzzy Controller

For NC feeding servo system, when the deviation is large, the main task of control system is to eliminate the deviation, then the weighting coefficient of deviation in the control rules; and when the deviation is small, system has been close to the steady state, which requires the variation of deviation in the control rules should be bigger. According to the above ideas, a rule factor are introduced for each deviation. The multi-rule factor control rules are formed, which can modify the rule factor under different controlled state to meet requirements of the control system. When the domain of deviation E, variation of deviation EC and controlled variable U are taken as $\{-3, -2, -1, 0, 1, 2, 3\}$, control rules with multi-rule factor can be expressed as:

$$U = \begin{cases} \alpha_0 E + (1-\alpha_0)EC & E = 0 \\ \alpha_1 E + (1-\alpha_1)EC & E = \pm 1 \\ \alpha_2 E + (1-\alpha_2)EC & E = \pm 2 \\ \alpha_3 E + (1-\alpha_3)EC & E = \pm 3 \end{cases} \qquad (\alpha_0, \alpha_1, \alpha_2, \alpha_3 \in [0,1])$$

2.3   Design of Fuzzy Controller

Fuzzy controller is proportion-integration type two-dimensional structure. Input variable is deviation $E = R - Y$ and variation of deviation $EC = E(k) - E(k-1)$. Output variable is controlled variable U. Fuzzy subset of E, EC and U are all defined as $\{NB, NM, NS, ZE, PS, PM, PB\}$. Universe of fuzzy sets is $[-1,1]$ and membership function uses symmetric triangular inhomogeneous distribution. According to feeding servo system actual performance requirements and system input-output characteristics, fuzzy control rules of fuzzy controller is designed and shown in table 1.

Table 1. Fuzzy control rule table of fuzzy controller output variable U

EC \ E	NB	NM	NS	ZE	PS	PM	PB
NB	NB	NB	NB	NM	NS	NS	ZE
NM	NB	NM	NM	NM	NS	ZE	PS
NS	NB	NM	NS	NS	ZE	PS	PM
ZE	NB	NM	NS	ZE	PS	PM	PB
PS	NM	NS	ZE	PS	PS	PM	PB
PM	NS	ZE	PS	PM	PM	PM	PB
PB	ZE	PS	PS	PM	PB	PB	PB

2.4   Simulation Experiment

The following is step response simulation experiment about feeding servo system based on self-adaptive multi-rule factor fuzzy controller. Model for transfer function is $G(s) = 2478.5/(s^2 + 101.7s + 2612)$. Figure 2 is simulation program which is designed with simulink toolbox of Matlab, the simulation results prove that self-adaptive multi-rule factor fuzzy controller is better than normal fuzzy controller.

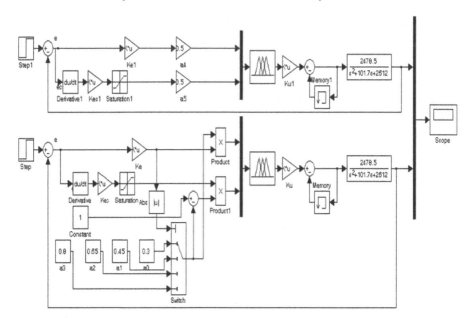

Fig. 2. Simulink experiment

3   Summary

From the above analysis, we can conclude that self-adaptive multi-rule factor fuzzy controller is better than normal fuzzy controller, which system overshoot is lesser, response time is short, adapt the controlled object parameter variation, can better satisfy rapidity and accuracy of NC feeding servo system.

References

1. Zhang, G.L.: Fuzzy control and Matlab. Xi 'an jiaotong university press, Xi 'an (2002)
2. Liao, B.Y.: Modern mechanical dynamics and its engineering application. China Machine Press, Beijing (2004)

Design of the Driving Circuit Based on CPLD in Transmitter of Transient Electromagnetic Instrument

Dongge Han, Guangmin Sun, Dequn Zhao, and Bo Peng

Department of Electronic Engineering, Beijing University of Technology,
100124, Beijing, China
handongge@emails.bjut.edu.cn, gmsun@bjut.edu.cn

Abstract. A driving circuit in transient electromagnetic instrument is designed based on CPLD. In the circuit, the transmitting waveforms are generated by CPLD and the synchronization is implemented by wired. In addition, the gate driver IR2110 for the power switch devices is employed to control the high-power MOSFET for driving the H-bridge. Compared with the reference circuit, the generated waveform of the proposed circuit has short turn-off time and high current. The testing results show that the proposed circuit has a better performance on driving the coils. The transmission delay is reduced and the measuring accuracy is further improved to deepen the detecting depth.

Keywords: CPLD, Transient Electromagnetism, Driving Circuit, IR2110, Self-detection.

1 Introduction

TEM (Transient Electromagnetic Method) is widely used in the fields of metal mining exploration, underground metal pipelines detection, groundwater or environmental pollution and engineering exploration. The detecting range of the instrument is 10m to several km [1]. However, it could not meet the demand for the deep-structure applications. On the other hand, recently there is a great demand for electrical measuring instruments in our country, but the key technologies are mastered by several countries [2], so it is mainly rely on imports for us. Therefore, to design and realize the multi-functional TEM devices is necessary.

For the designed transient electromagnetic transmitter, the current signals below 20A are usually used as the emission signal. The line synchronous and quartz synchronous are performed with the μs order of turn-off time for the transmitted waveform. There are some technical flaws by using these design methods. Firstly, the signals are easy to be overwhelmed in the noise. Secondly, the long transmission time delay and low measuring accuracy could not make the signal received efficiently. Thirdly, the low detecting depth could not meet the need for deep-structure exploration.

There are several factors to effect the detecting depth and accuracy for TEM, such as the turn-off time of the transmitted waveform, the transmitted current, electromagnetic noise, the power sensitivity and the electrical property. Therefore, the detecting depth and accuracy would be tested under certain conditions. In the paper, the low-delay time and high accuracy transmission methods is utilized to design the driving circuits in the views of turn-off time and current value.

D. Jin and S. Lin (Eds.): Advances in FCCS, Vol. 2, AISC 160, pp. 443–449.
springerlink.com   © Springer-Verlag Berlin Heidelberg 2012

2   Framework of the Proposed Driving Circuit

The detecting depth and accuracy of the system has a consanguineous relationship with the different factors of the current amplitude for the transmitter, turn-off time, waveform and the first sampling time after turning off etc. The design for the performance of transmitter is important in the development of the TEM devices. Improving the current amplitude of transmitter and shortening the turn-off time is efficient to enhance the system performance. Therefore, in this system, the maximum current of the driving circuit is 210A, the turn-off time is 900 ns and the waveform of transmitting signal can be selected artificially.

The driving circuit for TEM transmitter is constituted by logical control part, power driving part and detecting part. While transmitter working, it should synchronize with the receiver at first. After getting the synchronization signal, CPLD will generate the transmitted waveform. The signal is delivered to the power driving part through the photocoupler. For getting better performance, the transmission needs good robustness and low delay to reduce the distortion in the end of receiver. In the power driving part, a special driving chip is applied to drive the high-power MOSFET to make the circuit simple and stable. Moreover, to keep the circuit working safely and efficiently, the current of the transmitter, power supply and temperature of MOSFET should be detected. The results will be delivered to CPLD to determine whether to modify the parameters. The structure diagram of the proposed circuit is shown in Fig.1.

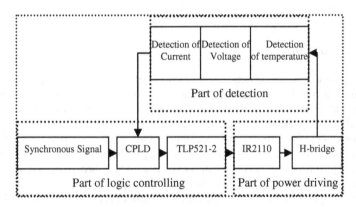

Fig. 1. Structural Block Diagram of Hardware Circuit

3   Circuit Design

3.1   Part of Logic Controlling

In early main control circuit of transmitters, relay interlock circuit is very common. Although the structure was simple and stable, the size of circuit is overlarge and the response speed is low. In addition, the analog devices are applied as the timer circuit, which leads to the big temperature excursion and bad consistency of circuit [3].

With the development of the microprocessor, the main controlling of transmitter has been improved in the aspects of intellectualization and integration. The application of the CPLD has overcome the disadvantages mentioned above. With the real-time of hardware, programmability of software, friendly human-machine interface and the powerful function of communication, CPLD could enhance the circuit design flexibility.

To enhance the security of transmission, the main control circuit must be reliable. As the large power of transmitter, its microwave power can reach several megawatts. Furthermore, many high-power switches and nonlinear electronic components may become strong interference sources. According to that, the main controlling circuit based on CPLD is so sensitive that it may lead to misoperation. Therefore, the feature of anti-jamming should be considered in design of system.

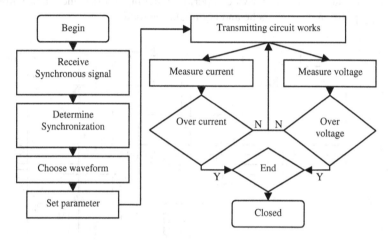

Fig. 2. Flow chart of logic controlling circuit

The part of CPLD is programmed by Verilog–HDL language, realizing the signal synchronization, overcurrent and overvoltage protection, as well as alarm indication. It can check the working state of each module, while alarming at the abnormal state. As system working well, the current, voltage and working mode information can be displayed. The H-bridge circuit is driven by a dual-driving signal to make it output the predetermined waveform. The system flowchart is shown in Fig.2.

3.2   Part of Power Driving

The power driving module, using the chip of IR2110, is isolated with H-bridge to protect the front of the circuit by TLP521-2 [4]. The IR2110 is a driver with high voltage, high speed, high independence and low side referenced output channels for power MOSFET and IGBT. Proprietary HVIC and latch immune CMOS technologies enable monolithic construction to ruggedize. Logic inputs are compatible with standard CMOS/LSTTL output. The output characteristic of drivers is a high pulse current buffer stage designed for minimum driver cross-conduction [5]. Propagation delays are matched to simplify the usage in high frequency applications. The floating channel can be used to drive a N-channel power MOSFET or IGBT in the high side configuration which operates up to 500 or 600 v.

In Fig.3, the driving voltage of gate of Q1 is clamped to 0V when the MOSFET is turned-off, and the driving signal makes VT1 to be on-state when the upper MOSFET of the H-bridge is on-state. The voltage of upper MOSFET is pulled to low voltage when VT1 is turned-off, and VT2 is on-state because of the high level of gate when the upper MOSFET is turned-off. In this way, the distortion on the gate driving voltage will be greatly reduced because the current produced by miller effect stream through VT2 [6]. H-bridge is formed by MOSFET SS70N10A. These N-Channel enhancement mode power FETs are produced using Fairchild's proprietary, planar stripe, DMOS technology. This advanced technology has been especially tailored to minimize on-state resistance, provide superior switching performance, and withstand high energy pulse in the avalanche and commutation mode.

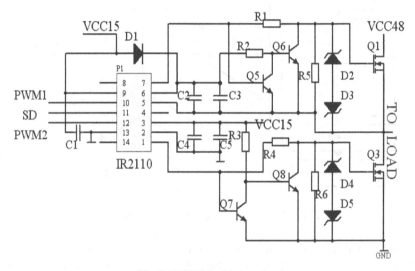

Fig. 3. IR2110 in driving circuit

4   Self-detection and Control of the Circuit

4.1   Detection and Control of Current and Voltage

Self-detection and control circuit of current and voltage is mainly formed by sampling resistance and A/D converter. Transmitting current is a pulse signal. Methods for detecting pulse current include sampling resistance, current transformer and hall element. Detection precision of the last two methods is not accurate, and not easy to be thousandth. The first method can assure a high detection precision, but the value is reduced because the sampling resistance is in series connection. So it needed collect a high-power calculated according to emitted current. AD1674 is selected as A/D converter. The 12 bits successive comparison A/D converter produced by American AD Company can reach up to the type conversion time of 10 ms, and single channel maximum sampling rate is 100 kHz, so it can satisfy the design requirements of the transmitter. There is a tri-state output buffer circuit in chip which can be directly linked with all kinds of typical 8 or 16 bit MCU together. Thus, a high precision of voltage reference and clock circuit within AD1674 is assured, and it is convenient that the chip finishes A/D conversion without any plus circuit and the clock signal. The internal structure of AD1674 is more compact, with the higher level of integration and working performance, and can make a better design plate area [7]. Thus, the cost is greatly reduced and the reliability of the system is improved.

Converting the current signal to the voltage signal which is to supply the comparator constructed by TL084, a 5 Ω high-power resistance is located. According to the measurement requirement, it needs adjusting the power resistance located in front end of the circuit to set the input and output voltage. The output is connected with the SD end of IR2110. The circuit will output the high level voltage when over-current. For the SD end of IR2110, it is valid in the high level voltage. When the SD is in high level, IR2110 output terminal is closed to protect the system circuit. The output of protecting circuit is connected with the alarm circuit so that it will be alarmed while outputting high level voltage.

4.2   Detection and Control of Temperature

As shown in Fig.4, the drain current decreases gradually with the increase of the temperature. To guarantee the transmitter work normally and safely, it is necessary to monitor the temperature of field-effect tube. The system uses a DALLAS temperature sensor.

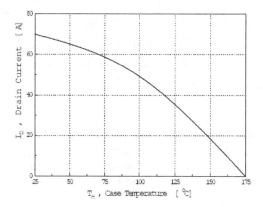

Fig. 4. Max Drain Current vs. Case Temperature

There are two kinds of methods for connecting CPLD with DS18B20 in hardware. For one thing, it is to pick up the external power source, GND grounding, and make I/O connected with CPLD's I/O connected; and the other is to use a parasitic power supply, UDD, GND grounding, connecting I/O with CPLD I/O. Both for the internal and external power supply mode, I/O will pick up a 5 KΩ resistance. The visiting DS18B20 by CPU contains 3 steps: 1) initializing the DS18B20 2) ROM operating 3) memory and data operating. Every step must follow the operation order and communication protocol in DS18B20 [8].

5   Experimental Results

To suppress the LF electromagnetic noise, pneumatic noise, remaining signal of last sampling period, interference of industrial power and its harmonics, the transient electromagnetic instrument system uses the bipolar pulse current whose duty cycle is 1:3 for the excitation source. The system observes transient characteristics of the secondary field in interval time between positive and negative pulse. When the value of input transient signal is negative, the switch will turn backward to keep the value of the signal which comes into the sampling integrator staying positive. Obviously after this processing, it will achieve the cumulative effect for the useful signal. However, for the slowly-decayed noise, its sampled value in the positive half-cycle will be offset by the one in the negative half-cycle, and the output interference is greatly reduced. The circuit uses ultra-fast recovery diode, whose responding time is in 100 ns at the current work state of 150A, to reduce the impact of coil on the power driving circuit and also to enhance the steep characteristics of transmitting waveform. The system connects with a 0.2Ω load, which can transmit the square wave under the 48V power supply. The transmitting current value is 192.5A. The testing result shows that the turn-off time of the wave is 900 ns. In table 1, it shows that the transmitting current of EM-42 which performs better among the reference circuits is about 100 A and its detecting depth can reach 1000 m.

Table 1. The comparison table of main parameters in different system

Type	Emitting current	Detecting depth(M)	Time range	Application for Exploration
EM-42[1]	100 A	<1000	N×10-1ms~Ns	oil and gas fields
EM-37 [1]	<20 A	100-1000	N×10-2~ N×102ms	metal、 coal
WDC-2[1]	<10 A	100-1000	N×10-1~ N×10ms	Mineral deposits surveys
Proposed Circuit	<200 A	N×1000	Nμs~N×10-2ms	oil and gas fields and deep structure

In the same condition, the output current of our proposed driving circuit can reach up to 210 A in theory. And the detecting depth is much deeper than the reference circuits listed. Furthermore, the receiver can receive data after several nano-seconds when the transmitter generates the transmitting wave, in order to increase the receiving accuracy.

6   Conclusion

In this paper, we proposed a driving circuit for transient electromagnetic transmitter with high detecting accuracy and large detecting range. The transmitting signals are generated based on CPLD and meet the requirement of the circuit. To transmit synchronization signal by Using max485 or GPS, it can also meet the requirement of multi-receivers working at the same time. Furthermore, it combines the advantage of IR2110 and H bridge circuit. Comparing with EM-42, the proposed circuit can generate the waveform with the advantages of short turning-off time, large current, little power supply and the simple power circuit. The testing results show that the proposed circuit can drive the power coil better, shorten the time delay of transmitting, improve the accuracy of measurement and extend detecting range.

References

1. Niu, Z.: Principle of Transient Electromagnetic Methods, pp. 106–119. Central South University of Technology Press (1992)
2. Kong, F.: Application of CPLD to Radar Transmitter. Ship Board Electronic Counter Measure, 100–103 (2008)
3. Chen, Z., Dong, H.: Development of Novel Transient Electromagnet Transmitter. Progress in Exploration Geophysics, 444–447 (2004)
4. Lee, J.B., et al.: Airborne TEM Surveying With a SQUID Magnetometer Sensor. Geophysics, 739–745 (2002)
5. Christensen, N.B.: A Generic 1-D Imaging Method for Transient Electromagnetic Data. Geophysics, 438–447 (2002)
6. Wang, Z., Lin, J., Yu, S., Ji, Y., Zhou, G.: ATTEM: An Instrument System Using Transient Electromagnetic Pulse for Subsurface Imaging. In: Proceedings of the IEEE Instrumentation and Measurement Technology Conference, pp. 2231–2234 (2006)
7. Wu, G.: The Development of Low Noise Transient Electromagnetic Methods Receiver Based on Equivalent Sampling Technology, pp. 3–4. Jilin University, China (2002)
8. Li, J.: Study on Multifunction Electromagnetic Transmitter, pp. 5–7. Chongqing University, China (2009)

A UHF RFID Reader Design Based on FPGA

Liyan Xu[1,*], Lingling Sun[2], and Xiaohong Zhang[2]

[1] College of Electrical Engineering, Zhejiang University, Hangzhou 310027, China
[2] College of Electronic Information, Hangzhou Dianzi University, Hangzhou 310018, China
zjuxuliyan@163.com, sunll@hdu.edu.cn, xhzhang@hdu.edu.cn

Abstract. ISO/IEC 18000-6C has been used in different applications and comes in myriad of forms today. The UHF RFID reader based on the ISO/IEC 18000-6C standard has been discussed in this paper. The reader's key issues of RFID reader's architecture and baseband design are studied. The proposed FPGA based baseband platform could implement physical layer and tag-identification layer. This form of baseband design can efficiently reduce the design time and development cost, FPGA could meet the complex requirement and flexiable apply to different standards.

Keywords: RFID, reader, UHF, ISO/IEC 18000-6C protocol.

1 Introduction

RFID technology shows a continuous growth in various application fields[1], like commerce, logistics, medical science, security, access control etc. Based on carrier frequency, RFID can be classified into low frequency (LF), high frequency (HF), ultra high frequency (UHF), and microwave. The corresponding frequency range and standards are 125KHz (ISO 18000-2), 13.56 MHz (ISO 18000-3), 860-960 MHz (ISO 18000-6), 2.45GHz (ISO18000-4), respectively. The RFID, which operation in UHF band from 840MHz to 960MHz, shows potential in low cost, data transfer robustness, high efficiency and high throughput[2].One of key fact to benefit from such technology is a low cost and high performance UHF RFID reader[3].

Many schemes about the design of UHF RFID reader have been discussed. such as the scheme[4] uses a microcontroller unit (MCU) as main controller, the scheme[5] uses a DSP chip as main controller. The baseband signal processor decides the performance of the reader system. A scheme[6] using ARM as main controller and FPGA as encoder and decoder. In this paper, a UHF RFID reader based on FPGA has been designed and it is compatible with ISO/IEC 18000-6C Standard. FPGA could meet the complex requirement and flexible apply to different standards. The paper is organized as follows. First, ISO/IEC 18000-6C protocol are briefly reviewed .Then the paper describes main blocks in a UHF RFID system. Detailed design of baseband is given .Finally the experimental results and a summary and conclusion are provided.

---

* Corresponding author.

D. Jin and S. Lin (Eds.): Advances in FCCS, Vol. 2, AISC 160, pp. 451–456.
springerlink.com © Springer-Verlag Berlin Heidelberg 2012

2   Specification Analysis and Architecture Choice

ISO/IEC 18000-6C UHF RFID standard[7] defines a communication protocol, operating frequency 860 ~ 960MHz. The protocol outlines the air interfaces and commands between RFID readers and RFID tags. The air interfaces mainly include the physical layer and the tag identification layer .The physical layer describes details of the communication mode between the reader and tags, which including data encoding and modulation, anti-collision algorithm, access control and the key technical aspects of privacy protection. The communication between the reader and tags are half duplex. In the UHF RFID system, the reader sends information to one or more tags by modulating an RF carrier using amplitude shift keying (ASK) with a pulse-interval encoding (PIE) format. Tags receive their operating energy from the reader's unmodulated RF carrier. The tags respond by modulating the reflection coefficient of its antenna, thereby backscattering an information signal to the reader. The tag's data encoding format, which selected in response to reader commands, is either FM0 or Miller-modulated.

The designed RFID reader is mainly composed of two parts: the analog front end and the digital baseband. The analog front-end includes the carrier signal synthesis module, modulator, power amplifier, signal isolation devices, filter circuits, demodulator, signal amplification circuits and so on. The mainly functions of digital baseband are processing data, communicating with application software, executing the commands from the application system software and completing quick real-time communication with tags. It's block diagram is shown in Fig.1.

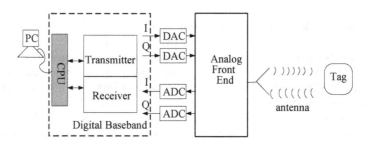

Fig. 1. Block diagram of the RFID system

The analog front-end could uses the Homodyne system theory[8] to avoid the mirror interference. Digital baseband outputs control signal for controlling frequency synthesizer to produce local oscillator (LO) carrier. If the reader sends a command, The baseband part accomplishes signal's ASK modulation and signal goes through the Digital Converter(DAC) ,then signal becomes analog signal. The analog front-end accomplishes carrier modulation. modulated carrier is sent to attenuator and power amplifier. Antenna transmits signal to tags and receives backscatter signals from tags. The received RF signal from the antenna is demodulated to low frequency (LF) signal with low-pass filters filtrating the high frequency signal. LF signal goes through amplifier and Analog to Digital Converter (ADC), and becomes baseband signal.

3   The Baseband Design Implementation

The FPGA implements CPU, transmitter and receiver function. There are the physical layer and the tag-identification layer to compatible with ISO/IEC 18000-6C.The physical layer is described in verilog HDL language and implemented in FPGA. The tag-identification layer is described in C language and implemented in NiosII core which embedded in FPGA. The baseband architecture based is showed in Fig.2.NIOSII core uses avalon bus to control DDR SDRAM memory, Flash memory, UART controller, Ethernet controller and the transmitter and receiver module. NiosII performs all commands and controls. This architecture is an advantage for implementing various kinds of RFID standards by changing the software of NiosII core. It efficiently reduces the design and development time and cost, and achieves flexible and efficient development. The baseband architecture is showed in Figure 3.

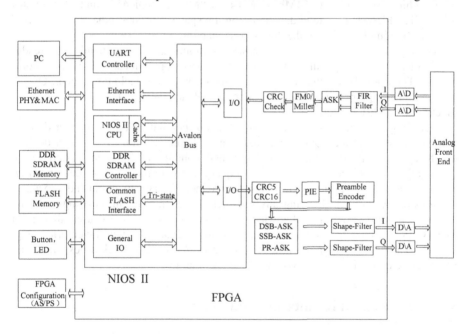

Fig. 2. The baseband architecture

FPGA is the system controller with NiosII core, which is used for controlling ADC, DAC, encoding, decoding, modulating, demodulating. If processor wants to send a command, CRC generator computes and adds either the CRC-5 or the CRC- 16 for the commands. According to ISO 18000-6C protocol , CRC5 is used in "Query" command , and CRC16 is used in others commands .Commands which are logical data sequences with CRC are encoded to pulse interval encoding (PIE) symbols. Commands shall begin with either a preamble or a frame-sync. The modulator converts up the baseband signals to double sideband amplitude shift keying (DSB-ASK), single sideband amplitude shift keying (SSB-ASK) or phase reversal amplitude shift keying(PR-ASK) modulation. To satisfy the RF envelope of the ISO/IEC 18000-6

Type C protocol, output signal of the encoder passes through the pulse shaping filter before DAC. After RF demodulating the tags' backscattered signals and ADC converting, The tag's signal first passes through FIR filter .then it is sent to demodulator. Demodulation mode is ASK or PSK. it could be decided by tag type. After that, bit synchronization circuit extracts clock information to synchronize the data and sends the data to decoder .The baseband receiver decoder is composed of a preamble detector and a decoder which is designed to decode both of FM0 and Miller encoding types. Except the general decoding procedure, the baseband decoder also detects collision occurrences and verifies whether there is a bit error or not by the CRC decoder. If CRC check passes without an error, the serial format data will be shifted to parallel format and be sent to processor.

Altera EP3C16Q240 FPGA is used in the design to ensure rapid availability and low cost. EP3C16Q240 FPGA devices can support complex digital systems on a single chip .We use a 64Mbits MT48LC8M8A2 SDRAM and 128Mbytes Intel28F128 FLASH to extend data and program spaces. Besides, ethernet and UART interfaces are implemented in the digital baseband circuit.

The tag-identification layer processes communication commands. A reader communicates to tag populations using three basic operations: Select, Inventory, Access. There are 14 commands to communicate with tag, including Select, Query, Query_adjust, Query_rep, Ack, Nak,Req_rn, Read, Write, Kill, Lock, Access, Block_write and Block_erase.

According to select command, a particular tag population based on user-defined criteria will be selected. Query command initiates an inventory round and decides which tags participate in the round query and contains a slot-count parameter Q. Query_adjust command repeats a previous query and may increment or decrement Q, but does not introduce new Tags into the round. Query_rep command repeats a previous query without changing any parameters and introducing new tags into the round. Ack command is to acknowledge the tag. Nak command causes all tags in the inventory round to return to arbitrate without changing their inventoried flag. Rq_rn command is to obtain a new RN16. Read, Write, Kill, Lock,Block_write and Block_erase commands are to write or erase tags memory.

4   Experimental Results and Summary

The waveform of transmitting signal from reader and the waveform of backscatter signal from tag could be captured by the spectrum analyzer. The communication process between the reader and tag is shown in Fig.4. The experimental result demonstrated that the reader could correctly send commands and receive data from the tag.

This paper presents the design of UHF RFID reader, which is compatible with ISO/IEC 18000-6C Standard. The digital baseband design has been described in details. The baseband circuit based on FPGA. Which according to the physical layer and the tag-identification layer to compatible with ISO/IEC 18000-6C.The physical layer is described in verilog HDL language and implemented in FPGA. The tag-identification layer is described in C language and implemented in NiosII core which embedded in FPGA. The architecture is not only applicable for being implemented in ISO/IEC 18000-6C protocol, but also applicable for other RFID standards. It could be flexible adjusted to implement physical layer of different protocol. The NiosII core in FPGA performs all commands and controls and could achieves rapid, flexible and efficient development.

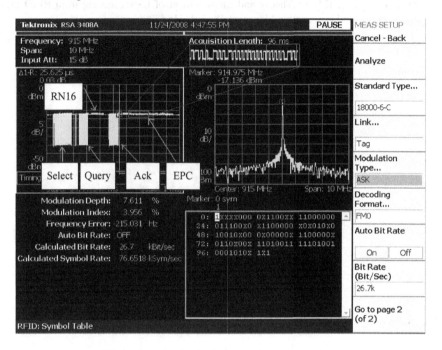

Fig. 4. Communication between the reader and tag

Acknowledgments. The work was supported by Zhejiang Provincial Technological Plan Foundation of China (GK110908002).

References

1. Finkenzeller, K.: RFID Handbook, 2nd edn., pp. 7–9. Wiley (2004)
2. Wang, W., Lou, S., Chui, K.W.C., Rong, S., Lok, C.F., Zheng, H., Chan, H., Man, S., Luong, H.C., Lau, V.K., Tsui, C.: A single-chip UHF RFID reader in 0.18um CMOS process. IEEE J. Solid-State Circuit 43(8), 1741–1754 (2008)
3. Karthaus, U., Fischer, M.: Fully integrated passive UHF RFID Transponder with 16.7uW minimum RF input power. IEEE J. Solid State Circuits 38(10), 1602–1608 (2003)
4. Pang, X.D., Yao, X.S., Liang, C.P.: Design and Realization of a Highly Integrated UHF RFID Reader Module. Microwave and Millimeter Wave Technology 3, 1506–1508 (2008)
5. Zhang, J.M., Lai, S.L., Chen, Y.T.: The Application of DSP Sampling and Identifying in UHF RFID Reader. Science Technology and Engineering 6(8), 956–959 (2006)

6. Li, H., Mou, X.: A New Implementation of UHF RFID Reader. In: TENCON 2009 -2009 IEEE Region 10 Conference, pp. 1–4 (2009)
7. ISO/IEC 18000-6C: Information technology-Radio frequency identification for item management-Part 6: Parameters for air interface communications at 860MHz to 960MHz, final draft[S] (2004)
8. Nikitin, P.V., Rao, K.V.S.: Theory and Measurement of backscattering from RFID tags. IEEE Antennas and Propagation Magazine 48, 212–218 (2006)

Modulation Recognition for Multi-component PSK Signals Based on Cyclic Spectral Envelope

ZhiBin Yu[1,*], ChunXia Chen[2], and NingYu Yu[3]

[1] School of Electrical Engineering, Southwest Jiaotong University,
Chengdu 610031, China
[2] Mechanical Engineering Department,
Chengdu Electromechanical College, Chengdu 610031, China
[3] The Second Artillery Engineering College, Xi'an 710025, China
zbinyu@126.com, yu_ccx@sohu.com, yuningyu1983@sohu.com.

Abstract. In order to recognize the modulation signals when the single-channel receiver works under multi-component signal (MCS) environment, a novel approach which can directly extract the feature invariants of each independent component is proposed for the multi-component PSK signals. Firstly, the difference between the modulation methods for MCS and single signal is analyzed. Furthermore, the spectral correlation function of MCS is deduced, the method of the feature extracted is proposed. Simulation results verified that the proposed algorithm can extract the features of each independently components without disturbance, and it can effectively identify arbitrary combination of dual-signal in the given signal set.

Keywords: Signal Processing, Cyclostationary, Feature extraction, Phase Shift Keying, Modulation Recognition.

1 Introduction

With the increasing complexity of the electromagnetic environment, the intercepted probability of time-frequency overlapped signals (Multi-source signals are partly overlapped in time domain and frequency domain, i.e. MCS) increases rapidly in single-channel receiver. But because of the complexity of MCS, the overlap part of intercepted signals is usually ignored in signal processing. Therefore, the false alarm rate is increased in fact. Due to the important study significance, analysis technology of MCS is concerned by more and more researchers in recent years [1], [2], [3], [4].

In recent researches, there are two thinks to consider the recognition of the signals under MCS. On the one hand, using the single-channel classification algorithm, the components of MCS can be obtained [3], [4], and the independence component be

---

* Corresponding author.

D. Jin and S. Lin (Eds.): Advances in FCCS, Vol. 2, AISC 160, pp. 457–462.
springerlink.com   © Springer-Verlag Berlin Heidelberg 2012

identified. But, these methods need a large of priori information. On the other hand, the feature invariants of the signal are directly extracted to identify the MCS [1], [2]. But this method is only valid when the time-frequency overlapped degree of signals is lower. In this paper, aiming to the problem of the recognizing multi-component PSK signals, an abstracted method of the feature invariant is proposed to recognize the signals based on the cyclostationary principle. Simulation results verified that the proposed algorithm can extract the features of each independently components without disturbance, and it can effectively classify arbitrary combination of dual-signal in the given set.

2   The Signal Model

Suppose the intercepted signals are two mutual independent multi-component PSK signals. The set of signals is {BPSK\QPSK\MSK}. The carrier frequencies of signals are in the same frequency domain, and their code rates are not equal. The signal model can be given as

$$r(t) = s_1(t) + s_2(t) + w(t),$$ (1)

where $r(t)$ is the intercepted MCS. $w(t)$ is the Gauss white noise and its standard deviation is σ_w^2 ; $s_1(t)$ and $s_2(t)$ indicate respectively the single independent component signals of the multi-component PSK signals and them can be expressed as

$$s_i(t) = A_i \left[I(t)\cos(2\pi f_{ci}t + \phi_{0i}) - Q(t)\sin(2\pi f_{ci}t + \phi_{0i}) \right],$$ (2)

with

$$I(t) = \sum_n a_{in} g_i \left(t - nT_{ci} - t_{0i} \right), Q(t) = \sum_n b_{in} g_i \left(t - nT_{ci} - t_{0i} \right),$$

where i expresses the ith component signal, and A_i is the amplitude of the ith signal and its carrier frequency, time-wide of codes, initial phase and the initial time are $f_{ci}, T_{ci}, \phi_{0i}$ and t_{0i}. The value of a_{in} and b_{in} are -1 and +1, respectively. $g_i(t)$ is the rectangle pulse. For the BPSK, if $\phi_{0i} = 0$ and $t_{0i} = 0$, $Q(t) = 0$. Thus, for MSK signal, we have

$$\begin{cases} I(t) = \sum_n a_{in} g_i \left(t - nT_{ci} - t_{0i} - T_{ci}/2 \right) \cos\left(\pi t / T_{ci} \right) \\ Q(t) = \sum_n b_{in} g_i \left(t - nT_{ci} - t_{0i} \right) \sin\left(\pi t / T_{ci} \right) \end{cases}.$$ (3)

The cyclic spectrum analysis of the signals is in the Literature [5], [6].

3   The Feature Extraction Method

3.1   The Cyclic Spectrum Analysis of the MCS

According the Eq. 1, the autocorrelation function of MCS can be obtained as

$$
\begin{aligned}
\hat{R}(t,\tau) &= E\left[r(t)r^*(t-\tau)\right] \\
&= E\left\{\left[s_1(t)+s_2(t)\right]\left[s_1(t-\tau)+s_2(t-\tau)\right]^*\right\} \\
&= E\left[s_1(t)s_1^*(t-\tau)\right]+E\left[s_2(t)s_2^*(t-\tau)\right]+E\left[s_1(t)s_2^*(t-\tau)+s_2(t)s_1^*(t-\tau)\right],
\end{aligned}
\tag{4}
$$

here $E\left[s_1(t)s_2^*(t-\tau)+s_2(t)s_1^*(t-\tau)\right]=0$, and $w(t)=0$. So, the Eq. 4 can be rewritten [7]

$$
\hat{R}(t,\tau) = \hat{R}_{s1}(t,\tau)+\hat{R}_{s2}(t,\tau).
\tag{5}
$$

The Fourier transformation of the Eq. 5 is

$$
\hat{S}_r^\alpha(f) = \hat{S}_{s1}^\alpha(f)+\hat{S}_{s2}^\alpha(f).
\tag{6}
$$

Namely, the cyclic spectrum of MCS is equal with the linear sum of its component signals' cyclic spectrum. In the theory, the MCS can be classified when the cyclic frequency set of the component signals are different each other. Therefore, the two component signals $s_1(t)$ and $s_2(t)$ can be separated by using the cyclic spectral envelope of signals.

3.2   The Feature Extraction

In this paper, based on the independence and symmetry of cyclic spectral correlation for the signals, the feature extracted can be obtained as

$$
F_1 = \frac{\left|\hat{S}_r^{\frac{2}{T_{c1}}}(f=f_{c1})+\hat{S}_r^{-\frac{2}{T_{c1}}}(f=f_{c1})\right|}{\left|\hat{S}_r^{\frac{1}{T_{c1}}}(f=f_{c1})+\hat{S}_r^{-\frac{1}{T_{c1}}}(f=f_{c1})\right|},
\tag{7}
$$

$$
F_2 = \left|\frac{\hat{S}_r^{2f_{c1}+\frac{1}{T_{c1}}}(f=0)+\hat{S}_r^{2f_{c1}-\frac{1}{T_{c1}}}(f=0)+\hat{S}_r^{-2f_{c1}+\frac{1}{T_{c1}}}(f=0)+\hat{S}_r^{-2f_{c1}-\frac{1}{T_{c1}}}(f=0)}{2\left(\hat{S}_r^{\frac{1}{T_{c1}}}(f=f_{c1})+\hat{S}_r^{-\frac{1}{T_{c1}}}(f=f_{c1})\right)}\right|.
\tag{8}
$$

Where the extracted feature invariants F1 and F2 are used to recognize the component $s_1(t)$ and the component $s_2(t)$. For the F1, the cyclic frequency set of component $s_1(t)$ is $\alpha=\pm1/T_{c1},\pm2/T_{c1},\pm2f_{c1}\mp1/T_{c1}$. According the Eq.6, the $\hat{S}_r^\alpha(f)=\hat{S}_{s1}^\alpha(f)$ can be got. Thus, based on the Eq. 7 and Eq. 8, the values of the

Table 1. The values of the feature invariants

Modulation type	F1	F2
BPSK	$\left\| \dfrac{Q^2\left(2f_{c1}+\dfrac{1}{T_{c1}}\right)+Q^2\left(\dfrac{1}{T_{c1}}\right)}{Q^2\left(2f_{c1}+\dfrac{1}{2T_{c1}}\right)+Q^2\left(\dfrac{1}{2T_{c1}}\right)}\right\|$	$\left\| \dfrac{Q^2\left(2f_{c1}+\dfrac{1}{T_{c1}}\right)+Q^2\left(\dfrac{1}{2T_{c1}}\right)}{Q^2\left(\dfrac{1}{2T_{c1}}\right)}\right\|$
QPSK	$\left\| \dfrac{Q^2\left(2f_{c1}+\dfrac{1}{T_{c1}}\right)+Q^2\left(\dfrac{1}{T_{c1}}\right)}{Q^2\left(2f_{c1}+\dfrac{1}{2T_{c1}}\right)+Q^2\left(\dfrac{1}{2T_{c1}}\right)}\right\|$	0
MSK	∞	∞

feature invariants can be shown in the Table 1 by using the cyclic spectral correlation of the component signal as the input.

Where $Q(f)=\sin(\pi fT_{c1})/\pi f$. From the Table 1, the multi-component PSK signals can be separated based on the F1 and F2.

4   Simulation Experiments

In this paper, the MCS (this is made of two component signals $s_1(t)$ BPSK and $s_2(t)$ QPSK) is used to test the valid of the proposed approach. The carrier frequency of $s_1(t)$ and $s_2(t)$ are respectively 2000Hz and 2048Hz, and the code speed of two components are also 563Hz and 512Hz, respectively. The sample frequency is 8192Hz. The two components are overlapped completely in time domain, and in frequency domain their overlapped degree is more 90%. The number of the sample points is 8192, and the number of the code is more 500.

The Fig.1 shows the various relationships of the feature value and signal-to-noise ratio (*SNR*). From the Fig.1, for the BPSK and QPSK, the value of the feature F1 is less 1. Furthermore, the value of F2 is 1 for BPSK while the value of F2 is 0 for QPSK. So, based on the feature F1 and F2, we can recognize the BPSK and QPSK in MCS.

When the signal power ratio is 1:1 and 1:2(Literature [9]), the experiment results in the Monte Carlo simulation for 200 times are shown in Fig.2. From the Fig.2, the proposed method is more effective than existing method. But the data need enough length in this paper. So, the time complex degree is larger.

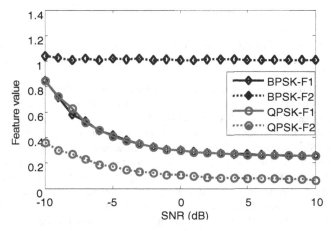

Fig. 1. The relationship of the feature value and *SNR*

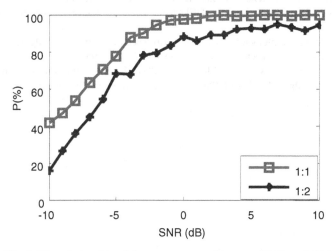

Fig. 2. The relationship of the correct recognition rate (*P* %) and *SNR*

5   Conclusions

Recognizing the time-frequency overlapped signal in single-channel receiver is a powerful challenge in electronic countermeasures. In this paper, an approach for judging directly the modulation type is presented to recognize BPSK and QPSK based on the cyclic spectrum analysis of the intercepted signals. Simulation results verified that the proposed algorithm can extract the features of each independently components without disturbance, and it can effectively recognize arbitrary combination of dual-signal in the given set.

Acknowledgments. This paper was supported by a grant from the National Natural Science Foundation of China (No. 60971103) and the Fundamental Research Funds for the Central Universities (No.SWJTU11BR026).

References

1. Spooner, C.M.: Classification of Co-Channel Communication Signals Using Cyclic Cumulants. In: Proc. of the 29th Asilomar Conference on Signals, Systems and Computers, pp. 531–536 (1995)
2. Spooner, C.M.: On the Utility of Sixth-Order Cyclic Cumulants for RF Signal Classification. In: Proc. of the 34th Asilomar Conference on Signals, Systems, and Computers, Pacific Grove, USA, pp. 890–897 (2001)
3. Yeste-Ojeda, O.A., Grajal, J.: Cyclostaionarity-based Signal Separation in Interceptors Based on a Single Sensor. In: IEEE Radar Conference, Rome, Italy, pp. 169–174 (2008)
4. Christensen, M.G., Jakobsson, A.: Optimal Filter Designs for Separating and Enhancing Periodic Signals. IEEE Trans. on Signal Processing 58(12), 5969–5982 (2010)
5. Gardner, W.A., Brown, W.A., Chen, C.K.: Spectral Correlation of Modulated Signals: Part II- Digital Modulation. IEEE Trans. on Commun. 35, 595–601 (1987)
6. Vucic, D., Obradovic, M.: Matrix-Based Stochastic Method for the Spectral Correlation Characterization of Digital Modulation. Electronics and Energetics 11(3), 271–284 (1998)
7. Jin, Y., Ji, H.B.: Influence of Stationary Noise on Cyclic-Autocorrelation Based PSK Symbol Rate Estimation. Journal of Electronics & Information Technology 30(25), 505–508 (2008)
8. Eric, A.: On the Implementation of the Strip Spectral Correlation Algorithm for Cyclic Spect-rum Estimation, Technical Note, Ottawa, Canada (1995)
9. Chen, H.W., Zhu, L., Wu, L.N.: Modulation Classification Algorithm for Jamming Signal Based on Cumulant. Journal of Electronics & Information Technology 31(7), 1741–1745 (2009)

Simulation on Inverter Fault Detection and Diagnosis of Brushless DC Motor

QiaoLing Yang[1] and HaiPing Zhang[2]

[1] College of Electrical and Information Engineering, Lanzhou University of Technology, 730050, Lanzhou, China
[2] Tianhua Research Institute of Chemical Machinery and Automation, 730050, Lanzhou, China
{winds-qiaoer,hapi1009}@163.com

Abstract. Power devices and its control circuit in inverter of BLDCM's speed regulation system is the weak link most likely occurs fault. Fault feature extraction is the key to fault diagnosis. The output voltage of BLDCM's inverter is the most sensitive characteristic parameter, which can reflect inverter's work state directly. Extracting it's spectrum by using improved windowed STFT, fault detection and diagnosis can be achieved. Simulation results demonstrate the effectiveness of this method.

Keywords: Inverter, BLDCM, Fault detection, Fault feature, Voltage spectrum, Power devices.

1 Introduction

Brushless DC motor(BLDCM) has simple structure, high efficiency, good speed performance etc. It's applications in the industrial field become more and more widespread with the rapid development of power electronics technology. Power devices and its control circuit in the inverter of motor's speed regulation system is the weak link most likely occurs fault. The problem about it's reliability has not been resolved completely[1]. New research shows that[2] faults of power converter is 82.5% of the whole drive system. So the fundamental problem to improve the safe operation and reliability of the system is prevent inverter's faults effectively.

Most systems take derating design or use redundant components or circuits in parallel way to reduce inverter's fault, what will make the power supply's cost too high, and only apply to the place where space conditions allow. In addition, there are others, such as based on analytical models method, based on knowledge method and based on signal processing method[3]. They have advantages and disadvantages, and can not be an ideal solution. In this paper, output voltage signals of inverter is regarded as the characteristic parameter. Extracting it's spectrum by using improved windowed STFT algorithm. Fault can be detected. Aritmetic operations of STFT is smaller than others, what can save computing time effectively. Simulation results demonstrate the effectiveness of this method.

D. Jin and S. Lin (Eds.): Advances in FCCS, Vol. 2, AISC 160, pp. 463–469.
springerlink.com   © Springer-Verlag Berlin Heidelberg 2012

2   Fault Mode and Characteristics Analysis

Figure1 is a BLDCM inverter drive system. Fault of inverter mainly by the following:1)one power devices open-circuit;2)one power devices short-circuit;3) two power devices in same arm open-circuit at the same time;4)one phase of motor open-ciecuit; When the system is running, power devices work in high-frequency switching status. It has larg power loss and most likely occurs fault[4]. The most common is open-circuit and short-circuit fault. No matter what kinds, extracting its features and find the location quickly and accurately is critical for fault treatment.

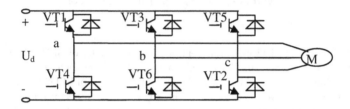

Fig. 1. BLDCM's inverter drive system

Output voltage reflects inverter's work status directly, it is the most sensitive characteristic parameter. Ideally, voltage in each-phase is symmetrical. Waveform of them is same. When a fault occurs, they become asymmetrical. Motor works on unbalanced power supply. Runing status of motor driving system can be detected effectively by analyzing the inverter's output voltage signal. During normal runing, inverter's output voltage in each phase Can be expressed as the following equation:

$$\begin{cases} u_a(t) = U \sin(\omega_0 t + \delta) \\ u_b(t) = U \sin(\omega_0 t + \delta - 2\pi/3) \\ u_c(t) = U \sin(\omega_0 t + \delta + 2\pi/3) \end{cases} \tag{1}$$

If make k_i equal to 1 or 0 to indicate the power tube is closed or disconnected, the voltage meet the following relationship:

$$\begin{cases} u_{ab} = k_a u_a - k_b u_b \\ u_{bc} = k_b u_b - k_c u_c \\ u_{ca} = k_c u_c - k_a u_a \end{cases} \tag{2}$$

where k_a、 k_b、 k_c is the switch function for a、 b、 c phase. k_i is dtermined by the modulation signal and carrier signal commonly. For SPWM inverter, it's double Fourier transform is[5]:

$$k_a = \frac{1}{2} + \frac{M}{2}\sin\omega_r t + \frac{2}{\pi}\sum_{m=1}^{\infty}\sum_{n=-\infty}^{\infty}\frac{1}{m}J_n(mM\frac{\pi}{2})\times\sin\left[(m+n)\frac{\pi}{2}\right]\sin(m\omega_c t + n\omega_r t)$$

$$k_b = \frac{1}{2} + \frac{M}{2}\sin(\omega_r t - \frac{2\pi}{3}) + \frac{2}{\pi}\sum_{m=1}^{\infty}\sum_{n=-\infty}^{\infty}\frac{1}{m}J_n(mM\frac{\pi}{2})\times\sin\left[(m+n)\frac{\pi}{2}\right]\sin\left[m\omega_c t + n(\omega_r t - \frac{2\pi}{3})\right]$$

$$k_c = \frac{1}{2} + \frac{M}{2}\sin(\omega_r t + \frac{2\pi}{3}) + \frac{2}{\pi}\sum_{m=1}^{\infty}\sum_{n=-\infty}^{\infty}\frac{1}{m}J_n(mM\frac{\pi}{2})\times\sin\left[(m+n)\frac{\pi}{2}\right]\sin\left[m\omega_c t + n(\omega_r t + \frac{2\pi}{3})\right]$$

(3)

where M is Modulation factor, ω_0 is angular frequency of modulate signal. ω_c is angular frequency of carrier signal, J_n is N-order Bessel function. So, u_{ab} is (4):

$$u_{ab} = \frac{\sqrt{3}}{2}MU\cos\delta - \frac{2U}{\pi}\sin(\omega_r t + \delta)\sum_{m=1}^{\infty}\sum_{n=-\infty}^{\infty}\frac{1}{m}J_n(mM\frac{\pi}{2})\times\sin\left[(m+n)\frac{\pi}{2}\right]\sin(m\omega_c t + n\omega_r t) +$$

$$\sin\left[m\omega_c t + n(\omega_r t - \frac{2\pi}{3})\right]\sin(\omega_r t + \delta - \frac{2\pi}{3}) + \sin\left[m\omega_c t + n(\omega_r t + \frac{2\pi}{3})\right]\sin(\omega_r t + \delta + \frac{2\pi}{3}) + \cdots$$

(4)

In this function, first part is DC component, second part is harmonics of carrier signal and others. If $\omega_c \gg \omega_0$, fault detection just need analyze low-frequency part. Low-frequency part of output voltage in each phase only contains DC component when inverter is runing normally. Either of the power switch failed, corresponding arm will not function properly. At this time, other low-frequency components will be contained. The following open-circuit fault occurs in VT1 tube, for example. A-phase no longer turn on. u_{ab} satisfy the following relationship:

$$u'_{ab} = k_a u'_a - k_b u'_b$$

(5)

It's no longer symmetrical. Negative DC component is included. Three-phase output currents sum to zero, A-phase and C-phase currents will contain negative DC component too[1]. Similarly, A-phase voltage become the following:

$$\begin{cases} u'_a = U_0/2 + \tilde{u}_a(t) \\ u'_b = -U_0 + \tilde{u}_b(t) \\ u'_c = U_0/2 + \tilde{u}_c(t) \end{cases}$$

(6)

where $\tilde{u}_a(t)$ 、 $\tilde{u}_b(t)$ 、 $\tilde{u}_c(t)$ are AC parts. u_{ab} can be written as follows:

$$u'_a = -\frac{U_0}{4} + \frac{\tilde{u}_b(t) + \tilde{u}_c(t)}{2} + \frac{M}{2}\sin(\omega_0 t - \frac{2\pi}{3})[\tilde{u}_b(t) - U_0] + \frac{M}{2}\sin(\omega_0 t + \frac{2\pi}{3})[\tilde{u}_c(t) + \frac{U_0}{2}]$$

$$+ \frac{2}{\pi}\sum_{m=1}^{\infty}\sum_{n=-\infty}^{\infty}\frac{1}{m}J_n(mM\frac{\pi}{2})\times\sin\left[(m+n)\frac{\pi}{2}\right]\sin\left[m\omega_c t + n(\omega_0 t - \frac{2\pi}{3})\right]\times(\frac{U_0}{2} + \tilde{u}_b(t)) +$$

$$\sin\left[m\omega_c t + n(\omega_0 t + \frac{2\pi}{3})\right](\frac{U_0}{2} + \tilde{u}_c(t))$$

(7)

It is clearly that when switch tube occurs open-circuit fault, there are not only DC component but also modulated signal and its harmonic frequency components in output voltage of each phase, what is not included when inverter is normal running. Therefore, we can select the voltage spectrum as a single-tube open-circuit fault characteristics.

3   Fault Characteristics Extraction

If the sampling frequency is an integer multiple of signal frequency, spectral leakage cann't occur when extracting the fault characteristics. But the fundamental frequency is fluctuant, it's difficult to keeping the relationship. Severe spectral leakage will affect the accuracy of signal's Fourier transformation[6,7].Windowed Fourier transformation for signal processing has become a widely used method. In this paper, an improved windowed short time Fourier transform(STFT) algorithm is used to obtain fault feature spectrum. The STFT of signal $x(t)$ is defined as follows:

$$STFT \ (f,\tau) = \int_{-\infty}^{\infty} [x(t)g(t-\tau)]e^{-j2\pi ft} \, dt \qquad (8)$$

which can be thought as Fourier transformation of $x(t)$ near time t. Concrete realization is devide the signal into a series of overlapping sub-segments by sliding the window function, assuming each segment is stable, and then make Fourier analysis. Resulting spectrum can characteriz time-frequency distribution of signal. Hardware implementation of discrete Fourier transform is complex. R. N. Bracewell has given the fast Hartley algorithm (DHT),which is defined as[8]:

$$X_H(k) = \sum_{n=0}^{N-1} x(n)\cos(kn2\pi/N) + \sum_{n=0}^{N-1} x(n)\sin(kn2\pi/N) \qquad (9)$$

For the real $x(n)$, $X_H(k)$ is real too.So,

$$X_H(N-k) = \sum_{n=0}^{N-1} x(n)\cos(kn2\pi/N) - \sum_{n=0}^{N-1} x(n)\sin(kn2\pi/N) \qquad (10)$$

The transformation between DFT and DHT is[9]:

$$X_F(k) = \frac{1}{2}[X_H(k) + X_H(N-k)] - \frac{j}{2}[X_H(k) - X_H(N-k)]$$

$$X_H(k) = \frac{1}{2}(1+j)X_F(k) + \frac{1}{2}(1-j)X_F^*(k) \qquad (11)$$

Deviding a signal sequence whose length N is 2m into two sequence whose length become N/2, that is[10]:

$$X_H(k) = \sum_{n=0}^{N/2-1} x(2n)cas(2kn2\pi/N) + \sum_{n=0}^{N/2-1} x(2n+1)cas[k(2n+1)2\pi/N] \qquad (12)$$

Where $cas(x) = \cos(x) + \sin(x)$, and $cas(x+y) = cas(x)\cos(y) + cas(-x)\sin(y)$, so

$$X_H(k) = \sum_{n=0}^{N/2-1} x(2n)cas(\frac{2\pi}{N/2}kn) + \cos(\frac{2\pi}{N}k)\sum_{n=0}^{N/2-1} x(2n+1)cas(\frac{2\pi}{N/2}kn)$$

$$+ \sin(\frac{2\pi}{N}k)\sum_{n=0}^{N/2-1} x(2n+1)cas(-\frac{2\pi}{N/2}kn) \qquad (13)$$

where $0 \le k \le N - 1$ .and $cas(-\dfrac{2\pi}{N/2}kn) = cas[\dfrac{2\pi}{N/2}k(\dfrac{N}{2} - n)]$ So,(13) can be writen as (14):

$$X_H(k) = X_{H2n}(k) + X_{H(2n+1)}(k)\cos(\dfrac{2\pi}{N}k) + X_{H(2n+1)}(\dfrac{N}{2} - k)\sin(\dfrac{2\pi}{N}k) \qquad (14)$$

where $X_{2n}(k)$ is DHT of $X(2n)$, and $X_{2n+1}(k)$ is DHT of $X(2n+1)$. Similarly, the following can be obtained:

$$X_H(k) = X_{H2n}(k) - X_{H(2n+1)}(k)\cos(\dfrac{2\pi}{N}k) - X_{H(2n+1)}(\dfrac{N}{2} - k)\sin(\dfrac{2\pi}{N}k)$$

$$X_H(k) = X_{H2n}(\dfrac{N}{2} - k) + X_{H(2n+1)}(k)\sin(\dfrac{2\pi}{N}k) - X_{H(2n+1)}(\dfrac{N}{2} - k)\cos(\dfrac{2\pi}{N}k) \qquad (15)$$

$$X_H(k) = X_{H2n}(\dfrac{N}{2} - k) - X_{H(2n+1)}(k)\sin(\dfrac{2\pi}{N}k) + X_{H(2n+1)}(\dfrac{N}{2} - k)\cos(\dfrac{2\pi}{N}k)$$

According to the above, spectrum of signal sequence can be obtained. It is clearly that aritmetic operations of STFT is small, what can effectively save computing time.

4   Simulation Resaults and Analysis

To confirming our findings, simulations on one power device open-circiut of BLDCM's inverter is performed.   Motor parameters are follows. rated speed is 3000r/min, stator windings R=0.356Ω, L=0.0028H, J=0.002kg.m2, p=1. Frequency of modulated signal is 60Hz. Both normal runing and fault status are simulated. The following are the resaults:

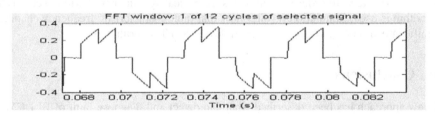

Fig. 2. Waveform of output voltage $u_{ab}(t)$

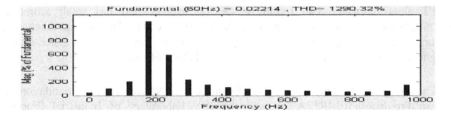

Fig. 3. Spectrum of $u_{ab}(t)$

Where $u_{ab}(t)$ is output voltage waveforms when inverter runing normally. when VTI occurs open-circuit fault, the waveform and spectrum of $u_{ab}(t)$ has significant changes. The following are waveform and spectrum of $u'_{ab}(t)$.

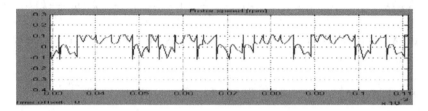

Fig. 4. Waveform $u'_{ab}(t)$

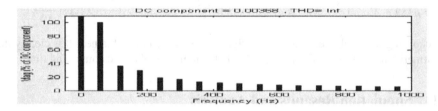

Fig. 5. Spectrum of $u'_{ab}(t)$

When single-tube occurs open-circuit fault, there are modulation signal and its harmonics components in output voltage spectrum. Compared fault signal with the normal signal, the spectral characteristics change significantly. Fault diagnosis can achieve accurately. In order to minimize spectral leakage, improved windowed STFT algorithm is used. Simulation results show that fault characteristic frequency changed significantly after transformed by the windowed STFT algorithm.

5   Conclusions

A new approach has been described what can detect and diagnose fault of BLDCM's inverter. When inverter has single-tube open-circuit fault ,the output voltage characteristics is different from normal runing. DC component is contains in it,which can be regard as fault feature.Extracting its spectrums by using improved windowed STFT algorithm,which can minimize spectral leakage and has small aritmetic operations,and comparing to normal spectrums,the fault can be detected and diagnosed accurately.So the output voltage spectrums can be the basis for fault detedte and diagnose in BLDCM.Simulation results validated this method.

Acknowledgements. This work was supported by the Natural Science Foundation of Gansu Province (1014RJZA024) and the Key Laboratory of Industrial Process advanced Control, Gansu Province (XJK0909).

References

1. Kastha, D., Bose, B.K.: Investigation of fault modes of voltage-fed inverter system for induction motor drive. IEEE Transactions on Industry Application 30(4), 1028–1038 (1994)
2. Wikstron, P.W., Terens, L.A., Kobi, H.: Reliability, availability, and maintainability of high-power variablespeed drive systems. IEEE Transactions on Industry Application 36(1), 231–241
3. Tao, C., ShanXu, D., Yong, K.: A survey of fault diagnosis technology for power electronics system. Electrical Measuement and Instrumentatioon 45(509), 1–7 (2008)
4. Tang, Q., Yan, S., Liu, S., et al.: Open-circuit fault diagnosis of transistor in three-level inverter. Proceedings of the CSEE 28(21), 26–32 (2008)
5. Jin, M.: Reserch on power electronics system of conducted interference modeling and prediction methods. Naval University of Engineering Press, WuHan (2006) (in Chinese)
6. Salvatore, L., Trotta, A.: Flap-top windows for PWM waveform processing via DFT. IEE Proc. Pt. B 135(6), 346–361 (1988)
7. Nuttal, A.: Some windows with very good sidelobe behavior. IEEE Trans. Acoustics, Speech, and Signal Processing 29(1), 84–91 (1981)
8. Bracewell, R.N.: Discrete Hartley transform. Oxford Univ. Press, Britain (1986)
9. Dee, H.S., Jeoti, V.: Computing DFT using approximate fast Hartley transform. In: Sixth International Symposium on Signal Processing and its Applications, vol. 8(13-16), pp. 100–103 (2001)
10. Bracewell, R.N.: The Fast Hartley Transform. Proc. IEEE (72), 1010–1018 (1984)

Static Pruning of Index Based on SDST

Lin Huo[*], Xianze Zou, Xiao Xing, and Ying Zhao

College of Computer and Electronic Information, Guangxi University,
Nanning, Guangxi, China, 530004
{nnxhy,hardhardstudy}@163.com, zouxianze@gmail.com,
oozhaoyingoo@126.com

Abstract. As an efficient index structure, Streamline Dynamic Successive Trees (SDST) is very suitable for Chinese information retrieval. Given the dependencies between terms in SDST and Chinese scenarios, we present a new bigram-centric pruning strategy and corresponding algorithm, utilizing our improved BM25 formula, to decide whether certain index information should remain in the index or not. This technology can be used to significantly reduce the size of SDST index, and still gives good enough answers. Thus, we make SDST index model more applicable in Chinese information retrieval.

Keywords: SDST, information retrieval, index pruning, index compression.

1 Introduction

Explosion of textual information urges the search engines to index collections of unprecedented size, which results in huge index that means not only occupation of large amount of disk space, but also more disk operations to access the index file. Thus, it's significant to utilize efficient compression scheme for index files.

Generally, there are two compression schemes: lossless compression and lossy compression. Under lossless scheme, the index files have a very compact representation by using efficient encoding mode. However, studies about the application of static index pruning, a lossy scheme, on Chinese information retrieval have been virtually nonexistent so far in terms of present references available. And there has been a lot of work about application of static index pruning on English information retrieval, with satisfactory results. Given the differences between English and Chinese, we choose another novel index model—SDST that is very suitable for Chinese information retrieval. Streamline Dynamic Successive Trees (SDST) incorporate merits of Inverted Files and Inter-Relevant Successive Trees model. The disadvantage of this model lies in high expansion ratio (ER) of its index. Thus, apply static pruning on SDST index has great theoretic value and practical sense.

2 Streamline Dynamic Successive Trees

Streamline Dynamic Successive Trees, denoted by SDST, is the data structure that is very suitable for implementing Chinese information retrieval. Lin Huo et al. [1]

---

[*] Corresponding author.

D. Jin and S. Lin (Eds.): Advances in FCCS, Vol. 2, AISC 160, pp. 471–476.
springerlink.com   © Springer-Verlag Berlin Heidelberg 2012

provided a complete description of this index model. In order to show our bigram-centric pruning strategy, a schematic drawing that illustrates this strategy based on SDST is introduced below. As shown in Fig. 1, *Root List* denotes a list contains all terms in root of SDST. *Leaf Lists* denotes a large amount of lists contain terms in leaf of SDST. In view of the dependencies between terms in *Root List* and *Leaf List*, that is, a term in *Leaf List* must directly depend on a term in *Root List*, we call such combination *(rt, lf)* bigram, denoted by *bi* . *Leaf Information Lists* denote a large amount of lists contain leaf information or index information. Each entry of *Root List* corresponds to a *Leaf List* and each entry of a *Leaf List* a *Leaf Information List*. Each entry of a *Leaf Information List*, denoted by *P((rt, lf), d)* or *P(bi, d)*, is a pointer that indicates the number d of document contains the current bigram *(rt, lf)*, absolute location of both *lf* and its successive term in the document d. $d_1, d_2, d_3, \ldots, d_N$ are document numbers and N is the total number of documents in collection. M is the number of terms that can be put into *Root List* and *Leaf List*. And the pruning of index based on SDST is a process to prune or delete such pointers in Fig. 1.

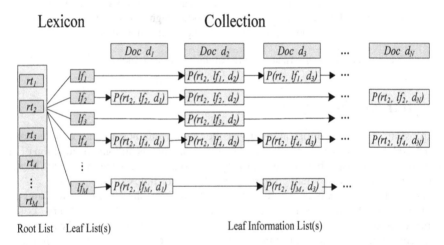

Fig. 1. A schematic drawing that illustrates the static pruning strategy

3   Bigram-Centric Pruning Strategy

D. Carmel et al. [2] introduced the concept of static index pruning to information retrieval. In their paper, they describe the static pruning method on inverted files under English context, and obtain satisfactory query performance. Given the dependencies between terms in SDST index, we present a new static index pruning strategy that follows a bigram-centric approach to decide whether certain index information should remain in the index or not. In order to improve precision at certain pruning level and adapt to Chinese scenarios, we add several improvements to BM25 weighting scheme by considering some key design features. The decision is made based on the bigram's contribution to the document's relevant score, utilizing our improved BM25 formula. And the initial form of BM25 [4] is:

$$S_{BM25}(i,d) = \sum_{i \in Q} \log \frac{(r_i + 0.5)/(R - r_i + 0.5)}{(n_i - r_i + 0.5)/(N - n_i - R + r_i + 0.5)} * \frac{(k_1 + 1)f_i}{K + f_i} * \frac{(k_2 + 1)qf_i}{k_2 + qf_i}. \quad (1)$$

where the summation over all terms (i) in the query Q; r_i is the number of relevant documents containing term i, n_i is the number of documents containing i, N is also the total number of documents in the collection, and R is the number of relevant documents for this query Q, f_i is the frequency of i in the document, qf_i is the frequency of i in the query, and k_1, k_2 are parameters whose values are set empirically. K is a factor and specifically, $K = k_1((1-b) + b * (bl/avgdl))$ where b is a parameter, dl is the length of the document, and $avgdl$ is the average length of documents in the collection. The first part of the Eq. 1's right reflects relevance feedback or just idf weighting if no relevance information is available; the second implements document term frequency and document length scaling, and third considers term frequency in the query.

Due to lack of relevance feedback information, it is reasonable that r_i (or r_{bi}) and R in Eq. 1 are set to zero, which changes the first part to $\log (N - n_{bi} + 0.5)/(n_{bi} + 0.5)$. The discerning reader could not fail to sense that if there is a bigram that the number of document contains it larger than $N/2$, the first part would be a negative value. In contrast of the general practice that directly deletes such terms or bigrams, we add 1.0 to $(N - n_{bi} + 0.5)/(n_{bi} + 0.5)$. Adding 1.0 in this way is a simple form of smoothing to avoid the possibility of negative weight. And the third part that considers term frequency in the query is unnecessary as the static index pruning is independent of any queries, which means we can set the third part to be a constant 1. There are two scaling parameters in the BM25 scoring function, i.e. b, k_1. We use the recommended values as b =0.75, and k_1=1.2, respectively. As for dl and $avgdl$, we define dl the number of document's unique bigrams that been put into SDST index and update dl and $avgdl$ after pruning because of their changeability during the pruning process. Thus, the final formula of BM25 weighting scheme employed in our paper to compute a document's relevant score is as follows:

$$S_{BM25}(bi,d) = \sum_{bi \in Q} \log(\frac{N - n_{bi} + 0.5}{n_{bi} + 0.5} + 1.0) * \frac{2.2 * f_{bi}}{f_{bi} + 1.2 * (0.25 + 0.75 * (dl/avgdl))}. \quad (2)$$

The scoring formula (2) works based on a 2-dimensional scoring table A, indexed by bigrams and documents. Table entries are set by the following Equation:

$$A(bi,d) = \log(\frac{N - n_{bi} + 0.5}{n_{bi} + 0.5} + 1.0) * \frac{2.2 * f_{bi}}{f_{bi} + 1.2 * (0.25 + 0.75 * (dl/avgdl))}. \quad (3)$$

We assume that $A((rt,lf),d)$ (or $A(bi,d)$) $=0$ if a bigram bi or (rt,lf) does not occur in d, and $A((rt,lf),d)>0$ otherwise. In our bigram-centric pruning strategy, we first compute $A((rt,lf),d)$, the bigram's contribution to the document's relevant score, for all pointers in SDST index and then delete those pointers with a low score. To guarantee search reliability, we set a pruning threshold for each row of table A and left the short rows unpruned.

4   Bigram-Centric Pruning Algorithm and Algorithm Analysis

According to bigram-centric pruning strategy, we offer the corresponding pruning algorithm as follows.

Input : I is the original or unpruned index files;

   k, μ are two user-specified parameter for controlling pruning level.

Output :   pruned index files at different pruning level.

$prune(I,k,\mu)$

   For each term rt in rootList of I

      For each term lf in leafList of rt

         Retrieve the pointers list $P_{(rt,lf)}$ corresponding to the (rt,lf)

         If $\left|P_{(rt,lf)}\right|>k$ $(k>1)$

            For each pointer $P((rt,lf),d)\in P_{(rt,lf)}$

               Compute $A((rt,lf),d)$; (By Equation (3))

               Sort and let z_t be the k-th highest score in row (rt,lf) of A

            $\tau_t=\mu*z_t$; $(0<\mu<1)$

            For each pointer $d\in P_{(rt,lf)}$

               If $A((rt,lf),d)\le\tau_t$

                  Remove the pointer $P((rt,lf),d)$ from $P_{(rt,lf)}$

For each bigram, the pruning algorithm needs to compute $A((rt,lf),d)$ and the time complexity is $O(N)$; Then sort the entries in table A and the general quicksort can reach the performance of $N*O(logN)$; The cost of remove operation is a constant, but it must traverse all pointers in index, so its complexity is $O(N)$; And thus the time complexity of the pruning algorithm is $O(NlogN*M^2)$. M, the number of terms in SDST index, is limited because we segment a Chinese character as a term in *Root List* and *Leaf List*. Thus, the time complexity of the pruning algorithm has the direct proportion with the document numbers N in the collection.

5   Experimental Evaluations

For the dataset, we use the Sougou corpus of Chinese categorized text which has about 18,000 text documents. We use the standard measures as in [3], i.e. mean average precision (MAP) and precision at 10 (P@10) and precision at 20 (P@20), to evaluate the retrieval performance for different pruning levels. The percent of the index is defined as the ratio of the number of pointers in the pruned index to that in the unpruned index. In all experiments, we do not use lossless compression method on SDST index. We create a sequence of pruned SDST index, where use varying values of k, μ in the pruning algorithm to reach the corresponding pruning level. During the query processing, we manually choose ten query strings to perform query on pruned index, and then average the values of P@10、P@20 and MAP.

Fig. 2 plots the precision at 10 and 20 of our search results at varying pruning level. As shown in Fig. 2, we can easily find out that at the pruning level of about 66%, we also receive the same P@10 as the full index. What is encouraging, we can obtain better P@10 and P@20 than the original index when the pruning level is low. Therefore, not very large pruning level of the SDST index is promising in improving the user experience. From the Fig. 2, we can also observe that the P@20 is more sensitive to the pruning than P@10, as P@20 experiences a larger decrease than P@10 at the same pruning level of the SDST index.

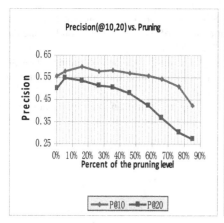 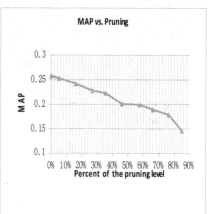

Fig. 2. Precision at varying levels   **Fig. 3.** MAP at varying levels

Fig. 3 plots the MAP of our search results at varying levels of pruning. As shown in Fig. 3, we can find out that the MAP experiences monotone decreasing along with the gradually rise of pruning level of index. That is, MAP is more sensitive or weak than both the P@10 and P@20. Theoretically speaking, the reason may well be that MAP considers the sequence of returned documents, but the first two measures does not.

6   Concluding Remarks

In this work we present our new bigram-centric pruning strategy and corresponding algorithm that reduce the size of SDST index by removing the least important pointers from the index, in terms of our improved BM25 formula. Our experimental results show that, in the Chinese context, we can retrieve satisfactory results even at higher pruning level. Therefore, we make SDST index model more applicable in Chinese information retrieval.

There are also many open questions. For example, we can consider the effect of many lossless compression methods after our lossy compression on SDST index. We can also consider more factors to reevaluate the bigrams' contribution to a document for the purpose of higher performance.

Acknowledgments. This work is supported by Projects of education department in Guangxi (20100712); Science and technology research project of science and technology agency in Nanning, Guangxi district(20100791).

References

1. Huo, L., Huang, J., Lu, Z., et al.: Research on Streamline Inter-relevant Successive Trees. Journal of Chinese Computer Systems (02) (2011)
2. Carmel, D., Cohen, D., Fagin, R., et al.: Static index pruning for information retrieval systems. In: Proceedings of the 24th International ACMSIGIR Conference on Research and Development in Information Retrieval, pp. 43–50. ACM, New York
3. Blanco, R., Barreiro, A.: Boosting static pruning of inverted files. In: Proceedings of the 30th Annual International ACMSIGIR Conference on Research and Development in Information Retrieval, pp. 777–778. ACM, New York (2007a)
4. Croft, W.B., Metzler, D., Strohman, T.: Search Engines: Information Retrieval in Practice. Addison Wesley (2009)

Development of Free Adjustable Function Generator for Drop-on-Demand Droplets Generation

Yan Wu[1] and Shengdong Gao[2]

[1] College of Information and Computer Engineering, Northeast Forestry University,
Harbin, 150040, China
ottp2002@163.com
[2] School of Mechatronics Engineering, Harbin Institute of Technology,
Harbin, 150001, China
sdgao@hit.edu.cn

Abstract. This paper discuss the generation of uniform droplets with DOD dsdroplet. DOD droplets generation system with an adjustable function generator based on PC and software of LabVIEW was setup. For comparing the droplets ejecting process, four type of adjustable signal were used to drive the DOD droplet generator. Finally, the experimental and simulated results of the droplets formation were discussed.

Keywords: drop-on-demand, uniform droplet, function generator, LabVIEW.

1 Introduction

Uniform droplets with stable ejection speed are required for many areas of research over the past 30 years. Their applications include DNA synthesis, microfluidics, combinatorial chemistry, combustion science and so on. The droplet generator can also be used for the development of particle sizing instruments, the calibration of optical measurement system and much more. In rapid prototyping manufacturing technology, metal parts, solder bump of integrated circuit array package and DNA-sensors have been applied with uniform droplets generator systems[1,2].

There are two types of uniform droplets generators, termed continuous droplets generation and drop-on-demand (DOD) droplets generation system. In continuous droplets generation, a stream of equally spaced droplets is formed by breaking up a cylinder liquid jet with a periodic disturbance introduced by a piezoelectric transducer. In drop-on-demand droplets generation, individual droplets are ejected by a pressure wave also introduced by a piezoelectric crystal only when needed [3,4]. So the DOD droplets generation is much more flexible than the continuous one. In the DOD droplets generation system, the formation of uniform droplets is depended on the driving signals. Adjustable driving signal will help to get stable, uniform droplets with DOD droplets generator [5].

In this paper, DOD droplets generation system with an adjustable function generator will be setup. Then the generation of the driving signal will be explained. Finally, some experimental and simulated results of uniform droplets formation are presented.

D. Jin and S. Lin (Eds.): Advances in FCCS, Vol. 2, AISC 160, pp. 477–481.
springerlink.com
© Springer-Verlag Berlin Heidelberg 2012

2   Experiment Setup

2.1   Droplets Generation System

Fig.1 shows a conceptual schematic of the DOD droplets generation system. The apparatus mainly consists of adjustable function generator, power amplifier, CCD camera, strobe light, and droplet generator with liquid supply.

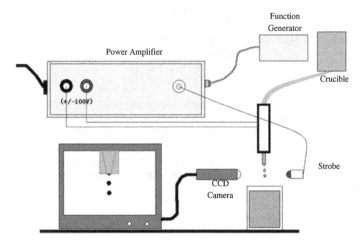

Fig. 1. Schematic of droplet generation system

For the reason that the generation of uniform droplets is influenced by signal types, driving signals with adjustable parameters are essential to droplet generator. In this paper, LabVIEW is applied as the functional generator, which converts digital signals to analog signals by Data Acquisition Card. Power amplifier was used to amplify the signal up to 100V which drives the droplet generator. For visualization of the formation of the uniform droplets, a CCD-camera is used in combination with a strobe for a transmitted light illumination of the droplet stream. The camera and strobe are placed on a separate traverse unit to enable changes of the position without disturbing the drop streams. The signal which drives the strobe is synchronized with the drive signal for the drop generators. As a result, the formation of the droplets can be recorded by the CCD-camera as image.

2.2   Free Adjustable Function Generator

Due to the influence of signal types to the formation of uniform droplets, it is important to generate different types of waveforms. The piezoelectric driving signal is generated with software (LabVIEW) and a ditital-analog converter. A PC is used to generate the signals and then the signals can be amplified with a power amplifier. The parameters of the waveform can be adjusted on the front panel of the software. Fig.2 shows the screenshot of the front panel of the function generator.

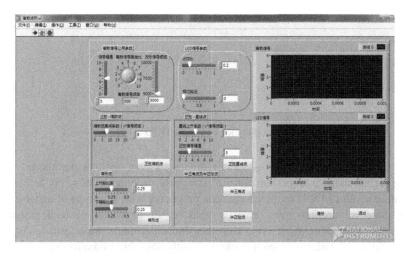

Fig. 2. The front panel of adjustable function generator programmed with LabVIEW

Four types of driving signals are programmed with LabVIEW (National Instruments), they're sinusoidal waveform, trapezoid waveform, sinusoidal-linear waveform, and sinusoidal-exponential waveform, shown in Fig.3.

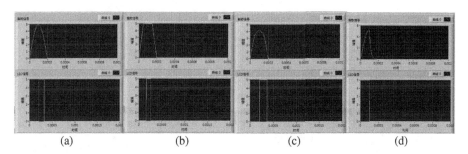

(a)	(b)	(c)	(d)

Fig. 3. Driving signals for piezoelectric with LabVIEW (a) sinusoidal waveform (b) trapezoid waveform (c) sinusoidal-linear waveform (d) sinusoidal-exponential waveform

3   Experiments and Simulation

Fig.4 shows a small droplet ejected from a DOD droplet generator with a nozzle of 100 micrometer in diameter, and the size of the droplet is also about 100 micrometer in diameter. Molten wax at 100°C was used as the droplet substrate. The driving signal shape in DOD droplet generator is a sinusoidal-exponential waveform. For the parameters used in experiments, a long tail is found after the droplet ejected from the nozzle. After a relative long flight distance, the tail coalesce with the main droplet to form a final droplet. In most cases, droplet with long tail or even satellite is needed to be avoided.

Fig. 4. A small droplet ejected from a DOD droplet generator

For the same DOD droplet generator, it has its own optimal parameter. So the driving signal shape is important for the process of droplet ejection. Fig.5 shows the simulation result of DOD droplet formation procedure with four type of driving signal shape. Satellite droplets may emerge with trapezoid waveform and sinusoidal-exponential waveform. It is found that the sinusoidal-linear waveform leads to the stable ejection of uniform droplets compared with other three waveforms.

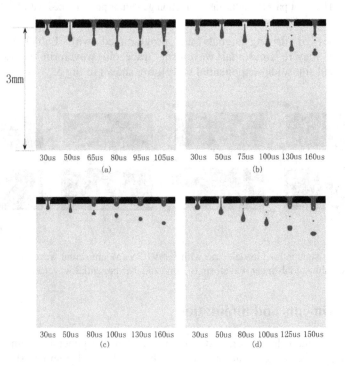

Fig. 5. DOD droplet formation procedure with different driving signal shape (a) sinusoidal waveform (b) trapezoid waveform (c) sinusoidal-linear waveform (d) sinusoidal-exponential waveform

4 Summary

The principle of the formation of uniform droplets with DOD droplet generator is the compression of a small chamber, so the liquid is ejected from the nozzle to form a small droplet under surface tension. Droplets ejected from the DOD droplet generator usually have the same size as the nozzle. The driving signal shape is an important parameter for the process of droplet formation. An adjustable driving signal can be used to optimize the droplet ejecting process. As a result, sinusoidal-linear waveform is much suitable for the DOD droplet generator built in this paper.

References

1. Lee, E.R.: Microdrop Generation. CRC Press, Boca Raton (2003)
2. Riefler, N., Schuh, R., Wriedt, T.: Investigation of a measurement technique to estimate concentration and size of inclusions in droplets. Meas. Sci. Technol. 18, 2209–2218 (2007)
3. Li, L., Saedan, M., Feng, W., Fuh, J.Y.H., Wong, Y.S., Loh, H.T., Thian, S.C.H., Thoroddsen, S.T., Lu, L.: Development of a multi-nozzle drop-on-demand system for multi-material dispensing. Journal of Materials Processing Technology 209, 4444–4448 (2009)
4. Laser, D.J., Santiago, J.G.: A review of micropumps. J. Micromech. Microeng. 14, 35–64 (2004)
5. Riefler, N., Wriedt, T.: Generation of Monodisperse Micron-Sized Droplets using Free Adjustable Signals. Part. Part. Syst. Charact. 25, 176–182 (2008)

The JIT Equipment Procurement Model with Reward System and Its Application in Nuclear Power Project

RuiYu Liu and Hui Guo

Economics and Management School, Wuhan University, Wuhan 430072, China
Liury0417@sohu.com, guohuihui577@163.com

Abstract. Proceeding from the particularity of equipment procurement in the nuclear power project, combined the reward system phenomenon exisiting in the supply chain of the nuclear power project, the complete information twostage Stackelberg master-slave game theory is applied to set up the equipment procurement model in the nuclear power project. By analysing and comparing with the results of the model, we finally conclude the scope of the application of the reward system, the relationship between the reward system and the delivery time, the parameters which affect the delivery time, and justify the JIT equipment procurement model with reward system existing in the nuclear power project in theory.

Keywords: equipment procurement, JIT model, reward system, Stackelberg master-slave game theory.

1   Introduction

Since 2004, China's nuclear power projects have developed rapidly. In order to achieve the scale effects of nuclear power project management and equipment procurement, the EPC pattern is adopted by more and more projects[1].And for any EPC project, the procurement is a connecting link between the preceding and the following. The EPC logical relationship can be seen in figure 1.

As everyone knows, the equipment procurement is the center of the procurement. As estimated, the investment proportion of the equipment procurement accouts for about 55% of the whole project[2]. Meanwhile, the supply of the nuclear power equipment is arranged moer closely under the EPC mode. So the delay of the schedule may bring about inestimable effect. To improve the effiency of equipment procurement management makes for reducing project cost and controlling project schedule.

In this article, the particularity of the nuclear power project's equipment procurement and the composition of both sides' cost are firstly analyzed. Then the complete information twostage Stackelberg master-slave game theory is applied to set up the JIT equipment procurement model with reward system. Lastly by analysing and comparing with the results, we finally conclude the scope of the application of the reward system, the relationship between the reward system and the delivery time, the parameters which affect the delivery time, and prove that this model can reduce the cost of both sides effectively in theory.

D. Jin and S. Lin (Eds.): Advances in FCCS, Vol. 2, AISC 160, pp. 483–488.
springerlink.com   © Springer-Verlag Berlin Heidelberg 2012

Fig. 1. The EPC logical relationship

2   Model Analysis

2.1   The Analysis of the Particularity of Nuclear Power Project's Equipment Procurement

For the equipment procurement of nuclear power project, the Supplier won't make the production plan until the General Contractor sent the demand order. In this MTO way of production, the JIT procurement mode can reduce the order response time as possible for the supplier and can improve the benefits of both[3].The supply chain management based on JIT procurement emphasizes the long-run winwin coorperation with the Supplier. This JIT management mode, which ensures communication and feedback about information in time, helps the Project Manager control the quality and schedule during the process of production, so as to make sure that the nuclear power project operates safely and continually.

In general, according to the schedule implemention carried by the Supplier, the Superviser of the nuclear power project would take some reward measures to the Supplier[4]. In this situation, the General Contractor would like to make over part of his income to the Supplier, in order to reduce extra losses caused by the delay in delivery. In this JIT model with reward system, just because of the existence of reward system, the General Contractor can be seen as a leader, who firstly decides the size of the variable "reward", the Supplier can be considered as a follower, who would consult the delivery time with the consideration of the reward size and his own ability of production and logistics[5].

2.2   The Analysis of the Composition of Both Sides' Cost

In the process of procurement, time is a principal factor of both sides' concern. As the procurement would be implemented exactly according to the purchasing schedule, if the delivery happens in advance, the inventory cost of the Contractor will increase; if not, the Contractor will burden the shortage cost. While the Supplier would make his production plan according to his own ability of production and logistics, if the agreed delivery time is too long, it will incease the inventory cost of the Supplier; if too short, the Supplier will burden the crash cost. So the length of delivery time derectly affects the cost and benefits of both sides.

As for the General Contractor, the procurment cost mainly consists of equipment cost、 reward cost and shortage cost. Considering the equipment cost which is decided by price and quality has no relation with the delivery time, the equipment cost is taken as a exogenous variable; the reward cost is just part of the Contractor's benefit, which is made over to the Supplier to ensure the timeliness of delivery; the shortage cost is the increased investment caused by the delay of delivery, which grows as the delay time goes up.

Here delivery in advance is not permitted. In order to simplify the model, the inventory cost which is produced during the process of the production and the daily productive cost are out of consideration, as they have no relation with the delivery time. So the cost of the Supplier only results from the order completed in advance or the late delivery. The former reason causes the inventory cost which is produced during the interval between the completion and the delivery, having a positive relationship with the delivery time. The later one causes the crash cost, which has a negative relationship with the delivery time.

As the production time is a random variable and obeys the exponential distribution, here is the first assumption that the order response time also obeys the exponential distribution. And another assumption is that the decision of both sides is based on complete information. The objective functions in this paper are minimizing the expected cost of both sides.

3   The Model and the Solution

3.1   The Definition of Related Variables and Parameters

T : The agreed order response time by both sides, in this period delivery in advance is not permitted;

t : The actual order response time, it is assumed as a random variable which obeys the exponential distribution;

$f(t)$: The density function of the variable"t", and

$$f(t) = \begin{cases} \lambda e^{-\lambda t} & t > 0 \\ 0 & others \end{cases}, \ \lambda > 0, \frac{1}{\lambda}$$ is the Supplier's average order response

time ;

C_1 : the General Contractor's unit shortage cost, $C_1 > 0$;

C_2 : the Supplier's unit inventory cost, $C_2 > 0$;

C_3 : the Supplier's unit crash cost, $C_3 > 0$;

a :   the unit reward given to the Supplier for in-time delivery;

EC_p : the expected cost of the General Contractor;

EC_s : the expected cost of the Supplier

3.2   The Establishment and the Solution of the Model

The objective functions are minimizing the expected cost of both sides.
So the function of the General Contractor can be expressed as follows:

$$\min EC_p(a) = a \int_0^{T(a)} f(t)dt + c_1 \int_{T(a)}^{\infty} (t - T(a))f(t)dt \qquad (1)$$

The function of the Supplier is:

$$\min EC_s(T) = c_2 \int_0^T (T-t)f(t)dt + c_3 \int_T^{\infty} (t-T)f(t)dt - a \int_0^T f(t)dt \qquad (2)$$

Solve the game process by backward induction, the equilibrium solutions are[6]:

$$T = \begin{cases} \dfrac{1}{2\lambda}\ln(\dfrac{c_2 + c_3 + c_1}{c_2}), & \text{if } c_1 > c_3 + \dfrac{c_3^2}{c_2} \qquad (3.1) \\[3mm] \dfrac{1}{\lambda}\ln\left(\dfrac{c_2 + c_3}{c_2}\right), & \text{if } c_1 \leq c_3 + \dfrac{c_3^2}{c_2} \qquad (3.2) \end{cases} \qquad (3)$$

$$a = \begin{cases} \dfrac{1}{\lambda}\left(\sqrt{c_2(c_2 + c_3 + c_1)} - c_2 - c_3\right), & \text{if } c_1 > c_3 + \dfrac{c_3^2}{c_2} \qquad (4.1) \\[3mm] 0 & \text{, if } c_1 \leq c_3 + \dfrac{c_3^2}{c_2} \qquad (4.2) \end{cases} \qquad (4)$$

$$EC_p = \begin{cases} \dfrac{\sqrt{c_2(c_2 + c_3 + c_1)} - c_2 - c_3}{\lambda}\left(1 - \sqrt{\dfrac{c_2 + c_3 + c_1}{c_2}}\right) + \dfrac{c_1}{\lambda}\sqrt{\dfrac{c_2 + c_3 + c_1}{c_2}}, & \text{if } c_1 > c_3 + \dfrac{c_3^2}{c_2} \qquad (5.1) \\[3mm] \dfrac{c_2}{\lambda} \cdot \dfrac{c_1}{c_2 + c_3} & \text{, if } c_1 \leq c_3 + \dfrac{c_3^2}{c_2} \qquad (5.2) \end{cases} \qquad (5)$$

$$EC_s = \begin{cases} \dfrac{c_2}{2\lambda}\ln(\dfrac{c_2 + c_3 + c_1}{c_2}) - \dfrac{\sqrt{c_2(c_2 + c_3 + c_1)} - c_2 - c_3}{\lambda}, & \text{if } c_1 > c_3 + \dfrac{c_3^2}{c_2} \qquad (6.1) \\[3mm] \dfrac{c_2}{\lambda} \cdot \ln\left(\dfrac{c_2 + c_3}{c_2}\right) & \text{, if } c_1 \leq c_3 + \dfrac{c_3^2}{c_2} \qquad (6.2) \end{cases} \qquad (6)$$

3.3   The Analysis of Model's Results

Conclution One: From the fourth formula, we can see that a is a positive number only

when $c_1 > c_3 + \dfrac{c_3^2}{c_2}$.That is to say the General Contractor won't provide the reward until his shortage cost is greater than the crash cost of the supplier. As for the nuclear power project, the cost of the delay of schedule is usually enormous, so it always stimulates the General Contractor to provide the Supplier with some reward, which can improve the possibility of in-time delivery. This conclution accords with the reality.

Conclution Two: Combining the formula 3.1 and 4.1, we can conclude that there is no direct relation between a and T. That means the General Contractor's reward won't affect the delivery time directly. While on the other hand, a and T both have a positive relationship with $\frac{1}{\lambda}$, which means that the reward should be payed according to the Supplier's ability of production and logistics. $\frac{1}{\lambda}$ can be reduced because of the reward, while T will also be smaller as $\frac{1}{\lambda}$ is reduced.

Conclusion Three: From the formula 3.1, we can see that T has a positive relationship with C_1、C_3、$\frac{1}{\lambda}$, a negative one with C_2. So if the General Contractor's shortage cost、 the supplier's crash cost and the supplier's average actual order response time are high, it will stimulates the Supplier to demand a longer delivery time. While if the Supplier's inventory cost is high, the Supplier would propose a shorter delivery time in order to reduce his expected inventory cost.

Conclution Four: We use V stands for the value of the formula, there is V(3.1)>V(3.2), V(5.1)<V(5.2), V(6.1)<V(6.2)[5]. The reasons is analysed as followed:

V(3.1)>V(3.2) If get reward, the Supplier would propose a longer delivery time to get a higher expected benefit. Without reward, the Supplier will make a independent decision. In this situation, the Supplier will take no accout of the General Contractor's shortage cost, proposing a shorter delivery time to minimize his own expected cost.

V(5.1)<V(5.2),V(6.1)<V(6.2) Due to the reward, the General Contractor's expected shortage cost will reduce rapidly, and the Supplier can get an amount of expected benefit.

4   Conclusion

By the analysis of result, we can see that there is an obvious reduction in the expected cost of both sides in the reward system. So it is rational for the General Contractor to provide the Supplier some reward stimuli, which not only reduces the total cost of the entire supply chain, but also achieves the win-win goal.

References

[1] Ding, X.-M.: The Search on Materials Purchasing Under the EPC Model, pp. 3–5 (2005)
[2] Sun, G.-Z.: The Strategy Analysis of CNNC's Equipment Procurement, pp. 14–16 (2009)

[3] Jahnukainen, J., Lahti, M.: Efficient Purchasing in Make-to-Order Supply Chains. International Journal of Production Economics 59, 103–111 (1999)
[4] Ping, M.: How to Control The Procurement Schedule of The Nuclear Power Project. Hei LongJiang Science Information (19), 103–104 (2009)
[5] Zhang, W.-Y.: Game and Information Economics, pp. 160–198. Shagnhai san-lian bookshop, Shanghai demotic publishing company, Shanghai, China (1998)
[6] Wang, Y.-Y., Shen, L., Li, B.-Y.: The Search on Several Kinds of In-Time Delivery Game Model. The Management Comment (19), 57–61 (2007)

Automatic Image Registration Based on the Vector Data of Non-linear Transformation

Yu Zhang[1,2] and Ling Han[1]

[1] Chang'an University, Xi'an 710064
[2] Geomatics Center of Shaanxi Province, Xi'an 710054
Sophi_82@163.com, hanling@chd.edu.cn

Abstract. For an urgent situation that the public need applications for geographic information in the process of informatization construction[1], we do an experiment on the technique of automatic image matching based on vector data of non-linear transformation[2], in order to propose an registration method of automatically getting control points for image matching from a security perspective. It can not only achieve a better registration effect between image and vector, but also save a lot of work on massively artificial selection of points, which provides valuable practical experience for the security of remote sensing image data processing.

Keywords: non-linear transformation vector, automatic matching of corresponding control points, image registration, Geographic Information Security.

1 Introduction

As society continues to promote information technology, government, industry and the general public increasingly need the application of geographic information. Departments of Surveying and Mapping chronically provide the paper maps and traditional GIS data with off-line mode, which has been unable to meet the quick access and application requirements of the geographic information for governmental management and decision, disaster prevention and mitigation, development of new industries, improving people's living resources and so on. In general, the different levels of users (government, industry, the public) have the different usage of geographic information. Basic geographic information data can be provided directly to a part of the special users or professional users. But, for the public users under internet environment, it needs the necessarily safe handling for basic geographic information and data through specific technology processing, such as content filtering, space accuracy decreasing and so on to provide in-line geographic information services. As to massive imaging resources, it must be handled by spatial transformation, pumping thin the image resolution, blurring the sensitive targets and so on for security in accordance with relevant provisions.

In this paper, through using the nonlinear transformation vector data as control data which was after the safe handling, it transforms the spatial coordinates of image based on the registration from image data to control vector data, which is technically feasible and can indirectly achieve security processing. The experiment studies the image matching technology which can automatically get corresponding control points in order to improve the traditional correction ways which have problems of workload, poor results and so on. It aims to provide reference to opening-up services of vast amounts of image data.

2   Introduction to Data Sources

In this experiment, it uses 1/4 ALOS image as the test area. The ALOS image is multi-spectral image and after crude correct with a total of four bands 1,2,3,4 and no RPC file. The resolution of image is 2.5meters and the coordinate system is WGS 84, UTM projection.

It uses the 1:50000 vector data which is after a safe handling of nonlinear transformation as the control vector. The control vector has big local deformation and the deformation is not regular. So, it has reached the aim of spatial location security. The contrast between vector data before and after nonlinear transformation is shown in Fig. 1.

Fig. 1. Comparison diagram of vector data before and after nonlinear transformation

3   Image Pre-processing

1. Synthesis the band 4,3,2 into false color image, then convert into the natural color(multi-spectral image);

2. Registration the multi-spectral image with band1 (pan-chromatic image);

3. Do image fusion between pan-chromatic and multi-spectral images using the principal component transform method and bilinear interpolation method.

4. Supported by Adobe Photoshop software, through artificial process, the image can highlight the readability and aesthetics in order to ease the public usage.

4   The research Methods

1. Traditional image registration

As the spatial coordinates of the control vector are random and arbitrary, so the image correction needs to find a greater stretching method. Commonly we use the image correction methods, such as: polynomial correction, IKNOS, Quick Bird and other correction model based on the image processing software ERDAS, ENVI, etc. The polynomial correction is to directly simulate the mathematical functions through the selected control points and to avoid the geometry process of images itself. IKNOS, Quick Bird and other correction model is through RPC (Rational Polynomial Coefficient) , digital elevation models and control points to establish a strict correction model, then do the overall orthorectified correction.

All these methods can't stretch images with a large degree of freedom, and is difficult to achieve a good fit with the control vector. By quadratic polynomial correction, the process and result are shown in Fig. 2 and Fig. 3.

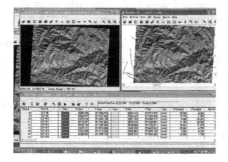

Fig. 2. Process of quadratic polynomial correction

Fig. 3. Result of quadratic polynomial correction

2.   Image registration based on automatically matching control points

Georeferencing is a registration module in ArcMap software[3,4]. The adjust algorithm in the module (Figure 4) is based on the principle of least square method, which can do nonlinear stretch on images with great arbitrariness. It can achieve a good match between images and vectors of non-linear transformation, but need manually

Fig. 4. ArcMAP registration Toolbar

select a large number of corresponding control points to support operations. For the applications involving large area, it will spend a lot of manpower and material resources.

3.   The principle of automatically matching control points

In this paper, the experiment realize the automatically extraction of corresponding control points based on a new algorithm, edit the control points into a file, then import the file into Georeferencing module to complete image correction.

The core of the image registration based on Georeferencing module is to find the corresponding control points. Through this algorithm, it can obtain the effective control points from vector data before registration (equivalent to the original image coordinates) and after registration (equivalent to the transformed image coordinates). Then, using the adjust tool in Georeferencing module to operate, it can complete the registration between image and vector. The specific steps of automatically matching the corresponding control points are as follows:

(1) Extract the line endpoint of vector data before and after nonlinear transformation, remove the pseudo nodes and the points between which the space is less than 3.5m, and respectively numbered them as the alternatively corresponding control points (red dot in Figure 5);

(2) In the vector data before and after nonlinear transformation, respectively choose the alternatively corresponding control points as the centers of circles, draw circles of which the radius is equal to 3m, get the intersections between the circles and the lines of the vectors (the blue dot in Figure 5), and make the center and intersections of a circle into a group. If the quantity of intersections of a circle is not equal to 3, remove this group;

(3) In every group, get the gravity ($X = \frac{(x_1 + x_2 + x_3)}{3}$, $Y = \frac{(y_1 + y_2 + y_3)}{3}$) of the triangle which is constructed by intersections (blue dots), calculate angle "a" between the north direction and the connection line of the alternatively corresponding control point and the gravity. Angle "a" is the final indicator of corresponding control points. It is shown in Fig. 5.

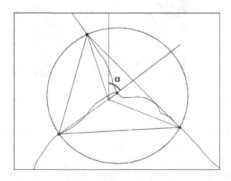

Fig. 5. Diagram of indicators for automatically selecting corresponding control points

(4) Compare the number of endpoints and the Angles "a" in the vector data before and after transformation, select the points which have the same number and the angles "a" are equal as the final corresponding control points.

5 The Experimental Results

Through running the program based on the algorithm above, we get a file of corresponding control points. Edit the file according to the file format (X coordinate of the original data; Y coordinate of the original data; X coordinate of the target data; Y coordinate of the target data) which is required by ARCGIS, then, import the file into the tool adjust in Georeferencing module. It is shown in Fig. 6.

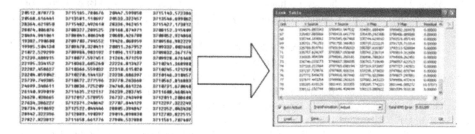

Fig. 6. Importing the file of corresponding control points into Georeferencing module

After re-sampling the image, the result is shown in Fig. 7:

Fig. 7. Overlap maps of result image and control vector

6 Conclusions

1. The experimental results of image registration based on automatically matching corresponding control points proved that the method of correction is right. It forms a practicable technology route of indirectly processing image data for security.

2. This experiment achieved a good result, saving a lot of manual work of selecting corresponding control points. It has high processing efficiency for big project with a

large amount of data items, and can provide valuable practical experience for security process of the massive remote sensing images.

3. As the algorithm is to match the intersections of linear features (such as: rivers, etc.), the match points are limited in quantity and uneven in distribution. Thus, the details of registration are not as good as the one of which the points are selected artificially. In order to improve the quality of the user experience, it can add some artificial control points to further improve the operation, and makes sense for the project with not very high accuracy requirements.

References

1. Information on, http://zgch1.enicp.net/index.asp
2. Yang, M., Wang, H.: The key Technology of Mapping for public based on Basic Geographic Information and data. Geomatics Technology and Equipment (2008)
3. Information on, http://library.columbia.edu/indiv/dssc/eds/georef.html
4. Information on, http://www.gissky.net
5. Wang, Z., Song, C., Jia, H.: Design and Implementation of Geographic Information Service System for Public. Science and Technology Information, 450–451 (2010)
6. Gong, J.: The Foundation of Geographic Information System. Science Press, Beijing (2001) (in press)
7. Hu, P., Huang, X., et al.: The Tutorial of Geographic Information System. Wuhan University Press, Wuhan (2002) (in press)

The Risk Management Analysis of Electric Power Engineering Project

JianFei Shen and Min Zhang

Economics and Management School, North China Electric Power University,
Changping Beijing 102206 People's Republic of China
{shenjianfei,zhangminphyllis}@163.com

Abstract. To improve the economic benefit is the goal of electric power enterprise pursuit of the goal, is also the basic guarantee for the survival and the development of the enterprise. Enterprise must fully realize all the kinds of risk, establish and improve the risk control mechanism, prevent the risk that occur in the development process of enterprise. So the risk management is particularly important. The better the risk management of engineering project, the better the performance of the corporation. This paper introduces the main source of risk management of the power engineering project and puts forward countermeasures.

Keywords: Electric Power Engineering Project, Risk Management, countermeasures.

1 Introduction

Nowadays the electric power market the competition is gradually fierce. If the company to achieve an invincible position in the competition, we must constantly improve project management, improve the economic benefit of enterprise. Enterprise must fully realize all the kinds of risk, establish and improve the risk control mechanism, prevent the risk that occur to the development process of enterprise, and to ensure the safe operation of electric power enterprise and the sustainable healthy development. Keeping good risk management is the main way to increase profitability of company, is also the foundation of enterprise development.

2 The Definition of Risk Management of Power Engineering Project

The definition of risk management of power engineering project is the management of risk that happened in the construction of power engineering project. That is the risk management personnel of the power engineering project have a recognition, forecasting, analysis, evaluation and effective treatment to the uncertainty of project that may lead to the loss. It is a scientific management method that can provide maximum security to finish the project successfully with the minimum cost. Risk management of electric power should run through the whole process of engineering project management. Electric energy is different from other sources of energy, each of establish of the power engineering project need to pass project, design, construction, operation stage.

In each stage, because of the conditions such as target, environment, personnel, cannot be exactly the same, so there is different between the difficulties to be overcome and the measures.

As well, there is different between the content of risk and difficulty, so the risk management exists in different stages of the project, and Implement the whole process of project management.

3   The Domestic and Foreign Research Status of Risk Management of Electric Power Engineering Project

Risk management problem was first originated from after World War I Germany. In 1931, the American Management Association first advocate the risk management, and research and studies various forms of risk management in the form academic conferences and research classes In later years. In 1963, "enterprise risk management" published in the U.S caused widespread attention in Europe and the United States. Thereafter, the trend of risk management is systematic, specialized, which make risk management become an independent discipline management. The establishment of Risk Management Association and the spread of Risk Management Education suggested that risk management has penetrated all spheres of society. In 1983, in the Risk and Insurance Management Association of America annual meeting, national experts and scholars gathered in New York, discussed and adopted the "101 Risk Management Guidelines ", which as general principles of risk management in various countries. This indicates that risk management has reached a new level. International Symposium held in Singapore in October 1986, which showed that the risk management movement has gone global and has become an international sport around the world.

The current situation and problems of Risk Management of Power Engineering project in China. Risk management technology was a dispensable tool in the modern project management, but in the late 70s, early 80s the project management theory and method has been introduced in China, whereas the risk management has not been introduced in China. However, since the mid-80s with the continuous development of China's economy, various foreign Risk management theory and the books were introduced to China, but also be applied to project management, especially in large power projects. For example, a project risk management plan has been developed in the Er Tan Power Project, the method of project risk management has been successfully applied in Three Gorges project. While risk management has gradually been adopted in China, and show a broad prospect in large project management, but there are still some problems, such as the risk identification, the error of risk assessment, backward methods of risk management.

4   The Characteristics of Electric Power Engineering Project and the Main Source of Risk

Compared to other construction projects, power project as a complex system engineering has its own distinctive characteristics. It needs Large amount of investment funds and it takes a long periods. Power projects are capital-intensive industry, which require huge amount of investment funds and Sophisticated professional technical. It needs one-time investment for power equipment and facilities and other fixed assets. As the power project is designed on the basis of specific geographical and environmental conditions, so equipment is manufactured according to standards, and equipment is product after signing the contract, which need long investment cycle and huge demand for funds.

It needs Long investment payback period. In general, because of the long period of construction project, the long duration of investment, the gradually recovery of funds, the investment is relatively slow effect. The equivalents funds have different values in different years, investors had to endure inflation and the risk of capital depreciation.

Technology-intensive, a larger number of supporting equipment, complex structure and properties, high quality and safety requirements, and ancillary equipment quality problems will directly affect the operating characteristics and operational levels of projects.

The benefit and risk is symbiosis. Direct purpose of power engineering construction is its economic and social benefits, but there is no benefit without risk, there is no risk without benefit. Each part of the mistake process of project construction may cause the generation of risk and then cause a loss. Therefore, it is necessary to enhance risk analysis and management, it directly affects the outcome of power engineering construction, which determine that whether it can achieve the requirements of technical design and whether it can meet its social role.

5   The Main Source of Risk of Power Engineering Project

According to the chronological order of development of power projects, the risk can be divided into three stages: the project risk of Pre-development phase, project risk of the implementation phase, the project risk of the production and operation phase. Each phase has different risk characteristics.

(1)The project risk of Pre-development phase: Before the formal construction of the project there has a long pre-development phase, including project planning, feasibility studies, engineering design, geological exploration, path selection and so on. During this period, there are many unknown and uncertain factors about the project. There is great risk investment of the project. There have two main factors that may impact on investment. The first is tendering and contracting. Power Transmission projects generally tend to select specific units to design general contracting by public tender, including survey and measurement of project, civil engineering design, installation engineering and communication process design and so on. And choose a project supervision unit and construction unit. If select the unit incorrect, it will result in test design quality and construction quality problems, which affect the operation of the project. The second is the project design. When the design units of power transmission project are confirmed, further in-depth design start according to the construction plan and the design contract of the feasibility study stage. If the investigation of geological conditions is not detailed and the technical and economic indicators do not meet the demand which will directly affect the construction quality, period and Investment returns of project.

(2)In the implementation phase, there are three major factors resulting in investment losses. The first is the construction progress period. Whether power construction projects can proceed smoothly as plan has a direct impact on investment efficiency, delaying period means the loss of funds which will not only increase the project cost but also increase the time of Investment income. The second is the problem of project quality and quality of equipment also exists. If there are quality problems on engineer and equipment, which will influence the progress of engineer as the equipment to be repaired or replaced, at the same time construction costs will be increased. The third is organization and management of project construction. The implementation of power construction projects involve purchase and install of materials and equipment, organization and deployment, implementation arrangements for the progress control of all aspects, and the coordination among human, financial, and material. If these aspects have some mistake, it will lead to investment loss.

(3)The project risk is high in the production and operation phase. Even though the project is completed and put into use, but if the project is not acquired adequate power supply in accordance with the original plan, the electricity demand of the region is still not satisfied; or forecast of demand is too high, which will result in excess waste of power resources. Line is also capable of withstanding the existence of the large flow peak Request or on line, communications equipment and civil works of the regular maintenance, repair and other aspects of risk. Production and operation phase of the project risk is a watershed project phase. It exists in a risk whether the line is able to withstand the peak flow of large requirements and the risk of lines, communications equipment and civil works for periodic maintenance. From this stage, the risk of the power company has gradually reduced as the recovery of funds.

There are some factors such as changes in economic environment and changes in market demand and supply that lead to investment risk. They will affect expected return of electric power project. When the power construction projects come into the normal production period, its production process is relatively fixed and the work is periodically repeated. At this stage, it's critical to establish reasonable and effective management system. After the completion of project, mismanagement of the project investment is a important reason for poor performance. Technological progress and technological innovation are also very important in the production period. There is also another important reason affecting investment return. For example, some power projects do not achieve the expected returns on investment and do not carry out technological transformation and do not adopt new technologies, do not improve product competitiveness and market share.

6   The Main Countermeasures in the Risk Management of Electrical Engineering

(1)Avoid the fatal risks. After fully opening the electrical construction and achieving the diversification of investment main bodies, the investors should avoid the projects which have too much risk when making decisions of electrical engineering projects. They also should pay enough attention to risk management, do detailed investigation and identification of geography and geology and prepare several sets of options in the preliminary work.

(2)Take active measures to control risk. The risk responsibilities could be specified In electrical engineering construction by decomposing the control targets of electrical engineering, so that the awareness of risk responsibilities could be enhanced. We should choose the best units of design, construction and supervision; strengthen the risk liabilities of design, supervision and adjustment; diversify the risks through the agreement of terms of the contracts; adhere to the strict measures of technology and management; strengthen the management of the project plan, the middle stage control and the management of middle decision to achieve effective control of project risks.

(3)Transferring risks through project insurance, Risk transfer can be achieved and risks can be transfered from the insured to the insurers through project insurance. The insured can get insurance benefits on the one hand and can strengthen the awareness of risk responsibilities and improve the effect of risk management on the other hand. Reasonable selective strategy should be taken when insuring electrical engineering projects. The first principles to be followed are selecting strong insurance companies which have good services and buying high level insurance which have full functions and can transfer risks in a reasonable way. As the electrical construction is complex, time-consuming and costly, the insurance mode that one insurance company is appointed as main Insurer and others are joined should be adopted by investors.

References

1. China Engineering Consulting Association. Large civil engineering project insurance, pp. 45–50. China Planning Publishing Social, Beijing (2001)
2. Root, L.F., Norman, G.: The management of construction risk, pp. 5–9. China Building Industry Press, Beijing (2000)
3. Chang, H.-C., Xia, G.: The system of project risk management. Wuhan University 9(7), 129–132 (2008)
4. Zhang, Z.: Project risk management and prevention measures. Railway Engineering Management 7(6), 34–39 (2009)
5. Xue, J.: About Power Management of project risk managers 24, 120–126 (2009)

The Coordination for Logistics Alliance of Manufacturer and TPL with Uncertain Demand

Ju-ning Su, Chuan-fang Shi, and Yan-qing He

School of Economics and Management, Xi'an University of Technology,
Xi'an 710054, China

Abstract. In the alliance of the manufacturer and TPL, the manufacturer logistics service would exist in supply short risk because the market demand which the manufacturer facing was uncertain. The coordination of logistics alliance was studied with uncertain logistics demand. Firstly, the decision behaviors of the manufacturer and TPL were analyzed in decentralized decision. Then the results between decentralized decision and centralized decision were compared, and found that decentralized decision couldn't reach the performances of centralized decision. The coordination mechanism based on cost-sharing was proposed. The analysis results of the coordination model indicated that the Pareto improvement of logistics alliance parties and the optimization of whole alliance could be achieved by designing appropriate cost-sharing ratio.

Keywords: logistics alliance, uncertain logistics demand, coordination, cost-sharing.

1   Introduction

As TPL occurs, many manufacturing enterprises either choose logistics outsourcing or make alliance with TPL enterprise. Alp, Erkip and Gullu studied the transportation contract parameter design between a manufacturer and a transporter[1]. Wu Qing and Dan Bin analysis the design of TPL coordination contract considering logistics service affecting the market demand change by using dynamic game model, designing the contract of service cost-sharing and combined contract of income-sharing and service cost-sharing[2]. Zhao Quan-wu, Zhang Qin-hong and Pu Xiang-zhi designed coordination contracts based on complete information and incomplete information considering shortage and damage to cargo occurred in the logistics operation of TPL[3]. Cui Ai-ping and Liu Wei proposed a coordination mechanism based on options contract to study the order of logistics capacity and investment decision between integrators and subcontractors from the viewpoint of the Stackelberg model[4].

The researches mainly focus on logistics outsourcing and logistics service supply chain，little research can be available about the logistics alliance formed by manufacturing enterprise and TPL. He Sheng-yu studied logistics alliance incentive scheme mainly based on manufacturing enterprise under certain logistics demand[5].

D. Jin and S. Lin (Eds.): Advances in FCCS, Vol. 2, AISC 160, pp. 501–506.
springerlink.com   © Springer-Verlag Berlin Heidelberg 2012

This paper extends the limitations of demand, to study the cooperation contract for logistics alliance of manufacturer and TPL under uncertain demand.

2   Problem Description

Study a logistics alliance consists of a manufacturing enterprise and a TPL enterprise, the manufacturer logistics service would exist in supply short risk because the market demand which the manufacturer facing was uncertain. The manufacturer chooses TPL by bidding, which undertakes all logistics tasks of the manufacturer. According to the unsured logistics demand D in single cycle(Probable Density Function and Cumulative Density Function is taken as $f(D)$ and $F(D)$), the manufacturer preorder a logistic size of d to TPL. Both party have agreement on that TPL not only offer d that the manufacturer preorder but also have an amount of extra logistics capacity(the logistics amount that extra logistics capacity can offered is Δ, it is called extra logistics amount). The decision variable for the manufacturer is d and the decision variable for TPL is Δ. Both are risk-neutral, equal status, information symmetrical. Other notations listed as follows: c_1: the per unit of logistics cost for TPL; c_2: the unit holding cost of residuary logistics capability; P: the logistics service price; g_1: the unit punishment that customer refer to the manufacturer when the logistics demand is unsatisfied; g_2: the unit punishment that the manufacturer refer to TPL when the logistics demand is unsatisfied; h: the unit award that the manufacturer give to TPL when real finished logistics amount is greater than the order amount; α: the logistics service level of TPL, when $D \le d + \Delta$, $\alpha = 1$; $D > d + \Delta$, $\alpha = \dfrac{d+\Delta}{D}$; M_c: the total logistics cost of the manufacturer; L_s: the total profit of TPL; T_c: the total logistics cost of logistics alliance. We assume that logistics demand obey to uniform distribution $D \in [a,b]$.

3   Decision Model in the Decentralized System

The decision of the manufacturer is:

$$\min_{d \ge 0} E(M_c) = \frac{1}{2(b-a)} \left\{ p \left[(d+\Delta)^2 - a^2 \right] + h\Delta^2 + (g_1 - g_2) \left[b - (d+\Delta) \right]^2 \right\}$$

The decision of TPL is :

$$\max_{\Delta \ge 0} E(L_s) = \frac{1}{2(b-a)} \left\{ (p - c_1) \left[(d+\Delta)^2 - a^2 \right] - c_2 \left[a - (d+\Delta) \right]^2 + h\Delta^2 - g_2 \left[b - (d+\Delta) \right]^2 \right\}$$

Under decentralized condition, the purpose of the manufacturer is to ensure the best order amount d according to demand amount D to minimize M_c. Whereas the purpose of TPL is to find the optimal extra storage Δ according to given d 、 g_2 and h. we get:

$$\Delta^* = \frac{g_1 b c_1 + g_1 b c_2 - g_1 b p - g_2 c_2 a - g_2 b c_2 - g_1 c_2 a - g_2 b c_1 - p c_2 a - 2 b g_2{}^2}{h(p + g_1 - g_2)}$$

$\because g_1 \geq g_2,\quad \therefore h(p + g_1 - g_2) > 0;\qquad \because p \geq c_1 + c_2,\quad \therefore g_1 b c_1 + g_1 b c_2 - g_1 b p \leq 0$

It can be concluded that $\Delta^* < 0$. That is to say, without considering the range of Δ, Δ that maximize $E(L_s)$ is less zero. But in the feasible region $[0, +\infty]$ of Δ, from $E(L_s)$ is concave with respect to Δ, we know that when $\Delta = 0, E(L_s)$ will be the largest. Therefore TPL may choose $\Delta^* = 0$.

Thus, we get equilibrium solution of Nash game in the decentralized system:

$$\Delta^* = 0, \quad d^* = \frac{b(g_1 - g_2)}{p + g_1 - g_2}.$$

Then the expected logistics cost of the manufacturer and the expected profit of TPL can be written as follows separately:

$$E^*(M_c) = \frac{1}{2(b-a)}\left\{ p\left[\left(\frac{bg_1 - bg_2}{p + g_1 - g_2}\right) - a^2\right]^2 + (g_1 - g_2)\left[b - \frac{b(g_1 - g_2)}{p + g_1 - g_2}\right]^2 \right\} \quad (1)$$

$$E^*(L_s) = \frac{1}{2(b-a)}\left\{ (p - c_1)\left[\left(\frac{bg_1 - bg_2}{p + g_1 - g_2}\right)^2 - a^2\right] - c_2\left[a - \frac{b(g_1 - g_2)}{p + g_1 - g_2}\right]^2 + h\Delta^2 - g_2\left[b - \frac{b(g_1 - g_2)}{p + g_1 - g_2}\right]^2 \right\} \quad (2)$$

The expected logistics service level is:

$$E^*(\alpha) = \int_a^{d^* + \Delta^*} f(D)dD + \int_{d^* + \Delta^*}^{b} \frac{d^* + \Delta^*}{D} f(D)dD = \frac{(d^* + \Delta^* - a) + (d^* + \Delta^*)\ln\frac{b}{d^* + \Delta^*}}{b - a} \quad (3)$$

4 Decision Model in the Centralized System

With the condition of logistics demand obey to uniform distribution $D \in [a, b]$, the purpose of decision in centralized system is:

$$\min E(T_c) = \frac{1}{2(b-a)}\left\{ c_1\left[(d + \Delta)^2 - a^2\right] + c_2(d + \Delta - a)^2 + g_1\left[b - (d + \Delta)\right]^2 \right\} \quad (4)$$

We can get $(d + \Delta)^* = \dfrac{g_1 b + c_2 a}{g_1 + c_1 + c_2}$ to make $E(T_c)$ minimize.

In the decentralized system, $d^* + \Delta^* = \dfrac{b(g_1 - g_2)}{p + g_1 - g_2}$, obviously

$d^* + \Delta^* \neq (d + \Delta)^*$. $E\left(T_c\left(d^* + \Delta^*\right)\right) > E\left(T_c\left((d + \Delta)^*\right)\right)$.

That is to say, the total cost of logistics alliance in the decentralized system is greater than that in the centralized system.

To motivate TPL hold extra logistics capacity, improve the service level of logistics alliance, the manufacturer can propose a coordination strategy based on cost-sharing: when the logistics demand is satisfied, the manufacturer takes cost-sharing to the holding cost of residuary logistics capability. When $D \leq d + \Delta$, the manufacturer share a proportion of k to the holding cost of residuary logistics capability. In this way, because of the risk that TPL takes is decreasing, the manufacturer can improve the punishment for TPL if the demand is not satisfied, we use g_2' to present the punishment of unfinished tasks by the manufacturer, $g_2' \geq g_2$.

5   Coordination Decision Model Based on Cost-Sharing

With the cost-sharing mechanism, the expected logistics total cost of the manufacturer and the expected profit of TPL can be written as follows respectively:

$$E'(M_c) = \frac{1}{2(b-a)}\left\{p\left[(d'+\Delta')^2 - a^2\right] + kc_2(d'+\Delta'-a)^2 + \left(g_1 - g_2'\right)\left[b - (d'+\Delta')\right]^2\right\} \quad (5)$$

$$E'(L_t) = \frac{1}{2(b-a)}\left\{(p - c_1)\left[(d'+\Delta')^2 - a^2\right] - c_2(1-k)(d'+\Delta'-a)^2 - g_2'\left[b - (d'+\Delta')\right]^2\right\} \quad (6)$$

By coordination strategy, can decision in the decentralized system for the manufacturer and TPL implement the performance of decision in the centralized system? After computation and analysis, we find voluntary cost-sharing strategy can't implement the optimization of logistics alliance total cost, that is the decision of logistics service order and extra logistics capacity storage made by the manufacturer and TPL from individual optimization can't get accord with the optimal decision of logistics alliance. However with the enforced cost-sharing strategy, the manufacturer can continue choose the order d^*, and also TPL must offer an extra logistics capacity storage of $(d + \Delta)^* - d^*$.

With the enforced cost-sharing strategy, the logistics order of the manufacturer $d'^* = \dfrac{b(g_1 - g_2)}{p + g_1 - g_2}$,   the   extra   logistics   storage   of   TPL

$$\Delta'^* = \frac{g_1 b + c_2 a}{g_1 + c_1 + c_2} - \frac{b(g_1 - g_2)}{p + g_1 - g_2}$$, Substitute them into (5) and (6), we will get the

expected cost of the manufacturer and the expected profit of TPL as follows:

$$E''(M_c) = \frac{1}{2(b-a)}\left\{ p\left[\left(\frac{g_1 b + c_2 a}{g_1 + c_1 + c_2}\right)^2 - a^2\right] + kc_2\left(\frac{g_1 b + c_2 a}{g_1 + c_1 + c_2} - a\right)^2 + \left(g_1 - g_2'\right)\left[b - \frac{g_1 b + c_2 a}{g_1 + c_1 + c_2}\right]^2\right\}$$

$$E''(L_s) = \frac{1}{2(b-a)}\left\{ (p - c_1)\left[\left(\frac{g_1 b + c_2 a}{g_1 + c_1 + c_2}\right)^2 - a^2\right] - c_2(1-k)\left(\frac{g_1 b + c_2 a}{g_1 + c_1 + c_2} - a\right)^2 - g_2'\left[b - \frac{g_1 b + c_2 a}{g_1 + c_1 + c_2}\right]^2\right\}$$

According to the participation constraints approached by coordination strategy
$E'^*(M_c) \le E^*(M_c)$ and $E'^*(L_s) \ge E^*(L_s)$, we can get the range of
cost-sharing ratio k .It can be proved that k exists , the specific expression will not
be showed here for the form is too complicated. We will calculate the range of k in
the part of computational study.

Then the expected logistics service level of TPL is:

$$E''(\alpha) = \int_a^{(d+\Delta)^*} f(D)dD + \int_{(d+\Delta)^*}^b \frac{(d+\Delta)^*}{D} f(D)dD = \frac{\left((d+\Delta)^* - a\right) + (d+\Delta)^* \ln \frac{b}{(d+\Delta)^*}}{b-a} \quad (7)$$

6   Computational Study

A car manufacturer's yearly logistics demand obey to uniform distribution which
range is $60000 \le D \le 100000$, $c_1 = 30000$, $c_2 = 20000$, $p = 50000$,
$g_1 = 200000$, $g_2 = 50000$, $g_2' = 150000$, $h = 2000$. We can get: before
coordination, $d^* = 75000$, $\Delta^* = 0$, $E^*(\alpha) = 0.914$; and after coordination,
$d^* = 75000$, $\Delta'^* = 9800$, $E'(\alpha) = 0.970$. According to the constraints of
coordination, we get the cost-sharing ratio $k : 0 \le k \le 0.317$.

7   Concluding Remarks

This paper studies the coordination for logistics alliance of manufacturer and TPL
with uncertain demand. Firstly, the decision behaviors of the manufacturer and TPL
were analyzed in decentralized decision. Then the results between decentralized
decision and centralized decision were compared, and found that decentralized
decision couldn't reach the performances of centralized decision. The coordination
mechanism based on cost-sharing was proposed, which would achieve the
coordination of logistics alliance through the sharing of the holding cost of residuary
logistics capability by the manufacturer and TPL. The analysis results of the
coordination model indicated that with voluntary execution of cost-sharing strategy
can't implement the optimization of whole alliance; with enforced execution of

cost-sharing strategy, we can find appropriate cost-sharing ratio interval, not only can we improve the logistics service level after coordination but also reach the goal of improvement of logistics alliance parties and the optimization of whole logistics alliance. Finally a numerical example was given to illustrate the related conclusions.

Acknowledgments. We acknowledge the support of Natural Science Foundation of Shaanxi (2011JM9004) and Scientific Research Program of Shaanxi Education Department (071C081, 11JK0168).

References

1. Alp, O., Erkip, N.K., Gullu, R.: Outsourcing logistics: Designing transportation contracts between a manufacturer and a transporter. Transportation Science 37(1), 23–39 (2003)
2. Wu, Q., Dan, B.: Third party logistics coordinating contracts with logistics market demand service dependent. Journal of Management Science in China 11(5), 64–75 (2008)
3. Zhao, Q., Zhang, Q., Bu, X.: Study on supply chain coordination based on logistics service quality with asymmetric information. Journal of Industrial Engineering/Engineering Management 22(1), 58–61 (2008)
4. Cui, A., Liu, W.: Study on capability coordination in logistics service supply chain with options contract. Chinese Journal of Management Science 17(2), 59–65 (2009)
5. He, S.: Research on some issues in the logistics alliance dominated by manufactory, vol. 9, pp. 112–119. Southwest Jiaotong University, Cheng Du (2006)

Comprehensive Analysis of Risky Driving Behaviors Based on Fuzzy Evaluation Model

Yulong Pei[1], Weiwei Qi[1], Xupeng Zhang[2], and Mo Song[1]

[1] School of Transportation Science and Engineering, Harbin Institute of Technology,
Harbin 150090, China
[2] Traffic Planning and Design Institute of Jilin Province, Changchun 130021, China
yulongp@263.net, {qwwhit,xiaobeibeiqj,07smhit}@163.com

Abstract. With the rapid development of the economy and society, people living in the city are faced with increasing pressures of life and working, thus leading to increasing risky driving behaviors, which severely threaten the driving conditions and the property safety of traffic participants. Therefore, it is a brand new start of traffic safety research to evaluate and sort different types of risky driving behaviors. Based on the statistics of traffic accident, we established a model base on fuzzy comprehensive evaluation to evaluate the influence on the vehicle damage under different risky driving behaviors. Through calculation, we can see that the risk degree of different risky driving behaviors based on the vehicle damage ranged in descending order as follows: risky lane changes, illegal occupation, frequent lane changes, illegal turning, inadequate longitudinal separation, rolling on markings, illegal u-turning, violation of signal and rolling on stop lines.

Keywords: risky driving behavior, vehicle damage, risk degree, fuzzy comprehensive evaluation.

1 Introduction

There exist two kinds of meanings in evaluating the risk degree of different risky driving behaviors: one is to calculate the risk degree of certain risky driving behavior, which is an important index of the longitudinal comparison for one kind of risky driving behavior; the other is to evaluate the risk degree of different risky behaviors relatively using the same criteria so as to compare with each other [1][2]. The basic ideas of evaluation is that the influence on vehicle damage of certain risky driving behavior is not only related to the frequency of traffic accident, it caused but also linked to other factors such as forms of the accident, levels of the accident, loss of damaged vehicles and so on [3][4]. Therefore, it's inappropriate to evaluate the influence of different risky driving behaviors on vehicle damage just according to the number of traffic accidents it caused, that is, we have need to adopt fuzzy comprehensive evaluation [5]. Besides, since there is no specific definition of the influence of each factor, it is relatively appropriate to adopt this method.

D. Jin and S. Lin (Eds.): Advances in FCCS, Vol. 2, AISC 160, pp. 507–512.
springerlink.com © Springer-Verlag Berlin Heidelberg 2012

2   Construction of the Evaluation Model

Estimation Subclass and Weighted Subclass
We take the number of lost vehicles, the forms of accidents and the level of accidents as three impact factors. The levels of accidents are divided into four levels including catastrophic traffic accident, serious traffic accident, general traffic accident and minor traffic accident. And the forms of accidents include head-on collision, side collision and others. An estimation subclass made up of a number of m evaluation results is shown as follows [6]:

$$U = \{u_1, u_2, \cdots u_m\} \tag{1}$$

By analyzing the weight distribution of each factor, we can build a weighted subclass expressed as weight vector as follows:

$$A = (a_1, a_2, \ldots a_n) \tag{2}$$

Weight Vector of Evaluation Factors
To compare the influence a set of n factors shown as $V = \{v_1, v_2, \cdots v_n\}$ on the target U and set their weight on the target, we take out two factors v_i and v_j each time, and the ratio of the two factors' influence on the target U is defined as C_{ij}, then we formed a pair wise comparison matrix $C = (C_{ij})_{N \times N}$. And A_r represents weight of different factors which can be solved by square-root method using formula (3) [7]:

$$A_r = \left(\prod_{j=1}^{n} C_{ij} \right)^{\frac{1}{n}} \tag{3}$$

For convenience, weights of different factors are normalized as follows:

$$A_r^0 = \frac{A_r}{\sum A_r} \tag{4}$$

The weights of the first level of evaluations are given by formula (3) and (4) as well as markings of a number of experts, and the result is shown in table 1. The

Table 1. Weights of the First Level of Evaluations

Evaluation Factors	$V_{\text{forms of accidents}}$	$V_{\text{levels of accidents}}$	$V_{\text{loss of damaged vehi}}$	A_r	A_r^0
$V_{\text{forms of accidents}}$	1	2	3	1.817	0.539
$V_{\text{levels of accidents}}$	1/2	1	2	1	0.297
$V_{\text{loss of damaged vehi}}$	1/3	1/2	1	0.550	0.164
Σ		—		3.367	1

Based on table 1, we defined the weight vector for the first level of evaluation factors: $A = (0.539, 0.297, 0.164)$.

Table 2. Weights of Different Forms of Accidents

Evaluation Factors	$V_{\text{head-on collision}}$	$V_{\text{side collision}}$	V_{others}	A_r	A_r^0
$V_{\text{head-on collision}}$	1	2	2	1.518	0.482
$V_{\text{side collision}}$	1/2	1	2	1	0.318
V_{others}	1/2	1/2	1	0.630	0.200
Σ		——		3.148	1

Based on table 2, we defined the weight distribution of different forms of accidents as follows: $A_{\text{forms of accidents}} = (0.482, 0.318, 0.200)$.

Table 3. Weights of Different Levels of Accidents

Evaluation Factors	$V_{\text{catastrophic traffic accident}}$	$V_{\text{serious traffic accident}}$	$V_{\text{general traffic accident}}$	$V_{\text{minor traffic accident}}$	A_r	A_r^0
$V_{\text{catastrophic traffic accident}}$	1	2	3	4	2.213	0.467
$V_{\text{serious traffic accident}}$	1/2	1	2	3	1.316	0.278
$V_{\text{general traffic accident}}$	1/3	1/2	1	2	0.760	0.160
$V_{\text{minor traffic accident}}$	1/4	1/3	1/2	1	0.452	0.095
Σ		——			4.741	1

Based on table 3, we defined the weight distribution of different levels of accidents as follows: $A_{\text{levels of accidents}} = (0.467, 0.278, 0.160, 0.095)$.

calculation results of weights of the second level of evaluations are shown in table 2 and 3.

Fuzzy Membership Function
Through evaluating the influence of different risky driving behaviors on the damaged vehicles, a fuzzy membership function can be defined by formula (5) [8]:

$$R(x) = \frac{x}{k+x} \quad x \geq 0, k > 0 \tag{5}$$

Where, k is the average number of traffic accidents led by the corresponding evaluation index of different risky driving behaviors; x is the number of accidents led by certain risky driving behaviors.

3   Sorting the Risk Degree of Different Risky Driving Behaviors

Based on the statistic analysis on the traffic accidents files of one province, we count the frequency of different levels of accidents, of different forms of accidents and the number of damaged vehicles led by different risky driving behaviors, the data is shown in table 4.

The estimation subclass we constructed is consisted of the influence of 9 kinds of risky driving behaviors, namely: $U = (u_1, u_2, \cdots\cdots u_9)$; And the evaluation factor set is $V = (v_1, v_2, v_3)$

Where, v_1, v_2, v_3 are the forms of accidents, the levels of accidents and the loss of damage vehicles.

Table 4. Statistics of the Traffic Frequency and Loss of Vehicles

Risky Driving Behaviors	Forms of Collision (times)			Levels of Accidents (times)				Loss of Vehicles (veh)
	Head-on	Side	Others	Catastrophic	Serious	General	Minor	
illegal turning	31	67	14	0	7	99	6	187
rolling on stop lines	0	2	7	0	1	8	0	15
illegal u-turning	8	42	5	0	2	51	2	93
frequent lane changes	22	41	15	1	2	75	0	124
risky lane changes	127	260	46	0	18	388	27	739
illegal occupation	146	96	61	0	20	273	10	447
inadequate longitudinal separation	9	4	174	0	2	181	4	331
violation of signal	7	18	0	0	0	25	0	49
rolling on markings	25	41	5	0	6	64	1	125
average	41.67	63.44	36.33	0.11	6.44	129.33	5.56	234.44

The impact on the estimation subclass is diverse from the levels of the accidents to the forms of accidents. To calculate the grade of membership of v_1 and v_2 to the target U, we need to evaluate v_1 and v_2. The weight vectors can be calculated:

$$A_{\text{forms of accidents}} = (0.482, 0.318, 0.200)$$

$$A_{\text{levels of accidents}} = (0.467, 0.278, 0.160, 0.095)$$

And the fuzzy membership functions can be calculated according to formula (5):

$$R_{\text{forms of accidents}} = \begin{bmatrix} 0.310 & 0 & 0.104 & 0.242 & 0.648 & 0.679 & 0.115 & 0.092 & 0.266 \\ 0.374 & 0.018 & 0.272 & 0.268 & 0.699 & 0.461 & 0.034 & 0.138 & 0.268 \\ 0.158 & 0.086 & 0.063 & 0.168 & 0.382 & 0.450 & 0.700 & 0 & 0.063 \end{bmatrix}$$

$$R_{\text{levels of accidents}} = \begin{bmatrix} 0 & 0 & 0 & 0.693 & 0 & 0 & 0 & 0 & 0 \\ 0.384 & 0.082 & 0.151 & 0.151 & 0.616 & 0.641 & 0.151 & 0 & 0.348 \\ 0.300 & 0.034 & 0.181 & 0.245 & 0.627 & 0.542 & 0.440 & 0.098 & 0.217 \\ 0.308 & 0 & 0.129 & 0 & 0.667 & 0.426 & 0.229 & 0 & 0.069 \end{bmatrix}$$

$$B_{\text{forms of accidents}} = A_{\text{forms of accidents}} \circ R_{\text{forms of accidents}}$$

$$= (0.300 \quad 0.023 \quad 0.149 \quad 0.235 \quad 0.611 \quad 0.564 \quad 0.207 \quad 0.088 \quad 0.266)$$

$$B_{\text{levels of accidents}} = A_{\text{levels of accidents}} \circ R_{\text{levels of accidents}}$$

$$= (0.184 \quad 0.028 \quad 0.083 \quad 0.405 \quad 0.335 \quad 0.305 \quad 0.134 \quad 0.016 \quad 0.138)$$

Together with v_3, their grade of membership to U consists of a fuzzy relation matrix R:

$$R = \begin{bmatrix} 0.300 & 0.023 & 0.149 & 0.235 & 0.611 & 0.564 & 0.207 & 0.088 & 0.226 \\ 0.184 & 0.028 & 0.083 & 0.405 & 0.335 & 0.305 & 0.134 & 0.016 & 0.138 \\ 0.302 & 0.033 & 0.177 & 0.223 & 0.631 & 0.508 & 0.433 & 0.102 & 0.224 \end{bmatrix}$$

Since we have considered the influence of different forms of accidents and levels of accidents on the estimation subclass, weight vector of influence on estimation subclass for v_1, v_2, v_3 can be estimated:

$$A = (0.539, 0.297, 0.164)$$

$$B = A \circ R = (0.266 \quad 0.026 \quad 0.134 \quad 0.284 \quad 0.532 \quad 0.478 \quad 0.222 \quad 0.069 \quad 0.200)$$

Finally, we draw a conclusion that the risk degrees of different risky driving behaviors based on the vehicle damage ranged in descending order as follows: risky lane changes, illegal occupation, frequent lane changes, illegal turning, inadequate longitudinal separation, rolling on markings, illegal u-turning, violation of signal and rolling on stop lines.

4   Conclusions

Based on the analysis on different risky driving behaviors, following conclusions can be got:

(1) The influence on vehicle damage of certain risky driving behavior is not only related to the frequency of traffic accident it caused, but also linked to other factors such as forms of the accident, levels of the accident, loss of damaged vehicles, and so on.
(2) When it comes to actual evaluation, if some evaluation factors are made up of several factors and it's difficult to define its weights and membership function, then we can turn to constructing a fuzzy comprehensive evaluation model.
(3) There exist certain limitations in the application of the conclusions, since relevant data are all collected from only one province. And the future key of the research lies in collecting data of risky driving behaviors on a national scale.

Acknowledgment. The study was supported by the National Natural Science Foundation of China (51178149).

References

1. Nasvadi, G.E.: Changes in self-reported driving behaviour following attendance at a mature driver education program. Transportation Research Part F: Traffic Psychology and Behavior 10(4), 358–369 (2007)
2. Iversen, H.: Risk-taking attitudes and risky driving behaviour. Transportation Research Part F: Traffic Psychology and Behavior 7(3), 135–150 (2004)

3. Shinar, D.: Aggressive driving: the contribution of the drivers and the situation. Transportation Research Part F: Traffic Psychology and Behavior 1(2), 137–160 (1998)
4. Larsen, L.: Methods of multidisciplinary in-depth analyses of road traffic accidents. Journal of Hazardous Materials 111(1-3), 115–122 (2004)
5. Chen, X., Qi, S., Ye, H.: Fuzzy Comprehensive Study on Seismic Landslide Hazard Based on GIS. Acta Scientiarum Naturalium Universitatis Pekinensis On Line 2(2), 1–6 (2007)
6. Wang, Y.: Systems Engineering, pp. 132–141. China Machine Press, Beijing (2003)
7. Li, S.: Engineering fuzzy mathematics and application, pp. 5–30. Harbin Institute of Technology Press, Harbin (2004)
8. Zhao, R.: Fuzzy Mathematics Mixing-up Evaluation Method of Danger Degree. Navigation of China 41(2), 40–45 (1997)

Cooperative R&D Analysis in Time-sensitive Supply Chain under Asymmetry Information

Ju-ning Su and Yan-qing He

School of Economics and Management, Xi'an University of Technology,
Xi'an 710054, China

Abstract. A supply chain with time-sensitive demand is discussed in the paper, which is consisted of a AM（assembly manufacturer）and a CS（component supplier）. The R&D price subsidy strategy in supply chain with symmetric information structure was first discussed. Then the R&D price subsidy strategy in supply chain under asymmetry information about CS's R&D ability was discussed. Some study conclusions which were drawn from a numerical example show that under symmetric information, the higher R&D ability of the CS, the less amount of price subsidy of the AM, and the profit of CS reduced too. But under asymmetry information, in order to stimulate the CS, the AM has to pay additional cost for the unknown R&D ability of the CS, this results in the increase of the amount of price subsidy, and the reduction of the AM's profit, while the profit of the CS increases due to holding private information. As a result, the whole supply chain coordination can not be achieved.

Keywords: asymmetric information, time-sensitive demand, R&D cooperation, price subsidy strategy.

1   Introduction

In the coordination of supply chain, the research on asymmetric information has already involved many aspects of purchasing, inventory, supplying, production, marketing and investment[1-2]. Some researches consider the demand under asymmetric information, such as references [3] considers the design of supply chain contract with symmetric information about market demand, references [4] builds a gaming model under two parties having asymmetric demand information; Some researches consider the cost under asymmetric information, references [5] studies how to design a contract if the cost of the buyer is asymmetric under random demand, references [6] studies supplier's optimal income-sharing strategy with the cost information of the retailer is hidden under asymmetric information.

The researches above don't considering the affect that time to whole supply chain. In an new economical environment that time competeveness becomes more common. Inputting new products faster than competitor can make enterprise more positive in competeveness and easier to get succeed in developing new products. Therefore, this paper attempts to analysis under the new product R&D input affects marketing-time

D. Jin and S. Lin (Eds.): Advances in FCCS, Vol. 2, AISC 160, pp. 513–518.
springerlink.com   © Springer-Verlag Berlin Heidelberg 2012

while marketing-time affects the demand situations, how to make decisions for assembly manufacturer and component supplier in supply chain and discuss the affect on them with asymmetric Information.

2   Problem Description

This paper studies a two level supply chain system consisting of AM (assembly manufacturer) and CS (component supplier). AM develops a new product and requests CS to research and develop a new component. The develop procedure of product consists of two parts: component R&D and product assembly. Correspondingly, R&D time of new product also consists of two parts: time of component R&D and time of product assembly. We assume that the R&D cycle of new product is mainly decided by the R&D time of component.

Notations and assumptions as follows: x —R&D input for component; t_m — production time of AM., t_m is a constant; $t_s(x)$ —component R&D time for CS; t —time that product put into market, $t = t_s(x) + t_m$; p —the market price of product; $Q(p,t)$ —the market demand for products; p_s —the marginal profit of the CS; p_m —the marginal profit of the AM; s —R&D cost for the AM; l — price subsidy per unit of product that AM gives to CS; π_s —the profit of the CS; π_m —the profit of the AM; π —the profit of the whole supply chain.

We assume $Q(p,t) = a - bp - gt$, a、b、g are constants greater than 0. We refer to references [7], and assume $t_s(x) = de^{-hx}$, d, h, both are constants greater than 0.

3   The Coordination Analysis of Price Subsidy Strategy under Supply Chain with Symmetric Information Structure

As benefit participator after successful R&D, in order to motivate CS, AM takes price subsidy strategy, that is AM offers some subsidy to CS's unit parts according to CS's R&D input. It also embodies the principle of supply chain: revenue-sharing and risk sharing. With symmetric Information, the AM decides the quantity of price subsidy so as to predict CS's R&D input. In price subsidy strategy, we form a Stackelberg game characterized by AM acts as leader and CS acts as follower.

3.1   The Decision of R&D Input for CS

When the price subsidy quantity of AM is l, the profit function of CS is:

$$\pi_s(l,x) = (p_s + l)[a - bp - g(t_m + de^{-hx})] - x \tag{1}$$

We know that the profit function of the CS is concave with respect to x, there exists x^* maximize CS's profit. We get :

$$x_s^*(l) = \frac{1}{h}\ln(\rho_s + l)ghd \qquad (2)$$

At this moment, time of product put into market is:

$$t^* = t_m + \frac{1}{gh(p_s + l)} \qquad (3)$$

At this time, the profit of CS can be written as follows:

$$\pi_s(l) = (\rho_s + l)[a - bp - g(t_m + \frac{1}{gh(\rho_s + l)})] - \frac{1}{h}\ln(\rho_s + l)ghd \qquad (4)$$

3.2   The Decision of Price Subsidy for AM

The profit function of AM is:

$$\pi_m(l) = (\rho_m - l)[a - bp - g(t_m + de^{-hx_s^*(l)})] - s \qquad (5)$$

We substitute (2) into (5) and simplify it, we get:

$$\pi_m(l) = (\rho_m - l)[a - bp - g(t_m + \frac{1}{(\rho_s + l)gh})] - s$$

We know AM's profit function is concave with respect to price subsidy l, there exists l^* that maximize AM's profit.

$$l^* = \sqrt{\frac{\rho_m + \rho_s}{h(a - bp - gt_m)}} - \rho_s \qquad (7)$$

4   The Coordination Analysis of Price Subsidy Strategy under Supply Chain with Asymmetric Information Structure

The above coordination of supply chain can be reached by assumption that both party has a complete information sharing, but it is difficult to make this come true in reality. The enterprise often hides some private information so as to get the optimal revenue. This paper assumes that other information in supply chain is symmetric and only R&D ability of CS is Asymmetric Information.

The R&D ability of CS h is complete known to itself, AM doesn' t know it completely. Only two possibilities can be obtained: one is $h = h_H$, its probability is $\theta (0 \leq \theta \leq 1)$; and another one is $h = h_L$, its probability is $1 - \theta$, $h_H \geq h_L$. Both party all know the above information difference. In this way, contract designing planning is written as follows P1:

$$\max_{(l_H, l_L)} \theta \left\{ (\rho_m - l_H)[a - bp - g(t_m + de^{-h_H x_H^*})] - s \right\} + (1 - \theta) \left\{ (\rho_m - l_L)[a - bp - g(t_m + de^{-h_L x_L^*})] - s \right\}$$

$$s.t \quad x_H^* \in \arg\max (\rho_s + l_H)[a - bp - g(t_m + de^{-h_H x_H})] - x_H \quad (8)$$

$$x_L^* \in \arg\max (\rho_s + l_L)[a - bp - g(t_m + de^{-h_L x_L})] - x_L \quad (9)$$

$$(\rho_s + l_H)[a - bp - g(t_m + de^{-h_H x_H^*})] - x_H^* \geq \pi_S^{\min} \quad (10)$$

$$(\rho_s + l_L)[a - bp - g(t_m + de^{-h_L x_L^*})] - x_L^* \geq \pi_S^{\min} \quad (11)$$

$$(\rho_s + l_H)[a - bp - g(t_m + de^{-h_H x_H^*})] - x_H^* \geq (\rho_s + l_L)[a - bp - g(t_m + de^{-h_L x_L^*})] - \frac{h_L}{h_H} x_L^* \quad (12)$$

$$(\rho_s + l_L)[a - bp - g(t_m + de^{-h_L x_L^*})] - x_L^* \geq (\rho_s + l_H)[a - bp - g(t_m + de^{-h_H x_H^*})] - \frac{h_H}{h_L} x_H^* \quad (13)$$

$\pi_S^{\min}$ is the reserved profit of CS. According to (8) and (9), we get:

$$x_H^* = \frac{1}{h_H} \ln (\rho_s + l_H) g h_H d \quad (14)$$

$$x_L^* = \frac{1}{h_L} \ln (\rho_s + l_L) g h_L d \quad (15)$$

In P1, it is easy to verify （11） and （12） are strict constraints, substituting (14) and (15) into the above planning problem and cancel (10) and (13), two loose constraints, then it transfer to P2:

$$\max_{(l_H, l_L)} \theta \left\{ (\rho_m - l_H)[a - bp - g(t_m + \frac{1}{g h_H (\rho_s + l_H)})] \right\} + (1 - \theta) \left\{ (\rho_m - l_L)[a - bp - g(t_m + \frac{1}{g h_L (\rho_s + l_L)})] \right\} \quad (16)$$

$$s.t$$

$$(\rho_s + l_L)[a - bp - g(t_m + \frac{1}{g h_L (\rho_s + l_L)})] - \frac{1}{h_L} \ln (\rho_s + l_L) g h_L d = \pi_S^{\min} \quad (17)$$

$$(\rho_s + l_H)[a - bp - g(t_m + \frac{1}{g h_H (\rho_s + l_H)})] - \frac{1}{h_H} \ln (\rho_s + l_H) g h_H d \geq$$

$$(\rho_s + l_L)[a - bp - g(t_m + \frac{1}{g h_L (\rho_s + l_L)})] - \frac{1}{h_H} \ln (\rho_s + l_L) g h_L d \quad (18)$$

According to (17), we get

$$l_L = \gamma^{\frac{-\text{lambertw}(-1,\frac{-\mu}{\gamma^\gamma})-\gamma}{}} - \rho_s$$

Among them, $\mu = h_L(a - bp - gt_m)$, $\gamma = 1 + Ingh_L d + h_L\pi_s^{\min}$,

lambertw() is a function, lambertw(x) implies the solution of the equation: $w * \exp(w) = x$, lambertw(k,x) stands for k branch of multi-valued function lambertw(), the specific value can be available by enter it in Matlab command window. Therefore planning problem can be also transferred to P3:

$$\max_{(l_H, l_L)}\left\{(\rho_m - l_H)[a - bp - g(t_m + \frac{1}{gh_h(\rho_s + l_h)})]\right\} \qquad (19)$$

s.t

$$(\rho_s + l_H)[a - bp - g(t_m + \frac{1}{gh_H(\rho_s + l_H)})] - \frac{1}{h_H}\ln(\rho_s + l_H)gh_H d \geq$$

$$(\rho_s + l_L)[a - bp - g(t_m + \frac{1}{gh_L(\rho_s + l_L)})] - \frac{1}{h_H}\ln(\rho_s + l_L)gh_L d \qquad (20)$$

From all the above, we know the solution to planning problem P3 is :

$$l_L = \gamma^{\frac{-\text{lambertw}(-1,\frac{-\mu}{\gamma^\gamma})-\gamma}{}} - \rho_s, \quad l_H = \max(l_{H1}, l_{H2})$$

5   A Computational Study

According to the actual meaning of relevant variables, we set values for the parameters in the model as follow: $\theta = 0.3$, $h_H = 0.009$, $h_L = 0.007$, $a = 200$, $b = 1$, $g = 8$, $d = 10$, $t_m = 1$, $\rho_s = 2$, $\rho_m = 8$, $s = 100$, $p = 12$, $\pi_s^{\min} = 300$. Substituting these parameters into the formula that are derived in former 2-3 section, we get under symmetric information: If $h = h_L = 0.007$, then $l^* = 0.817$, $x^* = 65.13$, $t^* = 7.34$, $\pi_s = 299.1$, $\pi_m = 828.7$, $\pi_m = 1127.8$. If $h = h_H = 0.009$, then $l^* = 0.485$, $x^* = 64.62$, $t^* = 6.59$, $\pi_s = 271.5$, $\pi_m = 916.7$, $\pi_m = 1188.2$.

Under asymmetric information, the contracts menu made by AM is $\{0.824 ; 0.797\}$. When the true R&D ability of the CS is h_L, then $x^* = 65.48$, $t^* = 7.32$, $\pi_s = 300$, $\pi_m = 828.7$, $\pi_m = 1128.7$. When the true R&D ability of the CS is h_H, then $x^* = 77.78$, $t^* = 5.97$, $\pi_s = 314.6$, $\pi_m = 910.4$, $\pi_m = 1125$.

6   Conclusions and Remarks

In a competitive environment base on time, there is a trend to consider the affect of time to the decision of supply chain. This paper studies a supply chain of the market demand of a product is constraint to time that the product put into the market. With the situation of R&D input of CS can shorten the R&D cycle of component, how to cooperate and analysis the AM's price subsidy strategies under symmetric information and asymmetric information respectively. This paper only analysis a coordinated model that both party cooperative and R&D in a two level supply chain and a certain demand, we expect to study further case that in multi-level supply chain and a random demand with more practical value.

Acknowledgments. We acknowledge the support of Natural Science Foundation of Shaanxi 2011JM9004) and Scientific Research Program of Shaanxi Education Department(11JK0168).

References

1. Gorbet, C., Groote, X.: A supplier's optimal quantity discount policy under asymmetric information. Management Science 46(3), 445–450 (2000)
2. Cao, J., Yang, C.-J., Li-Ping, Zhou, G.-G.: Design of supply chain linear shared-saving contract with asymmetric information. Journal of Management Science 12(2), 19–30 (2009)
3. Lau, A.H.L., Lau, H.S.: Some two—echelon style goods inventory models with asymmetric market information. European Journal of Operational Research 134(l), 29–42 (2001)
4. Zhou, Y.-W., Ran, C.-L.: Gaming in supply chain based on asymmetric demand information. Systems Engineering and Electronics 28(l), 68–71 (2006)
5. Ha, A.Y.: "Supplier-buyer" contracting: Asymmetric cost information and cut off level policy for buyer participation. Naval Research Logistics 48(l), 41–64 (2001)
6. Qiu, R.-Z., Huang, X.-Y.: Coordinating Supply chain with revenue-sharing contract under asymmetric information. Journal of Northeastern University (Natural Science) 28(s), 1205–1208 (2007)
7. Gilbert, M.S., Cvsa, V.: Strategic commitment to price stimulates downstream innovation in a supply chain. European Journal of Operational Research (150), 617–639 (2003)

Construction and Properties of Dual Canonical Frames of a Trivariate Wavelet Frame with Multi-scale

Yinhong Xia[1] and Xiong Tang[2]

Department of Mathematics Science, Huanghuai University, Zhumadian 463000, China
sxxa66xauat@126.com

Abstract. It is shown that there exists a frame wavelet with fast decay in the time domain and compact support in the frequency domain generating a wavelet system whose canonical dual frame cannot be generated by an arbitrary number of generators. We show that there exist wavelet frame generated by two functions which have good dual wavelet frames, but for which the canonical dual wavelet frame does not consist of wavelets, according to scaling functions.

Keywords: canonical frames, Gabor frames, Banach frames, frame operator, wavelet frame, Fourier transfer, mathematics modelling, tivariate, time-frequency analysis method, communication.

1 Introduction and Frame Operator

Information science is an interdisciplinary science primarily concerned with the analysis, collection, classification, manipulation, storage, retrieval and dissemination of information. Practitioners within the field study the application and usage of knowledge in organizations, along with the interaction between people, organizations and any existing information systems, with the aim of creating, replacing, improving or understanding information systems. Information science is often (mistakenly) considered a branch of computer science. Materials science is an interdisciplinary field applying the properties of matter to various areas of science and engineering. This scientific field investigates the relationship between the structure of materials at atomic or molecular scales and their macroscopicproperties. It incorporates elements of applied physics and chemistry. With significant media attention focused on nanoscience and nanotechnology in recent years, materials science has been propelled to the forefront at many universities Recently, wavelet frames have attracted more and more attention, just because they have excellent time-frequency localization property, shift-invariances, and more design freedom. Wavelet tight frames have been widely used in denoising and image processing. Tight frames generalize orthonormal systems. They preserve the unitary property or the relevant analysis ans synthesis operator. Frames are intermingled with exciting applications to physics, to engineering and to science in general. Frames didn't start in isolation, and even now in its maturity, surprising and deep connections to other areas continue to enrich the subject. The subjects are well explained, and they are all amenable to the kind of numerical methods where wavelet algorithms excel. Wavelet analysis is a particular time-or space-scale representation of signals which has found a wide range of applications in physics, signal processing and applied Mathematics in the last few years. The notion of frames was introduced by Duffin and Schaeffer[1] and popularized greatly by the work of Daubechies and her coauthors[2,3]. After this ground breaking work, the theory of frames began to be more widely studied both in theory and in applications[4-7], such as signal processing, image processing, data compression and sampling theory. The notion of Frame multiresolution analysis as described by [5] generalizes the notion of multiresolution analysis by allowing non-exact affine frames. However, subspaces at different resolutions in a FMRA are still generated by a frame formed by translates and dilates of a single function. This article is motivated from the observation that standard methods in sa-mpling theory provide examples of multirsolution structure which are not FMRAs. Inspired by [5] and [7], we introduce the notion of a generalized multiresolution structure (GMRS) of $L^2(R^3)$ generated by several functions of inte ger translates in space $L^2(R^3)$. We have that the GMRS has a pyramid decompoposition scheme and obtain a frame-like decomposition based on such a GMRS. It also lead to new constructions of affine frames of $L^2(R^3)$.

First, we introduce some notations. Let Ω be a separable Hilbert space. We recall that a sequence $\{\eta_t : t \in Z\} \subseteq \Omega$ is a frame for H if there exist positive real numbers C, D such that

$$\forall \lambda \in \Omega, \quad C \|\lambda\|^2 \le \sum_t \left| \langle \lambda, \eta_t \rangle \right|^2 \le D \|\lambda\|^2 \tag{1}$$

A sequence $\{\eta_t\} \subseteq \Omega$ is a Bessel sequence if (only) the upper inequality of (1) holds. If only forall $\lambda \in X \subset U$, the upper inequality of (1) holds, the sequence $\{\eta_t\} \subseteq \Omega$ is a Bessel sequence with respect to (w.r.t.) Ω. If $\{\eta_t\}$ is a frame, there exist a dual frame $\{\eta_t^*\}$ such that

$$\forall g \in \Omega, \quad g = \sum_t \langle g, \eta_t \rangle \eta_t^* = \sum_t \langle g, \eta_t^* \rangle \eta_t. \tag{2}$$

where C and D are called the lower frame bound and upper bound, respectively. In particular, when $C = D = 1,$ we say that $\{\xi_v\}_{v \in \Gamma}$ is a (nor- malized) tight frame in X. The frame operator $S : X \to X$, which is associated with a frame $\{\xi_v\}_{v \in \Gamma}$, is defined to be

$$S\varphi = \sum_{v \in \Gamma} \langle \varphi, \xi_v \rangle \xi_v, \quad \varphi \in X \tag{3}$$

Obviously, $\{\xi_v\}_{v \in \Gamma}$ is a frame in X with lower frame bound C and upper frame bound D if and only if S is well defined and $CI \le S \le DI$, where I denotes the identity operator on X. A sequence $\{f_t\} \subseteq X$ is a Bessel sequence if (only) the upper inequality of (1) follows. If only for all $\Upsilon \in \Lambda \subset X$, the upper inequality of (1) holds, the sequence $\{f_t\} \subseteq X$ is a Bessel sequence with respect to (w.r.t.) the subspace Λ. Let $\{\xi_v\}_{v\in\Gamma}$ be a frame in a Hilbert space X with lower frame bound C and upper bound D. Then its frame operator S is invertible and $C^{-1}I \le S^{-1} \le D^{-1}I$, that is to say, S^{-1} is also a frame operator. Moreover, we have the following result.

Theorem 1. Let $\{h_v\}_{v\in\Gamma}$ be a frame in X . Then the frame operators $S: X \to X$, and $S^{-1}: X \to X$ are self-adjoint operators, respectively.

Proof. For any $\phi, g \in X$, we have $\quad S\phi = \sum_{v\in\Gamma}\langle \phi, h_v\rangle h_v, \quad Sg = \sum_{v\in\Gamma}\langle g, h_v\rangle h_v.$

Then,

$$\langle S\phi, g\rangle = \left\langle \sum_{v\in\Gamma}\langle \phi, h_v\rangle h_v, g\right\rangle = \sum_{v\in\Gamma}\langle \phi, h_v\rangle\langle h_v, g\rangle$$

$$= \sum_{v\in\Gamma}\overline{\langle g, h_v\rangle}\langle \phi, h_v\rangle = <\phi, \sum_{v\in\Gamma}\langle g, h_v\rangle h_v> = \langle \phi, Sg\rangle$$

$$\langle S^{-1}\phi, g\rangle = \langle S^{-1}\phi, SS^{-1}g\rangle = \langle SS^{-1}\phi, S^{-1}g\rangle = \langle \phi, S^{-1}g\rangle .$$

Hence, the frame operators S and S^{-1} are all self-adjoint operators. Note that $\{S^{-1}\phi_v\}_{v\in\Gamma}$ is a dual frame of the frame $\{\phi_v\}_{v\in\Gamma}$ since

$$\phi = SS^{-1}\phi = \sum_{v\in\Gamma}\langle S^{-1}\phi, h_v\rangle h_v = \sum_{v\in\Gamma}\langle \phi, S^{-1}h_v\rangle h_v, \quad \forall \phi \in X; \qquad (4)$$

$$\langle \phi, g\rangle = \langle \sum_{v\in\Gamma}\langle \phi, S^{-1}h_v\rangle, g\rangle = \sum_{v\in\Gamma}\langle \phi, S^{-1}h_v\rangle\langle h_v, g\rangle \quad \forall \phi, g \in X \qquad (5)$$

The family $\{S^{-1}h_v\}_{v\in\Gamma}$ is said to be the canonical dual frame of the frame $\{h_v\}_{v\in\Gamma}$ and (4) is the canoonical representation of any ϕ in X using the frame $\{h_v\}_{v\in\Gamma}$ in X . Generally speaking, for a given frame $\{h_v\}_{v\in\Gamma}$ in X , there are many dual frames other than its standard dual frame. However, the standard dual frame $\{S^{-1}h_v\}_{v\in\Gamma}$ enjoys the below optical property(see[1]). For any element $\phi \in X$,

$$\sum_{v\in\Gamma}|\langle h, S^{-1}h_v\rangle|^2 \le \sum_{v\in\Gamma}|c_v|^2 \quad \text{whenever} \quad \phi = \sum_{v\in\Gamma}c_v h_v, \ \{c_v\}_{v\in\Gamma} \in \ell_2(Z).$$

In other words, the representation in (4) using the canonical dual frame has the smallest norm of the frame coefficients to represent a given element in X .

A wavelet frame is generated from several functions in $L^2(R^3)$ by dilates and integer translates. We say that $\{\psi_1, \psi_2, \psi_3\}$ generates a wavelet frame in $L^2(R^3)$ if $\psi_{l:j,k} : j, k \in Z, l = 1, 2, 3\}$ is a frame in $L^2(R^3)$. We say that $\{\psi_1, \psi_2\}$ and $\{\tilde{\psi}_1, \tilde{\psi}_2\}$ generates a pair of dual wavelet frame if each of them generates a wavelet frame and

$$\forall h, g \in L^2(R^3), \quad \langle h, g \rangle = \sum_{l=1}^{7} \sum_{j \in Z} \sum_{k \in Z^3} \langle h, \tilde{\psi}_{l:j,k} \rangle \langle \psi_{l:j,k}, g \rangle, \tag{6}$$

2   The Concept of Trivariate Generalized Multiresolution Structure

Let u be a positive integer, and $\Lambda = \{1, 2, \cdots, u\}$ be a finite index set. We consider the case of multiple generators, which yield multiple pseudoframes for subspaces of $L^2(R^3)$. In what follows, we consider the case of generators, which yield affine pseudoframes of integer grid translates for subspaces of $L^2(R^3)$. Let $\{T_v \phi_l\}$ and $\{T_v \tilde{\phi}_l\}$ $(l \in \Lambda, v \in Z^3)$ be two sequences in $L^2(R^3)$. Let Ω be a closed subspace of $L^2(R^3)$. We say that $\{T_v \phi_l\}$ forms a pseudoframe for the subspace Ω with respect to (w.r.t.) $\{T_v \tilde{\phi}_l\}$ $(l \in \Lambda, v \in Z^3)$, if

$$\forall f(x) \in \Omega, \qquad f(x) = \sum_{l \in \Lambda} \sum_{v \in Z^3} \langle f, T_v \phi_l \rangle T_v \tilde{\phi}_l(x). \tag{7}$$

Definition 1. We say that a Generalized multiresolution structure (GMS) $\{V_n, \phi_l(x), \tilde{\phi}_l(x)\}_{n \in Z, l \in \Lambda}$ of $L^2(R^3)$ is a sequence of closed linear subspaces $\{V_n\}_{n \in Z}$ of space $L^2(R^3)$ and $2u$ elements $\phi_l(x), \tilde{\phi}_l(x) \in L^2(R^3)$ such that (i) $V_n \subset V_{n+1}$, $n \in Z$; (ii) $\bigcap_{n \in Z} V_n = \{0\}$; $\bigcup_{n \in Z} V_n$ is dense in $L^2(R^3)$; (iii) $g(x) \in V_n$ if and only if $g(2x) \in V_{n+1}$ $\forall n \in Z$ implies $T_v g(x) \in V_0$, for $v \in Z$; (v) $\{T_v \phi_l(x), l \in \Lambda, v \in Z^3\}$ forms an affine pseudoframe for V_0 with respect to $\{T_v \tilde{\phi}_l(x), l \in \Lambda, v \in Z^3\}$.

A necessary and sufficient condition for the construction of an affine pseudoframe for Paley-Wiener subspaces is presented as follows. The filter banks associated with a GMS are presented as follows. Define filter functions $B_0(\omega)$ and $\tilde{B}_0(\omega)$ by the relaton $B_0(\omega) = \sum_{v \in Z} b_0(v) e^{-2\pi i \omega}$ and $\tilde{B}_0(\omega) = \sum_{v \in Z} \tilde{b}_0(v) e^{-2\pi i \omega}$ of the sequences $b_0 = \{b_0(v)\}$ and $\widetilde{b_0} = \{\widetilde{b_0}(v)\}$, respectively, wherever the sum is defined. Let $\{b_0(v)\}$ be such that $B_0(0) = 2$ and $B_0(\omega) \neq 0$ in a neighborhoood of 0. Assume also that $|B_0(\omega)| \leq 2$. Then there exists $f(x) \in L^2(R^3)$ (see ref.[8]) such that

$$f(x) = 2\sqrt{2} \sum\nolimits_{v \in Z^3} b_0(v) f(2x - v).$$ (8)

Similarly, there exists a scaling relationship for $\widetilde{f}(x)$ under the same conditions as that of b_0 for a sequence $\widetilde{b_0} = \{\widetilde{b_0}(v)\}$, i.e.,

$$\widetilde{f}(x) = 2\sqrt{2} \sum\nolimits_{v \in Z^3} \widetilde{b_0}(v) \widetilde{f}(2x - v)$$ (9)

3   The Canonical Dual Frame of a Wavelet Frame

In this part, we shall discuss the standard dual frame of a wavelet frame with two generators in general. Then we present the main results. Define the dilation and translation operators on $L^2(R^3)$ (see[6]) as follows:

$$D^j f(x) := 2^{3j/2} f(2^j x) \qquad \text{and}$$

$$T_k f(x) := f(x - k), \quad k \in Z^3, \quad f(x) \in L^2(R^3).$$

Suppose that $\{\psi_1, \psi_2\}$ generates a wavelet frame in $L^2(R^3)$. Let F denote its frame operator defined as follows: (for $h(x) \in L^2(R^3)$).

$$Sh = \sum_{l=1}^{7} \sum_{j,k \in Z} \langle h, \psi_{l,j,k} \rangle \psi_{l,j,k}.$$ (10)

It is evident to observe that $SD^j = D^j S$ for all $j \in Z$ and $D^j T_k = T_{4-jk} D^j$. We define the period $P(\{\psi_1, \psi_2\}) \in N \cup \{0\}$ of the wavelet frame generated by $\{\psi_1, \psi_2\}$ as follows:

$$\langle P(\{\psi_1, \psi_2\}) \rangle := \{k \in Z : S^{-1} T_{k+n} \psi_l = T_k S^{-1} T_n \psi_l, \forall n \in Z, l = 1, 2\}.$$ (11)

where $\langle x \rangle$ denotes the additive group generated by x.

Theorem 2[3]. Suppose that $\{\psi_1, \psi_2, \psi_3\}$ generates a wavelet frame in $L^2(R)$. For any nonnegative integer J, the following statements are equivalent:(i) There exist $2 \cdot 4^J$ functions $\widetilde{\psi}_1, \widetilde{\psi}_2, \widetilde{\psi}_{2 \cdot 3^J}$ such that they generate the canonical dual frame of the wavelet frame $\{\psi_{l:j,k} : j,k \in Z$ and $l = 1,2\}$; (ii) $\{F^{-1}[\psi_{l:Jk}] : k = 0,1,\cdots 2^J - 1$ and $l = 1,2\}$ and $\{\psi_{l:j,k} : k = 0,1,\cdots 2^J - 1$ and $l = 1,2\}$ generate a pair of dual wavelet frame in $L^2(R^3)$; (iii) $P(\{\psi_1,\psi_2,\psi_3\}) | 4^J$, where $P(\{\psi_1,\psi_2\})$ defined in (8). Moreover, when $\{\psi_1,\psi_2\}$ generates a wavelet frame which is a Riesz basis, conditions (i)-(ii) are also equivalent to(iv) $V_J(\{\psi_1,\psi_2\})$ is shift-invariant where $V_J(\{\psi_1,\;\;\psi_2\})$ is the closed linear span of $\psi_{l:j,k} : k \in Z,\quad j < J$ and $l = 1,2,3$. For any $v = (v_1,v_2,v_3) \in Z^3 := \{(\eta_1,\eta_2,\eta_3) : \eta_1,\eta_2,,\eta_3 \in Z\}$, the translation operator S is defined to be $(S_{va}\lambda)(x) = \lambda(x - va)$, where a is a pasitive constant real number.

Definition 2[3]. Let $\{S_{ta}\Upsilon,\; \iota \in Z^3\}$ and $\{S_{ta}\tilde{\Upsilon},\; \iota \in Z^3\}$ be two sequences in $L^2(R^3)$. Let G be a closed subspace of $L^2(R^3)$. We say that $\{S_{ta}\Upsilon,\; \iota \in Z^3\}$ forms an affine bivariate pseudoframe for G with respect to $\{S_{ta}\tilde{\Upsilon},\; \iota \in Z^3\}$ if the following condition is satisfied,

$$\forall\, \lambda(x) \in G,\; \lambda(x) = \sum_{\iota \in \mathbb{Z}^2} < \lambda, S_{ta}\tilde{\Upsilon} > S_{ta}\Upsilon(x) \tag{12}$$

4   Summary

In the paper, we prove that the frame operator and its inverse are all self-adjoint operators. We prove that the canonical dual wavelet frame cannot be generated by the translates and dilates of a single function.

References

1. Duffin, R.J., Schaeffer, A.C.: A class of nonharmonic Fourier series. Trans. Amer. Math. Soc. 72, 341–366 (1952)
2. Daubechies, I., Grossmann, A., Meyer, Y.: Painless nonorthogonal expansions. J. Math. Phys. 27(5), 1271–1283
3. Chen, Q., et al.: The characterization of a class of subspace pseudoframes with arbitrary real number translations. Chaos, Solitons & Fractals 42(5), 2696–2706 (2009)
4. Daubechies, I., Han, B.: Pairs of dual wavelet frames from any two refinable functions. J. Constr. Appro. 20, 325–352 (2004)
5. Daubechies, I., Han, B.: The canonical dual. frame of a wavelet frame. J. Appl. Comput. Harmon. Anal. 12, 269–285 (2002)

Basic Concept of Visual Informatics and Agricultural Application of Spectral Imaging as Its New Technology

Ming Sun and Yun Xu

Key Laboratory of Modern Precision Agriculture System Integration Research,
Ministry of Education China Agricultural University,
17 Tsinghua Dong Road, Beijing 100083, China
sunming@cau.edu.cn, sdxuyun@sohu.com

Abstract. Visual informatics, as a subject concept proposed in the early 90s, is a new interdisciplinary area to study theories, techniques and applications of imaging science systematically. Its research method correlates with mathematics, physics, physiology, psychology, information science, electronics and computer science. Its research scope overlaps with pattern recognition, computer vision, real-time image processing, computer graphics and virtual reality. Its research progress synchronizes with artificial intelligence, neural networks, genetic algorithms and fuzzy logic. Its development and application tie with bio-medicine, remote sensing, communication and document processing closely. This paper proposes some definite explanation and concept to revise the problem without unified definition of some terminologies. Then it introduces the concept and applications of a new technology in visual information---spectral imaging comprehensively. Finally, it put forward the key issues existing in the agricultural application.

Keywords: Visual informatics, Spectral imaging technology, Agriculture.

1   Basic Concept of Visual Informatics

Informatics. Informatics is a new subject which studies information acquisition, processing, transmission and regular use. Informatics is a comprehensive subject that aims to expand human information function and researches on information using computer technology as research tools.

Image science. Human brain, with the ability of perception, recognition, learning, association, memory and reasoning, is regarded as the most advanced bio-intelligence system among various biology systems. However human brain can perceive approximately 83% information from vision and the remaining from tactile, auditory, olfactory and other sensory organs. Therefore, vision, based on image, is an important way to obtain information. Image science includes a variety of image-related

Science in general. Researchers mainly use computer image processing technology to study digital image in the current.

Digital image. In the coordinate space a continuous image is converted to a discrete image to be processed by computer. Each unit in the image is called image element, pixel for short.

D. Jin and S. Lin (Eds.): Advances in FCCS, Vol. 2, AISC 160, pp. 525–533.
springerlink.com   © Springer-Verlag Berlin Heidelberg 2012

Visual informatics. The first conference of visual informatics[1] was held in Kuala Lumpur, Malaysia in November 2009. Visual informatics, a subject proposed in the early 1990s. Visual informatics, a new subject, proposed in the early 1990s, includes not only information and computer science but engineering information, medicine, health information science, its education and other related fields. It combines the use of basic principles and tools like mathematics, optics, computer science for effective extraction, storage, transmission, processing, analysis and interpretation of vision information in image science. In integrates the new theories, new methods, new algorithms, new tools and new equipments in the development of image science into a new comprehensive and new unified application frame. It becomes an overlap between computer vision, information visualization, real-time image processing, image information restoration, virtual reality, augmented reality, visual expression of information science, 3D graphics, multimedia integration, visual data mining, visual ontology and its service, and visual culture. Visual informatics is rich in content that can be divided into three levels, image processing, image analysis and image understanding according to the level of abstraction and different research methods.

Image processing. Image processing, a low-level operation to process large volumes of data in low abstract, mainly processes image in pixel level. It stresses image transformation and conversion. Image processing refers to a variety of science and technology in a broad sense and narrowly refers to image conversion to improve its visual effects for subsequent image analysis and understanding or saving storage space and transmission time for image coding. It includes image acquisition and acquisition, image enhancement, image smoothing, image enhancement, frequency conversion, geometric transform, image coding, image restoration.

Image analysis. Image analysis, middle-level with fewer volumes of data and higher degree of abstraction, mainly detect target image, extract characteristic and then use a more concise data format to describe images. The data, referring to detection result of target characteristic or detection based symbol description, covers the characters and inherent properties of the target in the image. Image analysis covers image segmentation, edge detection, morphological image processing, feature selection, texture analysis.

Image understanding. Image understanding, high-level with small amount of data and highest degree of abstraction, mainly focuses on the operation of image data or symbol and furthers the study on the properties and interaction of the image. It interprets the original content of the object through understanding of the meaning of images to guide and plan the action. Therefore its process and method is similar with human thinking. Image understanding stresses visual perception, knowledge representation and deeper understanding of some new theories, new tools and new technology like image recognition, image fusion and the use of wavelet transforms, neural networks, genetic algorithms, digital watermarking, spectral analysis, X-ray, video, remote sensing.

2   Agricultural Application of Spectral Imaging

Introduction. Spectral imaging refers to a new technology using multiple spectral channels to realize image acquisition, display, processing and analysis in terms of the features that sensitivity of target's spectral reflectance (absorption) rate differs in different bands. It uses a specific light source or filter equipment to select the range of the wave length of the light source, especially the wave length beyond visible light, and then enhance the different characteristics of the target parts of the image features in order to contribute to the quality of target detection. Spectral imaging, a new technology developed in the 1980s, is one of the important part of visual information in the modern imaging technologies. It combines traditional two-dimensional imaging and spectroscopy and integrates image analysis and spectroscopy[2]. Therefore this technology has special advantage in the detection of quality and safety of crops and agricultural products. Spectral image is a 3-D data block made up of a series of 2-D images with consecutive bands by the use of system in a particular spectral range. In each particular wavelength, spectral data can provide 2-D information, which can characterize the exterior features like size, shape and color of the objects, and different gray of the same pixel at different wavelengths provides spectrum information that features internal structure and ingredients. Therefore, the use of spectrum can not only give qualitative and quantitative analysis to the object to be detected but analyze its location. Spectral imaging can be divided to multi-spectral imaging, hyper-spectral imaging, hyper multi-spectral imaging technology according to spectral resolution. The following parts will primarily introduce widely used multi-spectral technology and hyper-spectral technology briefly.

Detection of crop nutrition level. Domestic and foreign scholars apply multi-spectral imaging in precision agriculture related study and primarily center around the image feature band on crop nutrition level as well as vegetation indices[3, 4]. Noh et al.[5] extract the green (G), red(R) and near infrared (NIR) feature parameters in the multi-spectral image(Duncan MS2100 (Duncan Tech, Auburn, CA, USA)). Neural network modeling, used to detect corn growing, predicts the value of chlorophyll content. The difference predicted value and the actual SPAD chlorophyll is 0.89. Jones et al.[6] uses multi-spectral camera (DuncanTech MS3100 (Auburn, CA)) to detect chlorophyll content of greenhouse spinach and extract green (G), red (R) and near infrared (NIR) channel image parameters then build the ratio vegetation index(NIR/R 和NIR/G) and normalized difference vegetation index (NDVI670 = (NIR-R)/(NIR+R) 和NDVI550 = (NIR-G)/(NIR+G)). Study shows the correlation between NDVI670 correlates and chlorophyll of single plant closely (R2=0.91).But the precision will reduce largely (R2=0.30) when the index is applied in the whole area.   Shibayama et al.[7] uses the visible red band (R) band (R) and near infrared (NIR) image to test and research on nitrogen content of field rice. The result of testing on different densities of near field plot (12m) shows that the correlation coefficient between R and field nitrogen absorption is up to 0.79. Similarly NIR is 0.81. But the value of N and NIR built normalized difference vegetation index (NDV) is not closely correlated with field nitrogen content. Miao et al.[8] uses hyper-spectral RS image to test corn growth. By analyzing the correlation between chlorophyll content SPAD and more than 30 vegetation indices, such as the ratio vegetation index (RVI), difference vegetation index (DVI), normalized difference vegetation index (NDVI), soil adjusted vegetation index (SAVI) and conversion-type vegetation index (TVI), the results show that hyper-spectral feature wavelength and RS vegetation index can detect 64% to 86% respectively 73%-88% field chlorophyll content. Based on field crop canopy reflectance spectroscopy and remote sensing images, Tian et al.[9] analyzed a variety of published vegetation indices to test the feasibility of rice leaf nitrogen concentration (LNCs), including R750/R710, R735/R720, R738/ R720, (R790-R720) / (R790 + R720), R434 / (R496 + R401) and R705 / (R717 + R491). Among them, the vegetation index R434 / (R496 + R401) and R705 / (R717 + R491) are largely associated with LNCs (R2 0.84 and0.83, respectively).Jiang and Sun[10] proposed a multi-spectral imaging based technology using the gray value of tomato leaves under the sensitive wavelength (532nm,610nm and 700nm) to predict the chlorophyll content. They establish a predictive model using multiple linear regression analysis, principal component analysis and partial least squares regression analysis method and get a good result. The correlation coefficient Rc2 and Rv2 are up to approximately 0.9. Liu et al.[11] use multi-spectral imaging to study on the content of cucumber leaf nitrogen and leaf area index detection. It shows that this method to predict tomato chlorophyll is effective and feasible, at the same time it set foundation for the development of crop growth detection equipment. Zhang et al.[12] use tri-band channel charge-coupled device (CCD) imaging technology, containing green (G), red (R) and near infrared (NIR), to have non-destructive test on rice leaves and seeds. The test results show that chlorophyll a and chlorophyll b, respectively, with G and NIR are significant linear correlated. What's more, nitrogen content of rice grain with G and R channel, normalized difference vegetation index (NDVI) shows a significant linear correlation. These results assist establishing a rice leaf chlorophyll and nitrogen content of the multi-spectral predictive model. Twenty one samples were used to test the model, in which seven significant linear correlated models have relative error RE (%) ranging from 9.36% to 15.7%, to achieve quick, accurate and non-destructive test on the content of grain leaf chlorophyll and rice nitrogen.

Nondestructive detection of agricultural products. Domestic and foreign scholars applied spectral imaging into nondestructive detection of agricultural products and made volumes of achievement. Lu[13] studied the feasibility to predict hardness of apple and soluble solids content using multi-spectral imaging technology. Firstly, scattering data, selected from five different bands in the range of 680-1060nm, as the input of neural network, predict the hardness of fruit and soluble solid content. The result shows that among the four bands combination 680, 880, 905 and 940nm, 680/940,880/905 and 906/940 can predict the hardness of fruit. Its correlation coefficient r = 0.87 and standard error of predication SEP=5.8N. 680/940 and 905/940 can predict the soluble solid content. Its correlation r = 0.77 and standard error of predication SEP=5.8%. Kleynen et al.[14] built a multi-spectral vision system, which includes four bands in the range of visible/near-infrared, to access the multi-spectral image of intact and defect apples. This system used pattern-matching algorithms to classify and study on the defect of apples (four categories: minor defect, serious defect, defect leading apple to be rejected, recent bruising). The result showed that serious defect and recent bruising have a higher rate of correct classification. Unay and Gosselin[15] studied automatic defect classification for Jonathan apples based on multi-spectral imaging technology and pointed out that the result of classification by sort supervision is the most accurate. And the specific level of identification error with the mean defect size and defect based full recognition of the measured error is more reliable. Xing et al.[16] studied on "Golden delicious" and "Jonagold" apples and used multi-imaging system to separate stems/fruit calyx region from the real defect parts. They selected several feature band by the principal component analysis of hyper-spectral images and then applied PCA and image processing technology in the multi-spectral image to recognize and separate stems/ fruit calyx region by analyzing the image characteristic of the first principal component. The results showed all stems/fruit calyx areas of Golden Delicious were rightly recognized, respectively, Jonagold were 98.3. Less than 3% bruising area was mistakenly recognized as stems/calyx area in both species. Lleó et al.[17] proposed two multi-spectral classification of maturity method for the red-meat soft peaches based on 3CCD camera (800nm,675nm,450nm) to obtain red and red/infrared images. Blasco et al.[18] firstly summarizes the application of near-infrared, ultraviolet and fluorescent computer vision systems in the detection of common defects of citrus fruit and proposed a fruit classification algorithm which combines spectral information of different wavelengths (including visible light) to classify the different types of fruit defects. The results showed that non-visible information can improve the detection and identification of some defect. Compared to color image, near infrared image can improve the accuracy of anthrax to 86% and fluorescent images can improve the accuracy of green mold from 65% to 94%. Peng and Lu[19-24] proposed an improved three parameters Lorenz equations to predict the hardness of Golden Delicious apple and the content of soluble solid. This equation made accurate prediction that hardness r is 0.894 with its standard error prediction SEP 6.14N and soluble solids r = 0.883, SEP = 0.73%. The coefficient of the test on the hardness of peaches was up to 0.998. The results above confirmed the feasibility of multi-spectral imaging in the bio-information test. Zhao et al.[25] managed to use hyper-spectral image combining with multivariate calibration methods to test the hardness of apple. The experiment extracted effective spectral information from the acquired image to create the prediction model of hardness of apple. In the modeling process, by comparing two multi-parameters, partial least squares (PLS) and support vector regression (SVR), the results showed that SVR model was better than PLS model. The modeling correlation coefficient of hardness prediction is 0.6808. Gamal et al.[26] did non-destructive tests on the moisture content of strawberries, total soluble solids and acidity based on hyper-spectral imaging. Gowen et al.[27] pointed that hyper-spectral imaging is an emerging tools of food security and security analysis. Yu et al.[28] applied spectral imaging technology in the study on non-destructive of nectarine inner quality. They collected the spectral images with different bands (632, 650, 670, 780 and 900 Bin), then implement the gray distribution of collected spectral to have Lorenz function fitting. Using the parameters from the fitting of nectarine sweetness and the Lorenz distribution function to create multiple linear regression and then built the best single wavelength, the best combination of dual-wavelength, the best three-wavelength combination and the best combination of four wavelength calibration functions. The correlation coefficient is: the best single wavelength r=0.789, the best dual wavelength r=0.893, the best three wavelength r=0.970 and the best four wavelength r=0.943. The test result showed it was feasible to detect the degree of nectarine sweetness using spectral imaging technology and set the technique foundation for the test on fruit inner quality. Guo et al.[29] proposed a method using multi-spectral imaging system to detect the orange sugar degree. They accessed the feature spectral wavelength of orange sugar derived from the orange reflectance spectral image then applied artificial neural system to build an orange sugar prediction model. The result, correlation coefficient R=01831 of the prediction model, proved the feasibility of hyper-spectral image for non-destructive orange sugar detection. Chen et al.[30] designed a spectral imaging based hyper-spectral system to collect data. They selected the feature images of three bands from amounts of data by the analysis of principal component then extracted six statistical moment based feature parameter, mean gray level, standard deviation, smoothness, third-order moments, consistency and entropy from each feature image. They extracted 8 principal factors to establish tea grade discriminant model based on back-propagation neural network based by principal component analysis of the 18 feature parameters of each sample. The overall recognition was ninety seven percents of training sample and ninety four percents of forecast. Xue et al.[31] studied on orange and used hyper-spectral imaging technology to detect pesticide residues in fruit surface preliminary. They collected the orange hyper-spectral images, whose bands are in the range of 625nm to 725nm then used principal component analysis (PCA) to obtain features wavelength images and used the third principal component images (PC-3) which got appropriate image processing to detect pesticide residues. The detection results showed hyper-spectral images can detect high concentrations of pesticide efficiently. Cai et al.[32] aimed at citrus rust and did principal analysis on the hyper-spectral images which got pretreatments. Then they optimized three feature wavelength bands 571nm, 652nm and 741nm to compose a new image block. The second step was principal analysis to obtain third principal image as the most appropriated citrus rust images. The last step was to complete digital feature extraction using median filtering, square root transformation, threshold segmentation and morphological operations. The results showed the correct ratio was up to 90% when using this algorithm. Zhao et al.[33] took a mixture of Chinese Coptis and Phellodendron Powder as an example, using multi-spectral imaging technology, to study on the fast detection on the adulteration. The result could not only supply a qualitative analysis of the seized goods but provide space positioning. Zou et al.[34] proposed a method combining hyper-spectral image and artificial neural network to clarify rapeseed species. They collected the hyper-spectral images, whose bands ranging from 400nm to 1000nm, of varieties of rapeseed species then used principal component analysis (PCA) to confirm feature wavelength. They extracted feature parameters based on the statistical methods of the combination between gray histogram and GLCM and establish rapeseed species clarification model to get the results that the rate of training model was 93.75% and the rate of forecasting was 91.67% based on the established neural artificial network. The results showed hyper-spectral imaging technology have good classification and identification effects on rapeseed species.

3   Summary

Spectral imaging has not only spatial resolution, but also the ability to distinguish spectra. Using spectral imaging can do not only "qualitative" or "quantitative" analysis, but also "locate" analysis even "regular" analysis. Because of the unique advantage, spectral imaging technology has enormous potential in the space remote sensing, crop and food quality testing. But the following key issues are still worth exploring in its applications.

(1) The spectral ranges of images applied in the agricultural image detection system mostly center on 400-1100nm, that is to say, from visible light to near-infrared spectral band. Spectral images of this range include the special chlorophyll and carotenoid absorption bands. And these two absorption bands are very useful on the detection of surface defects, damage or contamination of fruits and vegetables.

(2) New data analysis and new algorithms, such as independent component analysis, non-uniform second difference can be introduced to improve the effectiveness and accuracy of spectral image detection, to expand the robust of detection system and to gain unsupervised classification ability.

(3) It is difficult to detect rapidly because of spectral image containing large quantity of information and in need of a long period to obtain that information online. The solutions to this problem are as follows: firstly use spectral imaging to detect the most effective feature wavelength of the object under test then imply the spectral image system with several wavelengths into practical applications. This design of hyperspectral imaging system can greatly improve the detection efficiency, thereby achieving the purpose of testing of comprehensive quality of agricultural products online, fast, nondestructive.

Acknowledgement. This research was supported by the Special Fund for Agro-scientific Research in the Public Interest.

References

1. Badioze Zaman, H., Robinson, P., Petrou, M., et al.: Visual Informatics Bridging Research and Practice. Springer, Heidelberg (2009)
2. Zhou, Q., Zhu, D., Wang, C., et al.: Research progress in the detection of agricultural products/food by imaging spectrometry. Food Science 31(23), 423–428 (2010)
3. Tang, Q., Li, S., Wang, K.: Monitoring canopy nitrogen status in winter wheat of growth anaphase with hyperspectral remoate sensing. Spectroscopy and Spectral Analysis 30(11), 3061–3066 (2010)
4. Ju, Y.: Simple Discussion on nutritive diagnosis of nitrogen in corn plants. Contribution of Science and Technology to Development (8), 275 (2010)
5. Noh, H., Zhang, Q., Shin, B., et al.: A neural network model of maize crop nitrogen stress assessment for a multi-spectral imaging sensor. Biosystems Engineering 94, 477–485 (2006)
6. Jones, C.L., Maness, N.O., Stone, M.L., et al.: Chlorophyll estimation using multispectral reflectance and height sensing. Transactions of the ASABE 50, 1867–1872 (2007)
7. Shibayama, M., Sakamoto, T., Takada, E., et al.: Continuous monitoring of visible and near-infrared band reflectance from a rice paddy for determining nitrogen uptake using digital cameras. Plant Production Science 12, 293–306 (2009)
8. Miao, Y., Mulla, D.J., Randall, G.W., et al.: Combining chlorophyll meter readings and high spatial resolution remote sensing images for in-season site-specific nitrogen management of corn. Precision Agriculture 10, 45–62 (2009)
9. Tian, Y., Yao, X., Yang, J., et al.: Assessing newly developed and published vegetation indices for estimating rice leaf nitrogen concentration with ground- and space-based hyperspectral reflectance. Field Crops Research 120, 299–310 (2011)

10. Jiang, W., Sun, M.: Research on predicting modeling for chlorophyll contents of greenhouse tomato leaves based on multispectral imaging. Spectroscopy and Spectral Analysis 31(3), 758–761 (2011)
11. Liu, F., Wang, L., He, Y.: Application of multi-spectral imaging technique for acquisition of cucumber growing information. Optics Journal 29(6), 1616–1620 (2009)
12. Zhang, H., Yao, X., Zhang, X.: Measurement of rice leaf chlorophyll and seed nitrogen contents by using multi-spectral imaging. Chinese Science 22(5), 555–558 (2008)
13. Lu, R.: Multispectral imaging for predicting firmness and soluble solids content of apple fruit. Post-harvest Biology and Technology 31, 147–157 (2004)
14. Kleynen, O., Leemans, V., Destain, M.F.: Development of a multi-spectral vision system for the detection of defects on apples. Journal of Food Engineering 69, 41–49 (2005)
15. Unay, D., Gosselin, B.: Automatic defect segmentation of 'Jonagold' apples on multi-spectral images: A comparative study. Postharvest Biology and Technology 42, 271–279 (2006)
16. Xing, J., Jancso, P., De Baerdemaeker, K.J.: Stem-end/calyx identification on apples using contour analysis in multispectral images. Biosystems Engineering 96(2), 231–237 (2007)
17. Lleó, L., Barreiro, P., Ruiz-Altisent, M., et al.: Multispectral images of peach related to firmness and maturity at harvest. Journal of Food Engineering 93, 229–235 (2009)
18. Blasco, J., Aleixos, N., Gómez, J., et al.: Citrus sorting by identification of the most common defects using multispectral computer vision. Journal of Food Engineering 83, 384–393 (2007)
19. Peng, Y., Lu, R.: Prediction of apple fruit firmness and soluble solids content using characteristics of multi-spectral scattering images. Journal of Food Engineering 82, 142–152 (2007)
20. Peng, Y., Lu, R.: Analysis of spatially resolved hyperspectral scattering images for assessing apple fruit firmness and soluble solids content. Postharvest Biology and Technology 48, 52–62 (2008)
21. Peng, Y., Lu, R.: An LCTF-based multispectral imaging system for estimation of apple fruit firmness Part II. Selection of optimal wavelengths and development of prediction models. Transactions of the ASAE 49(1), 269–275 (2006)
22. Peng, Y., Lu, R.: Assessing peach firmness by multispectral scattering. Journal of Near Infrared Spectroscopy 13(1), 27–36 (2005)
23. Peng, Y., Lu, R.: Improving apple fruit firmness prediction by effective correction of multispectral scattering images. Postharvest Biology and Technology 41(3), 266–274 (2006)
24. Zhao, J., Chen, Q., Saritporn, V., et al.: Determination of apple firmness using hyperspectral imaging technique and multivariate calibration. Transactions ofthe CSAE 25(11), 226–231 (2009)
25. Gamal, E., Wang, N., Adel, E., et al.: Hyperspectral imaging for nondestructive determination of some quality attributes for strawberry. Journal of Food Engineering 81, 98–107 (2007)
26. Gowen, A.A., O'Donnell, C.P., Cullen, P.J., et al.: Hyper-spectral imaging an emerging process analytical tool for food quality and safety control. Trends in Food Science & Technology 18, 590–598 (2007)
27. Yu, X., Liu, M., Cheng, R.: A Study on non-destructive measurement of nectaring sugar content by means of spectralimaging. ACTA Agricultural Universitatis Jiangxiensis 29(6), 1035–1038 (2007)

28. Guo, E., Liu, M., Zhao, J.: Nondestructive detection of sugar content on navel orange with hyperspectral imaging. Transactions of the Chinese Society for Agricultural Machinery 39(5), 91–93 (2008)
29. Chen, Q., Zhao, J., Cai, J.: Estimation of tea quality level using hyperspectral imaging technology. ACTA Optica Sinica 28(4), 669–674 (2008)
30. Xue, L., Li, J., Liu, M.: Detecting pesticide residue on navel orange surface by using hyperspectral imaging. ACTA Optica Sinica 28(12), 2277–2280 (2008)
31. Cai, J., Wang, J., Huang, X.: Detection of rust in citrus with hyperspectral imaging technology. Opto-Electronic Engineering 36(6), 26–30 (2009)
32. Zhao, J., Pang, Q., Ma, J.: Spectral imaging technology applied to mixed powder of rhizoma coptidis and cortex phellodendri chinensis. Opto-Electronic Engineering 30(11), 3259–3263 (2010)
33. Zou, W., Fang, H., Liu, F.: Identification of rapeseed varieties based on hyperspectral imagery. Journal of Zhejiang University (Agricultural & Life Science) 37(2), 175–180 (2011)

Research of Recognition in ENA-13 Code Based on Android Platform

Jianhong Ma[1], Junxiao Xue[1], and Fengjuan Cheng[2]

[1] Software Technology School, ZhengZhou University, 450002, China
[2] College of information Science and Engineering, Henan University of Technology, 450002, China
majianhong@zzu.edu.cn, xuejx@zzu.edu.cn, yh@haut.edu.cn

Abstract. In order to solve the recognition of bar code based on image processing problems we put forward to use Android platform ENA-13 bar code recognition method. The method analyzed the basic principle of ENA-13 code recognition. Firstly we used the Otsu method for gray scale image to obtain a threshold for image segmentation. Then according to level the data of edge points distributed we determined the starting strip and realized barcode image preprocessing and data extraction. Finally, through calculating the standard deviation approach the most close to the pattern recognition of bar code was found out. And the results prove that the method of ENA-13 code recognition speed and accuracy can reach the practical level.

Keywords: Android, ENA-13 code, image preprocessing, image location, image recognition.

1 Introduction

With the development of the network and communication technology intelligent mobile phone has become people's daily consumer goods. In addition to the basic call function, intelligent mobile phone with PDA (personal digital assistant) as main function has become a new terminal node of the Internet. Android is an open-source platform developed by Google and the OHA(Open Handset Alliance)based on Linux2.6 smart phone operating system platform. October 2009, Google released the Android SDK (software developing kits) 2.0, including the Android emulator and a number of development tools for third-party software development space[1]. The simulator is very complete, users can use the keyboard input, mouse click emulator keystrokes, you can also use the mouse to click, drag the screen to manipulate. It's main screen achieved 320 * 480 pixels and colorfull. Therefore we choiced this emulator as our development environment. The bar code identification combined with Android mobile communications capabilities to eliminate the noise reduced image quality and other factors though digital image processing technology, not only increase the reliability of recognition results but also extends the range of bar code identification equipment.

D. Jin and S. Lin (Eds.): Advances in FCCS, Vol. 2, AISC 160, pp. 535–540.
springerlink.com   © Springer-Verlag Berlin Heidelberg 2012

2   The Basic Principles of EAN-13 Barcode Identify

EAN-13 commodity bar code consists of the 13 digit code bar code symbols from the left side of the blank area, start character, data character left, middle separator, the right data character, terminator, verifier, the right margin area.

EAN-13 is a (7, 2) code, in which each character a total width of 7 modules by two bar and two blank bar consists of alternating, and each empty width not more than 4 module. For example T is a character width, meet :$1 \leq Ci \leq 4$,Ci is integer, i=1,2,3,4, and $T = \sum_{i=1}^{4} C_i = 7$.

We use n as the current character unit module width, then   n=T/7.Make mi=Ci/n,i=1,2,3,4. By M1, M2, m3, M4 value can be coded. For example: if m1=1,m2=3,m3=1,m4=2,and bar's arrangement is bar-blank-bar-blank, It shows the current character encoding for 1000100, is on the right side of even character 7.If m1=3,m2=1, m3=1,m4=2,and bar's arrangement is   blank-bar-blank-bar, It shows the current character encoding for 0001011, is on the left side of   even character 9.

3   EAN-13 Barcode Image Recognition Method Overview

Bar code recognition mainly includes image sampling, image pre-processing, image recognition, as shown in figure 1. Images can be collected through the Android mobile phone camera, Image preprocessing mainly includes using digital image processing technique to remove noise, elimination of uneven illumination effect, fuzzy operation, Image segmentation is the location bar code, Bar code decoding is the bar code recognition, interpretation of bar code to carry the digital information, in the background database to find relevant information, in the Android mobile phone screen shows the price of goods, products, names and other information.

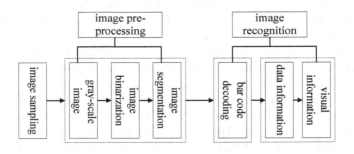

Fig. 1. EAN-13 barcode image recognition process

3.1   Image Pre-processing

Images obtained from cameras usually contain more noise, and there is the angle of tilt and geometric distortion, the purpose of image preprocessing is to eliminate the interference, the image of valuable information is extracted and used to calculate and improve the recognition rate[2].

3.1.1   Gray-Scale Image

The process of a color image into a grayscale image is known as gray-scale image processing. Grayscale images and color images of the same description still reflects the entire image of the whole and local color and brightness level of the distribution and characteristics. Gray-scale formula is as follows: I=0.3R+0.59G+0.11B(I is the subjective effects, R, G, B are red, green and blue components)

3.1.2   Image Binarization

Binary image segmentation is an important method. To transform into a gray scale image is only black and white 2 gray-scale images of the process is called binary. Binarization methods have a direct impact to the image processing capabilities, binarization threshold selection is crucial in the image binarization process, determine the effects of binary image [3]. The same image, different application requirements, select the threshold methods are not the same. Binarization threshold value which can be seen as a type (l) in the form of an operating function T:

$$T = T[x, y, p(x, y), f(x, y)] \tag{1}$$

$$g(x, y) = \begin{cases} 1, & f(x, y) > T \\ 0, & f(x, y) \le T \end{cases} \tag{2}$$

Which f (x, y) is the midpoint of the original image (x, y) the gray level, p (x, y) represents the point about the nature, such as to (x, y) as the center of the neighborhood, the average value.

After thresholding, the image g (x, y) defined in equation (2) above.

The tag 1(or any other suitable gray level) expressed the target sub-images, the tag 0 (or any other sub-image is not marked) expressed the background sub-images. According to the image content, we need to extract the contents and the relationships between background, Traditional binarization threshold into a fixed threshold, adaptive threshold and the quasi-adaptive threshold. Therefore, this paper puts forward a method for gray less than by Otsu method to determine the threshold value of the pixel processing.

Otsu threshold method is an adaptive threshold determination method, also called the Otsu method. It is based on the gray image characteristics, image can be divided into the background and objectives of part two. Background and objectives between-class variance the greater, indicating that constitute the difference between the two parts of the image the greater. Part target into the wrong background or part background into the wrong target will result in smaller differences between the two parts. Therefore, the maximum between-class variance means the division of the minimum probability of misclassification. For the image foreground (target)and background segmentation threshold is recorded as T,belongs to the foreground pixels accounted for the proportion of the whole image is recorded as $\omega 0$, its average gray is$\mu 0$, Background pixels accounted for the proportion of the whole image is recorded as$\omega 1$, its average gray is$\mu 1$, The total average gray level image denote by μ, class variance is denoted by g. Assuming the image size is M × N, the image pixel gray value is less than the threshold value T, the number of pixels recorded as N0, Pixel gray values greater than the threshold T number of pixels recorded as N1, there are:

$$\omega 0 = N0/(M \times N) \tag{3}$$

$$\omega 1 = N1/(M \times N) \tag{4}$$

$$N0 + N1 = M \times N \tag{5}$$

$$\omega 0 + \omega 1 = 1 \tag{6}$$

$$\mu = \omega 0 \mu 0 + \omega 1 \mu 1 \tag{7}$$

$$g = \omega 0 (\mu 0 - \mu)2 + \omega 1 (\mu 1 - \mu)2 \tag{8}$$

The equation (7) into equation (8), to be equivalent to the formula:

$$g = \omega 0 \omega 1 (\mu 0 - \mu 1)2 \tag{9}$$

Traversal method has been used to make the maximum between-class variance threshold T, is the request. The Otsu method is widely used, regardless of the image histogram has no obvious Shuangfeng, can get a satisfactory result, on the quality of segmentation is usually a certain security, can be said to be a stable segmentation.

3.1.3 EAN-13 Code Location

Many domestic and foreign scholars put forward the bar code positioning method, For example, based on the texture [4], using image block [5], using the discrete cosine transform (DiscreteCosine Transform, DCT), using JPEG2000 image compression, image corrosion, template matching, each have advantages and disadvantages. One dimensional bar code is characterized by parallel to the black and white according to certain rules of composition, black can be referred to as stripe, white can be referred to as blank, ideal bar code image of black and white border is obvious, has a strong edge characteristics.

Determine the characteristics of one-dimensional bar code as long as the horizontal direction of the strip and complete information can be decoded. Through the processed image into two binary images, if the bar code vertical or near vertical, current starting was the starting strip. If the bar code level or close to level, the starting strip was expanded by the distribution of initiation of the current of the upper and lower inner linear (such as how much, law and so on).If the bar code tilt ,no matter how the tilt Angle, as long as a pixel in the image across (the "horizontal" was a relative of the strip for the bar code) was taken, as long as all of the bar code in all of the stips to the full by line cutting, then, was sampled complete. The system up to 9 sampling, the first sampling from the bar code took the center of a pixel, and then the sample identification, if successful, would return the results, no longer continue, If it failed, then sampling. And, in the second start, every time in accordance with the 'up and down' sampling rotation sampling, every time sampling would have offset, closer to the image edge up and down. Whether the bar code was horizontal, vertical or inclined, using the above method can locate the bar code region. Location of bar code region could decode the recognition. Because the bar code on the uniform width blank and strip composition, even when tilted, the blank and strip width ratio constant, did not affect the result of decoding. On the tilt angle in 90 degrees near, did not need to be rotated to the vertical bar, just took each line separately decoding, then the results were compared. On approaching the level bar code, which must be rotated to the vertical or near vertical direction, to decode.

3.2   Recognition Algorithm Implementations on Mobile Phones

The system used a typical J2EE three layer structure: The Android platform mobile phone terminal application program as the entrance, and the server program to interact, we used Web server program by Java language, using Apache Tomcat 6 as the middle layer Web server software. On the deployment of the system server software: Which comprised a receiving module, identifying the core module, the database access module, the results generated module, MySQL database support.

Users through the Android mobile phone client sampled the bar code. Mobile phone terminal program bear in image acquisition and sent to the server to decode the task. In a mobile phone to find the installation " identifies the client ",we shoot bar code in frame. As shown in Figure 2. After taking the image, we compressed a certain degree of the image without affecting the recognition success rate in the premise, One could reduce the transmission cost, one can reduce decoding calculation pressure.

Fig.2. The sample interface of bar code at mobile phone terminal

Fig. 3. The successful decoding display results page

4   The End

This paper studied image gray, two values, bar code positioning, decoding and other image processing method in the application of one-dimension bar code recognition, and used the Android platform to achieve the EAN-13 code recognition, effective expansion of the bar code identification equipment. The next step of the research work is to improve the algorithm so that it can locate the reading of uneven illumination, fuzzy and other adverse conditions of commodity bar code image, and has better robustness and noise resistance

References

1. Liu, C.-P., Fan, M.-Y.: Light-weight access control oriented toward Android. Application Research of Computers 27(7), 2611–2613 (2010)
2. Zou, Y.-X., Yang, G.-B.: Research on image pre-processing for Data Matrix 2D barcode decoder. Computer Engineering and Applications 45(34), 183–187 (2009)
3. Hao, Y.-F., Qi, F.-H.: A Novel Machine Learning Based Algorithm to Detect Data Matrix. Journal of Image and Graphics 12(10), 1873–1876 (2007)
4. Jain, A.K., Chen, Y.: Bar code localization using texture analysis. In: Proceedings of the Second IEEE International Conference on Document Analysis and Recognition, pp. 41–44. IEEE Press, Washington, DC (1993)
5. Chai, D., Hock, F.: Locating and decoding EAN-13 barcodes from images captured by digital cameras. In: ICICS 2005: Fifth International Conference on Information, Communications and Signal Processing, pp. 1595–1599 (2005)

Short-Term Electric Load Forecasting Based on LS-SVMs

Fuwei Zhang[1] and Yanbo Li[2]

[1] College of Science, Changchun University of Science and Technology,
Changchun 130022, P.R. China
waywego@163.com
[2] Mathematics School, Jilin University, Jilin 130012, P.R. China
57458030@qq.com

Abstract. Short-term load forecasting of electrical power plays a crucial role in secured and economical operation of power systems. Least Square support vector machines (LS-SVMs) have been proposed as a novel technique to regress. In this paper, LS-SVMs are used for short-term load forecasting. The training sample sets are chosen and pretreated before every forecasting. Then the influence of the non-correlative and bad samples for the forecasting can be avoided. The effectiveness and the feasibility of forecasting of the employed method are examined using simulated experiments. The experimental results show the proposed method performs better than BP algorithm.

Keywords: Least square support vector machines, Regression, Load forecasting.

1   Introduction

Short-term load forecasting of electrical power plays a crucial role in secured and economical operation of power systems. It is important to forecast load variation rapidly and accurately for improving system safety and reducing the cost of generating electricity in practice. The forecasting has to be performed to adapt to the variation of energy demand as well as energy generation. A survey for the UK power system showed that the increase of 1% forecasting error would result in an increase of 10 million pounds in the operating costs of the system [1].

The obvious feature on load forecasting with lead-times, from a few minutes to several days, is to predict the future load according to the analysis of the past load data. Various techniques for power systems load forecasting have been proposed in the last few decades. In general, the previous methods for the load forecasting can be classified into statistical methods, intelligent systems, neural networks and fuzzy logic according to the techniques and models they used. Statistical methods regard the load pattern as time series signal analysis and forecast the value of the future load by employing the different time series analysis techniques [2]. Intelligent systems are considered which have appeared from the artificial intelligence based on mathematical expressions that came from the human experiences. Neural Networks has been considered the most promising area in artificial intelligence, and are based on human experience and on

D. Jin and S. Lin (Eds.): Advances in FCCS, Vol. 2, AISC 160, pp. 541–545.

links of input and output sets, learning or training concepts [3]. Fuzzy Logic models map an input variable set in an output using logic and the prediction values can be obtained using a reverse process named defuzzification [4].

SVMs have been proposed as a novel technique and applied successfully to classification tasks and more recently also to regression [5-8]. Suykens and Vandewalle [9] proposed a modified version of SVM called least squares SVM (LS-SVM), which resulted in a set of linear equations instead of a quadratic programming problem and is applied to time series prediction [10]. In this contribution, LS-SVMs are used in the field of electric load forecasting.

2   Theory of LS-SVMs for Regression Approximation

Consider a given training set $\{(x_i, y_i) \mid x_i \in R^n, y_i \in R\}_{i=1}^N$, where x_i is the input and y_i the corresponding target value. The LS-SVMs model has the following form:

$$y = w^T \phi(x) + b \tag{1}$$

where the nonlinear function $\phi(\cdot)$ maps the input into a higher dimensional feature space. For the function estimation, the optimization problems of LS-SVMs have the following form:

$$\min_{w,e_i} J(w,e) = \frac{1}{2} w^T w + \frac{\gamma}{2} \sum_{i=1}^N e_i^2 \tag{2}$$

subjected to the equality constraints:

$$y_i = w^T \phi(x_i) + b + e_i, \quad i = 1,2,\cdots,N \tag{3}$$

The Lagrangian corresponding to Eq. (2) can be defined as:

$$L(w,b,e_i,\alpha) = J(w,e) - \sum_{i=1}^N \alpha_i \{w^T \phi(x_i) + b + e_i - y_i\}, \tag{4}$$

where α_i $(i=1,2,\cdots,N)$ are Lagrange multipliers. Referring to Suykens and Lukas' work [11], the solution of the optimization problem (2) can be obtained by solving the following linear equations:

$$\begin{bmatrix} 0 & \vec{1}^T \\ \vec{1} & \Omega + \gamma^{-1}I \end{bmatrix} \begin{bmatrix} b \\ \alpha \end{bmatrix} = \begin{bmatrix} 0 \\ y \end{bmatrix}, \tag{5}$$

where $y = [y_1, y_2, \cdots, y_N]^T$, $\vec{1} = [1,1,\cdots,1]^T$, $\alpha = [\alpha_1, \alpha_2, \cdots, \alpha_N]^T$ and Ω takes the form as $\Omega_{kl} = \phi(x_k)^T \phi(x_l) = \psi(x_k, x_l)$ $(k,l = 1,2,\cdots,N)$ following Mercer's condition.

For the choice of the kernel function $\psi(\cdot,\cdot)$ one has several alternatives. Some of common kernel functions are listed in Table 1, where d and σ are constants.

After solving Eq. (5), the LS-SVM model for function estimation can be obtained as:

$$y = \sum_{i=1}^{N} \alpha_i \psi(x, x_i) + b,$$
(6)

It is observed that the Ω in Eq. (5) is only related to the input vector, the parameter γ, and the kernel function. The optimal parameters play a crucial role to build a compounds prediction model with high prediction accuracy and stability. The next section will give the method to obtain the optimal parameters.

3   Electric Load Forecasting Based-LS-SVMs

The past researches show that the future load depends on the past load, temperature, humidity and cloud cover. Considering the strong correlation between the load demands and weather variables, the past load and other factors [12], the load forecasting can be described as a mapping problem with relatively large scale and multi-characteristics. The mapping problem can be represented by the following regression function:

$$y_t = f(y_{t-1}, y_{t-2}, \cdots, y_{t-p}, w_t)$$
(7)

where $y_{t-1}, y_{t-2}, \cdots, y_{t-p}$ is the load before the load y_t, w_t denotes the weather vector consisting of temperature, humidity, wind speed and cloud cover.

According to the excellent characteristics of the LS-SVMs, they can be used as the regressors. The kernel function of the LS-SVMs is chosen as Gaussian kernel function. The user-prescribed parameters in the LS-SVMs are selected empirically.

Samples selection. The electric load fluctuates regularly according to weeks, seasons, and some other factors. The load curves have similar shapes under the similar condition. So according to the weather condition of the forecasting day, the data with similar weather to that of the forecasting day are chosen to form the training set every time. So the interference of the non-correlative samples for the forecasting can be avoided. The samples in the training set are arranged according to the natural time order.

Data Preprocessing. Because there exist some bad load data in the sample set due to some equipments or man-made reasons, and the bad data will result in relatively great error for load forecasting, the load data in the sample set need to be preprocessed before the forecasting as follows:

$$E_i = \frac{1}{N} \sum_{n=1}^{N} y_i^n, \quad V_i = \frac{1}{N} \sum_{n=1}^{N} [y_i^n - E_i]^2$$
(8)

$$S_i^n = \frac{|y_i^n - E_i|}{V_i}, \qquad \overline{y}_i^n = \frac{y_i^{n-1} + y_i^{n+1}}{2} \tag{9}$$

where y_i^n ($n = 1, 2, \cdots, N$; $i = 1, \cdots, 24$) denotes the load of the ith time of the nth day, and N the number of the days in the sample set, respectively. The load expectation E and variance V of the ith time in the sample set, namely, E_i and V_i can be obtained by Eq. (8). S_i^n in Eq. (9) is the deviating rate. y_i^n is considered as the normal load when $S_i^n < k$, otherwise the abnormal load, where k is a threshold given by experience. When y_i^n is abnormal, it should be substituted by $\overline{y}_i^n$.

Then the preprocessed samples are used to train the LS-SVMs. At last through the LS-SVMs the future load can be forecasted.

4   Experimental Results

In order to examine the performance and the efficiency of the LS-SVMs, we simulate the forecasting according to the data (hourly load and weather data) collected from a city in China.

The data are selected to perform the forecasting. We choose the data whose weather situation is similar to that of the forecasting day to form the sample set. The interference of the non-correlative samples for the forecasting can be avoided.

The Average relative error (ARE) and the root mean square error (RMSE) are used to evaluate the performance of SVMs. The RE and RMSE are calculated respectively by

$$ARE = \frac{1}{N} \sum_{i=1}^{N} |\frac{y_i - \hat{y}_i}{y_i}| * 100\% \tag{10}$$

$$RMSE = \sqrt{\sum_{i=1}^{N} (y_i - \hat{y}_i)} \Big/ N * 100\% \tag{11}$$

where y_i represents the actual load, $\hat{y}_i$ the predicted load and N the total number of data points in the test set, respectively.

Table 1 gives load forecasting error for one day. Figure 1 and Figure 2 show the curves of the load (including actual load and forecasting load) and the load forecasting error for one day. The experimental results show that the forecasting results are better than that of the BP algorithm.

Table 1. Load forecasting error for one day

	BP algorithm	LS-SVM
AR (%)	3.63	0.54
RMSE (%)	26.88	5.19

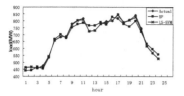 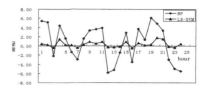

Fig. 1. Comparison of actual and forecasting load **Fig. 2.** Relative error curve of forecasting load

5   Conclusions

LS-SVMs have been proposed as a novel technique and applied more recently to regression. In this paper LS-SVMs are used for short-term load forecasting. According to the weather condition of the forecasting day, the data with similar weather to that of the forecasting day are chosen to form the training set every time. The interference of the non-correlative samples for the forecasting can be avoided. The preprocessed samples are used to train the LS-SVMs and the future load can be forecasted. Simulation results show that the LS-SVMs based method in this paper is efficient for the forecasting.

Acknowledgement. The authors are grateful to the support of the Open Fund Project from Key Laboratory of Symbol Computation and Knowledge Engineering of the Ministry of Education (93K-17-2010-K05).

References

1. Taskin, A., Guneri, A.F.: Economic analysis of risky projects by ANNs. Applied Mathematics and Computation 175, 171–181 (2006)
2. Vapnik, V.: The nature of statistical learning theory. Springer, New York (2000)
3. Cortes, C., Vapnik, V.: Support vector networks. Machine Learning 20, 273–297 (1995)
4. Wen, Y.F.: Materials experimental data analysis and application of support vector regression. Chongqing University Master's Thesis, 22–32 (2009)
5. Ward, S., Chapman, C.: Transforming project risk management into project uncertainty management. International Journal of Project Management 21, 97–105 (2003)
6. Raz, T., Michael, E.: Use and benefits of tools for project risk management. International Journal of Project Management 19, 9–17 (2001)
7. Zou, Z.H., Jiao, H.Y.: Based on least squares support vector regression for short-term water demand forecast. Water Supply and Drainage 34, 328–331 (2008)
8. Yan, G.H., Zhu, Y.S.: Support vector machine regression parameter selection method. Computer Engineering 35(suppl.), 218–220 (2009)
9. Vladimir, C., Ma, Y.Q.: Practical Selection of SVM Parameters and Noise Estimation for SVM Regression. Neural Networks 17, 113–126 (2004)
10. Sathiya, S., Keerthi, C.J.: Asymptotic Behavior of Support Vector Machines with Gaussian Kernel. Neural Computation 15, 1667–1689 (2003)

The Design and Implementation of OPC Sever for 3G Industrial Network

Ying Cai and Feng Liu

College of Computer and Information Science Southwest University,
swu, ChongQing, 400715
{liufeng,caixu}@swu.edu.cn

Abstract. This paper uses the idea of object-oriented software engineering to abstract field devices, variables, and the OPC tags of 3G industrial network as objects. And, it uses the theory of modular software engineering to design and implement OPC server for 3G industrial network. In order to accomplish the asynchronous communication of 3G network, event-based approach is used, and complete the data acquisition and write back of 3G industrial nodes. By the approach of using third-party dynamic link library, the processes of developing OPC sever are simplified, and help complete the development OPC server. After the validation of the standard OPC client-side, it suggested that the design and development of this OPC server meet the requirements of OPC specification.

Keywords: 3G Industrial Network, OPC Server, DLL.

1 Introduction

OPC technology is made and released by OPC Foundation, which is now one of the standard specification of data communication in industry network.OPC focus on the achievements of system integration and data exchange between hardware or software made by different manufacturers[1].The OPC Specification contains OPC Data Access Server, OPC Alarms and Events server, OPC server and OPC Historical Data Access server volume.There mainly there methods of developing OPC server: using COM provided by the MFC to develop OPC server, using ATL to develop OPC server and using OPC development tools[2].

3G Industrial node consists of three parts:MU103 communication module ,MSP430 processor module and traditional instruments.MSP430 MCU completes reading data from a traditional instrument and sent the date through MU103 module.MU103 can send data via both in SMS and TCP.The OPC server designed in this paper uses TCP protocal.

2 The System Architecture of OPC Server for 3G Industrial Network

The design of system architecture. In this paper, we develop the OPC server by using WTOPCSvr.dll.WTOPCSvr.dll provide us with a easy way of using API

D. Jin and S. Lin (Eds.): Advances in FCCS, Vol. 2, AISC 160, pp. 547–551.

functions to construct OPC data item. In this way, all the details of COM and OPC specifications are encapsulated, developers only focus on the reading of field instrumentations and do not need to concern the implementation details of OPC interfaces.WTOPCSvr.dll plays as a database,3G OPC server applications with the name and size of the structure marked and passed to the API function through the dynamic link library, dynamic link library record mark and construct OPC data item. The picture below shows 3G OPC server software architecture.

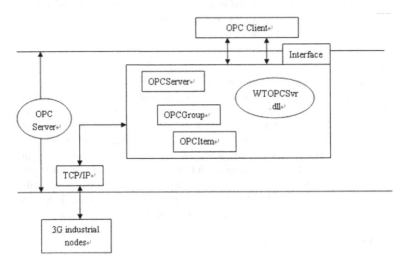

Fig. 1. 3G OPC server system architecture

The design of 3G OPC takes the client-server communication model, and the software system architecture includes an OPC client, OPC server and 3G industrial device notes. The data communication between OPC client and OPC server, OPC server and industrial device notes are on the basis of TCP/IP. The data source serves as OPC server, and it is responsible for providing the data used by the client, and the client is the user of these data.3G OPC server contains three COM objects: 3G OPCServer objects, 3G OPCGroup objects and 3G OPCItem objects. OPCServer object provides a way of accessing the source data, and served as the container of 3G OPCGroup objects.3G OPCGroup object provides the function of manipulating 3G OPCGroup objects for the OPC clients,and provides interfaces through IOPCServer and IOPCBrowseServerAddress..3G OPCItem object is at the bottom layer of the OPC model, and is included in the OPCGroup object. Each 3G OPCItem corresponds to a specific 3G industrial device monitoring data node[3].

The design of software architecture. OPC server software architecture can be divided into two parts: the first part is responsible for creating OPC, OPCItem queue management and updating data; the other part is responsible for reading and writing data through TCP/IP communication, by using listening and communicating Sockets to complete the design of 3G industrial nodes, and the data communication with the industrial nodes, as well as the write-back operation for the client.

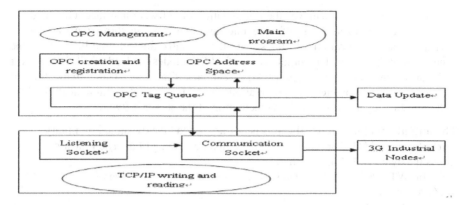

Fig. 2. 3G OPC server software architecture

3 The Implement of OPC Server for 3G Industrial Network

In this paper, we take the modular and object-oriented software engineering ideas for the development of 3G network OPC server software, and use object-oriented programming approach based on MFC to achieve all the functions. The server software includes: initialization module, communication module, OPC module, interface module[4].

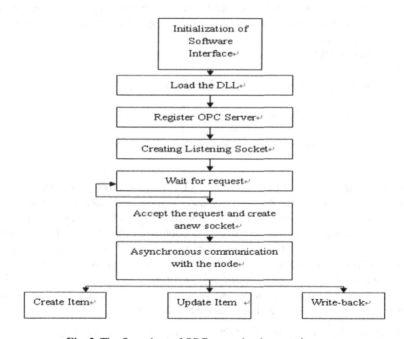

Fig. 3. The flow chart of OPC server implementation

Initialization module achieves the initialization of software interface, OPC and the sockets. The initialization of software interface includes the formation of window interface, the initialization of items list, the initialization of tree-like table, the generation of the menu; The initialization of OPC includes the load of dll, OPC server registration and the assignment of OPC callback functions. The initialization of sockets mainly complete the preparation work for socket communication. The following flow chart shows the implementation of the server software

The initialization of dll. In MFC,it is necessary to load the dynamic link library before been used,MFC provides static and dynamic loading of dll.The follow codes shows the dynamic loading method:

```
InitWTOPCsvr( (BYTE *)&CLSID_Svr, 100 );//Initialization of
WTOPCSvr.DLL
    UINT InitDriver(VOID)
    {
       CString DriverName;
       DriverName = "PA_DEV_DRV.dll";
        hDriverDll = LoadLibrary(DriverName);
       ......
    }
```

The registration of OPC server. It is necessary to register the OPC server before OPC clients connect to it,the dynamic link library (WTOPCSvr.dll) provides two API functions(UpdateRegistry( ) and UnRegisterServer ( )) for the registration. Both these two functions have a identifier (CLSID), which can uniquely identify the server and distinct form other registered servers. Server registration process as shown in the following program segment:

```
    int RegInitOPCSvr()
{   ..........
       CString  SvrName, SvrDescr, HelpPath;
       .......
       SvrName = "PbOpcSvr";
       SvrDescr = "OPC Server for Profibus";
       UpdateRegistry ((BYTE *)&CLSID_Svr, SvrName, SvrDescr, HelpPath);
    }
```

Reading and writing of data based on TCP/IP communication. The TCP/IP data communication launches by the sockets created and inherited from the class CasyncSocket. First of all, it will create the listening socket. When a listening socket accept a communication request from the client (3G industrial nodes), it creates an independent communication socket for this client. This communication socket inherited from CAsyncSocket and supports for asynchronous communication, and it will responsible for the communication between OPC server and 3G industrial nodes. The communication socket uses the message mapping mechanism to realize the function of sending and receiving message response[5].

OPC Item. The operations about OPC Item consists of three parts: create item, update item and OPC item write back. The operations on OPC based on the TCP/IP communication, and they are the core of OPC server. On the basis of completing communication and receiving field data, can function createTag ( ) and updateTag ( ) be called. It is necessary to specify the callback functions when registers OPC server. The write-back operation of data shared by OPC server and devices is complete by the callback functions.

4 Summary

3G OPC server take the object-oriented software engineering idea, and abstracted field devices, variables, and the OPC tags of 3G industrial network as objects, in this way ,will make it very convenient for data management and reuse. The 3G OPC server designed in this article can be detected by standard OPC client, which proved that the server is able to meet the regulatory requirements of OPC communication.

Acknowledgement. This research is supported in part by the ChongQing Information Industry Development Project under Grant NO.201016005.

References

1. Web Services for J2EE, http://www.w3.org/1999/XSL/Format/
2. OPC Foundation. OPC Data Access Custom Interface Standard Version 3.00 (2003)
3. Su, M., Wang, Z.: The Research and Implementation of OPC Sever Data Access. Computer Information 22(3-1), 11–13 (2006)
4. Lai, Y., Luo, J.: The Design and Implementation of OPC Client Based on. NET. Computer Knowledge and Technology 07(45-47) (2010)
5. Li, S., Wangjin, Xiao, W.: The Application of OPC Technology in Supervision and Control System in SCADA System. Marine Electric & Electronic Engineering 03(23-25) (2010)

The Biorthogonality Property of the Trivariate Multiwavelet Wraps and Applications in Education Technology and Materials Science

Lanran Fang

School of Science, Nanyang Institute of Technology, Nanyang 473000, P.R. China
zxs123hjk@126.com

Abstract. The rise of wavelet analysis in applied mathematics is due to its applications and the flexibility. A sort of multiwavelet wraps with multi-scale dilation factor for space $L^2(R^3, C^\nu)$ is introduced, which is the generalization of multivariate wavelet wraps. An approach for designing a sort of biorthogonal multiwavelet wraps in three-dimensional space is presented and their biorthogonality property is characterized by virtue of iteration method and time-frequency analysis method. The biorthogonality formulas concerning these wavelet wraps are established. Moreover, it is shown how to obtain new Riesz bases of space $L^2(R^3, C^\nu)$ from these wavelet packets.

Keywords: Education management, education theory, biorthogonality property, trivariate multiwav elet wraps, Riesz bases, Fourier transfer, iteration method.

1 Introduction

Teaching and learning are two major aspects of education. Since everyone needs to learn even after graduation, it is in particular important to learn during the university studies how to learn. Teachers should not only teach the course materials, but should also guide the students towards active learning methods. I strongly believe that, to be a good teacher, one should be able to teach students how to learn by themselves. Materials science is an interdisciplinary field applying the properties of matter to various areas of science and engineering. This scientific field investigates the relationship between the structure of materials at atomic or molecular scales and their macroscopic properties. It incorporates elements of applied physics and chemistry. With significant media attention focused on nanoscience and nanotechnology in recent years, materials science has been propelled to the forefront at many universities. It is also an important part of forensic engineering and failure analysis. Materials science also deals with fundamental properties and characteristics of materials Although the Fourier transform has been a major tool in analysis for over a century, it has a serious laking for signal analysis in that it hides in its phases information concerning the moment of emission and duration of a signal. Multiwavelets can simultaneously possess many desired properties such as short support, orthogonality, symmetry, and

D. Jin and S. Lin (Eds.): Advances in FCCS, Vol. 2, AISC 160, pp. 553–559.

vanishing moments, which a single wavelet cannot possess simultaneously. This suggests that multiwavelet systems can provide perfect reconstruction, good performance at the boundaries (symmetry), and high approximation order (vanishing moments). Already they have led to exciting applications in signal analysis [1], fractals [2] and image processing [3], and so on. Vector-valued wavelets are a class of special multiwavelets. Chen [4] introduced the notion of orthogonal vector-valued wavelets [5], Wavelet packets, owing to their nice characteristics, have been widely applied to signal processing [1], code theory, image compression, solving integral equation and so on. Coifman and Meyer firstly introduced the notion of univariate orthogoinal wavelet packets. Yang constructed a-scale orthogonal multiwavelet packets which were more flexible in applications. It is known that the majority of information is multi-dimensional information. Shen [5] introduced multivariate orthogonal wavelets which may be used in a wider field. Thus, it is necessary to generalize the notion of multivariate wavelet wrapts to the case of multivariate vector-valued wavelets. The goal of this paper is to give the definition and the construction of biorthogonal vector-valued wavelet packets and construct several new Riesz bases of space $L^2(R^3, C^v)$.

2　The Preliminaries on Multiple Multiresolution Analysis

We begin with some notations. Set $Z_+ = \{0\} \cup N$, $s, n, v \in N$ and $s, n, v \geq 2$, $Z^3 = \{(z_1, z_2, z_3): z_r \in Z, r = 1,2,3\}$, $Z_+^3 = \{(z_1, z_2, z_3): z_r \in Z_+, r = 1,2,3\}$. For any $Y, Y_1, Y_2 \subset R^3$, denoting $mY = \{mx: x \in X\}$, $X_1 + X_2 = \{x_1 + x_2: x_1 \in X_1, x_2 \in X_2\}$, $X_1 - X_2 = \{x_1 = \{x_1 - x_2: x_1 \in X_1, x_2 \in X_2\}$. There exist m elements $\mu_0, \mu_1, \cdots, \mu_m$ in Z_+^3 by finite group theory such that $m = \det(M)$, $Z^3 = \bigcup_{\mu \in \Gamma_0} (\mu + MZ^3)$; $(\mu_1 + MZ^3) \cap (\mu_2 + MZ^3) = \emptyset$, where $\Gamma_0 = \{\mu_0, \mu_1, \cdots, \mu_m\}$ denotes the set of all different representative elements in the quotient group $Z^3/(mZ^3)$ and μ_1, μ_2 denote two arbitrary distinct elements in Γ_0, M is a 3×3 matrix Set $\mu_0 = \underline{0}$, where $\underline{0}$ is the null of Z_+^3. Let $\Gamma = \Gamma_0 - \{\underline{0}\}$ and Γ, Γ_0 be two index sets. By $L^2(R^3, C^v)$, we denote the aggregate of all vector-valued functions $\Upsilon(x)$, i.e., $L^2(R^3, C^v) := \{\Upsilon(x) = (\gamma_1(x), \gamma_2(x), \cdots, \gamma_v(x))^T: \gamma_l(x) \in L^2(R^3), l = 1, 2, \cdots, v\}$, where T means the transpose of a vector. For $H(x) \in L^2(R^3, C^v)$, $\|H\|$ denotes the norm of vector-valued function $H(x)$, i.e.,

$$\|H\| := (\sum_{l=1}^{v} \int_{R^3} |h_l(x)|^2 \, dx)^{1/2}$$

In the below * means the transpose and the complex conjugate, and its integration is defined to be

$$\int_{R^3} H(x)dx = (\int_{R^3} h_1(x)dx, \int_{R^3} h_2(x)dx, \cdots\cdots, \int_{R^3} h_v(x)dx)^T.$$

The Fourier transform of $H(x)$ is defined as $\widehat{H}(\gamma) := \int_{R^3} H(x) \cdot e^{-ix \cdot \gamma} dx$, where $x \cdot \gamma$ denotes the inner product of real vectors x and γ. For $G, H \in L^2(R^3, C^v)$, their symbol inner product is defined by

$$[G(\cdot), H(\cdot)] := \int_{R^3} G(x)H(x)^* dx, \tag{1}$$

Definition 1. We say that a pair of vector-valued functions $H(x), \widetilde{H}(x) \in L^2(R^3, C^v)$ are biorthogonal, if their translations satisfy

$$[H(\cdot), \widetilde{H}(\cdot - n)] = \delta_{\underline{0}, n} I_v, \quad n \in Z^3, \tag{2}$$

Definition 2. A vector-valued multiresolution analysis of the space $L^2(R^3, C^v)$ is a nested sequence of closed subspaces $\{U_\ell\}_{\ell \in Z}$ such that (i) $U_\ell \subset U_{\ell+1}, \forall \ell \in Z$; (ii) $\bigcap_{\ell \in Z} U_\ell = \{0\}$ and $\bigcup_{\ell \in Z} U_\ell$ is dense in $L^2(R^3, C^v)$, where 0 denotes an zero vector of space R^v; (iii) $\Upsilon(x) \in U_\ell \Leftrightarrow \Upsilon(Mx) \in U_{\ell+1}, \forall \ell \in Z$; (iv) there exists $F(x) \in U_0$, called a vector-valued scaling function, such that its translates $F_n(x) := F(x-n), \quad n \in Z^3$ forms a Riesz basis of subspace U_0.

Since $F(x) \in Y_0 \subset Y_1$, by Definition 2 there exists a finitely supported sequence of constant $v \times v$ matrice $\{A_n\}_{n \in Z^3} \in \ell^2(Z^3)^{v \times v}$ such that

$$F(x) = \sum_{n \in Z^3} A_n F(Mx - n). \tag{3}$$

Equation (6) is called a refinement equation. Define

$$m\mathcal{A}(\gamma) = \sum_{n \in Z^3} A_n \cdot \exp\{-in \cdot \gamma\}, \quad \gamma \in R^3. \tag{4}$$

where $\mathcal{A}(\gamma)$, which is a $2\pi Z^s$ periodic function, is called a symbol of $F(x)$. Thus, (4) becomes

$$\hat{F}(M\gamma) = \mathcal{A}(\gamma)\hat{F}(\gamma), \quad \gamma \in R^3. \tag{5}$$

Let $U_j, j \in Z$ be the direct complementary subspace of Y_j in Y_{j+1}. Assume there exist m functions $\psi_\mu(x) \in L^2(R^3, C^v), \mu \in \Gamma$ such that their translates and dilates form a Riesz basis of X_j, i.e.,

$$X_j = \overline{(span\{\Psi_\mu(M^j \cdot -n) : n \in Z^3, \mu \in \Gamma\})}, \quad j \in Z. \tag{6}$$

Since $\Psi_\mu(x) \in U_0 \subset Y_1, \mu \in \Gamma$, there exist m sequences of constant $v \times v$ matrice $\{B_n^{(\mu)}\}_{n \in Z^4}$ such that

$$\Psi_\mu(x) = \sum_{n \in Z^3} B_n^{(\mu)} F(Mx - n), \quad \mu \in \Gamma. \tag{7}$$

By implementing the Fourier transform for the both sides of (9) , we have

$$\hat{\Psi}_\mu(M\gamma) = \mathcal{B}^{(\mu)}(\gamma)\hat{\Phi}(\gamma), \quad \gamma \in R^3, \quad \mu \in \Gamma. \tag{8}$$

$$\mathcal{B}^{(\mu)}(\gamma) = \frac{1}{m} \sum_{n \in Z^v} B_n^{(\mu)} \cdot \exp\{-in \cdot \gamma\}, \quad \mu \in \Gamma. \tag{9}$$

If $F(x), \tilde{F}(x) \in L^2(R^3, C^v)$ are a pair of biorthogonal vector-valued scaling functions, then

$$[F(\cdot), \tilde{F}(\cdot - n)] = \delta_{0,n} I_v, \quad n \in Z^3. \tag{10}$$

We say that $\Psi_\mu(x), \tilde{\Psi}_\mu(x) \in L^2(R^3, C^v), \mu \in \Gamma$ are pairs of biorthogonal vector-valued wavelets associated with a pair of biorthogonal vector-valued scaling functions $F(x)$ and $\tilde{F}(x)$, if the family $\{\Psi_\mu(x-n), n \in Z^3, \mu \in \Gamma\}$ Is a Riesz basis of subspace X_0, and

$$[F(\cdot), \tilde{\Psi}_\mu(\cdot - n)] = [\tilde{F}(\cdot), \Psi_\mu(\cdot - n)] = 0, \quad \mu \in \Gamma, \ n \in Z^3. \tag{11}$$

$$X_j^{(\mu)} = \overline{Span\{\Psi_\mu(M^j \cdot -n) : n \in Z^3, \}}, \mu \in \Gamma, j \in Z. \tag{12}$$

Similar to (5) and (9), there exist 64 finite supported sequences of $v \times v$ constant matrice $\{\tilde{A}_k\}_{k \in Z^3}$ and $\{\tilde{B}_k^{(\mu)}\}_{k \in Z^3}$, $\mu \in \Gamma$ such that $\tilde{F}(x)$ and $\tilde{\Psi}_\mu(x)$ satisfy the refinement equations:

$$\tilde{F}(x) = \sum_{k \in Z^3} \tilde{A}_k \tilde{F}(Mx - k),$$

and

$$\tilde{\Psi}_\mu(x) = \sum_{k \in Z^3} \tilde{B}_k^{(\mu)} \tilde{F}(Mx - k), \quad \mu \in \Gamma.$$

3   The Biorthogonality Propoty of a Class of Multiwavelet Wraps

Denoting by $H_0(x) = F(x)$, $H_\mu(x) = \tilde{\Psi}_\mu(x)$, $\tilde{H}_0(x) = F(x)$, $\tilde{H}_\mu(x) = \tilde{\Psi}_\mu(x)$, $Q_k^{(0)} = A_k$, $Q_k^{(\mu)} = B_k^{(\mu)}$, $\tilde{Q}_k^{(0)} = \tilde{A}_k$ $\tilde{Q}_k^{(\mu)} = \tilde{B}_k^{(\mu)}$, $\mu \in \Gamma$, $k \in Z^3$, $M = 14I_v$. For any $\alpha \in Z_+^s$ and the given vector-valued biorthogonal scaling functions $H_0(x)$ and $\tilde{H}_0(x)$, iteratively define, respectively,

$$H_\alpha(x) = H_{14\sigma+\mu}(x) = \sum_{k \in Z^3} Q_k^{(\mu)} H_\sigma(14x - k), \tag{13}$$

$$\tilde{H}_\alpha(x) = \tilde{H}_{14\sigma+\mu}(x) = \sum_{k \in Z^3} \tilde{Q}_k^{(\mu)} \tilde{H}_\sigma(14x - k). \tag{14}$$

where $\mu \in \Gamma_0$, $\sigma \in Z_+^3$ is the unique element such that $\alpha = 14\sigma + \mu$, $\mu \in \Gamma_0$ follows.

Definition   3.   We   say   that   two   sets   of   multiple   functions $\{H_{14\sigma+\mu}(x), \sigma \in Z_+^3, \mu \in \Gamma_0\}$ and $\{\tilde{H}_{14\sigma+\mu}(x),\qquad \sigma \in Z_+^3,\quad \mu \in \Gamma_0\}$ are multiwavelet wraps with respect to a pair of biorthogonal multi-scaling functions $H_0(x)$ and $\tilde{H}_0(x)$, resp., where $H_{14\sigma+\mu}(x)$ and $\tilde{H}_{14\sigma+\mu}(x)$ are given by (13) and (14), resp..

Applying the Fourier transform for the both sides of (18) and (19) yields, respectively,

$$\hat{H}_{14\sigma+\mu}(\gamma) = Q^{(\mu)}(\gamma/14)\hat{H}_\sigma(\gamma/14), \quad \hat{\tilde{H}}_{14\sigma+\mu}(14\gamma) = Q^{(\mu)}(\gamma)\hat{\tilde{H}}_\sigma(\gamma), \quad \mu \in \Gamma_0,$$

Lemma   1[4].   Assume   that $\{\tilde{H}_\beta(x), \beta \in Z_+^3\}$ and $\{\tilde{H}_\beta(x), \beta \in \quad Z_+^3\}$ are multiwavelet wraps with respect to a pair of multiple functions $H_0(x)$ and $\tilde{H}_0(x)$, resp. Then, for $\beta \in Z_+^3, \mu, v \in \Gamma_0$, we have

$$[H_{14\beta+\mu}(\cdot), \tilde{H}_{14\beta+v}(\cdot - k)] = \delta_{0,k}\delta_{\mu,v}I_v, \ k \in Z^3. \tag{15}$$

Theorem 1. If $\{H_\alpha(x), \alpha \in Z_+^3\}$ and $\{\tilde{H}_\alpha(x), \alpha \in Z_+^3\}$ are vector-valued wavelet packets with respect to a pair of biorthogonal scaling functions $H_0(x)$ and $\tilde{H}_0(x)$, then for any $\alpha, \sigma \in Z_+^3$, we have

$$[H_\alpha(\cdot), \tilde{H}_\sigma(\cdot - k)] = \delta_{\alpha,\sigma}\delta_{0,k}I_v, \ k \in Z^3. \tag{16}$$

Proof. When $\alpha = \sigma$,(16) follows by Lemma 1. as $\alpha \neq \sigma$ and $\alpha, \sigma \in \Gamma_0$, it follows from Lemma 1 that (16) holds, too. Assuming that α is not equal to β , as well as at least one of $\{\alpha, \sigma\}$ doesn't belong to Γ_0 , we rewrite α, σ as $\alpha = 14\alpha_1 + \rho_1$, $\sigma = 14\sigma_1 + \mu_1$, where $\rho_1, \mu_1 \in \Gamma_0$. Case 1. If $\alpha_1 = \sigma_1$, then $\rho_1 \neq \mu_1$. (16) follows by virtue of (13), (14) as well as Lemma 1 , i.e.,

$$[H_\alpha(\cdot), \widetilde{H}\sigma(\cdot - k)]$$

$$= \tfrac{1}{(2\pi)^3} \int_R \widehat{H}_{4\alpha_1 + \rho_1}(\gamma) \widehat{\widetilde{H}}_{4\sigma_1 + \mu_1}(\gamma)^* \cdot e^{ik\cdot\gamma} d\gamma = \tfrac{1}{(2\pi)^3} \int_{[0,2\pi]^3} \delta_{\rho_1,\mu_1} I_\gamma \cdot e^{ik\cdot\gamma} d\gamma = 0 \ .$$

Case 2. If $\alpha_1 \neq \sigma_1$, order $\alpha_1 = 14\alpha_2 + \rho_2$, $\sigma_1 = 14\sigma_2 + \mu_2$, where $\alpha_2, \sigma_2 \in Z_+^s$, and $\rho_2, \mu_2 \in \Gamma_0$. Provided that $\alpha_2 = \sigma_2$, then $\rho_2 \neq \mu_2$. Similar to Case 1, (16) can be established. When $\alpha_2 \neq \sigma_2$,we order $\alpha_2 = 14\alpha_3 + \rho_3$, $\sigma_2 = 14\sigma_3 + \mu_3$, where $\alpha_3, \sigma_3 \in Z_+^4$, $\rho_3, \mu_3 \in \Gamma_0$. Thus, after taking finite steps (denoted by κ), we obtain $\alpha_\kappa \in \Gamma_0$, and $\rho_\kappa, \mu_\kappa \in \Gamma_0$. If $\alpha_\kappa = \sigma_\kappa$, then $\rho_\kappa \neq \mu_\kappa$.

$$8\pi^3 [H_\alpha(\cdot), \widetilde{H}\sigma(\cdot - k)] = \int_{R^3} \widehat{H}_\alpha(\gamma) \widehat{\widetilde{H}}_{\sigma_1}(\gamma)^* \cdot e^{ik\cdot\gamma} d\gamma$$

$$= \int_{[0,2\cdot14^\kappa\pi]^3} \{\prod_{l=1}^\kappa \mathcal{Q}^{(\rho_l)}(\tfrac{\gamma}{14^l})\} \{\sum_{u\in Z_3} \widehat{H}_{\alpha_\kappa}(\tfrac{\gamma}{14^l} + 2u\pi) \cdot \widehat{\widetilde{H}}_{\sigma_\kappa}(\tfrac{\gamma}{14^l} + 2u\pi)^*\} \{\prod_{l=1}^\kappa \widetilde{\mathcal{Q}}^{(\mu_l)}(\tfrac{\gamma}{14^l})\}^* \cdot e^{ik\cdot\gamma} d\gamma$$

$$= \int_{([0,2\cdot14^\kappa\pi]^3} \{\prod_{l=1}^\kappa \mathcal{Q}^{(\rho_l)}(\gamma/14^l)\} \cdot O \cdot \{\prod_{l=1}^\kappa \widetilde{\mathcal{Q}}^{(\mu_l)}(\gamma/14^l)\}^* \cdot \exp\{-ik\cdot\gamma\} d\gamma = 0 \ .$$

Therefore, for any $\alpha, \sigma \in Z_+^3$, result (16) is established.

Definition 4. Let $\{U_n, \phi, \tilde{\phi}\}$ be a given GBMS. We say that the GBMS has a pyramid decomposition scheme if there are band-pass functions $\hbar_l, \tilde{\hbar}_l \in L^2(R^3)$, $l \in J$ such that $\forall \Gamma(x) \in L^2(R^3)$,

$$\sum_{v\in Z^3} \langle \Gamma, \tilde{\phi}_{1,va} \rangle \phi_{1,va} = \sum_{v\in Z^3} \langle \Gamma, S_{va}\tilde{\phi} \rangle S_{va}\phi + \sum_{l\in J} \sum_{v\in Z^3} \langle v, S_{va}\tilde{\hbar}_l \rangle S_{va}\hbar_l \quad (17)$$

Theorem 2[4]. Let $\phi(x), \tilde{\phi}(x), \hbar_l(x)$ and $\tilde{\hbar}_l(x), l \in J$ be functions in $L^2(R^3)$. Assume that conditions in Theorem 1 are satisfied. Then, for any function $f(x) \in L^2(R^3)$, and any integer n,

$$\sum_{z} \langle f, \tilde{\sigma}_{\dots} \rangle \phi_{\dots}(x) = \sum_{z}' \sum_{z}'' \sum_{z} \langle f, \tilde{\hbar}_{\dots} \rangle \hbar_{\dots}(x) \ . \quad (18)$$

Furthermore, for arbitrary $f(x) \in L^2(R^3)$, it follows that

$$f(x) = \sum_{l=1}' \sum_{z}'' \sum_{z} \langle f, \tilde{\hbar}_{\dots} \rangle \hbar_{\dots}(x) \ . \quad (19)$$

References

1. Telesca, L., et al.: Multiresolution wavelet analysis of earthquakes. Chaos, Solitons & Fractals 22(3), 741–748 (2004)
2. Iovane, G., Giordano, P.: Wavelet and multiresolution analysis: Nature of ε^{∞} Cantorian space-time. Chaos, Solitons & Fractals 32(4), 896–910 (2007)
3. Zhang, N., Wu, X.: Lossless Compression of Color Mosaic Images. IEEE Trans. Image Processing 15(16), 1379–1388 (2006)
4. Chen, Q., et al.: Biorthogonal multiple vector-valued multivariate wavelet packets associated with a dilation matrix. Chaos, Solitons & Fractals 35(2), 323–332 (2008)
5. Shen, Z.: Nontensor product wavelet packets in $L_2(R^s)$. SIAM Math. Anal. 26(4), 1061–1074 (1995)

Global Stability Analysis on the Dynamics of an SIQ Model with Nonlinear Incidence Rate

Xiuxiang Yang[1], Feng Li[1], and Yuanji Cheng[2]

[1] Department of Mathematics, Weinan Normal University, Weinan 714000 Shaanxi, China

[2] School of Technology Malmo University, SE-205 06 Malmo, Sweden

yangxiuxiang2000@yahoo.com.cn, lifeng5849@163.com,
yuanji.cheng@mah.se

Abstract. An SIQ epidemic model with isolation and nonlinear incidence rate is studied. We have obtained a threshold value R and shown that there is only a disease free equilibrium point when $R < 1$, and there is also an endemic equilibrium point if $R > 1$. With the help of Liapunov function, we have shown that disease free- and endemic equilibrium point is globally stable.

Keywords: SIQ model, Isolation, nonlinear infectious rate, threshold, equilibrium point, global stability.

1 Introduction

In this paper, we shall as well exploit the Lyapunov direct method to study the global stability analysis on the dynamics of an SIQ model with nonlinear incidence rate. Assuming that the population has a very short immunity period, which is negligible, and whence there is a class of recovered and the models being studied here are of SIQ type. Assume that the total population is divided into: S = S(t) susceptible, I(t) infected, Q(t) isolated and R(t)recovered classes at time t. Let A be the constant migration rate of the population, and d be natural death rate of each kind, α_1 and α_2 be the infected death rate of infectious and isolated classes respectively, and δ be the transition rate from infected to isolated classes, r, ε be immunity's lose rate of infected respectively isolated classes. We assume that the constants d, δ, A are positive, and $r, \varepsilon, \delta, \alpha_1, \alpha_2$ are nonnegative . Thus we consider the

$$\begin{cases} S'(t) = A - \dfrac{\beta IS}{1+mI} - dS + rI + \varepsilon Q \\ I'(t) = \dfrac{\beta IS}{1+mI} - (r+\delta+d+\alpha_1)I \\ Q'(t) = \delta I - (\varepsilon+d+\alpha_2)Q \end{cases} \quad (1)$$

562   X. Yang, F. Li, and Y. Cheng

we see that N satisfies: $N'(t) = A - dN - \alpha_1 I - \alpha_2 Q \le A - dN$, thus

$$N(t) \le N_0 e^{-dt} + \frac{A}{d}(1 - e^{-dt}) \qquad (2)$$

It follows from the above estimates that the solutions of initial value problem for system (1) exist on $[0, +\infty)$, $D = \{(S,I,Q) \ R_+^3 : S,I,Q \quad 0, S + I + Q \square A/d\}$ is invariant subset of (1).

2   Local Stability of Equilibrium

To find the equilibrium of system (1), we get the equilibrium $P_0(A/d, 0, 0)$ and $P_1(S_1, I_1, Q_1)$, let's define the threshold number for (1) $R_1 = \frac{d(r + \delta + d + \alpha_1)}{\beta A}$, Then we have that P_1 is in R_+^3 if and only if $R_1 < 1$.

Here $S_1 = \frac{(r + \delta + d + \alpha_1)}{\beta}(1 + mI_1)$,

$Q_2 = \frac{\delta I_1}{\varepsilon + d + \alpha_2}$, $I_1 = \frac{\beta A - d(r + d + \delta + \alpha_1)}{md(r + d + \delta + \alpha_1) + \beta(d + \alpha_1 + \frac{\delta(d + \alpha_2)}{\varepsilon + d + \alpha_2})}$

Theorem 1. For the system (1), the infection-free equilibrium point P_0 is locally asymptotically stable if $R_1 > 1$, If $R_1 < 1$, then the endemic equilibrium point P_1 is locally asymptotically stable and P_0 becomes unstable.

Proof. First we consider the system (1) and the infectious free equilibrium P_0, The Jacobian matrix $J(P_0)$, Which has eigenvalues:

$\lambda_1 = -(d + \varepsilon) < 0$, $\lambda_2 = \frac{\beta A}{d} - (r + \delta + d + \alpha_1)$, $\lambda_3 = -(\varepsilon + d + \alpha_2) < 0$,

When $R_1 < 1$, then λ_2 is also negative, and hence P_0 is locally stable.

In the P_1, $\beta S_2 = (r + d + \delta + \alpha_1)(1 + mI_2)$. Let $d_i = d + \alpha_i, i = 1, 2$

According to $J(P_1)$, the associated characteristic equation is $\lambda^3 + b_1\lambda^2 + b_2\lambda + b_3 = 0$,

$$b_1 = \beta I_2 + d(1 + mI_2) + mI_2(r + \delta + d_1) + (\varepsilon + d_2)(1 + mI_2) > 0$$

$$b_2 = (\beta I_2 + d + dmI_2)mI_2(r + \delta + d_1) + (\beta I_2 + d + dmI_2)(\varepsilon + d_2)(1 + mI_2)$$
$$+ mI_2(r + \delta + d_1)(\varepsilon + d_2)(1 + mI_2) + \beta I_2(\delta + d_1 - rmI_2) > 0$$

$$b_3 = mI_2(\varepsilon + d_2)(1 + mI_2)(r + \delta + d_1)(\beta I_2 + d + dmI_2)$$
$$+ \beta I_2((\varepsilon + d_2)(1 + mI_2)(\delta + d_1 - rmI_2) - \delta\varepsilon(1 + mI_2)^2) > 0$$

To verify the Hurwitz condition $b_1 b_2 - b_3 > 0$, we first rewrite:
$$b_1 = c_{10} + c_{11}mI_2,$$

$$b_2 = c_{20} + c_{21}mI_2 + c_{22}(mI_2)^2 \quad b_3 = c_{30} + c_{31}mI_2 + c_{32}(mI_2)^2 + c_{33}(mI_2)^3 ,$$

Where $c_{10} = d + \varepsilon + d_2 + \beta I_2$, $c_{11} = d + r + \varepsilon + d_2 + \delta + d_1$,

$c_{20} = d(\varepsilon + d_2) + (\varepsilon + d_2 + \delta + d_1)\beta I_2$,

$c_{21} = rd + (r + \delta + d_1)(\varepsilon + d_2) + (d + \beta I_2)(\varepsilon + d_2 + \delta + d_1)$,

$c_{22} = (r + \delta + d_1)(\varepsilon + d_2) + d(r + \varepsilon + d_2 + \delta + d_1)$,

$c_{30} = (\varepsilon d_1 + \delta d_2 + d_1 d_2)\beta I_2$,

$c_{31} = d(r + \delta + d_1)(\varepsilon + d_2) + 2(\varepsilon d_1 + \delta d_2 + d_1 d_2)\beta I_2$,

$c_{32} = 2d(r + \delta + d_1)(\varepsilon + d_2) + (\varepsilon d_1 + \delta d_2 + d_1 d_2)\beta I_2$,

$c_{33} = d(r + \delta + d_1)(\varepsilon + d_2)$.

Then get
$$b_1 b_2 - b_3 = c_{10}c_{20} - c_{30} + (c_{10}c_{22} + c_{11}c_{20} - c_{31})mI_2 +$$
$$(c_{10}c_{22} + c_{11}c_{21} - c_{32})(mI_2)^2 + (c_{11}c_{22} - c_{33})(mI_2)^3$$
$$C_0 =: c_{10}c_{20} - c_{30} > 0, C_1 =: c_{10}c_{22} + c_{11}c_{20} - c_{31} > 0 ,$$
$$C_2 =: c_{10}c_{22} + c_{11}c_{21} - c_{32} > 0, \quad C_3 =: c_{11}c_{22} - c_{33} > 0 ,$$ So P_1 is locally asymptotically stable.

3   Global Stability of Equilibrium

Theorem 2. For the system (1), If $R_1 > 1$, then the disease free equilibrium P_0 of (2) is globally asymptotically stable; when $R_1 < 1$, then P_0 is unstable, and the endemic equilibrium point P_1 is globally asymptotically stable.

Proof. we can easily proof the disease free equilibrium P_0 of (1) is globally asymptotically stable same as theorem. Following discuss asymptotic stability of P_1, we again change variable $N = S + I + Q$

$$\begin{cases} N'(t) = A - dN - \alpha_1 I - \alpha_2 Q \\ I'(t) = \dfrac{I}{1+mI}[\beta(N-I-Q)-(r+\delta+d+\alpha_1)(1+mI)] \\ Q'(t) = \delta I - (\varepsilon+d+\alpha_2)Q \end{cases}$$

$x = N - N_1, y = I - I_1, z = Q - Q_1$, then we deduce

$$\begin{cases} x'(t) = -dx - \alpha_1 y - \alpha_2 z \\ y'(t) = \dfrac{y+I_2}{1+m(y+I_2)}(\beta x - (\beta + m(r+d+\delta+\alpha_2))y - \beta z) \\ z'(t) = \delta y - (\varepsilon+d+\alpha_2)z \end{cases} \tag{3}$$

We have used the relation $(r+\delta+d+\alpha_1)(1+mI_2) = \beta S_2 = \beta(N_2 - I_2 - Q_2)$

Choose   the   Liapunov   function

$$V = \frac{w_1 x^2}{2} + w_2 (y - I_2 \ln(1+\frac{y}{I_2})) + \frac{my^2}{2}) + \frac{w_3 z^2}{2} + \frac{(x-z)^2}{2}$$

$$\left.\frac{dV}{dt}\right|_{(3)} = w_1 x(-dx - \alpha_1 y - \alpha_2 z) + w_2 y(\beta x - (\beta + m(r+d+\delta+\alpha_2))y - \beta z)$$

$$+ w_3 z(\delta y - (\varepsilon+d+\alpha_2)z) + (x-z)(-dx - (\delta+\alpha_1)y + (\varepsilon+d)z)$$

$$= -d(1+w_1)x^2 + (w_2\beta - w_1\alpha_1 - \alpha_1 - \delta)xy + (\varepsilon + 2d - \alpha_2 w_1)xz - w_2(\beta + m(r+d+\delta+\alpha_2))y^2$$

$$+ (\alpha_1 + \delta + w_3\delta - w_2\beta)yz - (\varepsilon + d + (\varepsilon + d + \alpha_2)w_3)z^2$$

If we choose $w_1, w_2, w_3 > 0$ $w_1 = (\varepsilon+2d)\big/\alpha_2, w_2 = (w_1\alpha_1 + \alpha_1 + \delta)\big/\beta, w_3 = w_1\alpha_1\big/\delta$,

$$\left.\frac{dV}{dt}\right|_{(3)} = -d(1+w_1)x^2 - w_2(\beta + m(r+d+\delta+\alpha_2))y^2 - (\varepsilon+d+(\varepsilon+d+\alpha_2)w_3)z^2$$

Which is clearly negative definite. Hence, $P_1(S_1, I_1, Q_1)$ is globally asymptotically stable.

References

1. Hethcote, H.W., van den Driessche: Some epidemiological models with nonlinear incidence rate. J. Math. Biol. 29, 271–287 (1991)
2. Li, M.Y., Muldowney, J.S.: Global stability of the SEIR model in epidemiology. Math. Biosci. 125, 155–164 (1995)
3. Ruan, S., Wang, W.: Dynamical behavior of an epidemic model with a nonlinear incidence rate. J. Diff. Equa. 188, 135–163 (2003)
4. Korobeinikov, A., Maini, P.K.: Nonlinear incidence and stability of infectious disease models. Math. Medicine and Biology 22, 113–128 (2005)
5. Kyrychko, Y.N., Blyuss, K.B.: Global properties of a delayed SIR model with temporary immunity and nonlinear incidence rate. Nonlinear Analysis: Real world Applications 6, 495–507 (2005)
6. Feng, Z.L., Thieme, H.R.: Recurrent outbreaks of childhood diseases revisited: The impace of isolation. Math. Biosci. 128, 93–130 (1995)
7. Gerberry, D.J., Milner, F.A.: An SEIQR model for childhood diseases. Mathematical Biology 59, 535–561 (2009)
8. Castillo-Chavez, C., Castillo-Garsow, C.W., Yakubu, A.A.: Matheamtical models of isolation and quarantine. JAMA 290, 2876–2877 (2003)
9. Xiao, D., Ruan, S.: Global analysis of an epidemic model with nonmonotone incidence rate. Math. Biosci. 208, 419–429 (2007)
10. Chinviriyasit, S., Chinviriyasit, W.: Global stability of an SIQ epidemic model. Kasetsart J., (Nat. Sci.) 41, 225–228 (2007)

A MCL-Based Localization Algorithm for Target Binary-Detection of Mobile Nodes

Jia Li, Xiaopeng You, Jianfa Xia, and Huaizhong Li[*]

College of Physics and Electronic Information Engineering, Wenzhou University,
Zhejiang China
hli@wzu.edu.cn

Abstract. This paper mainly studied the mobile node localization and tracking in Wireless Sensor Networks (WSNs). Monte Carlo Localization algorithm (MCL) is widely for mobile node localization and tracking. However, MCL doesn't consider the influence of different anchor nodes. Thus, we proposed a binary-detection Monte Carlo Localization algorithm (BD MCL), which combines binary-detection with MCL. Our algorithm mainly uses the time of target detection as the node weights in sampling process to enhance the influence of nearer anchor nodes. Simulations show that this new algorithm exhibits better performance than traditional MCL with regards to motion forecast and localization precision.

Keywords: WSN, anchor node, sensor node, MCL, localization, binary-detection.

1 Introduction

Wireless Sensor Network (WSN) concerns with wireless communication technology, sensor technology, embedded computing and distributed information processing. Recently, WSN, which requires highly cross-disciplinary and integrated knowledge, attracts extensive international attentions and research interests [1]. WSN has the features of self-organizing, low power, and low cost, it has broad application prospects in environmental monitoring, remote medical care, military and other fields.
Localization technique is one of the hot research spots in WSN. Presently, most of the WSN localization algorithms are aiming at static situations.

As the dynamic sensor networks have very complicated behavior, research involving dynamic sensor networks becomes more difficult. There are some research projects which concern with the dynamic sensor networks, for example, Zebra Net is one of the typical application projects. Mobile node tracking, precise localization and accurate description of trajectory are typical research focuses in recent years for dynamic WSNs. Unlike general static networks, research results for dynamic sensor networks can be applied to biological behavior research, medical monitoring, intelligent toys, car alarm, urban traffic management and other civilian and military areas.

---

[*] Corresponding author.

D. Jin and S. Lin (Eds.): Advances in FCCS, Vol. 2, AISC 160, pp. 567–572.
springerlink.com   © Springer-Verlag Berlin Heidelberg 2012

A research trend for dynamic sensor networks is to make the best use of movement information, and to design an algorithm for dynamic sensor networks or mobile node localization and tracking.

2   Research Status

If we simply apply a static nodes localization algorithm to the mobile nodes, although it also can updated to locate from time to time, however the mobility will lead to lower localization precision, it brings difficult for accuracy. There is one exception for Monte Carlo Localization (MCL) [2].MCL could use node mobility to improve localization precision, without other more cost.

In Ref. [3], the author first used Monte Carlo method to solve the mobile node localization. Its main principle is using a weighted sample set to show the possible distribution of nodes moving. After that it has been received extensive attention and improved a variety of derived methods. For example, Bagsio put forward MCB(Monte Carlo Localization Boxed) algorithm [4], which limited the sample area in a sample box made by beacon nodes box and sample box to improve the sampling rate. In Ref. [5], Marcelo combined MCL with RSSI, studied the RSS-based MCL, it used different precision orientation tracking sensors to obtain azimuth information, and then combined with MCL to achieve more precise localization. Other researches such as Multi-hop-based MCL [6], E-CDL [7], DRL etc., are all based on DV-Hop. Even recently a new type APIT-MCL [8] appeared.

In this paper, we pull in the concept of binary- detection [9].It can determine whether the target appeared within the detection range and choose the most suitable anchor nodes.; By collaboration with multiple sensor nodes to tracking, we can better improve the accuracy; Even more important, we applied binary-detection to Monte Carlo sampling period and used the detection time as node weights to improve the impact of some anchor nodes and calculated the target localization better.

3   MCL

As one of the most important method for target tracking, MCL has been widely used in state estimation. There are three main stages: initialization, prediction, filtration [10].

Initialization: select N sample points from probable nodes, and set initial weights. Prediction: estimate the current position based on a prediction sample set and motion models. In current time the position of the target can be constraint in a circle, with a center on the last estimated localization, and a radius of maximum velocity.

Filtration: according to the perceived data, target nodes screen some unable samples. If the remaining samples be small, then repeat the prediction period and filtration stages, until a sufficient number of samples collected.

Typically, samples are taken as uniform distribution; each sample has the same weight$1/N$, thereinto N represents the number of samples. Finally, we added the weighted samples as the node position estimation.

In MCL algorithm, many samples fall on an area with smaller location posterior density. Those value has smaller effect for localization, thus they reduced the tracking accuracy.

4   Optimization Algorithm

4.1   Concept of Binary-detection

Unlike other detection methods, which depend on determining distance to the target or the angle of arrival of the signal, this approach only requires the sensor be able to determine if an object is somewhere within the maximum detection range of the sensor. Thence, it has only two statuses: the target either within the range of anchor nodes, or not.

Fig.1 shows the sensor module. The dot stands for an anchor node. Its detection distance is R. The target can be detected within (R-e), otherwise out of (R+e) there is no detection. Specially between (R-e) and (R+e) it can be probably detected. Usually we delimit e = 0.1R [9].

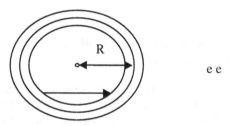

Fig. 1. Binary-detection module

Binary sensor is simple, convenient and low costs, its greatest advantage is ascertaining and recording the duration for mobile node in its detection range. The longer the duration, the nearer its distance from the sensor, the more accurate the data, therefore the larger weight.

4.2   Arithmetic Statements

In this paper we brought in binary-detection [11], applied it to Monte Carlo sampling phase, took the detection time as the anchor node weight to improve its impact factor, so that calculated the sensor node location accurately.

We placed binary sensors in the conventional anchor nodes to fabricate our new type anchor nodes, assumed it used in a two-dimensional plane. As a precondition all sensor nodes would be moving in uniform motion, and we ignored the error (including time error). The network was acquiesced in grid topology.

The new binary-detection based MCL algorithm has basic steps the same with classic MCL, Sampling algorithms is the main difference.

Assuming there is an anchor node marked, with a communication radius. It detected a mobile node with a start time stamp, and an end time stamp, after which the mobile node was not detected at least one monitoring period. Between the mobile node can be detected without any interrupt, so the length of detection time is, we call it [12].

Each particle has a different weight delimited by the detection time .The principle is shown in Fig.2 blew: for different anchor nodes detect the same moving node in the same time, the same speed is the longer the detected time, the larger the journey L. There is a triangle. According to the Pythagorean Theorem, it has smaller vertical distance. That shows it is closer to this sensor center, the data is more accuracy, this anchor node should have greater weight.

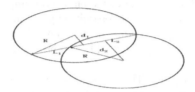

Fig. 2. Particle weights diagram

4.3   Concrete Steps

(1) Initialization
While K=0, choose N samples from all possible nodes, put them in a set, set their initialize weights, i=1, 2, …N;
(2) Prediction
While K=1, 2,…,N,   stands for the possible location in the last time , then in current time this target node will in a circle with the center   and the radius is . The probability of its location is given by a uniform distribution as follow:

$$p(\ell_t | \ell_{t-1}) = \begin{cases} \frac{1}{\pi v_{max}^2} & d(\ell_t, \ell_{t-1}^i) < v_{max} \\ 0 & d(\ell_t, \ell_{t-1}^i) \geq v_{max} \end{cases} \qquad (1)$$

In that, stands for the distance between.
(3) For those predicted location, they should first meet the filtration conditions. All coincident samples set their weights on the basis of the target detection duration. is a set collecting detected anchor nodes,, each   has a. is the value recorded by our new anchor nodes. is an integer associated with. Here we do a conversion: magnify each with a same ten times, hundred times, or even more multiples, choose the integer part as, rounding without decimal to reduce the computation, the new weights, so we implement weights influence.
(4) If samples are not enough, repeat step (2) and (3), until we collect enough samples.
(5) After get all the particle weights, we also need to normalize the weights until the total weights is 1. Sample weights normalized like this:

$$w_t = \frac{w_t^i}{\sum_{t=1}^{N} w_t^i} . \qquad (2)$$

(6)So the predicted location of sensor node will be:

$$x_t = \sum_{t=1}^{N} x_t^i w_t^i . \qquad (3)$$

$$y_t = \sum_{t=1}^{N} y_t^i w_t^i . \qquad (4)$$

Here we gave different anchor nodes for different weights, avoided excessive use of low-quality samples effectively and led to higher localization accuracy.

5   Simulation and Analysis

Single moving target localization is the basis of multiple target localization. In order to prove the effectiveness of this new algorithm, we use MATLAB to simulate one moving target from path prediction, anchor nodes density, error, and moving speed. Compare with classic MCL algorithm the results show as below.

Set the size of simulation scenario a square 300m×300m without any obstacle. Choose Random-Waypoint model as node movement. Take no account of the moving direction, all nodes have a same communication radius of 20m. Maximum moving velocity is. With a total of 200 nodes, the results show as below.

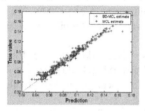
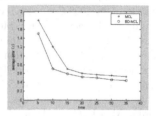

Fig. 3. path prediction different from true value

Fig. 4. Algorithm comparison

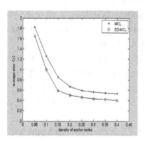
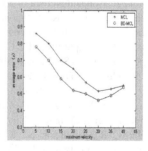

Fig. 5. Density of anchor nodes

Fig. 6. Moving speed

From the figures we can see clearly, new algorithm works better for trajectory prediction, Collaborative multi-sensor tracking can further reduce the error. Binary-detection MCL is more suitable for intensive anchor nodes situation. But there is a apparent issue, as speed increases, the detection time difference was not obvious, the impact of weight is so small that the error nearly has a convergence with MCL. Time error may even lead to worse error. How to bring in time weight better is a problem.

6   Conclusions

Considering the nearer anchor nodes could detect longer and have larger influence in practice, we brought in binary-detection in sampling process in this paper. Through simulation and experiments, this new BD-based-MCL has good scalability, effectively decreased localization error and improved the accuracy, especially work well when anchor nodes are densely density. Next step, we will consider more about the influence of changes in network topology and time synchronization, and how to set the weights better, even further we can add pre-monitor link to optimize energy. Ultimately we wish to achieve optimization Monte Carlo algorithms in applications.

References

1. Liu, Y., Wu, S., Nian, X.: The architecture and characteristics of wireless sensor network. In: Proc. of the 2009 Computer Technology and Development, p. 561 (2009)
2. Bagsio, A., Langendcen, K.: Monte Carlo Localization for Mobile Wireless Sensor Networks. Ad Hoc Networks, 718 (2008)
3. Hu, L., Evans, D.: Localization for Mobile Sensor Networks. In: The 10th Annual International Conference on Mobile Computing and Networking, Philadelphia, p. 45 (2004)
4. Wang, W.D., Zhu, Q.X.: RSS-Based Monte Carlo Localization for Mobile Sensor Networks. IET Communications, 673 (2008)
5. Martins, M.H.T., Chen, H., Sezaki, K.: OTMCL: Orientation Tracking-Based Monte Carlo Localization for Mobile Sensor Networks. In: Proceedings of the 6th Annual International Conference on Networked Sensing Systems, p. 151 (2009)
6. Chang, T.-C., Wang, K., Hsieh, Y.-L.: Enhanced Color-Theory-Based Dynamic Localization in Mobile Wireless Sensor Networks. In: Proceedings of the IEEE Wireless Communication and Networking Conference, Kowloon, China (2007)
7. Hsieh, Y.-L., Wang, K.: Efficient Localization in Mobile Wireless Sensor Networks. In: Proceedings of the IEEE International Conference on Sensor Networks, Ubiquitous, and Trustworthy Computing. IEEE Computer Society, USA (2006)
8. Wang, J., Fu, J.-Q.: Study on localization algorithm for wireless mobile node based on APIT and particle filter. Transducer and Microsystem Technologies, 1000–9787 (2011)
9. Li, P.: Wireless sensor network technology. Metallurgical Industry Press, Beijing (2011)
10. Luo, J., Chen, M.-M., Zegn, F.-Z., Li, R.-F.: Weighted Sampling Monte-Carlo Approach for Target Tracking. Computer Science 35(11A) (2008)
11. Mechitov, K., Sundresh, S., Kwon, Y., Agha, G.: Cooperative tracking with binary-detection sensor networks. In: Proc 1st Int'l Conf. on Embedded Networked Sensor Systems (2003)
12. Chao, Z.: Research of Tracking Technology on Wireless Sensor Network and the Application in Transport System. JiLin university, JiLin (2001)

ApRTSP: An Accurate and Prompt Remote Technical Support Platform for Equipment Maintenance of Large Ships

Hu Chen

Naval University of Engineering, Wuhan, China 430031
chen_nue@126.com

Abstract. This paper proposes a novel architecture for remote technical support platform that is used for equipment maintenances in large ships. In this architecture, we integrate some new techniques such as RFID, actuator and actor sensors for the detection of equipment abnormal status, voice information hiding in wireless communications for maintenance operations, and cloud index for fast response and solution localization. We call the architecture as ApRTSP, and its advantages are highly robust, low delay, and highly accurate.

Keywords: Equipment Maintenance, Remote Technical Support, Wireless Network, System Architecture.

1 Introduction

There exist a large number of equipments in large ships, which may present random failure or experience some abnormal performances. To maintain the availability and functionality of the equipments, we need to guarantee the following: faults and errors can be detected in real-time, the corresponding solution can be obtained instantly, and the problem can be solved correctly. The first step relies on the monitoring devices on the board, namely, the devices used for the detection and signal collection. The second step relies on the response team for faults and errors. Such team could be on-site, or remotely. The last step relies on the manually processing on the broad. Such process is related to the timeliness of the first step and the correction of the second step [1,2,3].

Since in large ships the number of relevant equipments is much more than that in usual ships, the fault detection and response in large ships present much more challenges, for example, the problem detection should be accurate and the solution localization should be fast. In other words, we need to discover the faults or potential faults instantly, the scope of the faults can be traced to a rational domain, and the fault diagnose can be conducted instantly, efficiently, and correctly.

Currently, some researches address the maintenance techniques for normal ships [4-8]; however, researches on large ships are seldom explored. Especially, a bunch of remote support techniques are not addressed systematically. In this paper, we propose a novel Remote Technical Support Platform (RTSP) for dealing with the aforementioned challenges. In RTSP, we propose to use RFID and wireless sensor networks to fast localizing and detecting faults in equipments. We also propose a secure communication architecture and remote support platform to accelerate the assurance, accurateness, efficiency of the solution discovery.

The contribution of this paper is as follows: 1) We propose a novel network and communication architecture for fault detection remotely by using RFID-WSNs hybrid techniques; 2) We propose a secure communication for fault signal transmission by using information hiding in the frequent traffics such as voice so as to save the communication resource and avoid the key management cumbersome. 3) We propose a novel platform for maintenance solution localization in large database by a virtual cloud index that has high performance and high availaiblity.

The rest of the paper is organized as follows: the Fault Detection section will address the new techniques for equipment problem detection. The Secure Communication section will propose a new architecture for secure guidance. The Cloud Index section will explore the techniques for fast guidance localization. The Conclusion section concludes the paper.

2   RFID-WSN Based Fault Detection

As in the large ship, there exists a large number of equips. They need to be fast traced, located and status detected. In other words, the equipments should be known its working status, fault risks and last maintenance results. Certain equipment is difficult to detected, for example, the devices are inner the large equipment. Some of them are expensive to detected, for example, some devices can only be detected by special operations. To address such challenges, we propose a new technique based on RFID and active sensors.

Traditional maintenance system on broad usually relies on the monitoring circuits and multiple sensors. They can collect the status data of equipments, including fault number, fault code, fault graph, fault sound, and equipment arguments et al.. Such data will be displayed on the equipment monitors, and be reported to the upper support center. The support operators then use the Portable Maintenance Aid (PMA) system and Interactive Electrical Technical Manual (IETM) to solve the faults [8-10].

There exist two weaknesses in the traditional systems: 1) The faults from the operational mechanics are not electrically equipped. To collect the fault information, extra circuits will have to be deployed in some manually operated devices. Especially, such objects may not be able to access from outside; 2) The traditional system can only collect the faults where the monitoring system are enabled. However, some faults can not be predicted upon deployments. In other words, the monitored signals are pre-defined, and the incremental enhancement may experience difficulties. Therefore, we propose to use RFID and Wireless Sensor Networks (WSNs) in the monitoring system.

The RFID is equipped with physical objects. It can return the device codes upon reader's reading. Some equipment that cannot be able to be accessed outside will be attached RFID tag. If some of them perform abnormally, the tag may be damaged. Thus, the reader can localize the fault by finding the absence of some tags. Moreover, the RFID tag can be collected automatically, so that the physical damages can be discovered instantly, without deploying any embedded circuits.

WSNs will also be deployed for monitoring some environment parameters, such as temperature, humidity, smoke, and pressure. They can be deployed in the location where the circuits are difficult to deploy. WSNs can finally upload the monitoring data into the control center.

Therefore, the supplement via RFID-WSNs can facilitate the monitoring system on broad to support real time feedback from more inner and inaccessible equipments.

3   Information Hiding in Voice

After the fault collection, maintenance operators on broad may not be able to totally solve the problems. Thus, they may have to consult the experts on the base or on the center. It arises a problem, how to secretly transport the questions to the base.

Normally, the ship will communicate with the base by satellite systems. The communication system will be encrypted by designated equipments such as dedicated encryption machine. However, it may result in two shortcomings: One is the maintenance communications will occupy communication resources such as bandwidth. The other is the key management issue between the base and ships may be cumbersome. Therefore, we propose a lightweight method without encryption, information hiding in voice.

The rationale of the idea is from following observations: the voice communications occur frequently. If the maintenance information can be hided in the normal voice communications, the consultancy of experts will not consume a dedicated communication channel. Thus the communication can be conducted in normal channels.

4   Cloud Index

In the consultancy of the experts in the base, the experts may also retrieve the aided information from the maintenance database. It thus confronts one challenge: how to localize the solution in a limited delay. It thus requires a high efficient index system to locate the solution in database.

We propose a Remote Technical Support Platform (RTSP) to empower the large data center and high performance retrieval response. As the equipment maintenance data may be distributed in different bases, the data center has two functions: to backup the distributed data in different bases; to construct an index to support the fast retrieval. Traditional RTSP has a Remote Diagnosis Support System (RDSS). It provides some decision support for the fault feedback and response. Such a system relies on the data mining or artificial intelligence techniques to fasten the solution discovery. However, it encounters the problems such as the accuracy may not be acceptable and response delay may be an overhead. To improve the performance of data mining and intelligent decision, we need to build the analysis and inference from a very large database.

The proposed RTSP will utilize cloud computing techniques such as virtualization, and high performance computing. It is not a naive application of cloud computing, in contrast, we propose a new technique called Cloud Index (CI). It is a dynamically generated index with semantic representation of data, called semantic meta data.

The data comprise of maintenance support information and such information will be organized by a tree with short depth. For building CI, we will index the data with layered index, and the generation relies on an automatic robot program. Such program can generate meta data and build the index of them as a regular routine. The index should be updated periodically to maintain the freshness.

The construction of CI is a virtue index and the storage of index is distributed in different physical computers to provide high performance and high availability. Large database over cloud center is an undergoing research topic and current results are not fully satisfied the requirement in RTSP, since RTSP database has its own characteristics, such as data item may be multimedia, data item is not updated frequently, and so on. Thus, the cloud index should be tailored design for tackle those specialties so as to achieve the optimal overall performance.

Figure 1 depicts the components in ApRTSP framework.

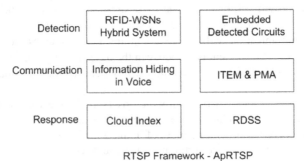

RTSP Framework - ApRTSP

Fig. 1. ApRTSP components

5   Summary

In this paper, we proposed a high level framework for the remote maintenance support platform. The proposed techniques address the current shortcomings or weakness of the existing solutions. It points out some critical problems in the current frameworks and highlight some promising techniques, such as RFID-WSNs based fault detection and monitoring, information hiding in voice for remote assistance communications, and cloud index tailed design for RTSP database. Such a framework may introduce some fresh arguments and discussions in the field of remote maintenance. We believe that such discussion will help further distinguishing the problems and foster new research directions in the field.

References

1. Flint, J.B.: A Vulnerability Analysis of 24 Electronic Chassis Found in Tube-Type Radar System. Naval Weapon Center NWC TP 5034, China Lake, CA (January 1971)
2. McCormick, N.J.: Reliability and Risk Analysis. Academic Press, San Diego (1981)

3. Recht, R.F.: Fragment Lethality, Naval Surface Weapons Center, NSWC TR 83-165, Dahlgren, VA (August 1983)
4. Teng, X., Cai, Z., Lin, S., Li, H., Zhang, Z.: Maintainability of ship and marine equipment. Journal of Shanghai Maritime University (1) (2007)
5. Peng, C.: Design and Realization of Distant Maintenance Technology Supporting Call Center. Ship Electronic Engineering (3) (2007)
6. Wenliang, Y.: Design and Implementation of Full Text Inspection in IETM Based on Web. Ship Electronic Engineering (2007)
7. Zhu, S.: On remote maintenance supporting system of war ship equipment. Journal of Naval University of Engineering (2004)
8. Fang, H.T.L.: LORA in the Ship Equipment Maintenance. Journal of the Naval Academy of Engineering (1999)
9. CoCoS, maintenance designed for maintenance excellence. MAN B&W Diesel A/S, Copenhagen, Denmark (2005)
10. Tarelk, O.W.: Improvement of ship mechanical equipment maintainability through design. Marine, Offshore and Technology, 91–98 (1994)

Finite Element Analysis in Vitro Expansion of Coronary Stents

Haiquan Feng, Xudong Jiang, Feifei Guo, and Xiao Wang

Mechanical Engineering College, Inner Mongolia University of Technology,
Inner Mongolia
fhq515@sohu.com, xudongjiang@sina.com,
feifeiyudian@126.com

Abstract. Coronary stent implantation has become an important method, of treating coronary heart disease. It has a pre-compression process when using the stent, which will be fixed in the balloon, and then be processing the expansion of operations. This research use the FE method, without considering the interaction of vessel stents and the circumstances under artery loading to simulate the three different materials (CoCr alloy, 316L stainless steel and Mg alloy), but it must has the same free pre-compression expansion process of the structure of the stent. By setting the mesh of the FE stent models, the boundary conditions, the loading, etc, and other key factors, The results demonstrate deformation mechanism, which can be used to calculate radical recoil rate and axial shortening rate. At last, the safety factors of stents are forecasted.

Keywords: coronary stent, mechanical properties, expansion deformation, finite element analysis.

1 Introduction

Coronary stent implantation is a method of treating coronary heart disease, which is developed on the basis of PTCA (Percutaneous Transluminal Coronary Angioplasty). When used, the stent will be crimped on the balloon matched with catheter, and then they get to the vascular lesion. When pressure is put in the balloon, the stent unfold and the balloon is pulled out, leaving the stent in the correct position, so it plays a role of supporting the vessel.

More and more people are beginning to study stent. Ni Zhonghua and others [1] proposed for numerical simulation method which was suitable to the coupling expansion of coronary stent, and compared the data with the vitro test data, verifying the rationality of this method. Qi Ming and others [2] studied the expansion of magnesium alloy stent. Wang Weiqiang, Liang Dongke and others [3] analyzed the dogboning phenomenon which is occurred in the instant expansion process, and tested the relation between the balloon and stent supporting and this phenomenon. Francesco Migliavacca and others [4] quantitatively analyzed the expansion and deformation of the BX Velocity stent and the axial and radial rebound effect without balloon, and compared with the experimental data.

D. Jin and S. Lin (Eds.): Advances in FCCS, Vol. 2, AISC 160, pp. 579–584.
springerlink.com   © Springer-Verlag Berlin Heidelberg 2012

In conclusion, currently many contents are limited to expansion and deformation process of the stent and balloon system, and the continuous process including pre-compression and expansion are seldom. The purpose of this paper is to compare deformed mechanism of different materials stents and obtain how affect stent properties of different materials, providing theoretical basis.

2   Materials and Methods

Geometric Model. This research take into account the interaction between the stent, balloon and crimping device in order to more accurately simulate the actual usage of stents. In order to analyze and compare better, three kinds of stents (magnesium alloy stent, stainless steel stent, and CoCr alloy stent) are simulated which have the same geometry but different materials. In this process of simulation, the stent plane structure is established in the software AutoCAD firstly, and then three dimensional model is build in the software Solidworks and grid in the software Hypermesh, and last it is import to software Abaqus to finish the relevant calculation. Considering that stent expansion is accompanied with material nonlinearity, geometry nonlinearity and connection nonlinearity, and stent structure is cyclical in axial direction, we only select a geometric unit to analyze. The geometric structure of stent is shown in Fig.1.

Fig. 1. The geometric structure of stent   **Fig. 2.** mesh model

Material Model. WE43 magnesium alloy, 316L stainless steel and L605 CoCr alloy are selected for stent material in this research. The elastic-plastic material properties of stents are test through the coaxial tensile experiment (see Table1), and their true elastic-plastic deformation follows the criteria of Von Mises and isotropic hardening.

Table 1. Stent material properties

parameter	CoCr-L605	Mg-WE43	SUS-316L
elastic modulus[GPa]	240	44.2	201
Poisson's ratio	0.3	0.27	0.3
nominal yield strength[MPa]	630	160	280
nominal strength limit[MPa]	1165	245	750

Boundary Conditions. Correct boundary conditions can not only avoid unnecessary rigid displacement of stent, but also reflect the real expansion process of stent. As is shown in Fig.2, boundary constraints are defined in the environment of cylindrical coordinates, where 'Z' direction represents the axial direction of the system, 'T' direction represents the angular direction and 'R' direction represent radial direction. As the friction parameter is unknown, it is defined no frictional contact between the crimp shell and stent, the balloon and stent.

Load Definition. Stent deformation process can be divided into four parts including compression, compression unloading, expansion and expansion unloading (see Fig.3). Exerting radial pressure upon the crimping shell in compression stage and making stent dilate from 1.55mm to 1.0mm. And then exerting reverse pressure upon the crimping shell, making stent have radial recoil; Exerting radial pressure upon the balloon during expansion stage and making stent's outer diameter get to 3.0mm. And then exerting reverse pressure upon it, making stent have radial recoil also. Especially, all the pressure here refers to the displacement.

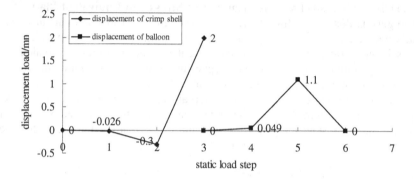

Fig. 3. Load displacement situation in Stent deformation process

3   Results

CoCr Alloy Stent Deformation Results. CoCr alloy stent is almost the most popular among all the alloy stent, so its deformation of the critical process is listed (see Fig.4). Though the other two stents are not given the deformation charts, the FE analysis turned out to be the same deformation process. Therefore, the most likely fractured location is the junction between supporting body and connecting body.

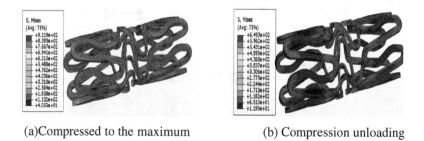

(a)Compressed to the maximum   (b) Compression unloading

Fig. 4. Deformation process of CoCr alloy stent

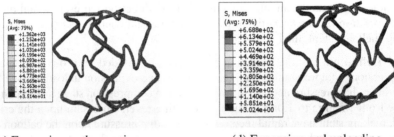

(c) Expansion to the maximum   (d) Expansion and unloading

Fig. 4. (*continued*)

Comparison Charts between the Equivalent Stress and Equivalent Plastic Strain in Dangerous Position of the Stent. Fig.5 (a) and Fig.5 (b) are curves about change of stent's equivalent stress and equivalent plastic strain in dangerous position. Three curves in Fig.5 (a) show that CoCr alloy stent has the maximum equivalent stress followed by stainless steel alloy and magnesium alloy has the minimum equivalent stress when these stents are expanded the same diameters, so it indicates CoCr alloy stent has the strongest resistance to external forces. Three curves in Fig.5 (b) demonstrate magnesium alloy has the maximum deformation followed by CoCr alloy stent and stainless steel alloy has the minimum deformation, but the difference of gradient is very small between three stents.

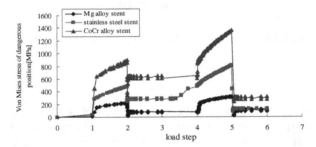

Equivalent stress chart of dangerous position

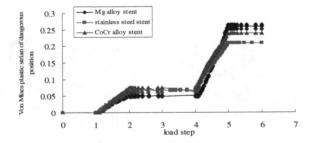

Equivalent plastic strain chart of dangerous position

Fig. 5. Stress and strain of Coronary Stent

Table 2. Stent deformation data

stent types	radial recoil ratio χ_2	axial shortening ratio χ_1	expansion uniformity χ_3
Mg alloy stent	6.82%	9.95%	3.89%
CoCr alloy syent	4.98%	10.61%	2.25%
stainless steel stent	3.32%	10.64%	0.65%

Table 3. Stent deformation data

stent types	Mg alloy stent	CoCr alloy stent	stainless steel stent
Von Mises stress of dangerous position[MPa]	317.7	1362	811.2
Von Mises plastic strain of dangerous position	0.26	0.24	0.21
Safety Factor	1.08	1.33	1.39

Analysis of Stent Deformation. In Table 2, we can see the radial recoil ratio and axial shortening ratio of stents, obviously, the radial recoil ratio of magnesium alloy stent is 6.82%, which is greater than the CoCr alloy stent (4.98%) and the stainless steel stent (3.32%), when stents are expanded to the same diameter. However, the axial shortening ratio is opposite. Safety factor (SF) is also one of the factors that can evaluate properties of the stents. It can be seen in Table 3 that safety factors of the three stents are meet the theoretical requirements, and the stainless steel stent has the maximum value followed by the CoCr alloy stent and the magnesium alloy stent has the minimum value.

4   Conclusions

This article simulates the pre-compression and free expansion of stents with   three kinds of materials by FE method and compares these results, getting some conclusions as followed: The key factors such as interaction, boundary conditions and loads are set to control solution. The differences in geometry deformation and stress distribution were compared, and stent's biomechanical properties are evaluated using axial shortening, radial elongation and static strength. What's more, Safety Factor is used to estimate stents safety. FEM can quantify some mechanical behavior of stents and optimize design to stents.

Acknowledgements. This work was financially supported by the National Natural Science Foundation (81160186), Inner Mongolia Natural Science Foundation (2011BS0708) and Inner Mongolia Municipal Education Commission (NJZY11077).

References

1. Ni, Z., Wang, Y., Chen, J.: Numerical simulation method of balloon-expandable coronary stents expansion mechanism. School of Mechanical Engineering 44(1), 102–108 (2008)
2. Qi, M., Si, C., Wang, W.: Stress change during the expanding process of magnesium alloy stents with different strut shapes and sizes. Journal of Clinical Rehabilitative Tissue Engineering Research 38, 7133–7135 (2010)
3. Wang, W., Liang, D.: The finite element analysis of the transitorily expanding process and design optimization of coronary stent system. Chinese Journal of Biomedical Engineering 24(3), 313–319 (2005)
4. Migliavacca, F., et al.: On the effects of different strategies in modeling balloon-expandable stenting by means of finite element method. Journal of Biomechanics 41, 1206–1212 (2008)

Research on the Coupling Expansion Mechanism of Balloon-Expandable Coronary Stent

Haiquan Feng, Feifei Guo, Xudong Jiang, and Qingsong Han

Mechanical Engineering College, Inner Mongolia University of Technology,
Inner Mongolia
fhq515@sohu.com, feifeiyudian@126.com, xudongjiang@sina.com

Abstract. The complete coupling expansion of stents includes constriction, free expansion and interactions with the artery vessel. This paper has used finite element method analyzing the static and dynamic continuous deformation process of stents with two kinds of materials (stainless steel-316L, CoCr alloy-L605) without considering plaque in vessel and studying the mechanism under periodic loading, especially. The key factors such as interaction, boundary conditions and loads are set to control solution. The differences in geometry deformation and stress distribution were compared, and the stents biomechanical properties are evaluated using axial shortening, radial elongation, static strength and fatigue strength. FEM can quantify some mechanical behavior of stents and design to stents.

Keywords: coronary stents, coupling expansion, finite element method, fatigue strength.

1   Introduction

Balloon-expandable stent will be fixed on the catheter matched with balloon at first, and when it reached the affected area by delivery system, the balloon expand, thus madding the stent unfolding and attaching to the lesion site, but the balloon must be pulled out. Because of the influence of stent's material, geometric size and expanding process, the unfolded stents will deform which can not only effect supporting properties but also make the blood dynamic characteristic change. So the research on the stent expansion process has been the hot topic in the interventional therapy field.

Since the stent technology has been applied, many researchers pay attention to analysis of the stent structure and performance, especially on balloon-expandable stent. Wei Wu and others [1] studied the expansion of a stent in a curved vessel and straight vessel, whose results indicated that stent can reduce lumen area of curved vessel but its maximum tissue prolapse was more severe. Walke and others [2] have combined finite element analysis with experimental method studying the biomechanical properties of stents in vessel. The expansion process of Medtronic S7 stent and Boston NIR stent was researched by Lally et al [3], moreover, the interaction with vessel was also analyzed. For analysis of vitro expansion of stents, research contents are mostly concentrated on the mechanism of balloon-stent system, and the pre-compression for stents is ignored;

D. Jin and S. Lin (Eds.): Advances in FCCS, Vol. 2, AISC 160, pp. 585–590.
springerlink.com   © Springer-Verlag Berlin Heidelberg 2012

for the analysis of coupling expansion of stents in the vessel, many researches are limited in the contact of balloon-stent-vessel; for fatigue strength of stents, the reports are seldom. This paper considers the pre-compression and interaction between balloon, stent and vessel when the coupling expansion of stents is researched, and uses the finite element method analyzing the change of stent's mechanical properties in the whole deformation process.

2   Model and Method

Geometric Model. The stent structure is tubular, and its outer diameter is Φ1.55mm with Φ1.41mm for the inner diameter, so the thickness is of 0.07mm. There are six symmetry supporting-bodies in circumferential direction, which are corresponded with connecting-bodies respectively, so its frame is thought as a typical closed-loop. Moreover, the crimp device, balloon and artery vessel are simplified as a thin shell in the premise of not influencing the stent deformation. In order to analysis and comparison easily, the CoCr alloy stent is with the same geometry as stainless steel stent, and the graphic model is constructed in the software CAD, and then imported to software Solidworks to build 3D model.

Material Model. The stent expansion process is accompanied with geometric nonlinearity and boundary condition nonlinearity, so it is simulated using commercial software ABAQUS in order to analysis stent deformation better. The elastic-plastic material properties of the two stents are test through the coaxial tensile experiment, and their true elastic-plastic deformation follows the criteria of Von Mises and isotropic hardening. The material properties of stents can be seen in Table 1.

Table 1. Stent material parameters

stent types	elastic modulus[GPa]	Poisson's ratio	nominal yield limit[MPa]	nominal ultimate tensile strength [MPa]
CoCr-L605	240	0.3	630	1165
SUS-316L	201	0.3	280	750

Meshing. Every part is meshed in software hypermesh, and Fig.1 shows the grid model. According to the characteristics of irregularity structure and large deformation theory, the element supports for a linear reducing integral unit C3D8R. What's more, a reliable calculated result often require sufficient number of units, so the supporting-body is arranged 4 layer units in the width and the connecting-body is of 2 layer units in the width, and they are of 4 layer units in the thickness. Transitions

Fig. 1. Mesh model

between supporting-body and connecting-body have a refinement mesh. For other models, quadrilateral element S4 is selected.

Contact and Boundary Conditions. There are three pairs of contact as crimp and stent, balloon and stent, stent and vessel of the system and standard surface to surface is used to treat contact problem. At the same time, since the friction coefficients are unknown, a frictionless contact type is stabled to simplify the model.

Loading. Two kinds of load are applied to the system: the displacement load acts on pre-compression and stent expansion, and we can terminate the calculation by controlling the constriction or expansion displacement value; the pressure acts on dynamic deformation in the vessel, we can get the deformation result by controlling pulsate pressure value and cycle number when we put sustained load on the stent inner surface and vascular wall.

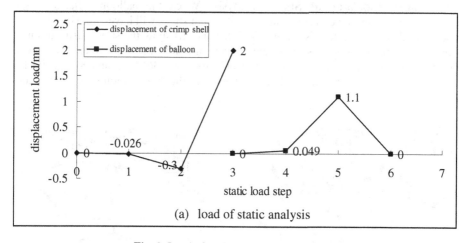

(a) load of static analysis

Fig. 2. Load of static and dynamic analysis

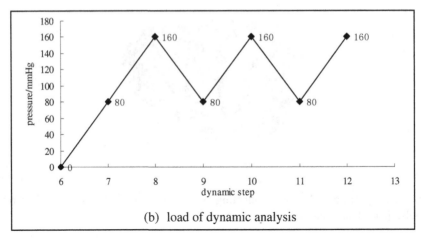

(b) load of dynamic analysis

Fig. 2. (*continued*)

Fig.2 (a) shows the change of displacement accompanied with load step in the static analysis, in which the positive value of displacement is said to be expansion process and the negative value indicates compression. The dynamic load exerted upon the stent is blood pressure, and it changes between120 ± 40mmHg. Fig.2 (b) qualitatively analyzed the change of dynamic load accompanied with step, while it is on the basis of static expansion in the actual simulation.

3   Results

Quantitative Indicator of Static Analysis. When the balloon inflates to the maximum diameter value in static analysis, the stent stress will have a maximum value (the inner diameter of stent is 3mm). Fig.3 (a) and (b) show the stress distribution of CoCr alloy stent and stainless steel stent when their diameters inflate to extreme value, in which the maximum stress value of CoCr alloy is 1366MPa and 815.3MPa for stainless steel stent. Moreover, Table2 demonstrates the deformation performance and safety factors of stents.

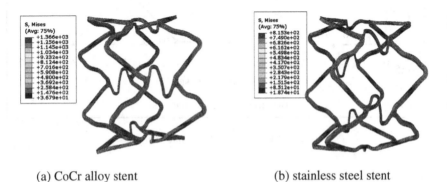

(a) CoCr alloy stent   (b) stainless steel stent

Fig. 3. The maximum static stress of stents

Results of Dynamic Analysis. Stent fatigue fracture may occur due to the continuous function of cycle alternative pulsating stress, so the loading of dynamic analysis can be used to simulate the fatigue loading. Fatigue life curve describes fatigue fracture strength of stent which has a quite life of 10 years, while the dynamic safety factor is a quantitative reflection of stent safety.

The dynamic safety factor $SF_{dynamic}$:

$$1 / SF_{dynamic} = \overline{\sigma}_{mean} / \sigma_{uts} + \overline{\sigma}_{alt} / \sigma_e \qquad (1)$$

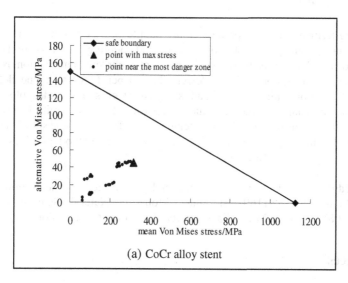

(a) CoCr alloy stent

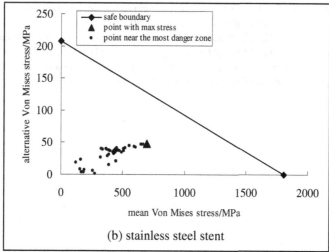

(b) stainless steel stent

Fig. 4. Fatigue security zone of danger position

Where σ_{uts} stands for the ultimate tensile strength, σ_e stands for the fatigue fracture limit, and $\overline{\sigma}_{alt}$, $\overline{\sigma}_{mean}$ represent equivalent alternative stress and mean Von Mises stress respectively.

Fig.4 (a) and (b) demonstrate stents fatigue curve of the most dangerous position and its adjacent area, and obviously, the most dangerous position can meet the security requirement, so the entire stent is in the security area during the dynamic analysis.

4   Conclusion

In the long-term application process, most people pay attention to the stent vitro expansion and the analysis of pre-compression and coupling expansion in the vessel is seldom, so a full comprehensive study of stent complete deformation is of some significance for stent optimization design. This paper focuses on the deformation mechanism of CoCr alloy stents and stainless steel stents under blood stream, quantitatively evaluates the biomechanical properties of stents in quite 10-year lifetime and obtains the conclusion that the two stents have sufficient fatigue strength to support the vessel. In this paper, the finite element modeling method provides a reliable scientific basis for stent evaluation.

Acknowledgements. This work was financially supported by the National Natural Science Foundation (81160186), Inner Mongolia Natural Science Foundation (2011BS0708) and Inner Mongolia Municipal Education Commission (NJZY11077).

References

1. Wu, W., Wang, W.-Q., et al.: Stent expansion in curved vessel and their interactions: A finite element analysis. Journal of Biomechanics 40, 2580–2585 (2007)
2. Walke, W., Paszenda, Z., Filipiak, J.: Experimental and numerical biomechanical analysis of vascular stent. Journal of Material Processing Technology 164-165, 1263–1268 (2005)
3. Lally, C., Dolan, F., Prendergast, P.J.: Cardiovascular stent design and vessel stress: a finite element analysis. Journal of Biomechanics 38(8), 1574–1581 (2005)

An Exploration to the Clinical and Engineering Cultivation Methods for Professional Medicine and Physics Majors

Zhang Ying[1], Wang Xuechun[1], Zhang Yantang[2], and Cheng Cheng[3]

[1] Department of Radiology, Taishan Medical University, China
[2] Taishan Medical University School Physician Hospital, China
[3] Radio-Diagnosis, China-Japan Friendship Hospital, China
{yzhang,xcwang,ytzhang}@tsmc.edu.cn,
chengcheng8000@163.com

Abstract. As medical physics is the merger of both medicine and engineering, the cultivation of its professional talents must adopt both the medical and engineering methods, construct the professional course system accordingly, and also form the teaching team of double-teachers so as to cultivate the professional talents on this major.

Keywords: combination of medicine and physics, medical physics, talent cultivation.

Contents

Taishan Medicine University launched the Department of Medical Imaging in 1985 and recruited its students on master degree, among which the direction of medical imaging technology is the only state-level master degree in China. Medical imaging major enjoys the great fame as the special major both on province level and on state level. Its laboratory acts as a model on provincial level. The clinical electronic appliance major was launched in 1989. In 2001, the biological medical engineering bachelor major was launched and its students were recruited. At present, this major is included into the special majors on provincial level. This college of radiology launched the medical physics major and began the exploration of courses system in 2005, so as to meet the urgent need for professional talents of this major from the society, which was based on its privilege of cultivating talents.

1 Construct the Course System Combining Both Medicine and Enginerring

Medical physics is a merger of courses which applies the basic acknowledge and methods of physics to human illness's precation, diagonosis, treatment and care-taking. It is also a professional course system to cultivate professional talents.

D. Jin and S. Lin (Eds.): Advances in FCCS, Vol. 2, AISC 160, pp. 591–595.
springerlink.com © Springer-Verlag Berlin Heidelberg 2012

1.1   The Principles for Designing the Course System

Apply the designing principle of cultivating both libertal talents and professional talents, and combine both medicine and engineering, handle the relations between basic acknowledge and major, theory and practice, major course and relevant courses, highlight tradition and modern education innovation and the characteristic of appliance graduate, construct the basic framework of the combined course system from two aspects both synchronically and diachronically. The courses are mainly divided into general classes, professional classes and elective classes.

1.2   The Framework of the Course System

The total number of compulsary classes is 38, with the total academic hours of 2810 and the total credits of 156. The total number of general classes is 10, whose academic hours reaching 828. The total number of elective classes is 6, the academic hours reaching 192 hours. The total credits are 20, adding both the compulsary and elective classes together.

General classes mainly include Marxism philosophical principle, Marxism political and economic theory, millitary training theory, Maoism ideas, Deng theory, ethics and basic laws, P.E., English, computer primaries and medical chemistry.

Professional primary classes mainly include advanced mathematics, primary biology, linear algebra, probability theory, engineering mathematics, Analog electronics technique, Principle of microcomputer and interface, Imaging section anatomy, computer programming language, An introduction to modern physics, Software design basis, Radiation physics.

Professional classes mainly include, Medical imaging diagnosis, Medical imaging physics, Overview of clinical medicine, Tumor radiation treatment for physics, Medical imaging equipment learn, Medical radiation biology, Clinical radiation oncology, Radiological diagnosis equipment quality guarantee, Radiation treatment equipment learn, Medical image processing.

Elective classes mainly include, bibliographic search, Biochemistry, Medical image Photography technology, Interventional radiology.

1.3   The Combination of Medicine and Engineering is the Basic Features

The basic features of course system lie in the following aspects: the cultivation of core courses and basic skills should care the requirements of the professional courses, highlight the mixture of medicine and physics, cultivate the all-round and practical talents; the designing of professional courses should make the two big majors of medicine and physics solid by emphasizing the merging courses and strengthen the courses of general medicine, electronics and basic computering.

So far the medical physics major has two Shandong Provincial superior elaborate courses, which are called respectively medical imaging appliance study and medical imaging diagnosis. Also the courses of electronics and radiological treatment study are excellent courses on school level.

2   Build a Professional "double-teacher" Teaching Staff

Medical physics which combines both medicine and science is a compound and applied undergraduate programme. Its requirement for the type and structure of the teaching faculty is different from traditional undergraduate programmes, which means it needs qualified lecturers, associate professors and professors who at the same time are engineers (technologist-in charge), senior engineers, and they must be able to combine science and medicine, have rich teaching and practical working experiences. This kind of teachers is called "double-teachers". There are not many "double-teachers" at present, we should develop and employ "double-teachers" actively from the actual situation of our university in order to build a "double-teacher" faculty. The approaches of our university are as follows: firstly, In order to build a "double-teacher" teaching staff which is of high comprehensive quality, practical ability, both comprehensive and professional, and combines medicine and science we strengthen the cultivation dynamics of our teachers, encourage young teachers to pursue further education, to get doctor's degrees, to take part in lesson planning meetings, academic meetings, to attend social practices, to fulfill their diplomas, to standardize their basic skills, to improve themselves in the practices. Secondly, the school selects teachers to take part in the construction of the internship base,and encourages the junior teachers to practice in the hospitals which are either attached to our school or internship bases, from which the teachers can enhance their selves, promote levels of profession and improve practical ability. Thirdly, the school hires senior medical professionals as part-time teachers from the attachment hospitals and internship bases, who are entitled with both theoretical and practical acknowledge. Meanwhile, the school attracts superior teachers and leaders in their fields from all over China and Shandong province. With effective measures, the school has achieved a hiring and measuring system both medically and technically to modulate teachers, and takes prority of the teachers who major in both medical study and technical study.

In radiation college, there are 59 professional teachers, there being 16 professors, 20 associate professors. There are 7 doctors and 30 masters. Radiation college has 1 national outstanding teacher, 1 school outstanding teacher, and 2 middle-aged backbones in school. Medicophysics teachers has a balance in age structure, professional title, knowledge structure and education structure. Radiation college pays attention to the training and improvement for professional teachers, emphasizing on in-service training and need to profession construction, adopting various forms to strengthen professional teaching staff construction. 1 teacher has been enrolled in Capital University of Medical Sciences for doctor degree, and 1 has taken part in the medical quality managing training, and 3 have taken part in medicophysics annual. They focus on teaching study, encouraging young teachers to combine with reality, taking a teaching reform and studying educating rules. Recently, medicophysics has won 12 study prizes, has set up 19 approval projects and has published 28 teaching thesises. Study has encouraged teaching and promoted teaching level.

3   Combine Medical Science and Industry, Combine Industry, Study and Research, Train Professionals of Medical Physical Science Major

When the medical and physical science major was opened , College of Radiology has regarded the combination of medical science and industry, combination of industry, studies and research as an important aim of construction of majors and training of talents, and have attached great importance to the combination of medical science and industry, the combination of industry, studies and research, strengthened the development of teaching and practice base, meet the requirements of training medical and physical science professionals, emphasized practical teaching, strengthed cooperation with industries and enterprises based on 8 attached hospitals to the school and 17 practical-teaching hospitals, established a new school mechanism of school-enterprise cooperation, complementing each other, coexisting and propering.

College of Radiology has cooperated with many corporations and research institions, such as Wandong Corporation which manufactures medical equipment, Dongruan Digital and Medical Corporationg, Shandong Xinhua Medical Equipment production company which mainly researches and produces rdiotherapy equipment. In recent years, Wandong Medical Equipment Corporation and Shandong Xinhua Medical Equipment Company have invested to our college 17 million yuan and 3 million yuan successively, for free to buy equipments. At present, there are more than 40 practices bases, most of which are third-level grade-A hospitals by all over the country, whose teaching faculties and equipments can adequately meet the needs of later teaching.

The medical and physical science major has medical equipment, imaging technology, medical engineering, physics, Jingong Practice Center and 5 laboratories, whose total value is 376 million yuan, and the equipoment is advanced. There are perfect and standard management systems in the laboratories. The laboratory equipments are in 100% good condition and have high efficiency. The one-time employment rate for the graduates of physics majors reaches 100%.

Pay attention to practical teaching. The practical teaching of medical physics mainly refer to the graduate internship and design of the post teaching, which includes graduate brochure, and designning. Students have rich and varied extracurricular activities. With the majority of students taking part in it, 12 out of them were awarded the prizes in the technical machine competition.

Through years of practices and exploration, we learned that medicine and engineering should be combined so as to cultivate all-round talents. In addition, a system of the two combinating majors and a double-teacher system should be constructed. With the combination of both medicine and engineering, together with the combination of both producing and researching, the professional and all- abled talents can be cultived.

References

1. Gu, M., Li, P.: Thesis on Chinese Researching Universitys' Status Quo of the Combination of Medicine and Engineering and Its Solution. Medical Education Exploration 5(9), 577–580 (2010)

2. Du, Y.: An Exploration to Chinese modes for cultivation of physics medicine talents. Medical Education Exploration 3(8), 228–230 (2009)
3. Wang, N.: Thesis on the Construction of Courses System for Biologically Medical Engineering Profession and the Assurance for the Qualified Cultivation of Applicant Talents. Xianning College Journal 2(29), 104–106 (2009)
4. Dong, Y.: Thesis on the Future of the Establishment of Medical Engineering Majors in Medical College and its Methods. Northwestern Medical Education 17(2), 229–230 (2009)
5. Department of Radiology, Taishan Medical University, Plan on the Cultivation of Applying Physics Professional Talents, Summary on the Construction of Medical Physics Teaching Team

Research on Decentralized Precise Support Mode and Related Strategies

Xiao Sun[1,2], Qigen Zhong[1], Zhiyi Cao[1], and Bin Liu[1]

[1] Ordnance Engineering College, Shijiazhuang, Hebei, 050003
[2] Science and Technology on Complex Systems Simulation Laboratory,
Beijing, 100101
xsunxiao@hotmail.com, {275770622,caozhiyi31,liubin957}@163.com

Abstract. A kind of decentralized precise support mode is proposed, which uses multiple local centers in different area to autonomous support. Under this mode the key information transmission will be prior to consider so that more global support requirement can be meet. Then the related strategies is discussed such as selection of local center, grade of support requirement, classification of information, incentive mechanism and so on. By a simulation example is shown to demonstrate essentiality and feasibility of decentralized precise support mode and related strategies.

Keywords: Precise Support, Decentralized, Multi-Center, Strategy.

1 Introduction

The development of information technology accelerates the appearance of precise support mode based on communication network. By accurate and detailed planning, building and using of resources with information technology, the resources and technology support are provided to afford combat operation in right time and right place, with right quantity quickly and efficiently to achieve the optimal effect-cost-ratio. That's to say, based on information technology, the precise supports mode builds an integrated support system, makes elaborate support plans to process and integrate information via command and control center and pushes forward the whole system to run systematically and orderly to use resources in its best way[1]. But the information network as material foundation has some inherent shortcomings: its vulnerability to interference and attack and its poor ability to conceal and to defense. If the information network is destroyed, especially when the control center can not be connected, the whole support system degenerates into the traditional imprecise support mode and will lose its functions substantially. So in most cases the communication network and information center have to be the protective emphasis.

When we study the information war, we must consider the world powers, which are fully capable of destroying our information infrastructure. So this paper studies the decentralized support mode and related strategies to develop precise support as much as possible in the environment that communication network is broken partly.

D. Jin and S. Lin (Eds.): Advances in FCCS, Vol. 2, AISC 160, pp. 597–602.
springerlink.com   © Springer-Verlag Berlin Heidelberg 2012

2   Decentralized Precise Support Mode

Since the role of command and control center is important, it will be the focus of attack and protection in war of both sides. In studying the equipment support, we have to pay close attention to the shortcomings of centralized information process and take effective measures to fill the gap.

The precise support system based on information technology is a network system at root. The theory of other network systems, such as society network[2], biology network[3], computer network[4] and so on, can be used for reference to the application of precise support system. Many network systems adopt multi-centered strategy[5] to share the burden of the master center and improve the viability of the whole system.

The basic idea of this paper is building multiple local centers in support system, as shown in Fig. 1. Any network nodes with information process capability can be selected as local centers, through which the associated nodes in network connect to the master center indirectly. When network situation is good, the master center takes up the task of information process and plan as a whole. Otherwise, the available local centers hold the information of associated support object nodes and support resource nodes, and coordinate their work spontaneously according to predefined support action rules or standards.

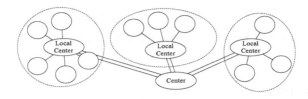

Fig. 1. Decentralized Precise Support Modes

3   The Related Strategies of Decentralized Support

The basic idea of decentralized support is simple; however, it needs strict support strategies to construct an integrated set of strategies to form a framework in order to effectively support military operations. The related strategies built in this paper, as is shown in Fig. 2, will be discussed one by one in the following section.

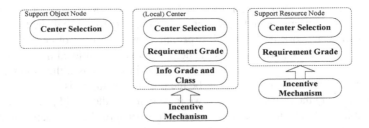

Fig. 2. Related Strategies of Decentralized Support

Selection of Local Center. Under default conditions, the local center in decentralized support system can be local commander nodes, which usually have more powerful information processing capability so that they are capable of achieving tasks of local center. If the local commander node is damaged, the nodes that have adequate information processing capability can be selected as local center. On this purpose these nodes of adequate information processing capability must install required software early. In some cases, the nodes can execute this procedure artificially.

Referring to other similar network systems, this paper proposed a simple executable center selection strategy. If a node can not connect to the original local center, the node searches a node with adequate information capability in the range of 90% max radio communication distance. If the node finds such a node, it connects to the latter as the new local center; if not, the node attempts to be the new local center. If the node cannot be the new local center, it will search again after a specific time.

Grade of Support Requirement. Grade of support requirement is the foundation for the local center to efficiently dispatch the support forces and resources. When a certain requirement occurs, it is required to grade it according to the task of combat element, the differentness of damage extent, the amount of resources consumed (ammunition, oil and ordnance stores) and so on. Then the local center disposes of the graded requirement depending on the current available resources and preconcerted support regulation. Normally the requirement of high grade is prior to fulfill. If the local center can not find enough resources, it attempts to demand more resources from the master center or other local centers.

The reasonable requirement grade standard and support regulation are essential to the decision making of local center. A feasible way is to firstly make basic items based on all factors, and then refine them repeatedly in actual combat practice or war game.

The Classification and Grade of Information. Since the local centers take up task of information processing and forwarding, an important part is the classification and grade of information in order to take different measure aiming at different kinds of information. When network environment is poor, this mode is beneficial to guarantee information will be transferred to the master center in accordance with its importance. The paper sorts the information in decentralized precise system into three kinds: original information, processed information and control information. The original information includes the state of combat element, the requirement from combat element, the state of support element, the amount of resources and so on, which is held by related local center and is the foundation of support. The processed information refers to the result achieved after the local centers analyze and integrate the original information, which shows the global state of element and the number of remaining resources in the local area. The processed information is smaller in size and more macro in view. According to the information, the master center can allocate and transfer the support element and resources to the managed area of the certain local center. The data delivering to master center is few, which decreases the pressure of master center in great measure. The control information is used to select the local center and control the entire system topology.

When grading the information, several factors should be considered. First, the information from more important element is relatively more important. Second, the information concerning requirement that can not be responded in the local center has higher grade. Third, the information about current deficient resources in the area is more important to the local center. Last but not least, the information that has not been processed for a long time has higher grade.

In wartime, according to the information grade strategy, information with high grade has a prior claim on transmission. While information with low grade will not be sent to master center until the communication capability is competent. At the same time the local center should record the autonomic decision, and report them later.

Incentive Mechanism. The support procedure of the local center and related resources is a self-organization behavior based on local information. To avoid the local center and support elements from profiting themselves, it is indispensable to formulate applicable incentive mechanism[6].

This paper proposes an effective evaluation method as a major reference to assess the elements' performance after war. The unit that continually has low evaluation will suffer punishment. This Incentive Mechanism enables support department to enhance support capability.

The effect evaluation will mainly concern the following factors: the absolute and relative amount of accomplished support task, response time of task, resource utilization, degree of its support to the local center, the accuracy of grading and so on.

4   Simulation Analysis

This paper simulated the support system of a typical tactical unit. In this case,   the number of nodes including command and control center, support object, resource is 56, the amount of resource type is 5, including ammunition, oil, repair power, spare part and vehicle. Restricted to the system scale, there are three level centers to autonomic support, the tasks that can not be disposed in current local center will be forwarded to the center higher level. Level 1 center is the lowest, level 3 center is master center. The roles of different level center are analyzed in the case.

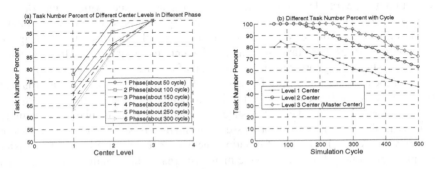

Fig. 3. Task Completed Proportion

Fig. 3(a) shows the percentage of tasks accomplished by different level centers in their own area in different phases. It is apparent that level 1 centers can accomplish most tasks and only a few tasks need the master node to dispose, which demonstrates the decentralized precise support mode proposed is feasible. Fig. 3(b), from another aspect, shows the real-time change curve of the percentage with simulation cycle.

The network resource requirement of different level centers in accomplishing tasks will be analyzed here. Table 1 shows the average search node number of different level centers to accomplish one task, the average hop count and delay from the support object node to resource node which can be found from different level centers. It indicates that higher level center managed more nodes and can search resource in a larger range, and the information amount for the search is more. Hence, the higher level center needs more network resource, which might lead to slow information transmission.

Table 1. Network Resource Requirement

Center Level	1	2	3
Average Search Node Number	4.74	10.2	17.79
Average Hop Count	1.89	4.33	6.0
Average Delay	175.57	346.75	465.67

From above analysis, the support efficiency and the network resource consumption in network-based precise support system are in inverse proportion. The goal of decentralized support mode is to solve the imbalance of the support efficiency and the network resource consumption. The research of precise support should not based on the assumption that network resource is sufficient for ever, but on selecting reasonable modes according to different network environment.

5   Conclusion

The decentralized precise support mode proposed in his paper aimed at facing the threat to information network in future war, fully considered how to guarantee the priority processing of more important information and provide supply support of combat elements by applying multiply level centers when network resources are limited In decentralized precise support system, besides the basic support mode, related strategies like grading strategy and incentive mechanism are required to form a complete set. The research and establishment of these strategies is a long-term procedure, which needs constant exploration in practice. Finally they had better to be written into wartime support regulation.

Acknowledgement. Supported by National Natural Science Foundation of China (60904071).

References

1. Kumar, S.: Connective technology as a strategic tool for building effective supply chain. International Journal of Manufacturing Technology and Management 10(1), 41–56 (2007)
2. Wang, X.F., Chen, G.: Complex networks: small-world, scale-free and beyond. IEEE Circuits and Systems Magazine (2003)

3. Canright, G., Deutsch, A., Babaoglu, O.: Design patterns from biology for distributed computing. ACM Transactions on Autonomous and Adaptive Systems (TAAS) 1(1), 26–66 (2006)
4. Xiao, S., Hui, W., Hao, W.: An adaptive topology management model in self-organizing overlay network. In: ICMSE 2007, Haerbin, Hei Longjiang (2007)
5. Kleis, M., Lua, E.K., Zhou, X.: Hierarchical peer-to-peer networks using lightweight superPeer topologies. In: 10th IEEE Symposium on Computers and Communications, ISCC 2005 (2005)
6. Wu, S.-P., Xu, X.-Y.: Incentive mechanism for two-echelon supply chain under asymmetric information. Computer Integrated Manufacturing Systems 14(3), 519–524 (2008)

Quasi RFID-WSNs Hybrid Object Networks for Fault Detection and Equipment Monitoring in Marine Ships

Hu Chen

Naval University of Engineering, Wuhan, China 430031
chen_nue@126.com

Abstract. The fault detection and monitoring of equipments in marine ships is a critical issue, since the maintenance operations and response need to be conducted timely. Traditional fault detection and monitoring system extensively rely on embedded circuits, which have to be deployed in advance. It thus encounters two inappropriate constraints such as the difficulties of incremental deployment, and detection of potential faults. In this paper, we propose a RFID-WSNs hybrid object networks for fault detection to enhance the accuracy and promptness of equipment monitoring. Hybrid object is a physical object with unique identity and appropriate sensing functions. The network of hybrid objects can detect the environmental parameters for the potential risks, especially the risks from the inner containers and inaccessible spaces. The identity can fast localize the potential risks on what physical objects.

Keywords: Equipment Maintenance, Remote Technical Support, Hybrid Networks, Fault Detection.

1   Introduction

There exist a large number of equipments in large ships, which may present Byzantine random failure or abnormalities. Since the marine ships are far from the base, to maintain the availability and functionality of the equipments is of foremost importance. It thus needs to guarantee the following requirements: faults and errors can be detected in real-time, and especially, potential risks can be predicated in advance [1, 2, 3]. The first step relies on the monitoring devices on the board, namely, the devices used for the detection and signal collection. The second step relies on the analysis of integrated information on-site.

Traditional monitoring systems on board usually highly rely on the embedded circuits. It imposes several undesirability as follows: the deployment of monitoring system is preloaded, not being able to incrementally redistribute; some physical objects are not suitable for the embedded circuits, for example, the operational mechanical handles, the inner solid actuators. The monitoring approaches indeed should be able to tackle the difference of monitoring objectives.

Furthermore, since in marine ships the number of relevant equipments is much more than that in normal ships, the fault detection present much more challenges, for example, the problem detection should be accurate and the solution localization should

D. Jin and S. Lin (Eds.): Advances in FCCS, Vol. 2, AISC 160, pp. 603–607.

be fast. In other words, we need to discover the faults or potential faults instantly, the scope of the faults can be traced to a rational domain, and the fault diagnose can be conducted instantly, efficiently, and correctly.

Currently, some researches address the maintenance techniques for normal ships [4-6]; however, researches on marine ships highly concentrate on embedded circuits. Based on aforementioned observations, in this paper, we propose a novel Hybrid Object Networks (HON) for dealing with the above challenges. In HON, we propose to use quasi RFID-WSNs hybrid object to fast localizing and detecting faults in equipments.

The contribution of this paper is as follows: 1) We propose a novel smart object call hybrid objects to monitoring the equipments in marine ships; 2) We propose a quasi RFID-WSNs networks to help the accurate environmental parameter gathering and fast problem localization. 3) We propose a novel architecture to address the shortcoming in the current detection and monitoring system for marine ships.

The rest of the paper is organized as follows: the Hybrid Object section will address the new techniques for equipment problem detection. The Quasi RFID-WSNs HON section will propose a new architecture for fault detection. The In-Networking Reliable Intelligence section will explore the techniques for fast on-site problem solving. The Conclusion section concludes the paper.

2   Hybrid Object

Since in the marine ships, there exist a large number of equipments. The detection of faults in the equipments has to be tailed to the characteristics of corresponding equipments. Traditional maintenance system on broad usually relies on the monitoring circuits and multiple sensors. They can collect the status data of equipments, including fault number, fault code, fault graph, fault sound, and equipment arguments et al.. Such data will be displayed on the equipment monitors, and be reported to the upper support center. The support operators then use the Portable Maintenance Aid (PMA) system and Interactive Electrical Technical Manual (IETM) to solve the faults [7-10].

In addition to the traditional embedded circuits, we observe that following objectives are not appropriated tackled:

Some physical equipment is integrated as a single and solid objective; they are difficult to deploy some embedded circuits. In this case, attaching a circuit chip may be the left choice.

Some monitoring parameters need to be collected incrementally. For example, some faults are estimated to be upcoming in some possible areas, the further monitoring response then are always required to deploy additionally for the interested details.

The detection needs to be conducted automatically, and the responses need to be instant. Some faults can be responded in time if the feedback logics are simple. Some faults need humanly confirmed and revoked manually.

The faults should be fast traced back and be located. It is important for the operators on board to process the fault by some computer aided devices such as PMA system and IETM.

To address above challenges, we propose a new technique based on RFID and active sensors.

Definition I. Hybrid Object (HO). Hybrid Object is an electrical device that has certain capabilities, such as wired and wireless communication interface, computation ability, memory ability, and so on. It is can be implemented by embedded devices such as Longxin or AMD processes and communication modules.

The Design of HO.

1) Communication ability. The communication interface card has wireless module and process unit. It can be provided by dedicated device providers. The software parts are more important. We propose a dedicate protocol stack. There exist physical layer and MAC layer in the stack. It can reference IEEE 802.15.4, IEEE 802.11n, and IEEE 802.3. We propose a dedicated MAC layer for the hybrid communication environments.

For the dedicated physical layer, we use dedicated preamble code. For the MAC protocol, we use revised CSMA/CA mechanism. It has different back off windows as the most HO that competes for the wireless channels are fixed, pre-deployed, and accounted. The back off window or timer could be improved in this scenario. As the topology is always fixed, the routing is always simply star routing and single hop transmission.

The data layer is above the MAC layer. It will carry two major fields. One is the identity of the monitoring objects; the other is the monitoring parameters. They are used for the object localization and monitoring parameter collection. Note that, HO is a supplemental of embedded circuits, so its functionality is targeting for the uncovered part of traditional circuits.

2) Computation and Storage ability. We propose to use highly reliable chip that may conduct double checks and smoothly failure migrations.

3   Quasi RFID-WSNs HON

Traditional monitoring networks consist of integrated circuits concentrating for the detection of equipment status. The detection confronts several limits:

1) The monitoring parameters usually from inner side of the objects, not outside of the environments. There exist certain locations that are not suitable for deploying wired monitoring lines. To enhance the flexibility, short range wireless network can be as a supplement.

2) The signals of detection impose reliability risks. That is, the circuits may be damaged in emergent situations, whereas the re-deployment experiences difficulties. To enhance the robustness, the incremental deployment is required.

Toe address above observations, we propose a quasi RFID-WSNs network, called Hybrid Object Network (HON). HON is deployed along the monitored equipments, and a network of HO. HON enables the wireless communications and support the flexible and robust deployment.

The topology of HON usually is start or chain topology. In start topology, the head is the core node and accumulate the data from the outer nodes. The connections between heads could be embedded circuits. In the chain topology, the data will be transferred hop by hop until to the core node.

The HON collects information and uploads them into central actuator nodes. Central actuators can be also a distributed network. Some actuators response locally, whereas some actuators response coordinately. That is, if the logic is simple, the decision at actuators will be locally. If the decision requires more information from other nodes, the decision will be made by upper layer or actuators.

HON is a reliable network, since we use following techniques in the design:

1) The topology is simple and most of them have only one hop. Thus, the connection can be maintained easily.

2) The redundancy is mandatory. Multiple nodes and multiple networks will be used for detection objective.

3) The hybrid architecture will help the robustness, since the failure of single architecture does not result in the failure of all architectures.

4) The response decision will be double checked. That is, only when two actuators make the same decisions, the response will make the final feedback.

4   In-Networking Reliable Intelligence

Traditional collection of detection and monitoring information will push to the operators to make the decision. To fasten the response, we propose to use a new technology called In-Networking Intelligence. The basic idea is that in the decision of the input should be an integrated collections, and the output should also be distributed. That is, there exists an in-networking logic process architecture in the HON, and that is a multiple input and multiple output paradigm.

The intelligence is based on the swarm intelligence, but the reliability of the intelligence is especially enhanced. Thus, the decision making relies on trust management. The detection signals are classified into different priorities. The higher one will be taken the higher weight. The computation is linear to decrease the delay. The smoothing techniques are also introduced for exploiting history. That is, the last data will take a weight in the final decision.

The process can be described as follows:

Suppose $n1, n2, \dots, np$ are p detection nodes. The $m1, m2, \dots, mq$ are q actuator nodes. Modeling in-networking process as Π, and denoted it as $(n1 \times n2 \times \dots \times np \rightarrow m1 \times m2 \times \dots \times mq)$. The decision path should be at least double. That is, suppose a path is denoted as $\prod ni \rightarrow \prod mj$, thus for each mj, there exist at least two $\prod ni$ such that $\prod ni \rightarrow \prod mj \in \Pi$.

Figure 1 presents the framework of Quasi RFID-WSNs Hybrid Object Networks.

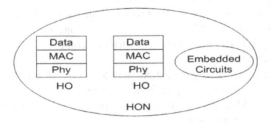

Fig. 1. Quasi RFID-WSNs Hybrid Object Networks

5   Summary

In this paper, we proposed a Quasi RFID-WSNs Hybrid Object Networks for the remote maintenance signal detection and monitoring. The proposed techniques firstly address the current shortcomings of the existing solutions. Next, we propose a device called hybrid object. It can trace back to the identity of monitored objectives, and collect the signals from the environment, which is an mandatory supplement for traditional embedded circuits. We also propose a quasi RFID-WSNs hybrid object networks for fault detection and monitoring, which enable the flexible deployment and incremental enhancement. A new in-networking reliable intelligence architecture is also proposed, as it enables trust evaluations and multiple decision paths in the responses. We argue that such discussions provide in-depth new directions in forthcoming approaches.

References

Dekker, R.: On the impact of optimisation models in maintenance decision making: the state of the art. Reliability Engineering & System Safety (1998)

Wang, Z., Xie, L.: Study on Equipment Maintenance Cost Budgeting Based on Risk Evaluation. Ship & Ocean Engineering (2008)

Yu, J.Y.W.X.F.: Design and Realization of the Maintenance Directed System of Ship Equipment on IETM. Ship Electronic Engineering (2008)

Wenliang, Y.: Design and Implementation of Full Text Inspection in IETM Based on Web. Ship Electronic Engineering (2007)

Peng, C.: Design and Realization of Distant Maintenance Technology Supporting Call Center. Ship Electronic Engineering (3) (2007)

Wenliang, Y.: Design and Implementation of Full Text Inspection in IETM Based on Web. Ship Electronic Engineering (2007)

Zhu, S.: On remote maintenance supporting system of war ship equipment. Journal of Naval University of Engineering (2004)

Fang, H.T.L.: LORA in the Ship Equipment Maintenance. Journal of the Naval Academy of Engineering (1999)

CoCoS, maintenance designed for maintenance excellence. MAN B&W Diesel A/S, Copenhagen, Denmark (2005)

Tarelk, O.W.: Improvement of ship mechanical equipment maintainability through design. Marine, Offshore and Technology, 91–98 (1994)

Confidence Interval Estimation of Environmental Factors for the Exponential Distribution under Type-II Censored

Feng Li[1,2]

[1] Department of Mathematics, Wei Nan normal University
[2] The Institute of Statistical Sciences and Social computing
Wei Nan, 714000, Shaanxi, China
lifeng5849@163.com

Abstract. The exact confidence interval estimations and the approximate confidence interval estimations of the environmental factors (EF) are investigated under type-II censoring date, the formulae to calculate the exact confidence limits and the empirical Bayes approximate confidence limits of EF are given. In order to investigate the accuracy of estimations, an illustrative example is examined numerically by means of Monte-Carlo simulation.

Keywords: environmental factors, Type−II censoring, exponential distribution, empirical bayes estimation.

1 Introduction

In reliability evaluation and accelerated life tests, reliability test information at various environments is often required. In most situations, it is difficult to use the laboratory failure data to estimate reliability at use conditions directly. If there are no failure data available at the field environment, we usually convert the laboratory test results into equivalent failure information at use condition. This equivalent information is used to estimate reliability at field conditions [1]. EF are powerful tools for solving this kind of problems. The first EF was introduced and defined for exponential lifetime distribution [2] and received some attention in reliability & quality engineering. EF for the s-normal, lognormal, inverse Gaussian, Binomial ,Gamma distributions are defined and their point & interval estimates are derived in [3-5]. But the Bayes estimation of EF was not addressed under Type-II censoring life test. In this paper, we investigate Empirical Bayes estimates of EF under Type-II censoring life test.

2 Exact Confidence Interval Estimation

Suppose the product life follow exponential life distribution $EXP(\lambda_i)$ $i = 1, 2$. Life time distribution under two different environment S_1 and S_2. Suppose that n_i ($i = 1, 2$) units are tested in environment S_i ($i = 1, 2$). When r_i ($i = 1, 2$) units failing, the test is stopped. We denote the failure moment in turn by $X_{i1} \leq X_{i2} \leq \cdots \leq X_{im_i}$. Thus $X_{i1}, X_{i2}, \cdots, X_{im_i}$ is a Type-II censored samples.

D. Jin and S. Lin (Eds.): Advances in FCCS, Vol. 2, AISC 160, pp. 609–613.

After the transformations of $Y_{ij} = \lambda_i X_{ij}$, $j = 1,2,\cdots,r_i$, $i=1,2$ $Y_{i1} \le Y_{i2} \le \cdots \le Y_{ir_i}$ is a Type-II censored sample from the standard exponential distribution since log function is an increasing function. Consider the following transformations

$$
\begin{cases}
Z_{i1} = n_i Y_{i1} \\
\cdots \\
Z_{ij} = (n_i - j + 1)(Y_{ij} - Y_{i,j-1}) \qquad i=1,2\ , j=1,2,\cdots,r_i \\
\cdots \\
Z_{ir_i} = (n_i - r_i + 1)(Y_{ir_i} - Y_{ir_{i-1}})
\end{cases}
$$

Thomas and Wilson (1972) showed that the generalized spacing $Z_{i1}, Z_{i2}, \cdots Z_{ir_i}$, are all independent and identically distributed as standard exponential. It can be seen that

$$
\sum_{j=1}^{r_i} Z_{ij} = 2\lambda_i [\sum_{j=1}^{r_i} X_{ij} - (n_i - r_i)X_{ir_i}] = 2\lambda_i T_{r_i} \sim \chi^2(2r_i) \qquad i=1,2 \tag{1}
$$

$$
\frac{2\sum_{j=1}^{r_1} Z_{1j} / r_1}{2\sum_{j=1}^{r_2} Z_{2j} / r_2} = \frac{r_2 \lambda_1 T_{r_1}}{r_1 \lambda_2 T_{r_2}} = \frac{r_2 \lambda_1 [\sum_{j=1}^{r_1} X_{1j} - (n_1 - r_1)X_{1r_1}]}{r_1 \lambda_2 [\sum_{j=1}^{r_2} X_{2j} - (n_2 - r_2)X_{2r_2}]} \sim F(2r_1, 2r_2) \tag{2}
$$

Where $\chi^2(2r_i)$ denotes a central chi-square distribution with r_i degree of freedom. Then the $1-\alpha$ confidence limit of k can be written as

$$
\frac{r_1 T_{r_2}}{r_2 T_{r_1}} F_{\alpha/2}(2r_1, 2r_2) \le \hat{k} \le \frac{r_1 T_{r_2}}{r_2 T_{r_1}} F_{1-\alpha/2}(2r_1, 2r_2) \tag{3}
$$

3   Approximate Confidence Interval Estimation

The likelihood function (see [6]) for the parameters λ_i is then

$$
L_i(x_{i1}, x_{i2}, \cdots x_{ir_i}) = \frac{n_i!}{(n_i - r_i)!} \lambda_i^{r_i} \exp\{-\lambda_i T_{r_i}\} \qquad i=1,2 \tag{4}
$$

where
$$
T_{r_i} = \sum_{j=1}^{r_i} x_{ij} + (n_i - r_i)x_{r_i}
$$

The joint likelihood function of S_1 and S_2 is

$$
L(x_{ij}; \lambda_1, \lambda_2) = L_1 L_2 = \prod_{i=1}^{2} \frac{n_i!}{(n_i - r_i)!} \lambda_i^{r_i} \exp\{-\lambda_i T_{r_i}\}
$$
$$
= \frac{n_1! n_2!}{(n_1 - r_1)!(n_2 - r_2)!} \lambda_1^{r_1} \lambda_2^{r_2} \exp\{-\lambda_1 T_{r_1} - \lambda_2 T_{r_2}\} \tag{5}
$$

we know λ_i is a random variable with a prior density function with density function

$$\pi(\lambda_i \mid \beta_i) = \beta_i \exp(-\beta_i \lambda), \quad \beta_i, \lambda_i > 0$$

Applying Bayesian theorem, combining the likelihood function and prior density, we obtain the posterior density of in λ_i the form

$$h_i(\lambda_i \mid x_{ij}) = \frac{g(x \mid \lambda)\pi(\lambda \mid \beta)}{\int_0^{+\infty} g(x \mid \lambda)\pi(\lambda \mid \beta)d\lambda} \qquad j = 1,2,\cdots,r_i, \quad i = 1,2 \qquad (6)$$

$$= [\beta_i + T_{r_i}]^{r_i+1}\Gamma(r_i+1)^{-1}\lambda_i^{r_i}\exp\{-\lambda_i(\beta_i + T_{r_i})\}$$

From [6], we can get

$$2(\beta_i + T_{r_i})\lambda_i \sim Ga(r_i+1,2^{-1}) = \chi^2(2r_i+2)$$

$$\frac{\lambda_1(r_2+1)(\beta_1+T_{r_1})}{\lambda_2(r_1+1)(\beta_2+T_{r_2})} \sim F(2r_1,2r_2) \qquad (7)$$

Then the $1-\alpha$ confidence limit of k can be written as

$$\frac{(r_1+1)(\beta_2+T_{r_2})}{(r_2+1)(\beta_1+T_{r_1})}F_{\alpha/2}(2r_1+2,2r_2+2) \le \bar{k} \le \frac{(r_1+1)(\beta_2+T_{r_2})}{(r_2+1)(\beta_1+T_{r_1})}F_{1-\alpha/2}(2r_1+2,2r_2+2) \qquad (8)$$

As β_i is an unknown constant, In order to estimate β_i, we need to use the the maximum likelihood estimation approach. As the probability distribution of every unit X_{ij} is a exponential distribution $EXP(\lambda_i)$, the margin density function X_{ij} is

$$f_{X_{ij}}(x) = \int_0^\infty f(x\mid\lambda_i)\pi(\lambda_i \mid \beta_i)d\lambda_i = \int_0^\infty \lambda_i\exp(-\lambda_i x)\beta_i\exp(-\beta_i\lambda_i)d\lambda_i = \frac{\beta_i}{(x_{ij}+\beta_i)^2} \qquad (9)$$

Let density function $h_i(x)$ and distribution function $F(x_r)$ be $f_X(x)$ and $F_X(x_r)$ respectively in equation (1), where

$$1-F_X(x_{r_i}) = \int_{x_{r_i}}^{+\infty} f_X(x)dx = \int_{x_{r_i}}^{\infty} \frac{\beta_i}{(x+\beta_i)^2}dx = \frac{\beta_i}{x_{r_i}+\beta_i} \qquad (10)$$

From [6], the associated density function of $X_{i1}, X_{i2}, \cdots, X_{im_i}$ is

$$L_i = \frac{n_i!}{(n_i-r_i)!}[\prod_{j=1}^{r_i} f_{X_{ij}}(x_{ij})][1-F_{X_{ij}}(x_{ir_i})]^{n_i-r_i} \qquad (11)$$

$$= \frac{(\beta_i)^{r_i} n_i!}{(n_i-r_i)!}[\prod_{j=1}^{r_i} \frac{1}{(x_{ij}+\beta_i)^2}](\frac{\beta_i}{x_{ir_j}+\beta_i})^{n_i-r_i}$$

$$\frac{d\lg L_i}{d\beta_i} = \frac{r_i}{\beta_i} - 2\sum_{j=1}^{r_i}\frac{1}{x_{ij}+\beta_i} + (n_i-r_i)(\frac{1}{\beta_i}-\frac{1}{x_{ir_i}+\beta_i})$$

The equation $\dfrac{d\lg L_i}{d\beta_i} = 0$ has only one root and the expression is

$$\beta_i^{(l+1)} = r_i[2\sum_{j=1}^{r_i}\frac{1}{x_{ij}+\beta_i^{(l)}} - (n_i - r_i)\frac{x_{r_i}}{\beta_i^{(l)}(x_{i_{r_i}}+\beta_i^{(l)})}]^{-1} \qquad i=1,2 \qquad (12)$$

Thus when the degree of confidence $1-\alpha$ is given, confidence limits of k is written as

$$\frac{(r_1+1)(\beta_2+T_{r_2})}{(r_2+1)(\beta_1+T_{r_1})}F_{\alpha/2}(2r_1+2,2r_2+2) \le \tilde{k} \le \frac{(r_1+1)(\beta_2+T_{r_2})}{(r_2+1)(\beta_1+T_{r_1})}F_{1-\alpha/2}(2r_1+2,2r_2+2) \quad (13)$$

4   Simulation Study

For given parameters $\lambda_1 = 5, \lambda_2 = 2$ of exponential distributions under the environment S_1, S_2, the simulated average confidence length for EF after 1000 simulation runs of the Monte Carlo simulation method are presented in Table 1, using formulas of (4), (12)-(13).

The average length for the interval estimation of environmental factors with $EXP(\lambda_i)$ for $\alpha = 0.9$ and $\alpha = 0.95$

(n_1, n_2)	(r_1, r_2)	$\alpha = 0.9$		$\alpha = 0.95$	
		Exact confidence	Approximate confidence	Exact confidence	Approximate confidence
(10,15)	(2,3)	3.78	2.8682	5.427	3.9777
(10,15)	(5,8)	1.755	1.4649	2.2839	1.91
(10,15)	(8,8)	2.9945	2.279	3.9087	2.9794
(20,15)	(2,3)	2.9616	2.4873	4.4508	3.5682
(20,15)	(5,8)	1.3832	1.2735	1.8944	1.7293
(20,15)	(10,15)	0.9839	0.9155	1.264	1.1744
(30,25)	(5,8)	0.9014	0.8711	1.2098	1.1557
(30,25)	(10,15)	0.6059	0.5944	0.786	0.772
(30,25)	(15,15)	0.8805	0.8418	1.0874	1.0437
(50,50)	(15,15)	1.1375	1.0782	1.4698	1.3922
(50,50)	(20,18)	1.1281	1.0712	1.4449	1.3721

5   Conclusions

Based on the type-II censored, we obtain the exact confidence limits and the approximate confidence limits of the EF. The Monte-Carlo simulation was used to examine the result of EF estimation. The simulation results show the accuracy of the approximate confidence limits of EF are better than the exact confidence limits.

Acknowledgements. This work was supported partly by the National Statistics Foundation of the National Bureau of Statistics. (N0. 2011LZ030)

References

1. Elsayed, E.A.: Reliability Engineering. Addison-Wesley (1996)
2. Method for Developing Equipment Failure Rate K factors, AD-B001039. Available from NTIS, Springfield, VA 22161 USA
3. Wang, H.-Z., Ma, B.-H., Shi, J.: Estimation of environmental factor for the normal distribution. Microelectronics & Reliability 32, 457–463 (1992)
4. Wang, H.-Z., Ma, B.-H., Shi, J.: A research study of environmental factors for the gamma distribution. Microelectronics& Reliability 32, 331–335 (1992)
5. Elsayed, E.A., Wang, H.: Bayes & Classical Estimation of Environmental Factors for the Binomial Distribution. IEEE Transactions on Reliability 45(4), 661–665 (1996)
6. Kan, C.: The life distribution and the theory of reliability mathematics. Science Press, Beijing (1999)

Research of Automation Control System of Manipulator Based on PLC

Wanqiang Hu and Lihui Guo

College of Electrical and Information Engineering, Xuchang University,
Xuchang 461000, China
hwq@xcu.edu.cn, gloria1981.xu@yahoo.com.cn

Abstract. The structure, working principle and the hydraulic system theory of the industrial manipulator, based on PLC were introduced in the paper. Some important components of the system were designed and selected. The technological processes of the industrial manipulator were analyzed. The I/O addresses of PLC were distributed, and the control of chucking power of the manipulator was designed by employing the BP-PID control strategy.

Keywords: Industrial Manipulator, Hydraulic Component Selection, PLC, WinCC.

1 Introduction

The industrial manipulator is a kind of automatic equipment, which can simulate the hand gestures of human been and achieve the automatic motion of grabbing, carrying and manipulation, according to the preset sequences, tracks and demands. It is one of the most important equipments, which can realize the industrial production mechanization and automation [1]. The mechanical clutch is one of the most important components of the actuating mechanism, and its performances will affect the effect and efficiency of the manipulator. How to design and control the grabbing power of a MC more efficient and accurate is one of the most important problems must being confronted [2].

In this article, a kind of IM based on WinCC and PLC is introduced. It is one of the coordinate hydraulic driving type, which is composed of wrist slewing mechanism, arm stretching mechanism, arm rotary mechanism and arm elevator mechanism etc., and with 4 degrees of freedom: arm stretching, rotating, elevating, wrist rotating. The structure of the system is show in figure 1.

The operation sequences of the manipulator are: lifting arm falling from original position → fingers clamping → lifting arm rising → base fast forward rotating → base slow forward → wrist rotating → telescopic boom stretch out → fingers loosening → telescopic boom retracting; Until the machining is finished, telescopic boom stretching out → fingers clamping → telescopic boom retracting → base fast back rotating → base slow back → wrist rotating → lifting arm falling → fingers loosening → lifting arm rising to the original position and stopping, preparing for the next loop.

D. Jin and S. Lin (Eds.): Advances in FCCS, Vol. 2, AISC 160, pp. 615–619.
springerlink.com   © Springer-Verlag Berlin Heidelberg 2012

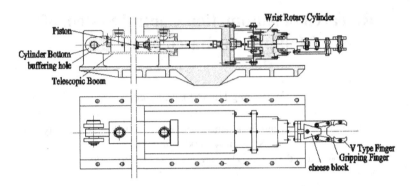

Fig. 1. The Structure Chart of Industrial Manipulator

2   The Chart of Hydraulic System

The hydraulic principle of the industrial manipulator is show in figure 2.

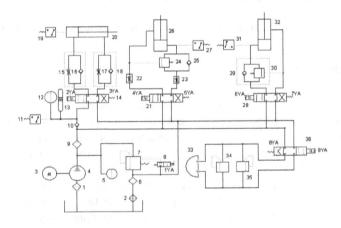

Fig. 2. The Hydraulic Principle of the Industrial Manipulator

There are 3 hydraulic cylinder mainly included in the hydraulic system: telescopic cylinder, lifting cylinder and gripping cylinder. A swing hydraulic motor is adopted to control the wrist. Single pump supplying is adopted by the hydraulic manipulator, while parallel supplying is adopted by arm stretching, wrist rotating and gripping motion. With the purpose of none mutual disturbances among multi-cylinder movements and realizing synchronous or nonsynchronous movements, meso-position "O" mode reversing valve is adopted.

3 The Selection and Verification of System's Part Significant Devices

Hydraulic Pumps. The max needing flow rate of the system's each executive component:

Clamping Cylinder: $Q_1 = 1.5L/min$, Telescopic Cylinder: $Q_2 = 15L/min$,
Lifting Cylinder: $Q_3 = 7.536L/min$, Rotation Cylinder: $Q_4 = 17.34L/min$.

So, $Q_p \geq kQ_{max} = 1.1 \times 17.34L/min = 19.074L/min \approx 20L/min$.

Displacement of the pump: $q = (Q_p/n_p) \times 10^3 = 13.6ml/r$.

Max pressure of each pumps while working: Clamping cylinder: $p_1 = 1MPa$, Telescopic Cylinder: $p_2 = 1.6MPa$, Lifting Cylinder: $p_3 = 2MPa$, Rotation Pump: $p_4 = 0.04MPa$.

So the max working pressure of the system: $p \geq p_{max} + \sum \Delta p = 2 + 0.4 = 2.4MPa$, $\sum \Delta p$ includes part losses of the oil liquid while flowing the flow valve and other components, and the path loss of pipe road etc. Its value is $0.4MPa$.

$5MPa$ can be selected as the system's rated pressure, for pressure storage.

According to the calculation above, YB-A16B pump is selected, whose displacement is 16.3ml/r, pressure5MPa, input power1.69kW, and rated rotation speed 1000r/min.

Pump Motor

The power of the pump motor is

$$N = \frac{p \cdot Q}{60\,\eta} = \frac{2.4 \times 17.34 \times 10^3}{60 \times 0.75} = 0.925\ kw \tag{1}$$

In the expression, p —Actual max working pressure of the pump; Q —Actual flow rate of the pump under the pressure; η —Total efficiency of the pump, valued0.65~0.75.

In consideration of the pressure loss, $1.5kw$ motor is selected according to the standard.

Calculation of the Oil Trunk Capacity. The Calculation of the oil trunk capacity is

$$V = m \cdot Q_p = (5 \sim 7) \times 19.074 = 95.37 \sim 133518L. \tag{2}$$

In the expression: V —Effective volume of the oil trunk(L); m —Coefficient, middle pressure system could be 5~7min; Q_p —liquid pump flow rate(L/min).

4 Control System Design

The gripping power control strategy of the manipulator is shown in figure 3. The neural network and PID combination control mode is mainly adapted. Make use of the self-learning characteristic, correct the PID controller's controlling parameters in time, and find the optimal P, I, D parameters under the PID control law [3,4].

BP-PID controller can be implemented in the WinCC configuration software. The global script editor provides a function interface of expanding system for users, where the project functions and movement functions, for system calling, can be written in C programming language by users.

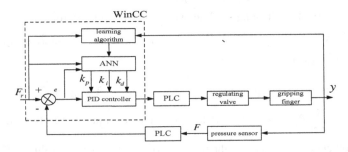

Fig. 3. The Controller's Principle Chart of Manipulator's Gripping Power

(1) Produce a project function ANN_PID() in WinCC global script, performing the neural network control algorithm. A section of the source program is shown as follows:

```
void ANN_PID()
{        int inputNum;// Actual Number of the Input Node
    int hideNum;// Actual Number of the hidden Level
    int trainNum;// Actual Number of Training
        // Obtain the Needing Variables of Control Algorithm Computation from
WinCC
        η =GetTagDouble( "LearnSpeed" );// Learning Efficiency
        α = GetTagDouble( "inertia" );// Obtain Inertial System
        e= GetTagDouble( "ErrorLevel" );// Obtain Accuracy Requirement
}
```

So, driven by the trigger signal, call the BP control algorithm once at regular time, and then export a group of PID real-time parameters, which will be transferred to PLC by transformation line, passing through the PLC analog output module, adjusting the control valve, and implement the control to gripping finger at last.

5 Conclusion

In this paper, a kind of industrial manipulator, based on PLC and WinCC, is presented. The hydraulic system of manipulator is designed. Part of components are designed and selected. The control of gripping power of the manipulation is realized successfully, by combining the .neural network and PID controller.

References

1. Wang, S., Xu, H.: Aplication of PLC in Manipulator. Journal of Changchun Insititute of Technology: Natural Science Edition (3), 45–48 (2008)
2. Gong, W., Hu, Q.: The Design of Manipulator PLC control system. Mechanical and Electrical Engineering Techhnology (11), 91–95 (2008)
3. Cai, W.: The Application of Human-computer Interface and PLC in Material Sorting System. Journal of Zhejiang Industry and Trade Polytechnic (3), 25–28 (2008)
4. Li, Y.: The Application of Step Motor Control in Industrial Manipulator Based on PLC. Science and Technology Information (18), 30–31 (2008)

Fault-Tolerant Integrated Navigation Algorithm of the Federal Kalman Filter

Jing Li[1], Jiande Wu[1,2,*], Junfeng Hou[3], Yugang Fan[1,2], and Xiaodong Wang[1,2]

[1] Faculty of Information Engineering and Automation, Kunming University of Science and Technology, Kunming 650093
[2] Engineering Research Center for Mineral Pipeline Transportation Yunnan, Kunming 650500
[3] Xuchang Cigarette Factory of China Tobacco Henan Industrial Co., LTD., Xuchang 461000
uutnight@qq.com, wjiande@gmail.com

Abstract. As the Federal Kalman filter has some good features such as flexibility, and good fault-tolerance, this paper proposes a federal Kalman filter design method and fault-tolerant structure. The structure uses the residual error between the local filter and reference filter for fault detection. In this paper, a simulation study of the integrated navigation system is done. The study shows that the algorithm is very simple, reliable, not only can quickly detect the fault of the external sensors and reference system, but also has good fault-tolerant. It can quickly detect and isolation fault, and let the integration of the system remain high precision.

Keywords: Federal Kalman filter, Fault-tolerant technology, Multi-sensor fusion.

1 Introduction

Kalman filter was proposed by Kalman in 1960. It is a linear minimum variance estimation recursive algorithm. It can estimate the random process of multi-dimensional (smooth, non-stationary). Since then, Carlson proposed a federal Kalman filter based on the decentralized filter, which has been widely used in navigation because of design flexibility, small amount of calculation, good fault-tolerance.

This paper proposes a design method of the Federal Kalman filter, and then detailed analyze the fault-tolerant technology and method of the federal Kalman filter. The final simulation shows that the method is effective.

2 Federal Kalman Filter

Kalman filtering for optimal multi-sensor data fusion has two ways: centralized filtering and decentralized filtering. The former is a filter to focus on all the information from the subsystems. In theory, a centralized Kalman filter can give the optimal state estimation, but it has the following problems [1]:

(1) A centralized Kalman filter has high state dimension, and the computation increase with the cube of filter dimensions. The computational burden is very heavy, and it can't guarantee real-time.

---

* Corresponding author.

(2) The more the number of the subsystem, the bigger the system failure rate. If only one of subsystem fails, the entire system will be contaminated. Therefore, the centralized Kalman filter has the bad fault-tolerant, and it is not conducive to fault diagnosis.

The above deficiency of Kalman filter makes the potential of the multi-sensor combined system to be fully unrealized. Parallel processing techniques, the importance of the system fault tolerance and a variety of new sensors makes the successful development of distributed Kalman filtering technique. A decentralized filtering system is one effective method to solve the state estimation of a big system, reduce the amount of computation, prevent the system numerical difficulties caused by high-order.

Decentralized filtering has been 20 years of development. Pearson[2] proposed the concept of decomposition and dynamic state estimation of the structure in 1971. Since then, Speyer, Willsky, Bierman, Kerr and Carlson all make the contribution to the decentralized filtering technique. Among them, Carlson[3] proposed Federal filter was paid attention because the use of information distribution eliminates the relevance of the sub-state estimates. It has small amount of calculation, good fault-tolerant and design flexibility. At the present time, US Air selected Federal filter as the basic "Public Kalman filter" algorithm [4].

The following Federal Kalman filter structure is designed:

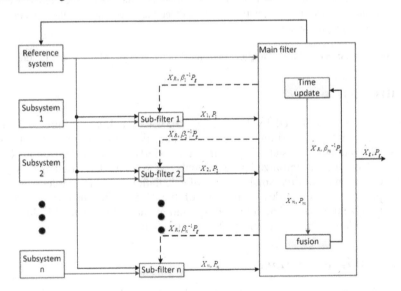

Fig. 1. The general structure of the federal Kalman filter

For the integrated navigation system, common reference system is generally the inertial navigation system in Figure 1, and its output is passed directly to the main filter on the one hand, on the other hand can be passed to the sub-filter as the observations. The local estimates $\hat{X}_i$ (public status) and its estimation error covariance matrix P_i of each sub-filter is sent to the main filter, and gets the globally optimal estimate by integrating with the estimated value of the main filter. Figure 1 also shows that the fusion global estimates $\hat{X}_g$ and their corresponding estimation error covariance matrix P_g of sub-filter and main filter is then amplified to $\beta_i^{-1} P_g (\beta_i \leq 1)$ as a feedback to the sub-filter (shown in dashed line), to re-estimates of the sub-filter set[1]:

$$\hat{X}_i = \hat{X}_g, P_{ii} = \beta_i^{-1} P_g (\beta_i \leq 1) \tag{1}$$

Meanwhile, the estimation error covariance matrix of main filter can also be reset to β_m^{-1} times $\beta_m \leq 1$ as the global estimation error covariance matrix. $\beta_i (i = 1, 2, \cdots, N, m)$ is called information distribution coefficient, and it is determined by the principle of information distribution. The different structures and different characteristics of the federal Kalman filter (fault-tolerance, precision and computation) can be gotten by using different β_i values.

The amount of information of state equation can be represented by the inverse of the process noise covariance matrix as Q^{-1}. Assuming the total amount of information Q^{-1} of the system process noise assigned to the sub-filters and main filter, the time update equation is:

$$P_{ii} = \Phi_{ii} P'_{ii} \Phi_{ii}^T + \gamma_i G_i Q G_i^T \tag{2}$$

$$P_{ji} = \Phi_{jj} P'_{ji} \Phi_{ii}^T = 0; P'_{ji} = 0 \tag{3}$$

The equation (3) shows that only $P'_{ji} = P_{ji(k-1)} = 0$ has $P_{ji} = P_{ji(k,k-1)} = 0$. That is to say, time update in each sub-filter is carried out independently, and there is not association among the every sub-filter.

Using above technology, the sub-filter estimates are not related to other sub-filter, and the observation update and time update can be carried out independently, so that the following formula (4) and (5) can be used for the optimal synthesis for the global integration of local estimates. The local estimates are sub-optimal, but the global fusion estimate is optimal.

$$P_g = [P_{11}^{-1} + P_{22}^{-1}]^{-1} \tag{4}$$

$$\hat{X}_g = [P_{11}^{-1} + P_{22}^{-1}]^{-1} (P_{11}^{-1} \hat{X}_1 + P_{22}^{-1} \hat{X}_2) \tag{5}$$

3   Federal Kalman Filter System-Level Fault Tolerance

As one major advantage of the federal Kalman filter, fault-tolerant design is an important way to improve the reliability of an integrated navigation system. The starting point of fault-tolerant design is changing the overall design of the system, rather than to improve the basic reliability of every component parts. The main method of fault-tolerant design is to make the system self-monitor, run through the monitoring system, real-time fault detection and isolation, and then take the necessary measures for the entire system still work safety[5].

The discrete system model with fault [6]:

$$\left.\begin{array}{l} X_k = \Phi_{k,k-1}X_{k-1} + \Gamma_{k,k-1}W_{k-1} \\ Z_k = H_k X_k + V_k + f_{k,\varphi}\gamma \end{array}\right\} \tag{6}$$

Where, γ is random vector which represent the size of fault, $f_{k,\varphi}$ is piecewise function.

$$f_{k,\varphi} = \begin{cases} 1, & k > \varphi \\ 0, & k < \varphi \end{cases}$$

Where, φ is the time when fault happen.

From the previous contents, each sub-filter is the Kalman filter, the residuals is:

$$r_k = Z_k - H_k \hat{X}_{k,k-1} \tag{7}$$

Where predictive value $\hat{X}_{k,k-1}$ is

$$\hat{X}_{k,k-1} = \Phi_{k,k-1}\hat{X}_{k-1} \tag{8}$$

When no fault, the residual r_k of Kalman filter is a zero-mean Gaussian white noise, and its variance is:

$$A_k = H_k P_{k,k-1} H_k^T + R_k \tag{9}$$

When fault, the residual r_k is not equal to zero. Therefore, it can determine whether the system is faulty though testing the mean of residual r_k.

For the following binary hypothesis to r_k:

$$H_0: \text{ no fault } \quad E\{r_k\} = 0, E\{r_k r_k^T\} = A_k$$

$$H_1: \text{ fault } \quad E\{r_k\} = \mu, E\{[r_k - \mu][r_k - \mu]^T\} = A_k$$

So, fault detection function is:

$$\lambda_k = r_k^T A_k^{-1} r_k \tag{10}$$

And $\lambda_k \sim \chi^2(m)$. m is the dimension of measure Z_k. The criteria of determine failure is:

$$\left.\begin{array}{l} \text{if } \lambda_k > T_D, \text{ fault} \\ \text{if } \lambda_k \le T_D, \text{ no fault} \end{array}\right\} \tag{11}$$

Where pre-set threshold T_D determined by the false alarm rate P_f. In this paper, $T_D = 13.82$.

For multi-system navigation study, GPS can be divided into GPS1 and GPS2, the divided method as follows: According to the principle of GPS positioning, four satellites can determine the point of the three-dimensional coordinates in the same place at the same time. With the increase in the number of spatial positioning satellites, four or more satellites will be observed in the same place at the same time. Assuming there are five satellites, (x_1, x_2, x_3, x_4) and (x_1, x_2, x_3, x_5) is two positioning signals, they are called GPS1 and GPS2. And such a choice is not unique, so that there is following simple guidelines. Guideline one: If GPS1 fault but GPS2 normal, navigational star x_4 fault; If GPS2 fault but GPS1 normal, navigational star x_5 fault; If GPS1 and GPS2 all fault, receiving machine fault.

Thus, a navigation system consists of three integrated navigation system is established: SINS, GPS1, GPS2. Now a fault-tolerant integrated navigation system is designed, shown in Figure 2.

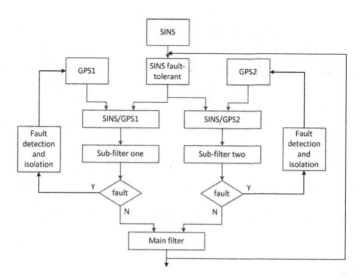

Fig. 2. Fault-tolerant system architecture of SINS/GPS1/GPS2

Guideline two:

If SINS/GPS1 fault, navigation system GPS1 fault;
If SINS/GPS2 fault, navigation system GPS2 fault;
So the following method is used for the fault detection of sub-filter.

If the subsystem SINS/GPS1 failure (corresponding to first sub-filter), then $\hat{X}_1$ is not correct, and therefore it can't be entered into the main filter, at this moment, the entire estimate of system error is:

$$\hat{X}_g = \hat{X}_2 \tag{12}$$

And the estimate of the output of failure subsystem is:

$$\hat{Z}_1 = H_1 \hat{X}_g \tag{13}$$

Similarly, SINS/GPS2 is also the case.

4 Simulation

A theoretical flight path is designed, the entire flight include the initial flight, takeoff, level flight, climbing stage, turning stage and final stage in level flight. Then the third direction in GPS2 sudden join the hard faults of 500m in 50s, and assuming that the signal is back to normal when the 100s. Without troubleshooting is shown in Figure 3. Where, the mean square error of the third direction is 146.86m. With troubleshooting is shown in Figure 4. Where, the mean square error of the third direction is 5.06m. From the figure we can clearly see that the fault-tolerant technology is effective and feasible.

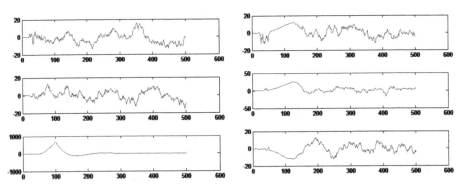

Fig. 3. The Position error without Troubleshooting

Fig. 4. The Position error with Troubleshooting

5 Conclusion

This paper presents a new fault detection structure and quantitative detection algorithm based on Federal Kalman filter, and simulation experimental results show that the fault-tolerant algorithm can detect the fault effect and little change in filtering accuracy.

Acknowledgments. This paper is supported by the foundation items: Project supported by Natural Science Foundation of China (Grant No. 51169007), Science & Research Program of Yunnan province (No. 2010DH004 & No. 2011CI017), Foundation of Kunming University of Science and Technology (No. KKZ3200903006).

References

1. Fu, M., Deng, Z., Yan, L.: Kalman filtering theory and its application in navigation system, 2nd edn. Science Press, Beijing (2010)
2. Pearson, J.D.: Dynamic decomposition techniques in optimization methods for large-scale system. McGraw-Hill, New York City (1971)

3. Carlson, N.A.: Federated filter for fault - tolerant integrated navigation systems. In: Proc. of IEEE PLANS 1988, Orlando, FL, pp. 110–119 (1988)
4. Loomis, P.V.W., Carlson, N.A., Berarducci, M.P.: Common Kalman filter: fault - tolerant navigation for next generation aircraft. In: Proc. of the Inst. of Navigation Conf., Santa Barbara, CA, pp. 38–45 (1988)
5. Qin, Y., Zhang, H., Shuhua, W.: Kalman filtering and navigation principles. Northwestern polytechnical university press, Xi'an (1998)
6. Liu, Z., Chen, Z.: A new fault detection structure and algorithm based on federal filter. Journal of Beijing University of aeronautics and astronautics 28(5), 550–554 (2002)

A New Formal Communication Protocol Model

ShengWen Gong

Qingdao University of Science & Technology, SongLing Rd. 99, 266071 Qingdao, China
qd_gsw@163.commail

Abstract. This paper presents the formal protocol model: SP-EFSM. We first present the classic Finite State Machine (FSM) model based on which SP-EFSM is defined; then the extension is introduced followed by a discussion of the relationship between two models. Finally a section is devoted to how communication protocol is modeled using SP-EFSM model.

Keywords: Finite State Machine, Communication Protocol, SP-EFSM.

1 Introduction

The Finite State Machine Model. FSM has been widely used in system specification of various areas, including network protocols, high level software design, real-time reactive systems, and logical circuits [1]. An FSM contains a finite number of states and produces output on state transitions after receiving an input symbol. It is often used to model control portion of a protocol. Validation and testing problems are well studied [2-4] for FSM.

Definition 1: A deterministic Finite State Machine (FSM) is a six-tuple $M = < S, s_{init}, I, O, \delta, \lambda >$, Where :

S is a finite set of states of size N ;

s_{init} is the initial state;

I is the finite input alphabet of size P ;

O is the finite output alphabet of size Q ;

$\delta : S \times I \rightarrow S$ is the state transition function;

$\lambda : S \times I \rightarrow O$ is the output function.

Note that state transition function is defined to be deterministic. Also, if δ and λ are total functions then the machine is complete otherwise the machine is partial.

The Extended Finite State Machine Model. Despite of its proven usefulness, FSM is not sufficient to model in an efficient way the data intensive protocols (such as TCP) which are very common in today's Internet. Extended FSM (EFSM) enriches FSM with variables as part of the state, and transitions with guard and actions. In general EFSM has the same power as Turing Machine and therefore can handle any complex protocols well, while when in practice we consider only variables with finite domains, EFSM is equivalent as FSM.

D. Jin and S. Lin (Eds.): Advances in FCCS, Vol. 2, AISC 160, pp. 629–634.

Each EFSM has an equivalent FSM representation, called its reachability graph, which takes the combination of EFSM state and variable values as state, and inherit all the transitions without guards and actions (I/O only). Because of the well known state explosion problem, reachability graph of any nontrivial EFSM will be large despite the invention of online minimization algorithms [5]. The definition of EFSM is as follows and we defer the discussion of reachability graph into next section together with the proposed new model.

Definition 2: An Extended Finite State Machine (EFSM) is a quintuple $M =< S, s_{init}, I, O, X, T >$, Where:

- S is a finite set of states;
- s_{init} is the initial state;
- I is the finite input alphabet of size P;
- O is the finite output alphabet of size Q;
- X is a vector denoting a finite set of variables with default initial values;
- T is a finite set of transitions. For each $t \in T$, $t =< s, s', i, o, p(x,i), a(x,i,o) >$ is a transition where s and s' are the start and end state, respectively; i and o are the input/output symbols; $p(x,i)$ is a predicate, and $a(x,i,o)$ is an action on the current variable values and input/output symbols.

2   The SP-EFSM Model

For purpose of modeling security protocols we further extend EFSM in the following ways.

- Parameterize input and output symbols in order to model the rich formats of modern protocol messages. Consequently the predicates and actions in the EFSM can refer to the input/output parameters;
- Add symbolic protocol message language with cryptographic primitives such as encryption/decryption/hash function. Both state variables and input/output parameters have value domain defined by this language. This extension is useful in order to cope with the rich semantics of security protocols.

Protocol Message. We define the security protocol message type as follows. First, there are three atom types: Int, Key and $Nonce$. A value of type Int is a non-negative bounded integer; a value of type Key ranges over a finite set K of keys and a value of type $Nonce$ ranges over a finite set N of nonces. K and N are not necessarily disjoint. Some keys are generated fresh and treated as nonce. Key type contains both symmetric keys and asymmetric key pairs. For the latter we use ku to represent a public key and kr for a private key. We treat Key and

Nonce as symbols in the sense that we are not concerned with their numerical values. In many practical protocols it is the case that some atoms are calculated from others, for example in SSL protocol the encryption key block is calculated using the master secret, which is in turn calculated from the pre-master secret and two nonces. We explicitly model such dependency relationship using a set of Derivation Rules. A derivation rule dr_i is of the form $dr_i = < N_{s1}, N_{s2}, N_{s3}, ..., N_{sk}) \rightarrow N_d >$, which means N_d could be calculated using $N_{s1}...N_{sk}$ together. The rules could be graphically represented by an AND-OR graph.

A protocol message is recursively defined as:

- An atom;
- Encryption of a message with a key;
- Secure Hash of a message;
- Concatenation of messages.
- A message can be represented by a string. For example $E(k_b,(k_a,H(n_a)))$ is a message that is formed by encrypting the concatenation of k_a and the hash of n_a with another key k_b. Given $A = < K, N >$, a set of keys and nonces, denote $L(A)$ as the message type and the set of messages formed using atoms in A. $L(A)$ is obviously infinite, and even if we restrict the number of atoms that a message contains, its size is exponential. There are some basic operations defined on message type. Let msg be a message, function $Elem(msg,i)$ calculates the i-th component in msg, and $D(k,msg)$ returns m' iff. $msg = E(k,m')$ when k is a symmetric key or $msg = E(k',m')$ when (k,k') is an asymmetric key pair. Both functions are partial and they are undefined for messages with incompatible formats. Define function $Encl : L(A) \rightarrow L(A)$ as for $\Omega \subseteq L(A)$, $Encl(\Omega)$ is the enclosure of Ω under functions $Elem()$, $D()$ and $E()$.

SP-EFSM Model. Now we can define a Symbolic Parameterized Extended Finite State Machine (SP-EFSM) model that uses finite (length-restricted) protocol message set $L(A)$ as input and output alphabet.

Definition 3: A Symbolic Parameterized Extended Finite State Machine (SP-EFSM) is a 7-tuple $M = < S, s_{init}, A, I, O, X, T >$, where:

- S is a finite set of states;
- s_{init} is the initial state;
- A is a finite set of atoms with certain derivation rules, and $L(A)$ is the set of messages formed using atoms in A;

- $I = \{i_0, i_1, ..., i_{P-1}\}$ is the input alphabet of size P; each input symbol i_k $(0 \le k < P)$ contains a parameter $\pi(i_k)$ of type $L(A)$.

- $O = \{o_0, o_1, ..., o_{Q-1}\}$ is the output alphabet of size Q; each input symbol o_k $(0 \le k < Q)$ contains a parameter $\pi(o_k)$ of type $L(A)$.

- X is a vector denoting a finite set of variables of type $L(A)$ with default initial values;

- T is a finite set of transitions; for $t \in T$, $t = < s, s', i, o, p(x, \pi(i)), a(x, \pi(i), \pi(o)) >$ is a transition where s and s' are the start and end state, respectively; $\pi(i)$ and $\pi(o)$ are the input/output symbols parameters; $p(x, \pi(i))$ is a predicate, and $a(x, \pi(i), \pi(o))$ is an action on the current variable values and parameters.

The meaning of predicate and action follows the tradition of state machine theory. There is an implicit semantic notion of protocol session in the model: A session variable SessionID is initialized with 0 and incremented each time the machine leaves state S_{init}. All atoms in set N are associated with a specific session and the derivation rules apply to atoms in the same session. In SP-EFSM we use parameterized input and output symbols to model the critical content of data packet. For instance, TCP data packets carry Sequence Number and Acknowledgement Number as parameters. Same all variants of EFSM, SP-EFSM model has the same computing power as Turing Machine, and it can be used to appropriately model network protocols.

3   Communication Protocol Specification

A protocol contains at least two principals. Each principal is identified uniquely. We use SP-EFSM to specify a principal in the communication protocol.

Definition 4: A communication protocol component is specified by a SP-EFSM with the following constraints: (1) It contains a variable id: the unique identifier of this component within the protocol system. (2) The input/output (I/O) alphabet is each divided into two subsets: internal (interaction within the component) and external (interaction with another component). External input (output) symbols carry an extra parameter to denote the message receiver (sender). (3) No transition contains both an external input and an external output.

According to (2) we denote an external input message received from M as $M_a?i$ and an output to M_a as $M_a?o$. The semantics of message sending/receiving follow the typical synchronous communication model such as in CSP: the I/O is executed as a rendezvous and simultaneously.

Definition 5: A communication protocol is specified as a set of protocol components: $M_{spec} = \{M_1, M_2, ..., M_C\}$, where each component has a unique id, and that they all share the same message type L.

Each component machine M_k represents a principal in the protocol system. It is possible that two machines are identical except for the identifiers, meaning there are symmetric peers in the protocol. Naturally we require that the sender/receiver parameter in Definition 4 is within range [1..C]. A trace from the specification is a set of ordered input/output events generated by the component machines. From the traces, an external event corresponds to a message exchange between two component machines, and they are observable to the network monitor. We define an observable trace (message trace) as a sequence of messages exchanged among components.

Definition 6: A message trace from a protocol M_{spec} with message type L is

$$tr = \{< s_1, r_1, msg_1 >, < s_2, r_2, msg_2 >, ..., < s_k, r_k, msg_k >\}$$, where the

messages are sequentially generated by M_{spec} with no other messages in between, and each triple contains the identifier of the sender machine s_i, receiver machine r_i and the message msg_i in L. Denote $Msg(tr)$ as the set of messages appeared in tr. Denote all message traces M_{spec} generates as $TR(M_{spec})$.

As an example, we show how the classic Needham-Schroeder-Lowe [6] two-way authentication protocol is modeled using SP-EFSM. There are two components machines for the initiator and responder, with id 0 and 1, respectively. Each of the components possesses a key pair and two nonces for each session (assuming there could be 3 valid identities). I/O symbols $\{Ask, Rpl, Cfm\}$ represent three types of message exchanged in the protocol. The message trace that leads to a successful run of the protocol is

$$< \{M_0, M_1, Ask[1, E(KU1, N01, 0)]\} >, < \{M_1, M_0, Rpl[0, E(KU0, N01, N10, 1)]\} >,$$
$$< \{M_0, M_1, Cfm[1, E(KU1, N10)]\} > .$$

4 Summary

This paper is concluded with a note on applying SP-EFSM and its simplified ancestors (including PEFSM, EFSM and FSM) in practical problems. SP-EFSM is most general and most powerful form in the family; however it is not always the appropriate one. For a particular protocol or problem, simpler models might be preferable. Furthermore, certain algorithms are not directly applicable on SP-EFSM, and we need to transform it to an equivalent FSM first.

References

1. Lee, D., Yannakakis, M.: Principles and methods of testing finite state machines - A survey. Proceedings of the IEEE, 1090–1123 (1996)
2. Lee, D., Chen, D., Hao, R., Miller, R.E., Wu, J., Yin, X.: A formal approach for passive testing of protocol data portions. In: 10th IEEE International Conference on Network Protocols (ICNP 2002), pp. 122–131. IEEE Computer Society (2002)
3. Lee, D., Netravali, A.N., Sabnani, K., Sugla, B., John, A.: Passive testing and its applications to network management. In: Proc. ICNP (1997)
4. Lee, D., Sabnani, K.: Reverse engineering of communication protocols. In: Proc. ICNP, pp. 208–216 (1993)
5. Lee, D., Yannakakis, M.: Online minimization of transition systems. In: STOC 1992: Proceedings of the Twenty-Fourth Annual ACM Symposium on Theory of Computing, NY, USA, pp. 264–274 (1992)
6. Needham, R., Schroeder, M.: Using encryption for authentication in large networks of computers. Communications of the ACM, 993–999 (1978)

The Quaternion Theory and Its Application on the Virtual Reality

Jianxin Gao, Hongmei Yang, and Yang Xiao

Hebei United University, Tangshan, 063009, China
gjx@heuu.edu.cn, yanghongmei@heuu.edu.cn, xy32@vip.sina.com

Abstract. Virtual reality simulation technology is widely used in robotics research, medical research, animation and many other fields.In the paper, a new interpolation technique based on the virtual human information was brought forward. The algorithm would automatically interpolate the information about the translational motion and rotation. And then the virtual human was animated more reality.

Keywords: Virtual Human, Human Body Model, Computer Graphics, Quaternion matrix.

1 Introduction

Virtual human animation technology is to redirect the three-dimensional (for short 3D) motion information we got to the 3D virtual human model, which makes the virtual human model to reproduce the action in computer technology and feedbacks the human motion 3D information.

People have done a lot to get the 3D information needed for driving virtual human motion in prior studies. But different degree problems exist in the three methods. The main problems are : that the manual point-by-point labelling method had the relatively large workload and long work period; automatic matching method often produces deviation and can not obtain the best matching results when the animation frames are more; motion capture methods (i.e. the method that paste sensors in each human joint point, and then automatically capture motion data), as a result of expensive and easily disturbed by the external environment and other factors, is not ideal [1].

Therefore, an interpolation method based on virtual human motion information were proposed in this article. This algorithm only needs to give 3D motion information of human body model in some animation key frames. Then the algorithm automatically interpolate between the transition information of frames. The algorithm use spline key frame interpolation method for the translational location shifts and quaternion interpolation method for the rotation angle variation.

2 Virtual Body Structure Analysis

2.1 Virtual Body Structures

Human motion complete by muscle contraction applied bones for the rod and bone connections for the fulcrum. The skeleton connected by bones forms the human

D. Jin and S. Lin (Eds.): Advances in FCCS, Vol. 2, AISC 160, pp. 635–640.

support, which support the soft tissues of the human body and the weight of the whole body. The muscle pulls skeleton rotating around the joints, so that the human body produces various sports [2]. All the limbs and even the whole body movements such as people's sitting, lying, walking are composed of the body base translation and the various rotary motion between the limbs. It is not generally a simple combination, but a coordinated motion of hands, arms, legs, feet and trunks in a complex way. In order to better describe the motion of human body, it is necessary to know the static description and the correlative motion information of the virtual human. One of the more commonly used method to solve the above problems is to use a virtual joint model, such as shown in figure 1. At the same time, we need to establish a virtual human tree structure, as shown in figure 2.

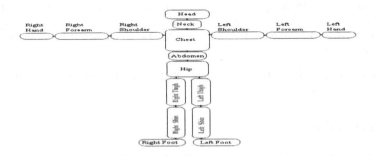

Fig. 1.

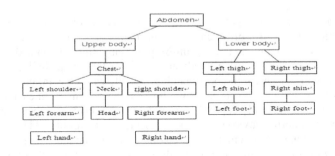

Fig. 2.

The illustrations show that: the head, left hand, right hand, left foot and right foot which are the ends of the tree, can move separately. Their parent nodes which are the neck, left forearm, right forearm, the left skin and the right skin, move with the son nodes. The limbs movement of a node will drive the movements of all its offspring nodes so as to make the virtual human motion according with the human movement.

2.2 Virtual Human Motion Control Technology

In order to make the virtual human animation more naturally and true, many techniques have been developed in computer animation. The more traditional motion control method is the parameterized key frame technology.

2.3 The Parameterized the Key Frame Technology

The parameterized key frame technology is the earliest method for virtual human motion control. In its early production process, the key frames are firstly designed by the senior animator, then the intermediate frames designed by the assistant animator. With the development of computer graphics, interpolation replaced the artists who design the intermediate frames. Intermediate frames can be generated by computer automatically. Any of the parameters for the interpolation effect of movement are so called parameter key frames [3].

2.4 The Interpolation Algorithm

Interpolation algorithm in the realization of generating the final animation by key frame technology has the very important status. The more simply interpolation method is linear interpolation method. But the animation produced by this method often came forth a rapid change. The reason for this is mainly that the movement object velocity is not continuous. In order to correct this phenomenon, spline interpolation method is used for the virtual human position changes and quaternion interpolation method for the human body joint direction changes in this paper.

2.5 Quaternion Interpolation

Set Q the four-dimensional vector space of a real number field. Its orthogonal basis

$(1,0,0,0)$、$(0,1,0,0)$、$(0,0,1,0)$、$(0,0,0,1)$ were expressed respectively by $e, i,$

j, k. The elements of Q are $q = [S, V] = a_0 e + a_1 i + a_2 j + a_3 k$. The scalar part

is $S = a_0$, and the vector part is $V = (a_1, a_2, a_3)$. q is called a quaternion [6].

Quaternion operation meets the addition exchange rule and addition combination rule. But the quaternion multiplication does not meet the exchange rule. In particular, the rotation θ around the unit axis n can be expressed as $q = [\cos\frac{\theta}{2}, n\sin\frac{\theta}{2}]$.

1. Transformation from Euler angle to quaternion

Located the Euler angles rotated around the x, y, z axis respectively as $\theta_1, \theta_2, \theta_3$ and the corresponding quaternion as $q_{\theta_1}, q_{\theta_2}, q_{\theta_3}$. By the type $R_q(p) = qpq^{-1}$ corresponds to quaternion:

$$q = q_{\theta3}q_{\theta2}q_{\theta3} = [\cos\frac{\theta_3}{2}, (0,0,\sin\frac{\theta_{31}}{2})][\cos\frac{\theta_2}{2}, (0,\sin\frac{\theta_2}{2},0)][\cos\frac{\theta_1}{2}, (\sin\frac{\theta_1}{2},0,0)]$$

Set unit quaternion $q = [\cos\frac{\theta}{2}, n\sin\frac{\theta}{2}] = [w, (x, y, z)]$, so

$$w = \cos\frac{\theta_1}{2}\cos\frac{\theta_2}{2}\cos\frac{\theta_3}{2} - \sin\frac{\theta_1}{2}\sin\frac{\theta_2}{2}\sin\frac{\theta_3}{2}$$

$$x = \cos\frac{\theta_1}{2}\cos\frac{\theta_2}{2}\sin\frac{\theta_3}{2} + \sin\frac{\theta_1}{2}\sin\frac{\theta_2}{2}\cos\frac{\theta_3}{2}$$

$$y = \cos\frac{\theta_1}{2}\sin\frac{\theta_2}{2}\cos\frac{\theta_3}{2} + \sin\frac{\theta_1}{2}\cos\frac{\theta_2}{2}\sin\frac{\theta_3}{2}$$

$$z = \sin\frac{\theta_1}{2}\cos\frac{\theta_2}{2}\cos\frac{\theta_3}{2} + \cos\frac{\theta_1}{2}\sin\frac{\theta_2}{2}\sin\frac{\theta_3}{2}$$

2. Transformation from quaternion to rotation matrix

According to the formula $R_q(p) = [0, (\cos\theta)r + (1 - \cos\theta)(n \cdot r)n + (\sin\theta)n \times r]$, the second rotation matrix determined by the quaternion $q = [w, (x, y, z)]$ is

$$M = \begin{bmatrix} 1 - 2y^2 - 2z^2 & 2xy - 2wz & 2xz + zwy \\ 2xy + 2wz & 1 - 2x^2 - 2z^2 & 2yz - 2wx \\ 2xz - 2wy & 2yz + 2wx & 1 - 2x^2 - 2y^2 \end{bmatrix}, \quad \text{expressed as}$$

$$\begin{bmatrix} M_{00} & M_{01} & M_{02} \\ M_{10} & M_{11} & M_{12} \\ M_{20} & M_{21} & M_{22} \end{bmatrix}.$$

3. Quaternion spherical linear interpolation

As a rotation transformation is mapped to a unit quaternion, the whole rotation group should be mapped to a unit sphere of the quaternion space. Thus the intermediate value curve after the interpolation should also be in this sphere. Considering the simplest case, two interpolations direct towards the key frame. If directly using the linear interpolation, the interpolation curve is not in the sphere. As a result the middle of the movement speeds. In order to ensure stable rotation, spherical linear interpolation must be adopted, that is, interpolation curve goes through the key-frame along the major arc.

Set q_1 and q_2 as unit quaternion, $u \in [0,1]$. The spherical first interpolation from q1 to q2 is $Slerp(q_1,q_2;u)$, it has the following formula:

$$Slerp(q_2,q_2;u) = \frac{\sin(1-u)\theta}{\sin\theta}q_1 + \frac{\sin u\theta}{\sin\theta}q_2, \text{ in which } \cos\theta = q_1q_2$$

3 The 3D Animation Generating Method Adopted in the Paper

The 3D animation regeneration method refers to given other animation roles to the captured 3D motion information (also known as the movement redirection) so as to producing animation impression. As the 3D scene is constantly changing, we can not directly apply the 3D motion information expressed in the Cartesian coordinates to the virtual human body. It needs to transform the 3D coordinates motion information into the rotational and translational information.

Each key frame has a series of control points corresponding to other key frame control point. The methods used in this paper are:

1. First of all, find out relative coordinates of each joint point corresponding to the gravity position of the virtual human, to eliminate the effect about the motion displacement to virtual human joint motion. Find the key points of each body's center of gravity relative to the virtual location of the relative coordinates

2. Applying inverse dynamics knowledge, find out the rotation angle of each joint relative to its" parent" node.

3. According to the position change of virtual human body gravity at different time, we can interpolate values between virtual human spatial location using a spline key frame interpolation method.

4. For the initial position of each joint on any two motion key frame and the rotation angle of the joints got by inverse dynamics, we interpolate values used the method of quaternion interpolation, thereby forming the intermediate picture.

5. Render and continuously play the model calculated in each frame attitude according to the real situation, and we get the virtual human model animation.

6. On the basis of this, the program itself can save the formed virtual human animation into a AVI file. It provides the necessary preparations to the latter part of the video compositing operations.

4 Summary

The common methods of implementation virtual human animation are manual labeling method, motion capture method and automatic identification method. This paper presents a interpolation method based on virtual human motion information, so as to reduce the workload of manual labeling, and to avoid high cost of motion capture and to make up for accuracy error of automatic recognition. The algorithm can automatically interpolate values between the translation information and the rotation information of human motion, which can better implement the virtual human animation.

References

1. Ranchordas, A.: Computer vision and computer graphics: theory and applications. CCIS, vol. xv, p. 275. Springer, Berlin (2009)
2. Adobe Systems, Adobe Creative Suite 4 design premium. Classroom in a book, vol. xii, p. 322. Adobe Press, Berkeley (2009)
3. Cruz-Monteagudo, M., Borges, F., Cordeiro, M.N.: Jointly Handling Potency and Toxicity of Antimicrobial Peptidomimetics by Simple Rules from Desirability Theory and Chemoinformatics. J. Chem. Inf. Model (2011)

A Design of Intelligent Parallel Switching Power Supply Based on Automatic Load Distribution

Xuezhong Ai, Ying Wang, Junfeng Zou, and Renyu Liu

College of Information and Control Engineering, Jilin Institue of Chemical Technology,
Jilin, China
Aixuezhong2006@163.com,
{36001814,350546270,349999195}@qq.com

Abstract. This article describes an intelligent switching power supply design. This switching power supply bases on push-pull mode, and it can work in multiple parallel mode . Besides, it can also automatically adjust according to the proportion of current distribution setting .To achieve the functionality of input-output isolation and automatic load distribution ,we make use of intelligent and digital control method .The parallel power system uses Fieldbus mode to information exchange, C8051F410 chip on each module as the core, DAC as the actuator to control switching power conversion unit .When all those units work, this switching power supply can realize stable voltage output and automatic load distribution. The article also gives the power module's software process through analysis the circuits .In the end, we have conducted Compliance test for the switching power supply designed.

Keywords: Switching Power Supply, Current distribution, System on Chip, Fieldbus.

1 System Solutions

This Switching power supply module is shown in Figure 1.This module uses the C8051F410 chip as the control core for the intelligent power supply, SG3524 to form push-pull power converter as the power output circuit, the OPA2335 and the HCPL7840 linear optocoupler to design isolated current sensing circuit, RS485 interface circuit for data exchange between the power supply module, LCD display and key circuit to conduct man-machine interface operation. The C8051F410 chip system controls the SG3524 power output circuit by adjusting the DAC0 output, and the output current measurement circuit output current sampled and feedback to the C8051F410 SoC ADC0, which makes the voltage stable at 8V. LCD display circuit and key circuit work as the man-machine exchange interface, and RS485 interface circuit works as the data transmission channel between the power modules. Then though Calculated the output voltage by the software, we can achieve automatically asssigning the load on each power module.

D. Jin and S. Lin (Eds.): Advances in FCCS, Vol. 2, AISC 160, pp. 641–646.

Fig. 1. Block diagram of intelligent switching power supply

2   Theoretical Analysis and Calculation

The calculation for transformer primary winding N1 turns of Dual-excited switching power. Dual-excited switching power transformer primary winding N1 turns is calculated as function (1) shows [1,2]. In the function , N1 is the least of the transformer primary coil winding   N1 or N2 turns; S is the transformer's magnetic core area (unit: square centimeters); Bm is the maximum flux density of the transformer core (unit: gauss); Ui is the voltage applied across the transformer primary coil winding N1 in volts; τ equals to Ton, named as the turn-on time of the control switch ,referred to as the pulse width or the width of the power switch conduction time.

$$N_1 = \frac{U_i \tau}{2 S B_m} \times 10^8 \tag{1}$$

The calculation for transformer's primary and secondary turns of Push-pull switching power supply. The push-pull switching power supply transformer's primary and secondary turns is calculated as function (2) below. In the function, N1 is the transformer primary coil winding turns N1 or N2; N3 is the number of turns for the transformer secondary coil; Uo is the DC output voltage; Ui is the DC input voltage; D is the duty cycle.

$$n = \frac{N_3}{N_1} = \frac{2 U_o (1 - D)}{U_i} \tag{2}$$

3   Circuit Design

The basic peripheral circuits on C8051F410 chip system. C8051F410 external circuit is shown in Figure 2 [3,4,5]. C8051F410 chip system includes JATG interface, power supply decoupling, reference filter, power-on reset, button interface, LCD interface and other basic peripheral circuits. Through DAC0 module, C8051F410 chip system outputs current signal, which work as the control signal of SG3524 power output circuit. The control circuits also test current output value through the ADC integrated on C8051F410 chip. And, buttons and LCD work as human-machine interface.

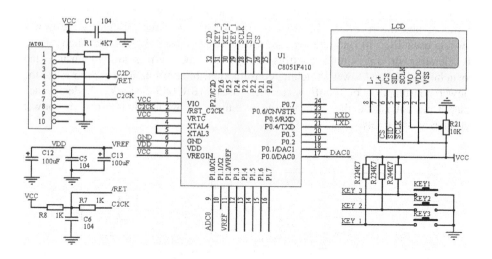

Fig. 2. The basic peripheral circuits on C8051F410 chip system

The DC-DC power output circuit based on SG3524. DC-DC power output circuit is shown in Figure 3 [6]. SG3524 ,a switching power tubes which controls power conversion circuit through the PWM signal , is the core of the power output circuit. Its pin 6 and 7 connect with 2.2K resistor and 0.01uF capacitor respectively, which determine that the oscillator frequency of approximately is 47KHZ. When the chips work on the push-pull output mode, the output pulse duty cycle ranges from 0% to 45%, and the frequency of oscillator pulse reduces to the half. So, resistance RT and capacitor CT determine the frequency of the power , such as type (3) below.

$$f = 1 / (R_T \cdot C_T)$$ (3)

Resistors R9 and R10 provide the voltage, and introduces inverting input of error amplifier through pin 2 ,;the negative terminal connect to feedback voltage controlled by MCU ; Pin 4 and 5 connect over-current sampling resistor, which determine the output power limit of the current value. In order to achieve frequency compensation, pin 9 Connect capacitance (C8) and resistor (R14) before to ground. Pins 11 and 14 connect with the gate of external power switch Q1, Q2 via switching diodes D1 and D2 respectively. R13, R14, Q3 and Q4 constitute the fast turn-off control circuit for the gate of FET IRF540, which aim at increasing the driving speed for the power switch.

Switching sides of the transformer is connected to the Schottky rectifier MBR2045, which is dual Schottky diode and the reverse voltage of is 45V, the maximum current allowed of is 20A, which form a full-wave rectifier circuit. And another group of secondary works as an auxiliary power supply, which provides front-end power for the isolated current sensing circuit via a LM7805 regulator.

In the switching power supply circuit, the efficiency depends on the power switches and the rectifiers. In order to improve the working efficiency, the power switch uses IRF540 and the typical resistance defines only 70mΩ. Then taking accelerating driving circuit controls the power tube on-ff status to ensure that switching speed and lower switching consumption. In addition, in order to reduce the power consumption of rectifiers, rectifier uses MBR2045 Schottky diodes. When output current is 4A, the voltage drop is only 0.3V, what can improve the power efficiency effectively.

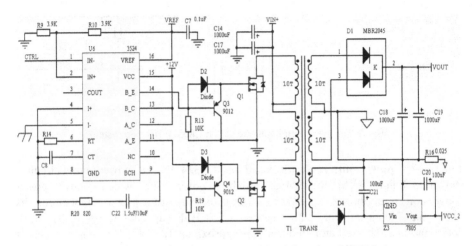

Fig. 3. The DC-DC power output circuit based on SG3524

Output current detection circuit. Output current detection circuit is shown in Figure 4.For the isolation between the power input and output, the output current detection circuit uses the linear optocoupler chip of 8-fold amplification, HCPL7840. pin 2 and pin 3 of HCPL7840 connect to the ends of output current sampling resistor respectively, and the differential output VOUT+ (pin 7) and VOUT- (pin 6) points were connected to the inputs of three times magnification differential amplifier, then the amplified voltage is sent to the ADC0 on C8051F410 SoC .The Current sampling resistor is R16 (Figure 3),when outputing current I, then the ADC0 output voltage of the output current detection circuit   is calculated as function (4), (5), (6) shows:

$$V_{OUT+} - V_{OUT-} = 8(V_{IN+} - V_{IN-}) \tag{4}$$

$$ADC0 = \frac{R_2}{R_6} \times (V_{OUT+} - V_{OUT-}) = 3 \times (V_{OUT+} - V_{OUT-}) \tag{5}$$

$$ADC0 = 3 \times 8 \times (V_{IN+} - V_{IN-}) = 24 \times Rs \times I \tag{6}$$

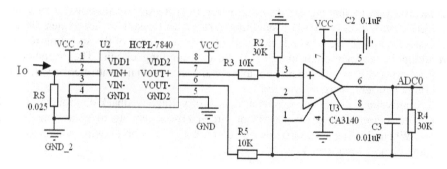

Fig. 4. Output current detection circuit

4 Experimental Testing and Analysis

Test Method. Two switching power supply modules of experiment output 8 Volts and its rated power is 16W, connecting the test circuit as Figure 5, we can get the test data of table 1.

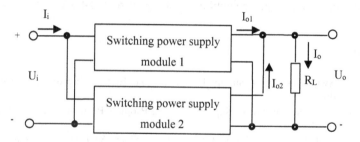

Fig. 5. The test pattern of two power supply modules forming parallel system

Table 1. Two modules assign current automatically according to I1: I2 = 1:1

Input current(A)	Output voltage(V)	I_{o1} (A)	I_{o2} (A)	I_o (A)	U_o (V)	P
0.47	23.8	0.25	0.26	0.51	8.08	—
0.65	23.8	0.52	0.53	1.05	8.07	—
0.86	23.8	0.74	0.74	1.48	8.06	—
1.09	23.8	0.94	0.93	1.87	8.04	—
1.55	23.8	1.49	1.51	3.00	8.02	—
1.92	23.8	2.02	1.99	4.01	8.01	70.2%

Data Analysis. Output current distribution error: There are two reasons for the output current distribution error: The sampling resistor will result in heat when flowing excessive output current, which causes the temperature drift of the sampling resistor, resulting in the measurement error of current sampling circuit; The overshoot of control process is not eliminated completely and the residual Control error will cause the output current distribution error.

Output voltage error: The temperature drift of voltage feedback control circuit will cause the output voltage measurement circuit errors, leading to controling output error; When flowing excessive output current, the resistor of wire will cause the output voltage drop on the load.

5   Conclusion

From the test data table we can get several conclusions following. When Load ranged from 0 to 4.0A output, power supply output voltage Uo changes between the 8.01 ~ 8.08V and can distribute the current as the proportion seted automatically. When the output power is 32W, the efficiency of the power can reach 70.2%. In addition, the power also designs load short-circuit protection and automatic recovery and the protection threshold current is 4.5A.

References

1. Pressman: Switching Power Supply Design Second Edition. Electronic Industry Press, Beijing (2005)
2. Maniktala, S.: Switching Power Supplies A to Z. People's Posts and Telecommunications Press, Beijing (2008)
3. Ai, X.-Z., Ji, H.-C., Zhang, Y.-L.: C8051F410 Pian Shang Xi Tong de Ni Bian Dian Yuan She Ji. Control and Instrumentations in Chemical Industry 37(8), 63–65 (2010)
4. Information on, http://www.xhl.com.cn
5. Ai, X.-Z., Jin, B.-T.: Ju You An Quan Jian Guan de Li Dian Chi Kuang Deng Zhi Neng Chong Dian Ji Shu De Yan Jiu. Control and Instrumentations in Chemical Industry 37(2), 100–101 (2010)
6. Zhai, Y.-W., Ai, X.-Z.: Electronic design and practice. China power press, Beijing (2005)

The Improving of Spread Spectrum Video Digital Watermarking Algorithm

Shiwei Lin[1] and Yuwen Zhai[2]

[1] College of Information and Control Engineering, Jilin Institute of Chemical
Technology, Jilin, China
13704406003@126.com
[2] Mechanical & Electrical Engineering College, Jiaxing University, Jiaxing, China
wanglei_new814@126.com

Abstract. Digital watermarking technology is that directly embedded a number of landmark information into the multimedia content through a certain algorithm, which does not affect the value and use of original content, and can not be aware of the perception system or note. MPEG-4 of the video structure is analyzed in the paper, we propose a spread spectrum-based video digital watermarking to improve the program, and gives application examples.

Keywords: spread spectrum, video digital watermarking, embedding algorithm, extraction algorithm.

1 Introduction

While digital watermarking technology has made rapid progress in recent years, but it focused on still images. The human eye has not been fully established including time-domain masking properties and more accurate visual model of, the video watermarking technology still lags behind the development of image watermarking. The other hand, attacking form of video watermarking appear, it made some unique requirements of video watermarking different from the still image watermark.

MPEG-4 of the video structure [1] is analyzed in the paper, we propose a spread spectrum-based [2] video digital watermarking to improve the program, and gives application examples.

2 Introduction of Video Watermarking Technology

2.1 Introduction of Digital Watermarking Technology [3-6]

In order to increase the attacker more difficult to remove the watermark, the watermark is currently making most of the program are used in cryptography encryption system to strengthen, in the watermark embedding, extraction using a key, or combination of several keys. The general method of Watermark embeds and extraction shown in Figure 1.

D. Jin and S. Lin (Eds.): Advances in FCCS, Vol. 2, AISC 160, pp. 647–651.
springerlink.com

2.2   The Several Aspects are Considered in Design of Video Watermarking Design

- Watermark capacity: the embedded watermark information must be sufficient to identify the multimedia content of the purchaser or owner.
- Imperceptibility: video data embedded in the digital watermark should be invisible or imperceptible.
- Robustness: the lower the video quality is not obvious of the watermark is removed is difficult under the condition that is not reduced obviously.

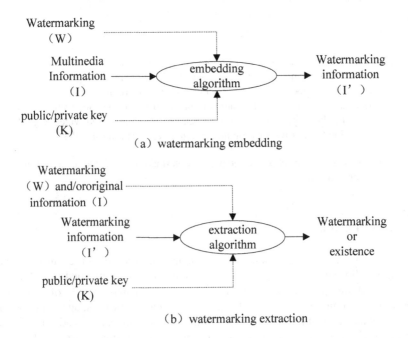

(a) watermarking embedding

(b) watermarking extraction

Fig. 1. The general method of Watermark embeds and extraction

- Blind detection: the watermark detection does not require the original video. The all the original video is saved is almost impossible.
- Tampering Tip: that the original data is tampered is sensitive to detect by the watermark extraction algorithm when multimedia content is changed.

2.3   The Scheme Selection of Video Watermarking

By analyzing the existing digital video codec system, MPEG-4 video watermark embedding and extraction program is divided into the following categories in current; it is shown in Figure 2.

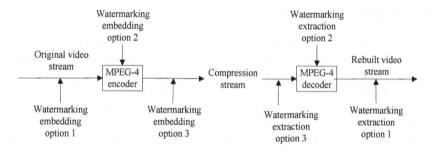

Fig. 2. Classification of MPEG-4 video watermarking

3   MPEG-4 Video Watermarking Implementation

Based on the above schemes, the improvement program is proposed for spread spectrum digital watermarking technology in MPEG-4 video coding system. The spread spectrum modulated watermark information embedded into the video stream IVOP (Intra Video Object Plane) in the chromo DC coefficients of DCT lowest. The scheme does not require full decoding, greatly reducing the computational complexity, improved real-time. At the same time the watermark is highly robust under the premise of the video.

3.1   Video Watermarking Algorithm and Implementation

In the MPEG-4 video, because IVOP the color sub-block DCT DC coefficient is always a presence in the video stream and a very robust argument, the watermark information by the program m sequence (longest linear feedback shift register sequence) embedded IVOP modulated chroma sub-block of DC DCT coefficients. This watermark does not affect the video quality in case of difficult to remove, so the robustness is strong enough. The program uses spread-spectrum approach to easily and efficiently detect the watermark, to resist various attacks and interference, security is good. Chroma key problem is that the DCT DC system is a visual system is very sensitive to the parameters, the program in the chroma DC coefficients of DCT watermark add the equivalent of adding small amounts of their interference, such interference must be below a certain threshold, the human eye's visual system to small changes in color video feel. After testing the watermark embedded in IVOP chroma DC coefficients of DCT potential energy to meet the minimum requirements.

3.2   Video Digital Watermark Embedding

Expansion of the pseudo-random sequence of length 255 (28-1), each watermark bit extensions through the pseudo-random sequence of modulation corresponding IVOP embedded into the DCT DC coefficient of the corresponding color (quantified, predicted the former), the lowest level, so does not affect the watermark information in the case of video effects is generally difficult to remove. Meanwhile, embedded in the lowest DC coefficient, the error caused is very small.

3.3   Video Watermarking Extraction

With the watermark information here inv_UDCij said chroma video IVOP DC DCT coefficients (before inverse quantization, DC, after calculating forecast) sequence; inv_ WMij that the watermark is detected extended modulation stream. Each IVOP chroma sub-block decoding to be an expansion in the modulation of the signal bit, each of 255 consecutive expansion modulated signal can be demodulated to get a bit watermark information, specifically as follows:

The original pseudo-random sequence with the same structure and be fully synchronized with continuous sequence of 255 to receive signals modulated extended sequence XOR, after a number of statistical operations recorded as OneCount. As the m-sequence autocorrelation function only two values (1 and -1 / (2n-1)), are two-valued autocorrelation sequences. Therefore, if the data has not been any attacks and interference, OneCount only two results: 255 or 0. When OneCount = 255, the resulting watermark bit is 1; when OneCount = 0, the resulting watermark information bit is 0. If the data is subject to attacks or interference, OneCount a variety of results. According to statistical analysis, when OneCount> 127, the resulting watermark bit is 1, and this 255 IVOP chrominance sub-block in (255-OneCount) sub-block attack or interference; when OneCount <127, the resulting watermark information bit is 0, and this sub-block 255 IVOP chrominance sub-blocks in OneCount attack or interference. In this way, you can count the total number of video IVOP chroma sub-block attack or interference, while strong to recover the original watermark information.

4   Result Analysis

The results showed that, m the length of the longer sequence, testing the better, but the watermark can be embedded in a corresponding reduction in the amount of information. The watermark is only embedded in the program video IVOP, do not modify PVOP and BVOP, of frame skipping and frame to remove offensive sound, because IVOP jump or can not be deleted. Also, because the watermark information is embedded in the DCT DC coefficient, while the DC coefficient of variation have a greater effect on the impact of the video, so take the watermark information embedded in the chrominance sub-block DCT DC coefficient of the lowest bit. This will not only make the watermark embedding computational complexity greatly reduced, as the MPEG-4 codec saves time, but also to achieve good visual effects, to the non-aware of. From a statistical point of view it would not increase the video stream. In addition, the watermark extraction without the original video. If the watermark information is not compromised, the program can accurately extract the watermark to the original full video; if the watermark is attacked, according to the nature of spread spectrum demodulator, the program can maximize the recovery of the original watermark information and the statistics are how many IVOP chromes sub-block attack.

References

1. Elbasi, E.: Robust multimedia watermarking: Hidden Markov model approach for video sequences. Turkish Journal of Electrical Engineering and Computer Sciences 18, 159–170 (2010)
2. Lee, M.-J., Im, D.-H., Lee, H.-Y., Kim, K.-S., Lee, H.-K.: Real-time video watermarking system on the compressed domain for high-definition video contents: Practical issues. Digital Signal Processing: A Review Journal 22, 190–198 (2011)
3. Chen, W.-M., Lai, C.-J., Chang, C.-C.: H.264 video watermarking with secret image sharing. In: IEEE Int. Symp. Broadband Multimedia Syst. Broadcast., BMSB (2009)
4. Yang, Y., Chen, G., Wang, Y.: A digital audio multi-watermarking algorithm based on improved quantization. In: Proc.-PACCS: Pac.-Asia Conf. Circuits, Commun. Syst. (2009)
5. Sleit, A., Abusharkh, S., Etoom, R., Khero, Y.: An enhanced semi-blind DWT-SVD-based watermarking technique for digital images. Imaging Science Journal 60, 29–38 (2012)
6. Cao, L., Men, C., Sun, J.: A double zero-watermarking algorithm for 2D vector maps. Harbin Gongcheng Daxue Xuebao/Journal of Harbin Engineering University 32, 340–344 (2011)

Medical Images Registration Method Integrating Features and Multiscale Information

Yang Yunfeng[1,2], Yang Hui[1], and Zhixun Su[2]

[1] School of Mathematical Science and Technology, Northeast Petroleum University,
Daqing, 163318, Peoples Republic of China
[2] School of Mathematical Sciences, Dalian University of Technology,
Dalian, 116024, Peoples Republic of China
yfyang1080@gmail.com, yanghui17012015@126.com,
zxsu@dlut.edu.cn

Abstract. Medical image registration is the key technology to the modern medical images and it is the basic issue of medical image fusion. A medical image registration method integrating features and multiscale information is proposed in the paper. Firstly, the edge feature images obtained by Canny operator are registered by Principal Axes method. Secondly, Harris corner detection method was used to extract the corners of the couple images. Then the corners of the couple images are registered by Iterative Closest Point Algorithm (ICP). Finally, the couple medical images were decomposed by wavelet transform, the mutual information is selected as similar measure, and only the low-frequency sub-images were selected as registration objects. Then, the registration process is begun from the coarse scale until the finest scale by Powell algorithm. Thus the medical images registration can be achieved through the above registration procedure. Experiment results show that this method not only can achieve a higher registration accuracy, but also can reduce the computational time greatly. The accurate registration results also can be achieved in the noisy environment.

Keywords: Registration, Edge feature, Multiscale, Mutual Information.

1 Introduction

Within the current clinical diagnosis, medical image is a vital component in the all kinds of diagnosis measures. Medical image can be used to obtain more information about patients, find the location of the tumor, monitor tumor's growing trend. All these works not only can help a clinical doctor to diagnose the disease accurately, but also can make a detailed treatment plan. Image registration technique is very important in these applications.

In general, the image registration method can be classed into approaches based on features [1,2,3] and approaches based on intensity information [8,9]. In addition, those approaches also can be classed into methods under space field and transform field. In the recent publications, wavelet transform is applied widely in the medical image registration [4,5,6]. The registration methods based on wavelet transform are the

D. Jin and S. Lin (Eds.): Advances in FCCS, Vol. 2, AISC 160, pp. 653–658.

approaches registering from rough registration to accurate registration, namely, registration is achieved through stepwise refinement process. The approaches based on features can't obtain ideal registration results in general for that the information used in the registration procedure is not enough. If more information used in the registration procedure, more accurate results could be obtained. But more information used in the registration process will reduce registration speed. In order to solve this problem, the intensity and feature information should be used reasonably. Based on this thought, a registration approach integrating the image's edge features, corners and multiscale information is proposed in the paper. The rough registration is achieved through the edge features and corner points. Then the accurate results can be obtained based on multiscale information by using Powell algorithm and mutual information.

2   Rough Registration Based on Features

A. The first rough registration based on edge features. The brain images can be seen as the rigid objects, and their edge features are clear. So the images' edge features are selected to accomplish the first rough registration. Canny operator is used to extract the couple medical images' edge in the paper. Canny operator was proposed in 1986, it can extract an image's edge clearly and accurately, and it has been widely used in the edge extraction of an image.

Set the coordinate of an image' edge pixel as $\{(x_i, y_i)|i = 1,2,\cdots N\}$, N denotes the number of pixel in the image's edge, then the center coordinate of it can be calculated by

$$\begin{cases} x = \dfrac{1}{n}\sum_{i=1}^{n} x_i \\ y = \dfrac{1}{n}\sum_{i=1}^{n} y_i \end{cases} \tag{1}$$

Where (x, y) denotes the image's center coordinate. Furthermore, the initial translation of the couple images can be calculated by

$$\begin{cases} \Delta x = x_f - x_r \\ \Delta y = y_f - y_r \end{cases} \tag{2}$$

Where (x_f, y_f) denotes the center coordinate of the floating image, (x_r, y_r) denotes the center coordinate of the reference image.

Based on expression (2), the rotation angle of the couple medical images can be calculated by the Principal Axes. Set the inertia matrix of an image's contour line is

$$I = \begin{pmatrix} u_{11} & -u_{12} \\ -u_{12} & u_{22} \end{pmatrix} \tag{3}$$

where $u_{11} = \sum_{i=1}^{N} (x_i - \overline{x})^2$, $u_{22} = \sum_{i=1}^{N} (y_i - \overline{y})^2$, $u_{12} = \sum_{i=1}^{N} (x_i - \overline{x})(y_i - \overline{y})$.

Based on the inertia matrix shows in (3), the long axes and the short axes of the couple medical images can be obtained for that they are the two eigenvectors of the inertia matrixes. Then, the included angles of the two long axes and the two short axes can be obtained respectively. The rotation angle of the two medical images is obtained by the average of the two included angle at last. Based on the initial translation value and the initial rotation angle, the first rough registration of the couple medical images is accomplished.

B. The second rough registration based on corner points. Although the first rough registration has been accomplished, there are also the larger errors between the couple medical images. In order to reduce the location error between the couple images, the feature points are used to achieve the registration further. Considering of the properties of Harris corner point extraction algorithm, it is selected to extract the corner points of the couple medical images. The main properties of Harris algorithm are accuracy, fast speed, and the ability of resisting noise. And it has been used in the image registration [2].

At this step, the corner points of the couple medical images which have been registered by the above step are extracted by Harris algorithm at first. Then, they are registered by ICP algorithm for that ICP algorithm is very effective to the registration of feature points.

The rough registration of the couple images has been accomplished after the step A and B stated above.

3   Fine Registration Based on Multiscale

A. The wavelet decomposition of an image. Wavelet transform has been widely used in the image processing after Mallat algorithm[7] was proposed for it 's local characters of time-frequency. An image can be seen as a 2D signal, so an image's wavelet transform can be obtained by the Mallat algorithm. In the wavelet frequency field, an image's edge feature information and detail information are distributed in high-frequency sub-images. When an image is decomposed by wavelet transform of k scales, $3k+1$ sub-images can be obtained, and they are $\{LL_k, HL_j, LH_j, HH_j\}$, $j = 1, 2, \cdots, k$. LL denotes the low-frequency sub-image, and   HL, LH, HH denote the high-frequency sub-images. An image's wavelet decomposition is showed in Fig.1.

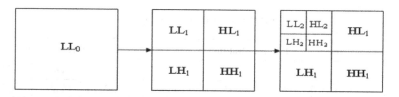

Fig. 1. The two level wavelet decomposition of an image

The low-frequency LL has the large part of energy of the initial image, and it's the approximation of the initial image, i.e. it contains the main information of the initial image. So the low-frequency sub-images in the wavelet decomposition are selected as the registration objects for that the more whole information of the initial image used in the registration procedure, the more accurate results can be obtained.

B. Registration based on the multiscale information. Mutual information (MI) is selected as the similar measure. MI is an important concept of information theory which is used to represent the statistics correlation of two sets of data which has been widely used in the image registration [8,9,10].

$$I(A,B) = H(A) + H(B) - H(AB) \tag{4}$$

Where $H(A) = -\sum_a P_A(a) \log P_A(a)$ denotes the entropy of image A, and the joint entropy is calculated by $H(AB) = -\sum_a P_A(a,b) \log P_A(a,b)$. The more large value of the MI is reached in the registration procedure, the more accurate registration results are obtained.

Powell algorithm is selected as the optimization search strategy. And in this registration procedure, the couple images are registered from the last level low-frequency sub-images to the first level low-frequency sub-images. And the couple initial images are selected as registration objects to accomplish the last step registration in the whole registration process.

4   Experiments and Results

The reference and floating images all consist of 256×256 and 8 bits grey-scale medical images. And all the images used in the experiments were supported by medical image data of Computational Geometry and Graphics Laboratory of Vanderbilt University. The couple images are all decomposed with two levels by wavelet transform. The experiment results are showed in Fig.2, Fig.3, Fig.4 and Fig.5. Fig.2 shows the registration result between two CT images, Fig.3 shows the result between two MR images, Fig.4 shows the result between CT images and MR images, and Fig.5 displays the result between two CT images under noisy environment. Table1 shows the parameters of the couple images before and after registration with the proposed approach in Fig.2, Fig.3 and Fig.4.

Table 1. Registration Parameters

	Parameters before registration			Parameters after registration			
	x	y	θ	x	y	θ	MI
Fig.2	15	20	15	14.94	19.97	15.00	2.53
Fig.3	15	20	15	14.91	19.95	15.01	1.99
Fig.5	15	20	15	14.96	19.95	15.07	1.67

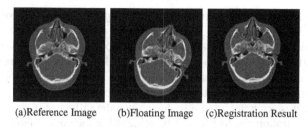

(a)Reference Image   (b)Floating Image   (c)Registration Result

Fig. 2. Registration Between CT Images

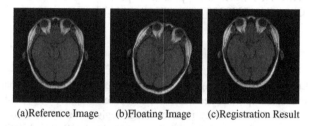

(a)Reference Image   (b)Floating Image   (c)Registration Result

Fig. 3. Registration Between MR Images

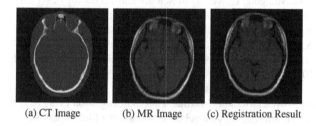

(a) CT Image   (b) MR Image   (c) Registration Result

Fig. 4. Registration Between MR and CT Images

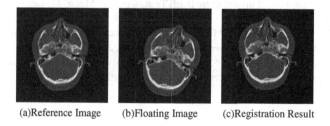

(a)Reference Image   (b)Floating Image   (c)Registration Result

Fig. 5. Registration under Noise Enviorment

5   Conclusion

A novel registration method of the medical images based on images' edge features, corner features and multiscale information is proposed in the paper. The images' features and multiscale information are used effectively and reasonably in the proposed method. Considering of the low-frequency sub-images of wavelet decomposition are the approximation sub-images of the initial image, so low-frequency sub-images are selected as registration objects. This not only can hold the whole characteristic of the initial image information, but also can cost the calculation. The experiments show that the proposed approach is an ideal medical image registration method. Because of Canny operator and Harris algorithm not only can be used to extract an image's edge and corner features accurately, but also can reduce the influence of noises effectively, the accurate registration results also can be achieved in the noisy environment.

Acknowledgment. This work is supported in part by the National Natural Science Foundation of China(No. U0935004, 60873181, 61173103).

References

1. Zhou, Y., Luo, S.: Medical Image Registration Based on Mutual Information of Feature Points. Journal of Computer-aided Design & Computer Graphics 14(7), 654–658 (2002)
2. Lv, X., Duan, H.: Multimodality medical image registration by mutual information and Harris corner detecor. Computer Engineering and Design 29(4), 998–1000 (2008)
3. Lv, X., Huang, X., Zhang, B.: A medical image registration method based on corner feature. Journal of InnerMongolia University of Science and Technology 29(1), 49–52 (2010)
4. Tang, B., Chen, T., Wang, Z.: Novel image registration method based on wavelet transform. Computer Applications 27(9), 2013–2015 (2007)
5. Quddus, A., Basir, O.: Wavelet-based medical image registration for retrievalap applications. In: International Conference on Biomedical Engineering and Informatics, pp. 301–305 (2008)
6. Sharman, R., Tyler, J.M., Pianykh, O.S.: A fast and accurate method to register medical images using Wavelet Modulus Maxima. Pattern Recognition Letters 21, 447–462 (2000)
7. Mallat, S.G.: Multifrequency channel decompositions of image and wavelet models. IEEE Transaction on Acoustics Speech and Signal Processing 37(12), 2091–2110 (1989)
8. Maes, F., Collignon, A., Vandermeulen, D., et al.: Multimodality image registration by maximization of mutual information. IEEE Transaction on Medical Imaging 16(2), 187–199 (1997)
9. Zhang, H., Zhang, J., Sun, J.: Medical image registration method based on mixed mutual information. Journal of Computer Applications 2(10), 2351–2353 (2006)
10. Liu, Q., Li, Y.: Improved sample method for medical image registration based on mutual information. Journal of Computer Applications 30(4), 947–949 (2010)

The Simulation Analysis of Fire Feature on Underground Substation

Xin Han, Xie He, and Beihua Cong

Shanghai Institute of Disaster Prevention and Relief, Tongji University, Shanghai 200092, China
{hanxin,bhcong}@tongji.edu.cn, sigvard@163.com

Abstract. Underground transformer substations constructed with non-dwelling buildings have a positive meaning to a resource-saving and environment friendly society. The fire safety is very important for the effective operation of these projects. By introducing the method of oxygen suffocation, this paper carries out simulation analysis of fire feature on underground substation. The corresponding fire protection strategy is also put forward.

Keywords: Underground substation, Totally enclosure, FDS, Fire feature, Fire protection strategy.

1 Introduction

With rapid development of the national economy and urban construction, there is a great demand for the increment of electric power. More and more transformer substations have been constructed in the downtown region of the city. On the other hand, owing to the shortage of land resources, some substations should be constructed in the underground space. Perhaps these underground substations need to be built with non-dwelling buildings. Hence, the fire safety is very important for the effective operation of these projects. By introducing the method of oxygen suffocation as well as with the help of software FDS, this paper carries out simulation analysis of fire feature on underground substation. In accordance with the simulated results, the related fire protection strategy is also put forward.

2 Construction Simulation Model

A certain project consists of three parts, including 220KV/110KV underground substations and a building of management center. The joint part of the construction consists of 5 floors, 4 basements. Among the construction, the 220KV/110KV underground substations are located on the second floor to the fourth floor. The EHV management center lay on the aerial parts. The garage servicing for the building of management center is located in the basement, between the underground substations and the center space.

This paper mainly carries out studies on the fire protection strategy for the underground 220KV main transformer room. The size of the room is 15m ×12m × 12.8m, locating underground of floor 2 and floor 3. The wall thickness of the room is 0.5m and the thickness of the bottom and the top surface is 1.5m, the material is concrete.

The upper part of the room sets two exhaust pipes and the lower part sets two air supply pipes, the size being 1m×0.5m×0.5m and the materials being stainless steel. The spacing between the top and bottom piping from the side wall is 0.5m. The spacing between the upper pipes from the top surface is 0.1m and that of the lower part of the pipelines is 0.06m. Each pipeline is uniformly arranged three vents, size being 1m×0.5m. The air flow, under normal circumstances, is 35000m3/h. When the heat detector reaches to 70°C, the vents could be shut down automatically.

The flue gas control valve is located in each pipe, size being 1m×0.5m×0.5m. The spacing between the valves from the external walls is 1m, material being stainless steel, etc. The valves could be tolerated at 260°C continuing 1.5 hours. The level of fire doors is the group A, size being 2m×2m, being located in the right and left from the side wall of 1.5m. The leak volume is 500m3/m2(h, and the material is steel.

3   Simulation Scenarios

The load of fire source is designed as 10MW, reaching the designed peak heat release rate at 60s. The simulation scenarios could be divided into four groups: Scenario 1 (small interface oil leakage when totally enclosure), Scenario 2 (large interface oil leakage when small leak), Scenario 3 (small interface oil leakage when totally enclosure), Scenario 4 (large interface oil leakage when small leak). The area of small interface oil leakage is 20m2, located around the main transformer and that of large interface oil leakage is 217m2, covering the entire room. The oil spill height in the room is on the ground. The software of FDS is adopted during the process of simulation analysis [5].

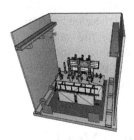

Fig. 1. Simulation model

4   Results Analysis

4.1   The Comparison of Fire Extinguished Time in Different Scenarios

Scenario 1: Fire is completely extinguished in 401s.
Scenario 2: Fire is completely extinguished in 515s.
Scenario 3: Fire is completely extinguished in 480s.
Scenario 4: Fire is completely extinguished in 635s.

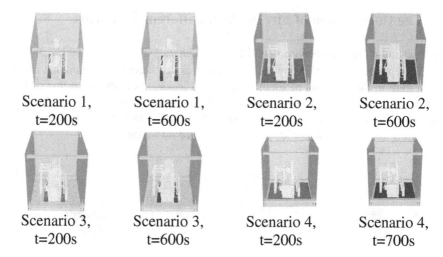

Scenario 1,   Scenario 1,   Scenario 2,   Scenario 2,
t=200s   t=600s   t=200s   t=600s

Scenario 3,   Scenario 3,   Scenario 4,   Scenario 4,
t=200s   t=600s   t=200s   t=700s

Fig. 2. HRRPUV in different scenarios

The comparison of Heat Release Rate in different scenarios

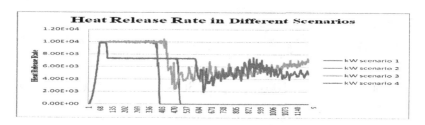

Fig. 3. Heat Release Rate in different scenarios

- Scenario 1: Fire reaches the peak value of 10MW at 60s, and then maintains 10MW from 60s to 398s, followed by plummeting from 398s to 401s, and finally being extinguished at 401s.
- Scenario 2: Fire reaches the peak value of 10MW at 60s, then maintains 10MW from 60s to 68s at a quite short period, and continues 7MW from 75s to 470s, among these two values being a violently decline. At the end, the fire is extinguished at 515s.
- Scenario 3: Fire reaches the peak value of 10MW at 60s, and then maintains 10MW from 60s to 460s. After that, the fire has a severe shock and decreases from 460s to 480s, which achieving the smallest value at 480s. However, the fire is not extinguished. On the contrary, it keeps shocking to a low temperature value.
- Scenario 4: Fire reaches the peak value of 10MW at 60s, and then maintains 10MW from 60s to 68s briefly, and maintains 7MW from 75s to 600s. There is a violent decline among these two values. In the burning process of the preceding, the burning results are strikingly similar to that of Scenario 2, although the duration of the lower value burning is longer than the former one. In the subsequent process, the fire has a shock with a slightly increase.

5 The Comparison of Fire Scale in Different Scenarios

The maximum fire temperature in different scenarios is 310℃, 260℃, 320 ℃, 270 ℃ separately. By analyzing the simulation results, it could be found that the fire temperature in the situation of small interface oil leakage is higher than that of large interface oil leakage when the environmental conditions of the room are the same. Similarly, the fire temperature when the room has small leak is higher than that when the room is totally enclosure with the same interface oil leakage.

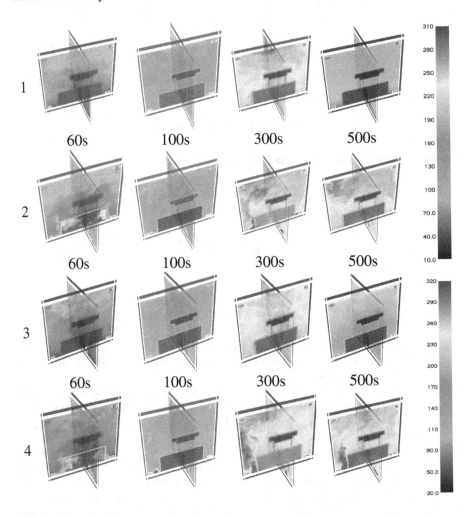

Fig. 4. The maximum fire temperature in different scenarios

6 The Comparison of Temperature of the Flue Gas Control Valve in Different Scenarios

The maximum temperature of the flue gas control valve in different scenario during the whole burning process is 26.5℃, 29℃, 34℃, 35℃ separately. It could be suggested that the maximum temperature of the flue gas control valve in small interface oil leakage is higher than that of large interface oil leakage when the environmental conditions of the room are the same. The maximum temperature of the flue gas control valve when the room has small leak is higher than that when the room is totally enclosure with the same interface oil leakage.

Table 1. The fire temperature of the flue gas control valve in different scenarios

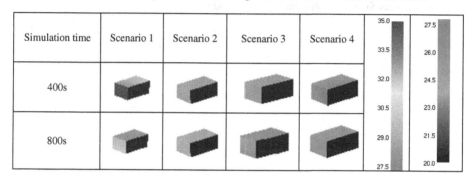

7 Conclusion

1. The environmental conditions of the room, i.e. small leak or totally enclosure, play a crucial impact on the room fire of underground substations. Fire could be extinguished because of the lack of suffocate oxygen when the room is totally enclosure. On the contrary, in the condition of small leak, the fire burn after a period of peak value and will not be extinguished. It would be maintained at a lower heat release rate, continuing to burn.

2. As for the same fire scale of 10MW, the room fire being totally enclosure is conductive to that of small leak with the same interface oil leakage. Therefore, it is suggested that the room should be constructed, as much as possible, being totally enclosure. On the contrary, the fire burning scale with large interface oil leakage is better than that with small interface oil leakage when the environmental conditions of the room are the same. Thus, it should be focused on normal operation of the spill valve and short circuit protection in case of fire.

3. During the simulation, the safety of the ventilation control valves has been a great degree of assurance. It is illustrated that in accordance with the current design, the control valve does not exist danger because of the high temperature deformation.

4. The fire could be effectively controlled at a later stage. If the fire extinguishing system has no longer been activated, the fire could also be quickly extinguished. It could probably be regarded as a reasonable and effective method to establish the corresponding fire protection strategy for underground substations.

Acknowledgments. The support of the Natural Science Foundation of China (Grant No. 50906063) is gratefully appreciated.

References

1. Cox, G., Chitty, R.A.: Study of the Deterministic Properties of Unbound Fire Plumes. Combustion and Flame (1980)
2. Heskestad, G.z.: Proc. 18th Inter. Symp. On Combustion, Pittsburgh PA (1981)
3. BSI Technical Committee FSH/24. Fire Safety Engineering in Building, Part1: Guide to the Application of Fire Safety Engineering Principles (1997)
4. McCaffrey, B.J.: Purely Buoyant Diffusion Flames: Some Experimental Results. NBSIR 79-1910, Washington, DC (1979)
5. Fire Dynamics Simulator (Version 5) User's Guide

The Design and Implementation of Soccer Robot Control System

Sheng-Jie Zhao and Chuan Wang

Henan Normal University, Xinxiang, 453007, Peoples R China
{shengjiezhidao,wangchuan20112012}@163.com

Abstract. Aiming at the problem that the intelligence is not high of the soccer robot control system, with analysis of energy materials, this paper can not only simplifies the design of the hardware and software of the control system but also enhances considerably its reliability and performance, which includes a core controller that based on S3C2440, a position loop based on LM629, and software strategy based on UC/OS-II. Experiments show that the new soccer robot controller features a quick response, high control precision and high servo rigidity; it also has a good self-adaptability.

Keywords: S3C2440, LM629, Control System, Soccer Robot.

1 Introduction

Robot soccer is put forward of multi-agent system developing platform[1], and it is a typical multi-agent robotic system in recent years[2]. The robot soccer has a variety of types, different types of games the limits of soccer robot structure design[3]. Mirosot football game the system by the robot general robot subsystem, vision subsystem, decision subsystem and communications subsystem four parts[4]. The bottom of soccer robot control system as a whole system of actuators, the performance of the whole system quality plays an important use[5]. At present, most soccer robot control system stability low-rise is bad, but also need sampling over a long period of time, real time is not strong,and intelligent is not high, which cannot meet the needs of the robot control system. Therefore, development of a kind of high performance robot bottom control system become the urgent needs of robot soccer fans.

In this paper ,a new soccer robot control system is design,which includes a core controller that based on ARM920T,a position loop based on LM629, and software strategy based on UC/OS-II.

2 Underlying the Overall Design of the Control System

The overall design of the control system is to receive the wireless communications subsystem over information, combined with their own cameras and sensors received information, after a decision strategies it makes a choice, and translates into drive DC

D. Jin and S. Lin (Eds.): Advances in FCCS, Vol. 2, AISC 160, pp. 665–670.
springerlink.com   © Springer-Verlag Berlin Heidelberg 2012

motor of the directive driven robot cars. According to the function of the robot execution, the task can be divided for hardware and software task.

This design scheme ARM920T kernel in the design S3C2440 chip is main control chip[6], motor driver adopted L298 device, and install a infrared, pressure sensors[7]; In addition to the expansion of wireless communication mode and video acquisition module, the overall design is shown in Fig.1.

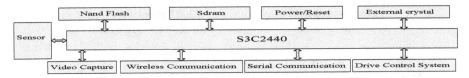

Fig. 1. The overall design

3   The Hardware Design of the Control System

Main processor. Main processor choose samsung by the production of S3C2440 microprocessor, this product is for hand-held devices and commonly used low power consumption, high performance application plan provides small size microprocessors, the internal use 16/32-Bit RISC structure and it has strong instruction set; The built-in 16KB instruction Cache and 16KB data Cache; Chip MPLL and UPLL can provide up to 400 MHZ clock frequency; 60 interrupt source, 24 external interface interrupt source; 5 a PWM channels; A32.768 KHZ real-time clock; Article 3 the DMA channel based on or terminal of the serial port channel; 16-Bit watchdog timer and so on many kinds of peripheral interfaces

Servo controller. LM629 as servo control regulator, in addition to accept ARM9 instruction set position, velocity, acceleration, 3 movement parameters and filter the PID parameters KP, KI, KD outside, and at the same time LM629 will output signal processing, the gain position, the digital PID operation PWM and reverse control after the output signal, the dc machines to drive chip.

LM629 have internal trapezoid speed planning generator and digital PID controller. The two functions cover most of the CPU workload. Set the position control mode, the position S3C2440 velocity, acceleration into LM629, LM629 use these data calculation trajectory. Motor running, may change the operation parameters, generating speed track. Set speed control mode, the motor need only acceleration and speed. LM629 internal PID controller using incremental PID algorithm, the control parameters KP, Kt, KD by S3C2440 are given.

Power supply circuit design. By S3C2440 chip voltage power supply characteristic, it is known that S3C2440 internal voltage power supply for 1.2V, external I/O voltage power supply for 3.3V. Here LM1117-18 chip selection provides 1.25V voltage, choose LM1117-33 chip provides 3.3V voltage. They characteristics are, output current is big, high precision, high stability, low power consumption. For an external voltage for 12V, so the LM2567 chips will 12V into 5 V voltage.

Communication module. Communication module USES LPC 2212 UART1 connected to a wireless communication module, mainly from the PC to receive the world coordinate system in position, velocity, instructions. Because of the LPC2212 UART can be as high as 115200 in wave frequency normal work, and it has the unique more bytes of FIFO structure, when the interrupt way accept data only by meeting trigger depth which will produce the interruption of the data are available, to a great extent, it can reduce the number of response to an interrupt processor, to improve operation efficiency of the processor. UART hardware has surveillance "overflow error, parity error, frame error" and so on the function of the mistakes, so software don't set the check code, and write convenient.

Sensor module and LCD module. Robots are installed on six ultrasonic ranging module, model PI D#615078-6500, used to detect 6 inches to 35 feet around the obstacles in information. This module interface articulated in 1_PC2212\bus. LCD module mainly used to display some of the system operation state, is through the bus interface connection in the main controller.

4   The Software Design of the Control System

System software is mainly including the program module and interrupt service procedure module. The main program module complete system of initialization and interrupt Settings, and an interrupt service routine is the main body of the software, and complete the system control function.

Initialize system clock. Outside of the 12 MHZ provide crystals, the points can be as high as 400 MHZ frequency. For peripheral equipment. S3C2440 internal clock logic control can provide CPU FCLK clock, and the internal bus AHP and the HCLK and PCLK APB clock. Therefore, in the peripheral equipment normal work before, it must first initial need clock.

RMPLLCON MPLL configuration register is used to will provide external clock frequency into need frequency MPLL and UPLL, the former is the kernel and internal bus provide a clock, the latter for USB devices provide clock.The dividing circuits of system clock is shown in Fig.2.

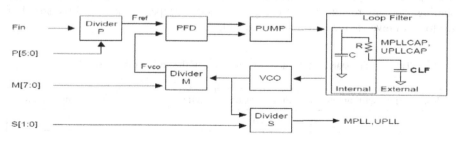

Fig. 2. The dividing circuits of system clock

Interrupt Close. Must turn off interrupts before the system enters normal operation, the register addresses of interrupt mask register MASK and SUSMASK are INTMASK = 0x4A0000008 and INTSUBMASK = 0x4A00001C. Set each bit of the register into"1", mask the interrupt.

The establishment of the memory. After the power supply circuit, the 4 KB Nand flash before the code will automatically be copied to the buffer 4 KB kernel, and then it began to run. But the problem is that, if our code is more than 4 KB, the code will be how to run? This is usually happened. The solution is to copy the code after 4 KB to SDRAM operation.

Therefore, to SDRAM [29] related register related configuration:

(1)   rBWSCON address: 0 x48000000 configuration width and waiting for bus state;
(2)   rBANKCON0~rBANKCON5 address: 0x48000004~0x48000018 configuration top five bank;
(3)   rBANKCON6, rBANKCON7 address: x4800001C 0, 0 x48000020 configuration 6, 7 bank;
(4)   rREFRESH address: 0x48000024 SDRAM refresh control register;
(5)   rBANKSIZE address: 0x48000028 SDRAM size capacity configuration register;

UC/OS-II transplantation. UC/OS-II operating system is based on task priority operation, so in distribution robot task, it must be made for each robot task and distribution. For most of the UC/OS-II [6] ANSI C code is to use language to write, so UC/OS-II of portability is very good. However it still need to use C and assembly language to write some relevant code processor. For UC/OS-II transplant modify it only OS_CPU. H, OS_CPU_C. C, OS_CPU_A. ASM three files.

5   The System Test and Analysis Results

With the adoption of the circuit which is based on ARM9 panel and transplantation after the operation system, might as well to watch the performance of the robot. Through field experiment mensuration, soccer robot around the time needed for starting, stopping round table 1. The given speed of Upper machine and the speed of the forward/reversal rotation is shown in table 2.

Table 1. The starting time and stopping time of the soccer robot around. (ms)

No.	Left starting time	Right starting time	Left stopping time	Right stopping time
1	13.1	13.0	14.3	14.5
2	12.8	13.1	14.6	14.4
3	13.0	12.8	14.5	14.7

Table 2. The given speed of Upper machine and the speed of the forward/reversal rotation

The given speed of Upper machine with forward rotation(m/s)	0.3	0.6	0.9	1.2	1.5	1.8	2.1
The speed of the forward rotation(r/s)	2.1	4.2	6.4	8.6	0.8	13.4	16.1
The given speed of Upper machine with reversal rotation(m/s)	0.3	0.6	0.9	1.2	1.5	1.8	2.1
The speed of the reversal rotation(r/s)	2.0	4.2	6.3	8.7	10.0	13.4	16.0

From table 1 and table 2 it is not difficult to see soccer robot starting time and the stopping time is short, the robot motor no matter are turning around the output of the wheel or reverse linear are all very good, two rounds of coordination or robot stable operation. The whole system design is reasonable.

6   Conclusion

Research and design on the S3C2440 chip, LM629 servo module, L298 is based on the power amplifier module, and transplant the UC/OS-II operating system to the bottom soccer robot control system, can not only meet the requirement of real-time soccer robot, but also for the buck to the robot transformation laid the foundations, realize the intelligence PID adaptive control, through the experiments it shows that the system has high accuracy and stability, system design is reasonable and feasible. The system is better to meet the game requirements.

Acknowledgement. In this paper, the research was sponsored by the Nature Science Foundation of China (Project No.61173071)and the Soft Science Project of Henan Province(No.112400440062,112400450305).

References

1. Xu, C.: Mechanical servo system based on fuzzy neural network for complex control. Control Engineering 17(2), 146–148 (2010)
2. He, Z.: LM629-based motor servo control system design. Mechanical Design and Manufacture (2), 40–42 (2009)
3. Chen, N., et al.: For the field of industrial and highly interconnected, TI launched a new Sitara ARM9 microprocessor. Global Electronics (5), 86–87 (2010)
4. Hiroki, K., Minoru, A., Yasuo, K., et al.: RoboCup: a challenge problem for AI and robotics. HirokiK. In: RoboCup-97: Robot Soccer World Cup, pp. 38–43. Springer, Berlin (1998)

5. Wu, C.-Y., He, L.-Y.: Design and Realization of Instructional RPPR-Robot. Research And Exploration In Laboratory, 26(10) (2007)
6. Yin, Z.: Application of FPGA control DC motor servo system. Inner Mongolia Science and Technology and Economy 177(23), 101–103 (2008)
7. Gao, J., Huang, X., Peng, G.: Design of Control System Based on DSP for Remote-brainless Soccer Robot. Computer Engineering and Applications (12), 19–21 (2006)

Decoupled Sliding Mode Control for the Climbing Motion of Spherical Mobile Robots

Tao Yu, Hanxu Sun, Qingxuan Jia, Yanheng Zhang, and Wei Zhao

School of Automation, Beijing University of Posts and Telecommunications,
Beijing 100876, China
yutaogzm@sohu.com

Abstract. Based on the equivalent control method and Lyapunov function technique, a decoupled sliding mode control approach is presented for stable control of the climbing motion of spherical mobile robots with sliding mechanics. At first the dynamic equations of the climbing motion of a spherical mobile robot are derived based on the Euler-Lagrange equation, and the equilibrium conditions of the climbing state are then analyzed. Combining the feature of the dynamic model, a decoupled sliding mode controller is proposed and the asymptomatic stability of the system is theoretically analyzed. Finally, the proposed control approach is applied to a spherical mobile robot and simulation results verify the validity of the proposed control approach.

Keywords: decoupled control, sliding mode control, climbing motion, spherical robot.

1 Introduction

A spherical mobile robot is a new type of mobile robot that appears in recent years, and it has a spherical shell to encompass all its driving mechanisms, control devices and energy sources. A spherical robot is usually characterized as having a simple, compact structure and high maneuverability. A spherical robot will inevitably encounter such situations as climbing slopes and traversing obstacles during its implementation of practical exploration missions. Climbing ability and obstacle surmounting capability have become important indicators for measuring the motion performance of a spherical robot.

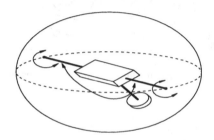

Fig. 1. Structure of a pendulum-driven spherical mobile robot.

D. Jin and S. Lin (Eds.): Advances in FCCS, Vol. 2, AISC 160, pp. 671–677.
springerlink.com   © Springer-Verlag Berlin Heidelberg 2012

The spherical shell enables a spherical robot to avoid a general class of obstacles. Compared with a wheeled mobile robot, the property of the contact between the sphere and the ground, which is an approximated point contact, makes it easier for a spherical robot to be subjected to outside disturbances while rolling along a slope, which adds the challenge to the control of climbing motion of a spherical robot. This approximated point contact may also cause another problem, i.e., when a spherical robot rolls ahead along a slope, if the turning motion is simultaneously conducted, the rolling without slipping constraint between the sphere and slope may be ruined. As a result, the robot cannot remain stable and will fall along the slope. We don't wish this phenomenon to occur, and only the rolling ahead motion of a pendulum-driven spherical mobile robot along a slope is considered in this paper. The structure of a pendulum-driven spherical mobile robot [1, 2, 3] is shown in Fig. 1.

The rest of this paper is structured as follows: In section 2, the Euler-Lagrange equation is used to derive the dynamics of the climbing motion of a spherical robot. In Section 3, the proposed decoupled sliding mode controller is described, and the stability of the system is simply proved. The proposed controller is utilized to control a spherical robot and simulation results are presented in Section 4. Finally, the summary is given in Section 5.

2   Dynamic Analysis

The system can be greatly simplified by taking advantage of its inherent geometric symmetry. By only considering the performance of the rolling ahead motion along the slope, we can remove the nonholonomic constraints normally associated with a rolling sphere in three dimensional setting. We may further reduce complexity by imposing constraint condition of no-slip and no bounce. In order to leave some generality, we assume the planar model to be rolling on an arbitrary incline of γ degrees. The idealized planar model of a pendulum-driven spherical mobile robot is shown in Fig. 2. In addition, the definitions for the model parameters are listed in Table 1.

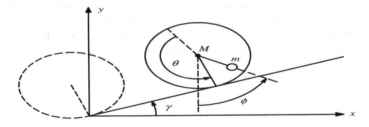

Fig. 2. Kinematic diagram of a pendulum-driven spherical mobile robot along a slope.

Table 1. Definitions for robot model parameters

M	m	I	l
mass of the outer shell	mass of the pendulum	moment of inertial of the shell	length of the link
R	g	γ	τ
radius of the shell	gravitational acceleration	inclination angle of the slope	torque applied to the pendulum

We assume that the mass of the link is negligible relative to the shell and eccentric mass, and we also model the eccentric mass as a point instead of a rigid body. We impose the assumptions to reduce the system to two coordinates: θ for sphere rotation and ϕ for pendulum rotation angle. Utilizing the Euler-Lagrange equation, the dynamics of the climbing motion of the robot can be described as

$$
\begin{bmatrix} (M+m)r^2 + I & mrl\cos(\phi-\gamma) \\ mrl\cos(\phi-\gamma) & ml^2 \end{bmatrix} \begin{bmatrix} \ddot{\theta} \\ \ddot{\phi} \end{bmatrix} + \begin{bmatrix} (M+m)gr\sin\gamma - mrl\sin(\phi-\gamma)\dot{\phi}^2 \\ mgl\sin\phi \end{bmatrix} = \begin{bmatrix} \tau \\ \tau \end{bmatrix}
$$

(1)

Consider the dynamic equations (1), two states of equilibrium can be easily derived. First, consider the case in which the robot sits stationary on the slope. In this limiting case, all angular velocities and accelerations reduce to zero. When we enforce this condition upon (1), they reduce to

$$
(M+m)gr\sin\gamma = mgl\sin\phi_o
$$

(2)

We use the notation such that ϕ_o denotes the equilibrium value of ϕ. It stands to reason that there exists a limiting value of slope γ after which the robot will be incapable of holding its position. To determine this operational boundary we solve for ϕ_o in the above equation. The resulting solution is

$$
\phi_o = \arcsin\left(\frac{(M+m)r}{ml}\sin\gamma \right)
$$

(3)

Clearly, γ must be bounded above and below to ensure an inverse sine operand less than unity:

$$
-\arcsin\left(\frac{ml}{(M+m)r} \right) \leq \gamma \leq \arcsin\left(\frac{ml}{(M+m)r} \right)
$$

(4)

The bounds associated with (4) correspond to stable node bifurcations at which the equilibrium solutions coalesce and disappear. This phenomenon is associated with the dynamic condition of whirling [4] in which the robot unsuccessfully attempts to either remain stationary or climb the slope.

The second condition for equilibrium is defined by assuming the robot maintains a constant velocity over constant-slope terrain. This condition can also be satisfied by $\dot{\phi}=0$. When this condition is enforced upon (1), we can find the equilibrium pendulum angle continues to satisfy (3).

3   Control Design

Consider the following coupling nonlinear systems which can be divided into two subsystems as

$$A:\begin{cases} \dot{x}_1 = x_2 \\ \dot{x}_2 = f_1(x) + b_1(x)u \end{cases} \qquad B:\begin{cases} \dot{x}_3 = x_4 \\ \dot{x}_4 = f_2(x) + b_2(x)u \end{cases} \tag{5}$$

where $x = [x_1 \ x_2 \ x_3 \ x_4]^T$ is the state vector, $f_1(x)$, $b_1(x)$, $f_2(x)$, $b_2(x)$ are nonlinear functions, and u is the control input.

Then we construct the following linear functions as sliding surfaces for the two subsystems [5, 6]

$$s_1 = c_1 x_1 + x_2 \qquad\qquad s_2 = c_2 x_3 + x_4 \tag{6}$$

where c_1 and c_2 are positive constants.

We define an intermediate variable z, which represents the information from subsystem A, and it is incorporated into s_2. Therefore, the sliding surface s_2 can be modified as

$$s_2 = c_2(x_3 - z) + x_4 \tag{7}$$

Here the intermediate variable z is related to s_1. For decoupling control, we define z as

$$z = \tanh(s_1/\Phi_z) \cdot z_U \tag{8}$$

where z_U is the upper bound of $abs(z)$, $0 < z_U < 1$, Φ_z is a positive constant, and $\tanh(\cdot)$ is the hyperbolic tangent function defined as follows

$$\tanh(s_1/\Phi_z) = \frac{e^{s_1/\Phi_z} - e^{-s_1/\Phi_z}}{e^{s_1/\Phi_z} + e^{-s_1/\Phi_z}} \tag{9}$$

Since z_U is less than one, z presents a decaying signal. As s_1 decreases, z decreases too. When $s_1 \to 0$, we have $z \to 0$, $x_3 \to 0$, and then $s_2 \to 0$, and the control objective will be achieved.

Differentiating (8), $\dot{z}$ can be calculated as

$$\dot{z} = \alpha\left(s_1, \Phi_z, z_U\right) \cdot \dot{s}_1 \qquad (10)$$

where

$$\alpha\left(s_1, \Phi_z, z_U\right) = \frac{\operatorname{sech}^2\left(s_1/\Phi_z\right) \cdot z_U}{\Phi_z} \qquad\qquad \operatorname{sech}\left(s_1/\Phi_z\right) = \frac{2}{e^{s_1/\Phi_z} + e^{-s_1/\Phi_z}}$$

Differentiating (7) and using (10), we can calculate $\dot{s}_2$ as

$$\begin{aligned}
\dot{s}_2 &= c_2\left(\dot{x}_3 - \dot{z}\right) + \dot{x}_4 \\
&= c_2 x_4 + f_2 - \alpha c_2\left(c_1 x_2 + f_1\right) + \left(b_2 - \alpha c_2 b_1\right) u
\end{aligned} \qquad (11)$$

The equivalent control u_{eq} can be obtained from $\dot{s}_2 = 0$, i.e.

$$u_{eq} = -\frac{c_2 x_4 + f_2 - \alpha c_2\left(c_1 x_2 + f_1\right)}{b_2 - \alpha c_2 b_1} \qquad (12)$$

We further assume the system control input u to have the following form

$$u = u_{eq} + u_{sw} \qquad (13)$$

Here u_{sw} is the switching control. Define the Lyapunov function candidate $V = \dfrac{1}{2}s_2^2$ and differentiate V with respective to time, we have

$$\begin{aligned}
\dot{V} &= s_2 \dot{s}_2 \\
&= s_2\left[c_2\left(\dot{x}_3 - \dot{z}\right) + \dot{x}_4\right] \\
&= s_2\left(b_2 - \alpha c_2 b_1\right) u_{sw}
\end{aligned} \qquad (14)$$

Then we choose the switching control u_{sw} as follows

$$u_{sw} = -\frac{\eta \operatorname{sgn}\left(s_2\right) + k s_2}{b_2 - \alpha c_2 b_1} \qquad (15)$$

where η and k are positive constants.

Theorem 1. Suppose that the system (5) is controlled by the control input described in (12), (13) and (15), then the sliding surface s_2 is asymptomatically stable.

Proof: Substituting (15) into (14), we can obtain

$$\begin{aligned}
\dot{V} &= s_2 \dot{s}_2 \\
&\leq -\eta\left|s_2\right| - k s_2^2 \leq 0
\end{aligned} \qquad (16)$$

Integrating both sides of (16), we have

$$V(t)-V(0)\leq \int_0^t \left(-\eta|s_2|-ks_2^2\right)d\sigma \qquad (17)$$

From (17), we can further obtain

$$V(t)=\frac{1}{2}s_2^2 \leq V(0)<\infty \qquad \lim_{t\to\infty}\int_0^t \left(\eta|s_2|+ks_2^2\right)d\sigma \leq V(0)<\infty \qquad (18)$$

From (18), we have $s_2 \in L_\infty$ and $s_2 \in L_2$. But from (16) we have $\dot{V}=s_2\dot{s}_2 <\infty$, then we can obtain $\dot{s}_2 \in L_\infty$. According to babalat's lemma, we have $\lim_{t\to\infty}s_2 =0$.

4 Simulation Study

We apply the proposed control scheme to a spherical robot to demonstrate the effectiveness of the controller. Define the state variables to be $x_1 =\theta$, $x_2 =\dot{\theta}$, $x_3 =\phi$, $x_4 =\dot{\phi}$, the dynamic equation of the climbing motion (1) can be converted into a canonical form described in (5) by adopting coordinate transformation and only using simple mathematical manipulations. The robot model parameters [3] and controller design parameters are listed in Table 2. We assume the robot to be executing a rest-to-rest maneuver along a slope of 15 degrees. The initial values of the system states are chosen as $x_0 =(0 \quad 0 \quad \phi_o \quad 0)^T$, and we choose the desired values of the system states $x^d =(\pi \quad 0 \quad \phi_o \quad 0)^T$.

Table 2. Robot model parameters and controller design parameters

M	m	I	l	R	g
17[kg]	14[kg]	0.6659[kg·m²]	0.2[m]	0.3[m]	9.81[m/s²]
c_1	c_2	z_U	Φ_z	η	k
0.5501	2.3272	0.9626	7.4822	2.7688	0.6769

The tracking performance of the proposed controller is shown in Fig. 3. We can find that not only the rotation angle of the sphere but also the rotation angle of the pendulum can reach their desired values in a short time. And before the robot reaches its desired position, the pendulum angle has already converged to its equilibrium value only after a few oscillations.

5 Summary

In this paper, we present a decoupled sliding mode control approach for pendulum-driven spherical mobile robots' climbing motion. The whole system is divided into two subsystems and two sliding surfaces are constructed through the state variables of the system. Then an intermediate variable, which represents the information of a sub-sliding surface, is incorporated into the other sub-sliding surface. The sliding mode control law of the system is finally derived by adopting Lyapunov stability theorem. Simulation results conducted on a spherical robot, which show that the proposed controller can achieve good regulation performance, verify the validity of the proposed sliding mode controller.

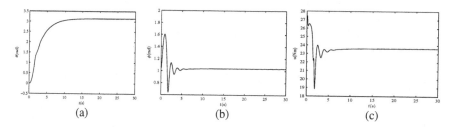

Fig. 3. Tracking results for the climbing motion. (a) sphere rotation angle; (b) pendulum rotation angle; (c) control input.

Acknowledgement. The authors wish to acknowledge the financial support provided by National Natural Science Foundation of China (No. 51175048).

References

1. Ming, Y., Deng, Z.: Dynamic modeling and optimal controller design of a spherical robot in climbing state. Chinese Journal of Mechanical Engineering 45, 46–51 (2009)
2. Liu, Z., Zhan, Q.: Motion control of a spherical mobile robot for environment exploration. Acta Aeronautica Et Astronautica Sinica 29, 1673–1679 (2008)
3. Xiao, A.: The research on a novel spherical mobile robot. Beijing University of Posts and Telecommunications, Beijing (2005)
4. Abbott, M.S.: Kinematics, dynamics and control of single-axle, two-wheel vehicle. Virginia Polytechnic Institute and State University, Blacksburg (2000)
5. Lo, J.C., Kuo, Y.H.: Decoupled fuzzy sliding-mode control. IEEE Transactions on Fuzzy Systems 6, 426–435 (1998)
6. Lin, C.M., Chin, W.L.: Adaptive hierarchical fuzzy sliding-mode control for a class of coupling nonlinear systems. Int. J. Contemp. Math. Sci. 1, 177–204 (2006)

Author Index